# RENEWING OUR RIVERS

# RENEWING OUR RIVERS

## STREAM CORRIDOR
## RESTORATION
### IN DRYLAND REGIONS

—— Edited by ——

### MARK K. BRIGGS & W. R. OSTERKAMP

THE UNIVERSITY OF
ARIZONA PRESS

TUCSON

The University of Arizona Press
www.uapress.arizona.edu

ISBN-13: 978-0-8165-4148-5 (paperback)

Cover design by Leigh McDonald
Cover photograph of San Pedro River, SE Arizona © Brooke Bushman/The Nature
Conservancy
Interior design and typesetting by Sara Thaxton in 10/13.5 Warnock Pro (text),
Hardwick WF, and Payson WF (display)

Library of Congress Cataloging-in-Publication Data
Names: Briggs, Mark K. (Mark Kendig), 1961–editor. | Osterkamp, W. R., editor.
Title: Renewing our rivers : stream corridor restoration in dryland regions / edited by
    Mark K. Briggs and W. R. Osterkamp.
Description: Tucson : University of Arizona Press, 2020. | Includes bibliographical refer-
    ences and index.
Identifiers: LCCN 2020022348 | ISBN 9780816541485 (paperback)
Subjects: LCSH: Stream restoration—Handbooks, manuals, etc. | Arid regions ecology. |
    LCGFT: Handbooks and manuals.
Classification: LCC QH541.5.S7 R48 2020 | DDC 333.91/62153—dc23
LC record available at https://lccn.loc.gov/2020022348

Printed in the United States of America
♾ This paper meets the requirements of ANSI/NISO Z39.48-1992 (Permanence of Paper).

*Dedicated to our brothers and sisters around the world
who are working tirelessly to improve stream conditions for
the benefit of humans and native species.*

*The real cure for our environmental problem is to understand that our job is to salvage Mother Nature.*

—JACQUES YVES COUSTEAU

*Heaven is under our feet as well as over our heads.*

—HENRY DAVID THOREAU

# CONTENTS

## 6. Implementation: Putting Your Stream Corridor Restoration Plan into Action     **213**

## 7. Monitoring the Results of Stream Corridor Restoration     **313**

# ACKNOWLEDGMENTS

The publication of *Renewing Our Rivers* has been an amazing, diverse, and rewarding decade-long journey involving fifty-five authors from four countries: Australia, Mexico, Spain, and the United States. There are many people we want to acknowledge and thank. First and foremost are the authors of the guidebook, who never flinched from their support, and who contributed research findings, case studies, and experiences from all over the world. At the start of this journey, funding in support of the 2010 Tri-national Restoration Lessons Learned Conference—the spark that led to the restoration guidebook—was provided by Karl Flessa (from the University of Arizona, via the Research Coordination Network: Colorado River, National Science Foundation Award 0443481) and Michele Girard (then with the National Park Service, now with Cuenca de los Ojos). We are indebted to Osvel Hinojosa-Huerta (then with Pronatura Noroeste), Carlos Sifuentes (with CONANP), Pat B. Shafroth (with the U.S. Geological Survey), and Francisco Zamora (with the Sonoran Institute), who voiced great encouragement after the conference to do more than a final report but to produce a restoration lessons-learned publication for widespread distribution. We also want to thank Karin Krchnak (then with the World Wildlife Fund, now with the World Bank Group), who supported the effort during a period when we most needed outside encouragement.

Especially important were the time, energy, and support of Sarah LeRoy, Amy McCoy, and Daniel Bunting—three of the guidebook's lead authors. Sarah LeRoy, lead author of the climate-adaptation chapter, managed to enlist some of the highest-level climate scientists in the world. Amy McCoy was one of our strongest supporters throughout the process; we probably met with her two hundred times during the course of this journey, and her positive and encouraging insights were critical not only to finishing her chapter but to completing the entire guidebook. And many thanks to Daniel Bunting who was lead author on two of the guidebook's longest chapters that, all told, involved twenty-nine of the guidebook's fifty-five authors.

Finally, there are three heroes we want to acknowledge. When drafts of the chapters were completed and combined, we hired Mary Black (retired from the University of Arizona) as an editor. The contract covered maybe half of the time she needed to go through our drafts. She never stopped, and her edits were critical to getting the guidebook completed and presented for publication. When we receive complements on how well the book is written, we say "Mary Black." Eduardo González (from Colorado State University) and Katharine Hayhoe (from Texas Tech University) took time to review the entire guidebook and update references. We cannot thank Eduardo and Katharine enough for their efforts and contributions. Lindsay White (of the Commonwealth Environmental Water Office) was part of this effort from the start and stayed with us through its completion. It is no understatement to say that without Lindsay's encouragement, insights, and technical input, the guidebook could not have been completed. Finally, it is with profound sadness that I note my co-editor, long-time friend, and mentor, Waite Osterkamp, passed in March 2020, only a few months before the completion of the guidebook. Waite spent his career with the U.S. Geological Survey and worked on rivers all over the United States. He was with us on this project from the onset, and the guidebook would not have been completed without his contributions and leadership. Waite was a kind gentleman, had a great sense of humor, loved practical jokes, and was a joy to work and be with. He is sorely missed. The world was a better place with Waite in it.

# ABBREVIATIONS

| | |
|---|---|
| ABCWUA | Albuquerque-Bernalillo County Water Utility Authority |
| ACIS | Applied Climate Information System (of the RCC program, U.S.) |
| ADEQ | Arizona Department of Environmental Quality (U.S.) |
| ADWR | Arizona Department of Water Resources (U.S.) |
| AEURHYC | Asociación Ecológica de Usuarios del Río Hardy-Colorado, A.C. (Ecological Association of Hardy-Colorado River Users) |
| ANCOLD | Australian National Committee on Large Dams |
| ANOSIM | analysis of similarity |
| ANOVA | analysis of variance |
| APHIS | Animal and Plant Health Inspection Service (of the USDA) |
| B-IBI | Benthic-Index of Biological Integrity |
| BLM | Bureau of Land Management (of the USDOI) |
| BOM | Bureau of Meteorology (AU) |
| Bosque NWR | Bosque del Apache National Wildlife Refuge |
| CAKE | Climate Adaptation Knowledge Exchange (of EcoAdapt) |
| CASC | Climate Adaptation Science Center (of the USGS) |
| CBI | Conservation Biology Institute (U.S.) |
| CCAST | Collaborative Conservation and Adaptation Strategy Toolbox |
| CCC | Community Center for the Conservation of the Santa Cruz River |
| CCRN | Cochise Conservation and Recharge Network |
| CCSM | Community Climate System Model |
| CDEC | California Data Exchange Center (U.S.) |
| CEA | Comisión Estatal del Agua |
| CEWO | Commonwealth Environmental Water Office |
| CFE | Comisión Federal de Electricidad (Mexico's Federal Electricity Commission) |
| CICES | Common International Classification of Ecosystem Services |

| | |
|---|---|
| CICESE | Centro de Investigación Científica y de Educación Superior de Ensenada, Baja California (Center for Scientific Research and Higher Education in Ensenada, Baja California, MX) |
| CILA | Comisión Internacional de Límites y Aguas |
| CLIMAS | Climate Assessment for the Southwest (a RISA) |
| CLICOM | Climate Computing project (of CICESE, SMN, INECC, and others, MX) |
| CMIP | Coupled Model Intercomparison Project |
| CMP | Coastal Management Plan |
| CNAP | California-Nevada Applications Program (a RISA) |
| CNRM | Centre National de Recherches Météorologiques |
| CONABIO | Comisión Nacional para el Conocimiento y Uso de la Biodiversidad (MX) |
| CONACyT | Consejo Nacional de Ciencia y Tecnología (Mexico's National Science Foundation) |
| CONAGUA | Comisión Nacional del Agua (Mexican National Water Commission) |
| CONANP | Comisión Nacional de Áreas Naturales Protegidas (Mexican Nation Commission for Natural Protected Areas) |
| CORDEX | Coordinated Regional Climate Downscaling Experiment |
| CRAVe | Climate Registry for the Assessment of Vulnerability (of EcoAdapt) |
| CRP | Texas Clean Rivers Program (U.S.) |
| CSIRO | Commonwealth Scientific and Industrial Research Organisation (AU) |
| CWA | Clean Water Act |
| DAP | Drought Action Plan (AU) |
| DBH | diameter breast height |
| DEM | digital elevation model |
| DFFA | Designer and Functional Flow Assessment |
| DWT | Colorado River Delta Water Trust (MX) |
| EC | electroconductivity |
| ELOHA | Ecological Limits of Hydrologic Alteration |
| EMP | Estuary Management Plan |
| ENM | environmental niche model |
| EPA | Environmental Protection Agency (U.S.) |
| ESA | Endangered Species Act (U.S.) |
| ESCC | Earth Systems and Climate Change (of the NESP) |
| ESDM | empirical statistical downscaling model |
| ET | evapotranspiration |
| EVI | Enhanced Vegetation Index |
| FEMAT | Forest Ecosystem Management Team (of the USFS) |
| G5 | Grupo Cinco, Ejido Miguel Hidalgo, Sonora, Mexico |
| GCM | global climate model |
| GHCN | Global Historical Climatology Network (of the NCEI) |
| GMT | global mean temperature |
| GNSS | Global Navigation Satellite System |
| GPCP | Global Precipitation Climatology Project |
| GPS | global positioning system |

| | |
|---|---|
| HEC | Hydrologic Engineering Center (of the USACE); HEC-RAS is its River Analysis System |
| IBWC | International Boundary and Water Commission (U.S. and MX) |
| IFIM | Instream Flow Incremental Methodology |
| IHA | Indicators of Hydrologic Adjustment |
| IMTA | Instituto Mexicano de Tecnología del Agua (Mexican Institute of Water Technology) |
| INECC | Instituto Nacional de Ecología y Cambio Climático (National Institute of Ecology and Climate Change, MX) |
| INEGI | Instituto Nacional de Estadística, Geografía, e Informática (MX) |
| INIFAP | Instituto Nacional de Investigaciones Forestales, Agricolas y Pecuarias (National Institute of Forestry, Agriculture and Livestock Research, MX) |
| IPCC | Intergovernmental Panel on Climate Change (of the United Nations) |
| KWRP | Kooragang Wetland Rehabilitation Project |
| LAI | leaf-area index |
| LAN | Ley de Aguas Nacionales (Mexico's National Water Law) |
| LCC | Landscape Conservation Cooperative (formerly of the U.S. Department of the Interior) |
| LiDAR | Light Detection and Ranging |
| LNMySR | Laboratorio Nacional de Modelaje y Sensores Remotos (National Laboratory of Modeling and Remote Sensing, MX) |
| LSA | Level of Sustainable Activity |
| MAR | multivariate autoregressive |
| MIROC | Model for Interdisciplinary Research on Climate |
| MODIS | Moderate-Resolution Imaging Spectroradiometer |
| MRG | Middle Rio Grande (in New Mexico, U.S.) |
| MRGCD | Middle Rio Grande Conservancy District (of New Mexico, U.S.) |
| NAM | North American monsoon |
| NAWQA | National Water-Quality Assessment Program (of the USGS) |
| NCA | National Climate Assessment (U.S.); NCA3 refers to the third assessment report, published in 2014; NCA4 to the fourth assessment reports, published in 2017 (vol. 1) and 2018 (vol. 2); NCA5 due out in 2022 |
| NCAR | National Center for Atmospheric Research |
| NCCARF | National Climate Change Adaptation Research Facility (AU) |
| NCEI | National Centers for Environmental Information (of the NOAA) |
| NCEP | National Centers for Environmental Prediction |
| NDVI | Normalized Difference Vegetation Index |
| NEPA | National Environmental Policy Act (U.S.) |
| NESP | National Environmental Science Programme (AU) |
| NGO | nongovernmental organization |
| NIDIS | National Integrated Drought Information System (U.S.) |
| nMDS | non-metric multidimensional scaling |
| NOAA | National Oceanic and Atmospheric Administration (U.S.) |
| NPDES | National Pollutant Discharge Elimination System (of the EPA) |

| | |
|---|---|
| NPS | National Park Service (of the USDOI) |
| NRCS | National Resource Conservation Service (of the USGS) |
| NSW | New South Wales, Australia |
| NWP | nationwide permit |
| NWR | National Wildlife Refuge (of the USFWS) |
| OPUS | Online Positioning User Service |
| OTA | Office of Technology Assessment |
| PCA | principal component analysis |
| PDSI | Palmer Drought Severity Index |
| PGP | general state programmatic permit |
| PHABSIM | Physical Habitat Simulation |
| PINCC | Programa de Investigación en Cambio Climático (Program for the Investigation of Climate Change, MX) |
| POD | Point of Diversion |
| PRISM | Parameter-elevation Regression on Independent Slopes Model dataset |
| PVC | polyvinyl chloride (or CPVC—chlorinated polyvinyl chloride) |
| R | R Foundation for Statistical Computing |
| RBI | riparian biotic index |
| RCC | Regional Climate Center (of the NCEI) |
| RCM | regional climate model |
| RCN-CRD | Research Coordination Network for the Colorado River Delta (U.S. and MX) |
| RCP | Representative Concentration Pathway |
| Reclamation | U.S. Bureau of Reclamation |
| REDESClim | Red de Desastres Asociados a Fenómenos Hidrometeorológicos y Climáticos (Network of Disasters Associated with Hydrometeorological and Climatic Phenomena, of CONACyT) |
| REW | RiversEdge West |
| RGB | Rio Grande / Río Bravo or Rio Grande/Bravo is the international reach of this river that forms the border between Mexico and the U.S., with the state of Texas on the left side of the river and the Mexican states of Chihuahua, Coahuila, Nuevo León, and Tamaulipas on the right. For purposes of this guidebook, 'Rio Grande' (not 'Rio Grande/Bravo') is used when referring to non-international parts of the river (such as the Rio Grande through the U.S. state of New Mexico). |
| RGP | regional general permit |
| RGSM | Rio Grande Silvery Minnow |
| RGVA | Rio Grande vulnerability assessment |
| RGWF | Rio Grande Water Fund |
| RISA | Regional Integrated Sciences and Assessments (of NOAA) |
| RMRS | Rocky Mountain Research Station (of the USFS) |
| RTK | real-time kinematic |
| SAVS | System for Assessing Vulnerability of Species |
| SCENIC | Southwest Climate and Environmental Information Collaborative (a tool of the WRCC and SW CASC) |

| | |
|---|---|
| SCIPP | Southern Climate Impacts Planning Program (of NOAA-RISA) |
| SD | standard deviation |
| SDL | sustainable diversion level |
| SEMARNAT | Secretaría del Medio Ambiente y Recursos Naturales (Mexican Natural Resources Agency) |
| SGM | Servicio Geológico Mexicano |
| SHPO | State Historic Preservation Office |
| SI | Sonoran Institute |
| SIGA | Subgerencia de Información Geográfica del Agua (MX) |
| SIMPER | similarity percentage |
| SJPIS | San José de Pandos Irrigation Society |
| SMN | Servicio Meteorológico Nacional (Mexican National Weather Service) |
| SOP | standard operating procedures |
| SPRNCA | San Pedro Riparian National Conservation Area (MX and U.S.) |
| SRCC | Southern Regional Climate Center (of the NOAA) |
| SRES | Special Report on Emission Scenarios |
| SRLCC | Southern Rockies Landscape Conservation Cooperative |
| SWAT | Soil and Water Assessment Tool |
| SWPPP | Storm Water Pollution Prevention Plan |
| SWQM | Surface Water Quality Monitoring Program (of the TCEQ) |
| TCCC | The Coca-Cola Company |
| TCEQ | Texas Commission on Environmental Quality |
| TIN | triangular irregular network |
| TNC | The Nature Conservancy |
| TNRIS | Texas Natural Resources Information System (U.S.) |
| UNAM | Universidad Nacional Autónoma de México (National Autonomous University of Mexico) |
| USACE | U.S. Army Corps of Engineers |
| USA-NPN | U.S. National Phenology Network |
| USDA | U.S. Department of Agriculture |
| USDOI | U.S. Department of Interior |
| USFS | U.S. Forest Service |
| USFWS | U.S. Fish and Wildlife Service |
| USGCRP | U.S. Global Change Research Program |
| USGS | U.S. Geological Survey (of the USDOI) |
| UTM | Universal Transverse Coordinate System |
| VA | vulnerability assessment |
| VI | vegetation indices |
| WARSSS | Watershed Assessment of River Stability and Sediment Supply (of the EPA) |
| WRCC | Western Regional Climate Center (U.S.) |
| WWA | Western Water Assessment (a NOAA-RISA) |
| WWF | World Wildlife Fund |
| YDP | Yuma Desalting Plant |

# RENEWING OUR RIVERS

# 1

# A Case for Stream Corridor Restoration

*Mark K. Briggs, Osvel Hinojosa-Huerta, W. R. Osterkamp, Patrick B. Shafroth,*
*Carlos Sifuentes, Lindsay White, and Francisco Zamora*

## INTRODUCTION

This guidebook is about designing and implementing stream restoration. Given the state of rivers throughout the world, the need for river restoration may be more urgent than ever. "Rivers in Crisis" announced the cover headline of the journal *Nature* (September 29, 2010). Sapped by dams, with their overallocated waters supplying the needs of irrigated agriculture and seemingly evergrowing, thirsty populations, once mighty rivers like the Rio Grande / Río Bravo (RGB) in Mexico and the United States, the Tarim River in the Xinjiang Autonomous Province, the Huang He (Yellow) River of China, the Indus in southern Asia, the Nile in Africa, the Murray-Darling in Australia, and the Colorado of the western United States, among many others, have deteriorated dramatically and, in some cases, rarely reach the sea. In a recent global study, water security and biodiversity indices for *all* of the world's major rivers have declined as a result of pollution, water diversion, impoundment, fragmentation, introduced species, and a variety of other factors (Vorosmarty et al. 2010).

Before jumping to the conclusion that saving a river necessarily equates to large-scale, top-down efforts, it is interesting to note that in the same study that described the global deterioration of rivers, Vorosmarty and colleagues (2010) noted that localized efforts that incorporate both the needs of humans and local wildlife may be the most effective way to improve river conditions. Indeed, one of the tenets for developing this guidebook is that comprehensive restoration efforts almost always start small, often with a group of individuals (sometimes just one person, maybe you!) who have meager resources and are focusing on streams of compromised condition that, in many cases, flow through complicated sociopolitical landscapes.

Given that you are reading the guidebook, a particular river or stream has likely attracted your interest. Maybe the stream is near where you live or recreate or is a stream that has become a concern due to its deteriorated condition. Picture in your mind what your stream looks like today. Now envision stream conditions as you'd like them to be in the future. What are the main improvements that you see and how would they benefit wildlife and people? And now, the $64,000 question: How do you get there from here? In an ideal, but improbable, world, a raft full of experienced stream restoration practitioners would float around the bend of your stream and ask if they could help you answer this question.

This guidebook is our attempt to bring the seasoned restoration practitioners to you. The intended audience for *Renewing Our Rivers* is broad and may best be described as community and student leaders, public officials, natural resource managers and scientists who want to initiate and potentially even lead an effort to bring back a damaged stream, yet who may not have stream restoration experience. Keep in mind that, although the involvement of seasoned restoration expertise is essential to successful stream restoration, just as important are people who care about it and are good thinkers, organizers, planners, and leaders.

A good analogy to this guidebook might be books on home construction for homeowners with little construction experience but who want to play a strong role in designing and constructing their home. Such books will not turn a homeowner into a general contractor. Yet they will provide a start-to-finish overview that will educate a homeowner on the main aspects of construction that, in turn, will allow them to understand the construction process, avoid pitfalls, select a good contractor, save money, ask intelligent questions, and so on, all of which are needed for long-term success.

Similarly, although *Renewing Our Rivers* will not make you a stream restoration expert, it will provide you start-to-finish guidance, resources, lessons, and tools required to help you bring back your favorite stream or river. The practices and principles that will help your project succeed will be applicable to flows and riparian corridors ranging from small ephemeral streams to larger rivers with significant year-round flows. By picking up this guidebook, you are inviting us to be part of your stream restoration journey. We are honored!

The impetus for assembling this guidebook came out of a 2010 "lessons learned" stream restoration conference in Tucson, Arizona, that was attended by restoration practitioners from dryland (arid and semiarid) parts of southeastern Australia, northern Mexico, and the southwestern United States. The 2010 conference was the spark. Over the next decade, a wealth of research on stream restoration was conducted, additional experts and authors were involved, and their experiences and lessons learned were gathered and organized into this guidebook. Guidebook authors include scientists, managers, conservationists, engineers, lawyers, business people, and citizens representing a variety of state and federal agencies, nongovernmental conservation organizations, academic institutions, and communities. The river raft of experience that is joining you for this restoration journey is a well-balanced and enthusiastic group of stream practitioners with decades of restoration experiences on rivers in a variety of settings.

The guidance we provide comes from our own stream restoration experiences as well as other stream corridor restoration efforts. Specific organizing questions that we address include the following:

- What were the main lessons learned from these stream corridor restoration experiences?
- What goes into the development of thoughtful and realistic stream corridor restoration goals and objectives?

- What are the main stressors driving the hydroecological decline of the streams that we included in our research, and what can realistically be done to minimize—or at least reduce—their impacts?
- What scientific research should be undertaken to fill critical information gaps regarding current conditions, trends, and reasons for changes that have occurred?
- How might we quantify and secure environmental flows?
- How should we incorporate climate change knowledge into restoration planning and implementation to be as certain as possible that restoration efforts will be effective, regardless of which climate change scenario ultimately manifests?
- How can small-scale restoration projects (e.g., site-scale projects that were implemented over a couple of years) be scaled up both spatially and temporally for greater impact?

Building on these interests, this guidebook investigates the process of developing stream restoration projects and identifies the elements that are key to success, the pitfalls to avoid, and the resources available to help. The aim is to bring together the hard-earned lessons from past restoration experiences—many of which are undocumented, buried in end-of-project reports stored in dusty file cabinets, or relegated to the memories of the people who did the work (Briggs 1996; Bernhardt et al. 2007; Kondolf et al. 2007)—in a manner that guides future stream restoration efforts and improves your likelihood for success.

Before moving on, let's pause here a moment to define a few terms. First, given that it is the central theme of the guidebook, we want to make sure we clarify what we mean by "restoration." In a general sense, restoration refers to the sentiment of not being content with the status quo and actively searching for responses that reverse declines in water quality, ecosystem services, and loss of critical freshwater habitat (Bernhardt et al. 2007; Rohwer and Marris 2016). As defined for this guidebook, restoration is any action meant to shift a river's deteriorated hydroecological condition toward a state characterized by enhanced resilience, decreased vulnerability to the impacts of stressors, improved quality habitat for native flora and fauna, and enhanced ecosystem services for people. In this sense, we use the term more broadly than it is used in many ecological texts, where it often means to reestablish the hydroecological conditions that existed prior to significant human disturbance (Murdock 2008; Osterkamp 2008). The broader definition we employ here aligns well with that put forward by other members of the restoration community, which define ecological restoration as "the process of assisting the recovery of an ecosystem that has been degraded, damaged, or destroyed" (SER 2004; McDonald et al. 2016). One of the advantages of using a broader definition of restoration in this guidebook is that it inherently includes multiple possible endpoints, from taking a river back to its predisturbance state, maintaining current conditions (i.e., holding the line on continued deterioration), or achieving some degree of ecological improvement that may include establishing novel river ecosystems.

Second, what do we mean by "stream corridor restoration"? By including "corridor" in the title we are emphasizing the importance of having a watershed perspective. Injuries to the stream are often caused by impacts outside the stream's immediate aquatic and riparian environment and, in order for stream restoration to be successful, practitioners need to identify and address those impacts. In other words, a stream ecosystem is more than a stream of water, more than a channel, it is a corridor that includes flood plain or adjacent land that is sculpted by the water and sediment that is conveyed through the stream from its contributing watershed. The myopic perspective that stream restoration is limited to what happens between the banks of a stream needs to be done away with. Understanding how and why water and sediment runoff are changing at the scale of the watershed is essential to restoration success. It is with that broad spatial perspective in mind that we use the phrase "stream corridor restoration." With that important point noted, we understand that "stream corridor restoration" is a bit of a mouthful and, where appropriate, we will shorten to "stream restoration" in parts of the guidebook where the broader intent conveyed by "corridor" is inherently understood.

Third, the terms "rivers," "streams," and "creeks," are all used to refer to water flowing on the Earth's surface and, depending on context, can be pretty much interchangeable. Scale-wise, creeks are often seen as the smallest, with streams being in the middle, and rivers being the largest (see the USGS Water Science School website: https://water.usgs.gov/edu/earthrivers .html). Meanwhile, "arroyo" is a term coming out of dryland regions that typically refers to smaller drainages (a creek or stream) that dry out after the wet season. To avoid debate and confusion, we use the term "stream" from this point forward unless we are referring to a specific drainage system. For example, we may use the phrase "river restoration" in the context of discussing restoration actions that are specifically associated with the Murray-Darling River.

And, finally, fourth, herein we define "drylands" as arid and semiarid areas that have an imbalance between natural water availability and the moisture needed to sustain human activities and prospering native ecosystems. Although arid and semiarid zones (receiving, respectively, less than 250 mm, and from 250 to 500 mm, of precipitation annually), where potential evaporation typically exceeds precipitation, are emphasized, many of the broad concepts that are highlighted in the guidebook are appropriate and can be effectively applied to tropical and temperate regions as well. That stated, it is the aspect of water deficiency that we use to define drylands that is an important thread woven throughout the guidebook and one that, in particular, is emphasized in the case studies and the lessons that are learned from them.

## GUIDEBOOK ORGANIZATION AND STRUCTURE

The assembly of your stream restoration program involves a variety of steps that will lead to implementation. These steps need to be understood and connected from beginning to end when planning your restoration effort (Bernhardt et al. 2007; Hawley 2018; Yochum 2018). Superficially, a restoration project can be equated to putting together an "assembly required" purchase (like a shelf bought at a hardware store) that requires numerous steps to be completed in order to realize design specifications. A stepwise approach provides a path for progress as various steps are considered, addressed, and completed. For stream restoration, first we develop restoration goals and objectives and then we develop restoration tactics, followed by implementation, and so forth.

Of course, the analogy between assembling a shelf and a stream corridor restoration project can only be taken so far. A shelf is an inanimate object, while a stream is a highly dynamic ecosystem that sustains people and wildlife. A stream trickles, then floods, adjusts its channel,

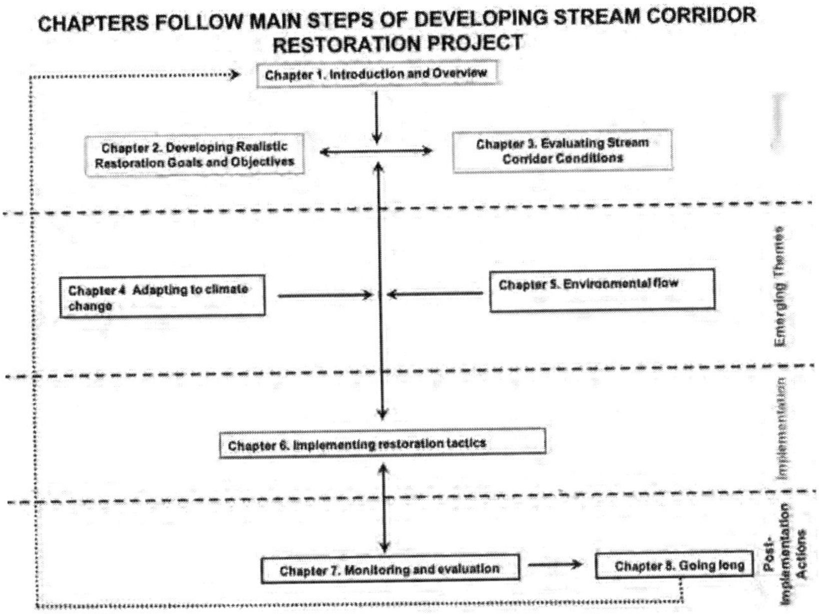

**CHAPTERS FOLLOW MAIN STEPS OF DEVELOPING STREAM CORRIDOR RESTORATION PROJECT**

**FIGURE 1.1** Flow chart depicting main phases and steps of developing a restoration program, and the guidebook chapters where the main steps are discussed. Double flow arrows indicate feedback loops between various steps. Adapted from Shafroth et al. 2008.

all the while affecting (and being affected by) the wildlife and people who live in or near it. For this reason, the steps of developing a stream corridor restoration project are rarely finalized with nary a glance back. Whereas steps for assembling a shelf are inherently unidirectional (step two cannot be undertaken until step one is completely finished), steps for developing a stream corridor restoration project are inherently iterative. The stream corridor restoration process is almost always a two-steps-forward and one-step-back (or sideways) dance that carries an umbra of frequent reassessment. Although iteration is inevitable, it is very helpful to have a sense of the direction forward and an awareness of what to be thinking about and reaching for. Providing that "sense of direction" is the overarching aim here.

This guidebook is structured so as to follow the general arc of developing a stream corridor restoration program (see figure 1.1):

i. Planning—This includes the formulation of restoration goals and objectives, as well as evaluating both the biophysical conditions of the stream corridor and the sociopolitical landscape that the stream passes through (chapters 1 to 3).

ii. Emerging themes and issues—Over the last several decades, two important themes have emerged as important considerations for developing stream corridor restoration programs: climate change and environmental flow (chapters 4 and 5).

iii. Implementation—Based on information gathered from previous chapters, the restoration plan developed for your stream should cover all aspects of who, where, when, and how identified restoration tactics will be carried out. Topical areas of focus in this chapter include riparian revegetation, eradication of nonnative plants, native fish conservation, and community-based efforts (chapter 6).

iv. Post-implementation—This phase focuses on monitoring, evaluation, and adaptive management, as well as considerations for the future and next steps (chapters 7 and 8).

In the context of introducing the organization and structure of the guidebook, we want to note that although we follow the arc of developing a stream restoration project, there are topics in every chapter (ministeps) that may need to be considered earlier on in the restoration process than is indicated by the chapter number. For example, chapter 7, which focuses on monitoring, is placed toward the end of the book, given its focus on monitoring, evaluating and maintaining the results of your stream restoration effort. However, monitoring of site conditions prior to implementing restoration tactics (which is the focus of chapter 6) is important in order to quantify conditions before and after implementation of restoration tactics. Another example is the importance of securing the legal permits you may need to implement your restoration tactics. This topic is covered in chapter 6 as part of implementing your restoration tactics, yet should be considered early in the restoration process given the time it can take to secure necessary permits. Given these temporal nuances, we encourage you to read through the entire guidebook (underline parts that are particularly relevant to your restoration situation and timeline, cross out parts that are not) and make an actionable schedule pertinent to your own restoration project.

Also, in the context of discussing the guidebook's organization, we want to acknowledge the importance we place on including citations. That is, central statements on a variety of stream restoration topics are followed by citations of authors whose work form a body of knowledge on the particular point that is being made. If you read scientific literature, you are used to seeing citations. They are par for the course. However, for other readers, we understand how text laden with numerous citations can be off-putting. If you are in this camp, we ask that you bear with us. Pass citations by if they concern topics that you are not particularly interested in. However, for topics you are interested in, please see citations as invitations that open doors for deeper dives into thematic areas that we were not able to adequately cover in the confines of a single guidebook. Indeed, entire books are written on some of the restoration topics that we were able to mention only briefly. A well-placed citation can be a key that opens up an entire world of literature, knowledge, and even potential collaborators.

## TAKING STOCK: RESTORATION PRINCIPLES GAINED FROM PAST STREAM CORRIDOR RESTORATION PROJECTS

### Top Eight Principles Learned About Stream Restoration

Setting the stage for the remainder of the guidebook, a top eight list of principles learned from past stream restoration efforts are summarized below that were distilled from the 2010 Lessons Learned Conference as well as from subsequent research—for example, standards and strategies for successful stream restoration (Palmer et al. 2005; Wohl et al. 2015; McDonald et al. 2016; Hawley 2018) and up-to-date investigations on climate change (IPCC 2018). The principles were selected based on their importance to the long-term success of stream corridor restoration.

With this list, we embark together on your stream restoration journey. As you read through the list and subsequent chapters, there will be topics and lessons more pertinent to your stream and restoration objectives than others, but all are nuggets gained from the hard-fought failures and victories of seasoned stream practitioners. Some may resonate with you immediately, others might seem a bit abstract or even irrelevant to restoring your stream. We encourage you to take in all the lessons and experiences, even the ones that at first glance may seem peripheral to your specific situation.

*Principle 1. Stop the bleeding, please!* Whether you are a first responder at a vehicular accident or a stream practitioner considering a damaged stream, the first rule is the same: remove the victim from harm. For the accident, remove victims from the road. For the stream practitioner, identify and eliminate (or reduce the impacts of) the main stressors that are causing the stream to decline. Stream ecosystems are resilient and can strongly bounce back without further restoration input after a main stressor is addressed (Briggs 1996; Srivastava and Vellend 2005; Shafroth et al. 2008; Côté and Darling 2010; González et al. 2017). Indeed, this may be the most important fork in the restoration road. If the major stressors associated with your target stream cannot be addressed, it is likely that you will be left working on the fringes of restoration, making only minor progress toward real stream improvement. On the other hand, if a main stressor is eliminated or its impacts are significantly reduced, there may be great potential for dramatic improvement of stream conditions at a much larger scale.

*Principle 2. Restoration success hinges on clear objectives and adequate planning.* Many of the practitioners involved in the case studies included in this guidebook note the cost of jumping too soon from planning their restoration response to implementing restoration tactics. You must plan before you build, and what goes into planning your restoration response is the central theme of chapter 2. Although planning your stream restoration response is an essential first step, an important key to success is connecting such planning actions as the setting of goals and objectives all the way through implementation and evaluation.

*Principle 3. You have to understand the processes that underpin the condition of your stream and its watershed.* Four key questions need to be answered to form the foundation for effective stream restoration design and implementation: (1) What are the current hydro-ecological processes that shape the condition of your stream? (2) How have these processes changed? (3) Why have these processes changed? and (4) How can the stressors that are driving changes in process be addressed? These key questions can only be addressed via a thoughtful assessment program that emphasizes an understanding of water and sediment flux in the stream corridor. Scorecard or cookbook evaluation approaches should be avoided. Chapter 3 is devoted to assessing stream corridor conditions, providing a framework that requires stream practitioners to put on their thinking caps as they address specific evaluation questions related to stream corridor conditions.

*Principle 4. Factor climate change into the stream restoration equation.* Human-induced climate change has already warmed the planet by 1°C, or 1.8°F (NCA4 2017). The Intergovernmental Panel on Climate Change (IPCC) warns the globe will continue to warm by another 0.5°C, or 0.9°F, in the next ten to thirty years, with greater changes at higher latitudes (IPCC 2018). Among the many ramifications of this rapid warming and other aspects of climate change (e.g., shifts in precipitation) are its impacts on streams. Climate change will affect flood and drought intensities, altering stream sediment and water fluxes, channel morphology, and water quality and temperature, among other parameters (Baron et al. 2002; Sabater and Tockner 2010), all of which will impact restoration objectives (i.e., what realistically can be realized) and the effectiveness of the tactics we put forward to achieve them (Perry et al. 2015). In this light, stream restoration planning has to be both "climate-informed" and "climate-adapted," such that goals, objectives, and strategies will be successful no matter which climate scenario might become reality. Background and sources of information on climate change as well as a framework for incorporating climate change into the stream corridor restoration planning process is presented in chapter 4.

*Principle 5. Protect streamflow.* Given the multitude of threats to streamflow, protecting or augmenting streamflow for environmental purposes ("environmental flow") is an essential component of long-term stream corridor restoration. This may be particularly relevant to dryland streams. For those of you contemplating small pilot projects with minimal budgets, developing an environmental flow program may not seem practical. Yet successful stream restoration is inherently a long-term process, and the risk of not protecting and/or augmenting streamflow may compromise the long-term viability of your project. Chapter 5 provides a framework, numerous sources of information, and case studies for developing an environmental flow program.

*Principle 6. The implementation of stream restoration tactics needs to consider the hard-fought lessons gained from past stream restoration efforts.* A critical threshold is crossed as you and your team go from planning your stream restoration response to implementing restoration tactics. There are important considerations to embrace that will enhance the likelihood for success, as well as pitfalls to avoid. Taking advantage of the hard-fought lessons gained from the implementation of past stream restoration efforts is critical. Chapter 6 reviews key lessons learned from the implementation of stream restoration projects that focus on enhancing community involvement, managing nonnative invasive plants, conducting native riparian revegetation, and implementing native fish conservation and reintroduction.

*Principle 7. Monitoring is essential to evaluate and report progress.* Lack of adequate funding, resources, and monitoring standards; limited information on how to implement monitoring; and the inherent challenge in isolating restoration effects in complex, dynamic systems have all been cited as reasons for the lack of efforts to monitor results of stream restoration efforts. When restoration results are not monitored and documented, it is a loss to the entire restoration community. In addition, monitoring is a prerequisite to adaptive management that allows stream practitioners to access progress toward restoration objectives and to address unanticipated changes and keep the project on course. Funders and practitioners alike need to demand that monitoring not only takes place but is an integral part of the stream restoration program. Questions regarding how stream practitioners will monitor progress toward achieving restoration goals and objectives should receive as much attention as questions that focus on what tactics will be used to bring back the deteriorated stream (i.e., the restoration tactics). Monitoring and evaluating stream restoration results is the theme of chapter 7.

*Principle 8. Strong political support, public recognition, and community participation can produce real and long-lasting results.* Stream systems all over the world have deteriorated as a result of human impacts that often have a long history in their respective watersheds—in some cases, hundreds of years. Minimal restoration budgets of less than decadal time frames can only do so much in the face of such monumental impact and momentum. Therefore, to be effective, the stream restoration response needs to be long-term, and having a long-term perspective should be a constant ambition for practitioners and funders alike who are looking to make a difference.

▪ ▪ ▪

To inspire a temporal perspective that goes beyond the life of a one-and-out grant, it is critical that we—as river restoration practitioners, river managers, scientists, and people living and working on rivers—actively foster sociopolitical support and community participation in river restoration by emphasizing the benefits of restoration to human well-being. To effectively renew our rivers, local citizens, community leaders, businesses, farmers, ranchers and many others need to be strongly and equally engaged with scientists, managers, and practitioners; all four elements are needed for successful river restoration on a grand scale (Bernhardt et al. 2007; Gonzalez et al. 2017; Hawley 2018). In chapter 8, the guidebook closes with strategies that will foster continued restoration of your stream well after your initial restoration project is completed.

So please continue to turn the pages as we travel together on this stream corridor restoration journey. The ultimate goal of this journey is to provide the steps, knowledge, and sources of information that will allow your restoration effort to be successful (to meet its objectives) and provide multiple benefits to people and wildlife far into the future. Your community and your progeny will appreciate your efforts. Now, hold on tight, we are about to take off.

## REFERENCES

Baron, J. S., N. L. Poff, P. L. Angermeier, C. N. Dahm, P. H. Gleick, N. G. Hairston, R. B. Jackson, C. A. Johnston, B. D. Richter, and A. D. Steinman. 2002. "Meeting Ecological and Societal Needs for Freshwater." *Ecological Applications* 12:1247–60.

Bernhardt, E. S., E. B. Sudduth, M. A. Palmer, J. D. Allan, J. L. Meyer, G. Alexander, J. Follastad-Shah, and L. Pagano. 2007. "Restoring Rivers One Reach at a Time: Results from a Survey of U.S. River Restoration Practitioners." *Restoration Ecology* 15 (3): 482–93.

Briggs, M. K. 1996. *Riparian Ecosystem Recovery in Arid Lands: Strategies and References.* Tucson: University of Arizona Press.

Côté, I. M., and E. S. Darling. 2010. "Rethinking Ecosystem Resilience in the Face of Climate Change." PLoS 8 (7). https://doi.org/10.1371/journal.pbio.1000438.

González, E., A. Masip, E. Tabacchi, and M. Poulin. 2017. "Strategies to Restore Floodplain Vegetation After Abandonment of Human Activities." *Restoration Ecology* 25:82–91.

Hawley, R. J. 2018. "Making Stream Restoration More Sustainable: A Geomorphically, Ecologically, and Socioeconomically Principled Approach to Bridge the Practice with the Science." *Bioscience* 68 (7): 517–28.

IPCC. 2018. "Summary for Policymakers." In *Global Warming of 1.5°C: An IPCC Special Report on the Impacts of Global Warming of 1.5°C Above Pre-industrial Levels and Related Global Greenhouse*

*Gas Emission Pathways, in the Context of Strengthening the Global Response to the Threat of Climate Change, Sustainable Development, and Efforts to Eradicate Poverty*, edited by V. Masson-Delmotte et al. Geneva: World Meteorological Organization.

Kondolf, G. M., S. Anderson, R. Lave, L. Pagano, A. Merenlender, and E. S. Bernhardt. 2007. "Two Decades of River Restoration in California: What Can We Learn?" *Restoration Ecology* 15 (3):516–23.

McDonald, T., C. D. Gann, J. Jonson, and K. W. Dixon. 2016. *International Standards for the Practice of Ecological Restoration—Including Principles and Key Concepts.* Washington, D.C.: Society for Ecological Restoration.

Murdock, J. N. 2008. "Stream Restoration." In *Applications in Ecological Engineering*, edited by S. E. Jørgensen, 113–19. Amsterdam: Elsevier.

NCA4. 2017. *Climate Science Special Report: Fourth National Climate Assessment.* Vol. 1. Edited by D. J. Wuebbles et al., 333–63. Washington, D.C.: U.S. Global Change Research Program. https://doi.org/10.7930/J0VM49F2.

Osterkamp, W. R. 2008. *Annotated Definitions of Selected Geomorphic Terms, and Related Terms of Hydrology, Sedimentology, Soil Science, Climatology, and Ecology.* U.S. Geological Survey Open-File Report 2008–1217. https://pubs.er.usgs.gov/.

Palmer, M. A., E. S. Bernhardt, J. D. Allan, P. S. Lake, G. Alexander, S. Brooks, J. Carr, et al. 2005. "Standards for Ecologically Successful River Restoration." *Journal of Applied Ecology* 42:208–17.

Perry, L. G., L. V. Reynolds, T. J. Beechie, M. J. Collins, and P. B. Shafroth. 2015. "Incorporating Climate Change Projections into Riparian Restoration Planning and Design Ecohydrology." *Ecohydrology* 8 (5): 863–79. https://doi.org/10.1002/eco.1645.

Rohwer, Y., and E. Marris. 2016. Renaming Restoration: Conceptualizing and Justifying the Activity as a Restoration of Lost Moral Value Rather Than a Return to a Previous State. *Restoration Ecology* 24 (5): 674–79.

Sabater, S., and K. Tockner. 2010. "Effects of Hydrologic Alterations on the Ecological Quality of River Ecosystems." In *Water Scarcity in the Mediterranean: Perspectives Under Global Change*, edited by S. Sabaterand and D. Barcelo, 15–39. Berlin: Springer Verlag.

Shafroth, P., M. K. Briggs, A. Sher, V. Beauchamp, and M. C. Scott. 2008. "Planning Riparian Restoration in the Context of Tamarix Control in Western North America." *Restoration Ecology* 16:97–112.

Society for Ecological Restoration (SER). 2004. *SER International Primer on Ecological Restoration. Society for Ecological Restoration.* Science and Policy Working Group, Washington, D.C. http://ser.org/resources/resources-detail-view/ser-international-primer-on-ecological-restoration.

Srivastava, D. S., and M. Vellend. 2005. "Biodiversity-Ecosystem Function Research: Is It Relevant to Conservation?" *Annual Review of Ecology, Evolution, and Systematics* 36:267–94.

Vorosmarty, C. J., P. B. McIntyre, M. O. Gessner, D. Dudgeon, A. Prusevich, P. Green, S. Glidden, et al. 2010. "Global Threats to Human Water Security and River Biodiversity." *Nature* 467:555–61.

Wohl, E., S. N. Lane, and A. C. Wilcox. 2015. "The Science and Practice of River Restoration." *Water Resources Research* 51 (8):5974–97. http://doi.org/10.1002/2014WR016874.

Yochum, S. E. 2018. *Guidance for Stream Restoration.* U.S. Department of Agriculture, Forest Service, National Stream and Aquatic Ecology Center, Technical Note TN-102.4. Fort Collins, Colo. https://www.fs.fed.us/biology/nsaec/assets/yochumusfs-nsaec-tn102-4guidancestreamrestoration.pdf.

# 2

# Stream Corridor Restoration

SOME ASSEMBLY REQUIRED

*Mark K. Briggs, Eduardo González, W. R. Osterkamp, Patrick B. Shafroth,*
*and Francisco Zamora*

## INTRODUCTION

In this chapter, we focus on planning your stream restoration project. You need to have a plan to know where you are going. "If you don't know where you are going, you'll end up someplace else" (Yogi Berra). Thoughtful planning at the outset provides the foundation for all actions that follow, increases the likelihood of avoiding costly mistakes, and ultimately provides for the success of your entire restoration project (Shafroth et al. 2008; Hermoso et al. 2012; Wohl et al. 2015). Much like the carpenter's maxim—take your time, measure twice, cut once—it is critical to take the necessary time during the planning phase to assemble the right people and gather the necessary data and information.

Of course, you should not spend all your time, funds, and energy on planning (eventually you need to implement!), but the more prepared you are when you complete the planning phase, the more likely your efforts will succeed. As with almost everything, there are bookend extremes to the amount of planning that is done, from doing no planning at all to a never-ending planning process. As the late Dr. Julio Alberto Carrera López (former director of Noreste y Sierra Madre Oriental, CONANP, Mexico) stated as part of planning binational restoration actions along the RGB (which constitutes over nine hundred miles of the international boundary between the United States and Mexico), "All we seem to do is meet and talk about how much we don't know. In this room, we know enough to begin!" As important as planning is to restoration success, knowing when you are sufficiently informed to begin implementing restoration tactics can be subjective. Certainly, the depth and length of planning will be determined by the scope of your project. The U.S. Army Corps of Engineers (USACE) spent millions of dollars and more than five years to plan and review restoration tactics for the Mississippi Delta (Orth et al. 2005). Most of the rest of us with project budgets in the tens of thousands of dollars don't have that luxury. Given the great caveats of varying project objectives and stream conditions, a rule of thumb is to divide your total project budget as follows: allocate one-quarter of the budget to planning activities, one-half to implementation, and one-quarter to monitoring. Moreover, particularly if you are operating on a small budget, keep in mind that effective information-gathering and planning can be accomplished inexpensively and provide great long-term dividends.

This chapter will focus on three key parts of planning your stream restoration response: (1) developing your restoration team to provide the energy and expertise you'll need to not only develop restoration goals and objectives, but all other aspects of your stream-restoration response; (2) developing thoughtful and realistic stream restoration goals and objectives; and (3) prioritizing your restoration response: how your goals and objectives can provide the basis for this and the criteria and strategies that can be used to do it. Developing restoration goals and objectives, as well as prioritizing the restoration response, will be greatly informed by another essential aspect of the planning phase: evaluating the conditions and trends within your stream corridor and its watershed, which is the focus of chapter 3.

## FORMING YOUR RESTORATION TEAM

Stream corridor restoration may begin with just you or a few interested individuals, but there are good reasons for creating a team. Foremost, it is nearly impossible to restore a stream on your own. Stream restoration is often a complex undertaking from both sociopolitical and hydroecological standpoints. This is not meant to discourage you but to emphasize the importance of forming a cohesive team to lead the effort and distribute responsibilities among people with diverse backgrounds and expertise.

Because most projects are conducted with limited funding and personnel, when forming your restoration team, strive for a group composition that is realistic and relevant. You may fail to recruit everyone you'd like, but forming a team with critical mass and diversity is essential. Team formation takes time, and changes will occur as your project matures, objectives expand, and funding for the project increases. For example, the inclusion of community leaders, politicians, and business leaders as integral team members may not occur until after you have completed a first-round restoration effort. That is, seeing tangible restoration results may be needed to inspire such important potential team members to leap on board your restoration journey. Chapter 8 discusses this issue and provides strategies for taking advantage of initial restoration results to garner the sociopolitical resources needed to carry your project far into the future. In addition, keep in mind that as you form your team there will invariably be an inner circle of team members who you confer with frequently and outer circle members who are consulted with more strategically.

Understand that some team members may have multiple areas of expertise; examples of the people, technical expertise, and types of representation that you may want on your team are highlighted below and divided into two categories: (1) the types of expertise that you may need, and (2) potential sources where such expertise may be secured from (i.e., the agencies, institutions, and organizations you may want to reach out to).

## Specific Types of Expertise That May Be Needed

- *Biologists*—Including one or more biologists on your team is usually essential. The biological expertise needed will be determined by the issues, species, and habitats that are central to the objectives of the stream restoration project.
- *Climate experts*—Climate change is occurring much more rapidly than even many scientists had anticipated, and adapting your stream restoration response to climate change is the focus of chapter 4. Thus, having an expert on your team from the start of program implementation may be necessary to evaluate the best available information on how best to adapt restoration goals, objectives, and tactics to rapidly changing climate conditions.
- *Consumptive and nonconsumptive water users*—Representatives of irrigation districts, mines, municipalities, businesses, anglers, boaters, and other water users are part of the broad "stakeholder and citizen" category below, but are singled out here because of the importance of securing their support and influence to protect water resources and developing win-win solutions for the environment and people.
- *Environmental flow experts*—If conditions describe dramatic changes in streamflow conditions, flow-management strategies that recognize the needs of native ecosystems may require input from someone with experience in quantifying and securing environmental flow.
- *Funders*—No matter where funding for your restoration project is coming from, including key personnel associated with those funding sources on your restoration team from the outset can provide a variety of benefits such as assuring the donor that funding is being used properly, strengthening trust and relations with the funder, technical input, and laying the foundation for additional funding (from the same funder or securing help from the current funder to identify additional funding opportunities).
- *Hydrologists and fluvial geomorphologists*—Stream channels convey water *and* sediment. A team member knowledgeable about how channels respond to changing fluxes of water and sediment at the watershed scale can provide the expertise necessary to develop realistic goals and objectives and to identify the science needed to fill information gaps.
- *Restoration specialist*—The team should include a specialist experienced in designing and implementing a restoration program. Planning and implementation are complicated processes that involve many steps. The restoration specialist must be able to guide the program from its inception to the identification of priority objectives and then on to subsequent steps. Having such expertise will help to ensure that key steps in the restoration process are not missed and that potential pitfalls are avoided, both of which will likely save time and money throughout the project.
- *Stakeholders and citizens*—Those who live and work along a stream are the ultimate stakeholders who will be most affected by both the deterioration of river conditions and the impacts of the restoration program that is developed. A strong stakeholder/citizen presence is essential for developing realistic project objectives that link environmental concerns with the needs of riverside communities. These individuals and groups will help you to address fundraising needs, provide site access, win public and political support, assist with monitoring, and provide long-term maintenance and management of restoration sites after the project is completed. Some may even become directly involved in the implementation of your restoration effort (see figure 2.1).

**FIGURE 2.1** Trained work crews composed of residents of several riverside communities along the RGB conduct a substantial part of the restoration work taking place along this important river that forms the international border between Mexico and the United States. Photo by Javier Ochoa-Espinoza, CONANP.

## Potential Sources of Expertise

- *Academic institutions*—College and university departments are sources of natural resource expertise and information, particularly if they are conducting relevant studies in your watershed. Investigations of natural resources, land-use change, socioeconomic conditions, and changing demographics can add great value to your restoration program. Professors and students should be considered for inclusion on your core team.
- *Consultants*—Stream restoration practitioners often develop contracts with consultants to help them carry out restoration tasks in a timely manner. Given the range of activities inherent in stream restoration (evaluating stream conditions, implementing restoration tactics, monitoring results, and everything in between), consultants can be in a single consulting firm or comprise a mixture of the types of individuals on this list. As was mentioned above, hiring locally can help strengthen community support and is recommended when possible (see figure 2.1).
- *Department of Agriculture*—Compared to land and water management agencies (mentioned below), the Department of Agriculture may not leap to mind as a federal agency that would provide restoration support. Yet agricultural departments in many countries can be a critical source of natural resource expertise and information, as well as funding. In Australia, the National Water Initiative is housed under the Department of Agriculture and focuses on the sustainable management of the nation's rivers, wetlands, estuaries, and aquifers (Kingsford et al. 2005). In Mexico, La Secretaria de Agricultura y Desarrollo Rural supports a variety of

rural development programs that can overlap with the objectives of a variety of environmental conservation initiatives, including stream restoration efforts. In the United States, the Natural Resource Conservation Service provides expertise and funds a variety of conservation initiatives.

- *Nongovernmental organizations (NGOs)*—NGOs that work in the region where your stream is located can assist with communication, coordination, and technical expertise. NGOs in the southwestern United States and northern Mexico include The Nature Conservancy, the Environmental Defense Fund, Sky Island Alliance, the Audubon Society, Pronatura, Profauna, and the Sonoran Institute. In Australia, the Australian Conservation Foundation, the Australian Wildlife Society, and the Australian Wildlife Conservancy are examples of well-established NGOs that work throughout the continent on a variety of natural resource conservation issues, including stream restoration. Another positive benefit of including NGOs in your restoration team is their inherent flexibility in working with a diversity of constituents (perhaps more so than many natural resource agencies) and may be able to facilitate cooperation among federal and state agencies, businesses, institutes, communities, the political realm, and others across state and country boundaries.

- *Land-management agencies*—Reaching out to land-management personnel may be essential if your watershed includes lands managed by federal or state land agencies. If the pertinent agencies have resource-management missions, they likely have science, management, and conservation programs that can be advantageous to your restoration project. Federal sources of information can fill data gaps and help to establish project implementation and monitoring. Both water and land-management agencies are often staffed with hydrologists, biologists, and related personnel who can provide expertise and guidance on many facets of your restoration program.

- *State agencies and local governments*—Many state and local government agencies have natural resource departments that generate similar products as do federal agencies. Each state of the western United States has a department that manages and regulates fish and game; five states and three territories of Australia maintain similar natural resource departments, such as the Department of Environment and Resource Management in Queensland. The northern states of Mexico—Sonora, Chihuahua, and Coahuila—also have natural resource departments, and many counties and municipalities in the United States support natural resource missions for land management, monitoring, and restoration activities (such as the Pima County Flood Control District in Arizona). Finally, many First Nation governments also have active water rights and natural resource departments that collect and maintain a variety of data as well as conduct stream restoration (e.g., the Colorado River Indian Tribes and Quechan Nation along the lower Colorado River, southwestern United States). Finally, many local governments—cities, towns, boroughs (delegaciones in Mexico), counties, parishes, and so on—have natural resource departments and expertise whose involvement can help not only to fill key information gaps but also to strengthen local support for your restoration effort.

- *Water management agencies*—In addition to land-management agencies, agencies that plan, develop, distribute, and manage water resources can be invaluable participants in your restoration effort. These include the Department of Environment and Bureau of Meteorology in Australia, La Comisión Nacional del Agua (CONAGUA) in Mexico, and the USACE and the U.S. Bureau of Reclamation (Reclamation) in the United States.

# DEVELOPING GOALS AND OBJECTIVES FOR STREAM CORRIDOR RESTORATION

There is nothing more important to the success of a stream restoration project than developing sound restoration goals and objectives. Restoration goals and objectives set the path for the entire restoration project. In essence, they provide the foundation that allows you and your team to develop appropriate and effective restoration tactics, project schedules, and roles (who does what), and determine how results will be evaluated (monitoring). All of which provide the basis for finalizing your budget, personnel, and equipment needs, among other considerations. For groups whose mission is to bring together a diversity of stakeholders to address stream decline (e.g., watershed partnerships), there may be no more important contribution to bringing back a damaged stream than developing strongly supported sound restoration goals and objectives.

In this section, we summarize the main considerations needed to develop sound restoration goals and objectives. In this context, it is important to underscore the iterative nature of stream corridor restoration. Although sound restoration goals and objectives are required to move confidently forward, they will also be informed by what you learn as your restoration project is implemented and as project results are monitored and evaluated.

It is important to note that restoration goals and objectives developed during the planning phase are not set in stone, never to be altered no matter the results of implementing restoration tactics. View stream restoration as an adaptive experiment with multiple possible outcomes (Woelfle-Erskine et al. 2012). What you learn from your stream restoration "experiment" as your restoration project is rolled out will provide additional information that will allow you to thoughtfully adapt and adjust your restoration goals and objectives to the hydroecological and socioeconomic realities of the landscape your stream passes through. That stated, it is essential at the onset to develop sound restoration goals and objectives that are based on the best information you have.

In our experience, stream restoration goals and objectives are either (1) developed at a broad spatial scale (e.g., watershed scale) and then used as a lens to identify locations in the watershed that have the environmental, land-use, socioeconomic, political, and other characteristics deemed most suitable for achieving them; or (2) based on the environmental, land-use, socioeconomic, and political characteristic of a specific part of the watershed (from sub-basin to site scale) (Shafroth et al. 2008). An example of the former might be using the results of a watershed-scale evaluation of streamflow, water quality, and other characteristics to identify stream reaches in your watershed that have characteristics most suitable for the reintroduction and long-term viability of native fish. An example of the latter would be an evaluation of biophysical, socioeconomic, and political conditions along a particular part of the basin (e.g., reach of a stream) that indicate how far local entities who want to improve conditions along their stream may be able to realistically move the restoration needle (i.e., how far the needle can be pushed toward wild and natural conditions).

It is also true that the development of restoration goals and objectives can be a mix of the two approaches. For example, restoration goals can be used as a filter that allows you to narrow geographical focus within a watershed to locations (stream segments, reaches, or sites) where restoration efforts would be most likely to achieve those goals. Then, stream restoration objectives are fine-tuned and adapted to the realities of on-the-ground conditions. For example, in the southwestern United States and northern Mexico, expanding riparian habitat for the yellow-billed cuckoo (*Coccyzus americanus*) has become a conservation priority in both countries (at least for the western segment of the yellow-billed cuckoo population). In this region, habitat for this threatened bird can include the establishment of both obligate riparian trees (e.g., cottonwoods [*Populus. Fremontii*] and willows [*Salix gooddingii*]) as well as such nonobligate riparian trees as mesquite (*Prosopis* spp.). If results of evaluation indicate low water availability and high soil salinity along key reaches of your stream (areas where biologists feel that expansion of yellow-billed cuckoo habitat would be critical), the restoration goal of expanding habitat for yellow-billed cuckoos could still be met by adjusting the restoration response at reach scale to an emphasis on establishing mesquite and other nonobligate riparian species over obligate riparian species.

Regardless of the starting point, stream restoration goals and objectives set expectations about the extent to which we can reverse trends in stream biophysical conditions and about how far we can realistically achieve more desirable conditions. This requires an understanding of the interplay of a host of biophysical, socioeconomic, and data-availability factors associated with the project, the stream corridor, and its watershed. These may include the hydroecological condition of the stream corridor and human factors such as political and community support, as well as legal and socioeconomic considerations (Shafroth et al. 2008; Beechie et al. 2010; Palmer et al. 2014b; Wohl et al. 2015; Jorda-Capdevila and Rodríguez-Labajos 2016; González et al. 2017a). We will begin with developing stream restoration goals and then move on to developing stream restoration objectives.

## Developing Restoration Goals

In this guidebook, goals are characterized as lofty and broad in scope and describe the long-term intentions (over multiple decades) of your stream corridor restoration program. Restoration goals can stem from a vision statement that describes desired conditions. The difference between vision and goal is that a well-defined goal describes the overall purpose of restoration, including such themes as improving natural processes, aesthetics/recreation, riparian habitat and floodplain conditions, biodiversity, and water quality. The goals that are selected and prioritized will underpin the finalization of restoration objectives and have profound consequences on how restoration is planned and implemented (Falk et al. 2006; Shafroth et al. 2008; McDonald et al. 2016).

If the goal is vaguely described, assumptions about the project's intent and expectations may be in question (Shafroth et al. 2008). During this first step of planning, agree upon a preliminary restoration goal, then vet it with your team, distribute it to others, and modify it based on comments. The restoration goal should also include a discussion of spatial scale, which can vary from small revegetation projects to watershed efforts that reconnect habitat at the landscape scale. Don't feel discouraged if your project is at site scale with a modest budget. Keep in mind that, although large-scale restoration goals and objectives at the watershed scale can be the most effective approach for habitat restoration (Bohn and Kershner 2002; Laub and Palmer 2009), realizing restoration at such a large spatial scale is often accomplished by building on restoration projects at the subwatershed scale (Wohl et al. 2015). Indeed, one of

> ## CASE STUDY 2.1
> Example of a Restoration Goal: Pecos River, New Mexico
>
> Along the Pecos River in New Mexico, the Pecos bluntnose shiner is one of two subspecies of *Notropis simus* (Cope) that is currently restricted to this tributary of the RGB (Hatch et al. 1985). Decline of the species is attributed to stream desiccation, impoundment, and competition with nonnative fish (Hatch et al. 1985; Brooks et al. 1991; Dudley and Platania 2007; Hoagstrom et al. 2008). The U.S. Fish and Wildlife Service was tasked to develop a recovery plan for the Pecos bluntnose shiner that includes the goal
>
> > to ensure survival of the Pecos bluntnose shiner in the wild and provide habitat essential for its recovery. This goal will be realized when sufficient protection is provided to ensure appropriate flows and water quality are maintained in designated critical habitat, and when additional populations of the shiner are established in suitable areas of historic occupancy.
>
> (U.S. Fish and Wildlife Service 1992) (see also Case Study 6.8).

the main purposes of pilot-scale projects is to foment awareness about the need to scale up (Gonzalez et al. 2017a). Regardless, begin at a scale that is realistic to you and your restoration team and broaden it as restoration progress is made (see Case Study 2.1).

## Developing Restoration Objectives

Moving past broad goals, we define restoration objectives as outcomes that last up to a decade and are tangible, quantifiable, and achievable (Shafroth et al. 2008). Restoration objectives describe the time frame and destination of your project and provide a foundation for identifying priorities and tactics. They are the means to assess success, which is how well the results of your project meet stated restoration objectives (Briggs 1996; Kentula 2000; Palmer et al. 2005; Wohl et al. 2015).

In the context of discussing restoration objectives, another term that is often used is "benchmark," which is used to compare progress toward stated goals or objectives relative to the desired time frame for accomplishment. As such, establishing reasonable benchmarks provides an effective way for managers to evaluate conservation and restoration efforts midstream (so to speak), allowing opportunities for course correction if results do not reveal adequate progress. In this sense, near-term benchmarks can be developed for multiple time periods (for example, at two-year increments within a decadal objective), with the assumption that the level of confidence and detail for the two-year restoration objective will be higher than for subsequent years (Hughes et al. 2005; McDonald et al. 2016). Assessing progress toward stated objectives at selected benchmarks may lead to modification of restoration tactics and/or the restoration objective itself, and is the heart of an adaptive-management program (see chapter 7).

So, what qualities does a sound stream corridor restoration objective have? In addition to being realistically achievable in a specified time frame, sound restoration objectives should be quantifiable (measurable), scientifically sound, economically defensible, and flexible (Briggs 1996; Palmer et al. 2005; Tear et al. 2005; Kondolf et al. 2007; Shafroth et al. 2008; McDonald et al. 2016). These qualities are discussed, below.

### Quality 1: Realistically Achievable

Restoration objectives should be realistically achievable. Having the objective to restore your stream corridor to its condition prior to significant human interference may be noble but unrealistic, owing to a variety of factors, including (1) a lack of appropriate reference systems, (2) rapid catchment change that makes comparisons to historic reference conditions inappropriate, (3) impacts of climate change on ecosystem processes and conditions, (4) the extent and distribution of invasive species, and (5) landscape-scale development and fragmentation of habitat and alteration of natural processes (Dufour and Piégay 2009; Thorpe and Stanley 2011).

Between the extremes of restoring a deteriorated stream to its wild, natural state and leaving the stream in its present deteriorated condition is a vast grey area where the realistic restoration objective for your stream resides. Addressing the following questions can help you and your team determine what is realistic:

- *Are the restoration objectives sustainable and resilient?* Restoration objectives must be self-sustaining and resilient to external perturbations to the extent that minimal follow up is required after completion of activities (Palmer et al. 2005; Miller and Hobbs 2007; McDonald et al. 2016). If reestablishing native species (as through revegetation of riparian plants or reintroducing native fish) is part of the restoration objective, is it likely that the reintroduced native species will survive to maturity and regenerate? If eradicating an invasive, exotic species is a central part of a restoration objective, will significant maintenance be required in the long run to keep the species at bay? Thoughtful responses to these types of questions require a strong knowledge of stream corridor biophysical conditions and trends (Gurnell et al. 2016), which will be the focus of chapter 3, as well as the physiological requirements and tolerances of the species that are central to the restoration objective (González et al. 2018).
- *Can the main drivers behind the deterioration of your stream be addressed?* Improvement of stream corridor conditions becomes more feasible when the causes of decline are reversed or eliminated (Hobbs and Norton 1996; Shafroth et al. 2008; Jones and Schmitz 2009; Laub and Palmer 2009; McDonald et al. 2016). Restoration objectives that describe returning a stream to near natural or wild conditions are not likely to be achieved (i.e., will be unrealistic) if such major stressors as river impoundment, flow diversion, or major upland land-use changes cannot be addressed. In such situations, objectives may need to be altered or the geographic focus moved to a stream reach (or another stream entirely) where stressors are having less of an impact or where there is greater opportunity to reduce or eliminate their impacts.
- *What level of socioeconomic and political support do you have for your restoration project?* Technically feasible and scientifically valid objectives cannot be realized without societal acceptance (Cairns 2000; Choi 2007; Gumiero et al. 2013). Community acceptance reduces socioeconomic and political constraints and may lead to long-term commitment and increased funding, personnel, and equipment (see Case Study 6.2). On the other hand, lack of sociopolitical

support may preclude conducting restoration and, at the very least, make pushing the needle toward reestablishing wild or natural stream conditions quite difficult (Hinojosa-Huerta et al. 2005; Miller and Hobbs 2007; Shafroth et al. 2008; Metcalf et al. 2015).

- *Is your time frame realistic?* The time required for restoration will vary with your objectives and the condition of the stream. Many years may be needed to improve conditions and reestablish native species in a stream system that has been highly altered. For example, several decades of concerted binational effort was needed to orchestrate the environmental pulse flow for the Colorado River Delta in northwest Mexico (see Case Study 5.5). As you consider what a realistic time frame may be for achieving your objectives, keep in mind that time frames depend on sources of funding that may have spans of less than a few years. A key challenge will be to work with the funding source to determine what can be accomplished within the allotted time. Extending a short restoration time frame even a few years may make an unrealistic restoration objective feasible by providing the time needed to gain or augment socioeconomic and political support or the time required for plants to become established and offer sufficient habitat for key wildlife species.
- *Are available resources commensurate with restoration objectives?* Knowledge of available resources—funding, staff, equipment, materials—is another key ingredient to gauging whether objectives are attainable or not. If the resources required to realize objectives exceed current availability, practitioners will have to adjust restoration objectives, scale, type of project, and/ or the time frame.

## Quality 2: Quantifiable

Restoration objectives must be quantifiable in a manner that allows measurement of progress toward their realization (Ruiz-Jaen and Aide 2005; Schroeder 2006; González et al. 2015). For a revegetation effort to establish a desirable riparian forest, would a few surviving trees of the thousands planted be interpreted as success or would much greater establishment rates be required? A quantifiable restoration objective provides the basis to answer this critical question by allowing monitoring of results to quantify progress toward realizing objectives (and, as such, provides a strong foundation for adaptive management). There are a variety of ways that this can be accomplished. Certainly, take advantage of experts on your team that are associated with your restoration objective. For example, if the restoration objective is to establish habitat for wildlife, experts of the focal species will likely be able to describe in quantifiable detail the habitat that is required (e.g., the density, diversity, and vertical complexity of the riparian plant community that provides habitat for a key bird species, per Van Riper et al. 2008).

Identifying reference sites may be useful. Reference sites can be used as model ecosystems to represent desired conditions at your restoration site, providing information that will help practitioners define restoration goals, determine the restoration potential of sites, and evaluate the success of stream restoration (Thorpe and Stanley 2011; McDonald et al. 2016). They offer a guiding image and provide a basis for comparing characteristics and quantifying change at your restoration site. In the context of stream restoration, reference sites do not necessarily have to be characterized by biophysical conditions that represent a "wild" state, or what the system was like prior to major human interference. The endpoint should, however, be site-specific and realistic and represent what is desired and can realistically be supported given the current biophysical and chemical condition of the stream corridor as well as the economic or sociopolitical landscape of your target stream.

Like control sites (see chapter 7), reference sites should have biophysical and climatic conditions that are similar to the restoration site. The difference is that reference sites are already characterized by desired conditions that practitioners feel can be realistically established at the target restoration site. As such, reference sites do not require restoration action to establish desired conditions. In comparison, control sites are typically in a similar degraded state as the site being restored. That is, channel conditions and biological assemblages should be less disturbed at reference sites than the targeted restoration site (Palmer et al. 2005; Morandi et al. 2014; González et al. 2015). Looking ahead, in chapter 4 we discuss the potential value of identifying reference sites (as well as reference time periods) that are characterized by climate conditions similar to those forecast for the region where your restoration project is located in future decades.

To select an appropriate reference site, the following factors need to be considered: similarities in elevation and watershed landscape; geology/soils; hydrology (frequency, magnitude, and duration of flow and its source—snowmelt/rainfall, gaining/losing stream); channel form (slope, sinuosity, streambank); and channel morphology (substrate, sediment load, alluvium) (Harrelson et. al. 1994). Other criteria for defining a reference site include community complexity, the presence or absence of exotics, floristic structure, and dominant land use (White and Walker 1997; Harris 1999; Dufour et al. 2019). Another benefit of examining reference sites is the potential to identify important ecological linkages or processes that need to be restored at your project site in order to realize restoration objectives (Williams 2011). In other words, reference sites can improve the design of your restoration project by highlighting differences that were not originally observed or accounted for.

Riparian preserves or conservation areas have been established in some watersheds and, depending on how well they are buffered from human impacts, could offer appropriate reference sites (Seastedt at el. 2008). For example, an area of Tumacácori National Historic Park protects just under two kilometers of the Santa Cruz River near Tucson, Arizona, and its rare cottonwood-willow and mesquite bottomland plant communities, two of the more endangered plant communities in the United States. The Santa Cruz River through Tumacácori was used as a reference site for a restoration project 65 km (40 mi.) downstream from where the Santa Cruz River passes through the San Xavier District of the Tohono O'odham Nation (a southwestern U.S. First Nation). Plant data collected at Tumacácori described plant diversity, percent cover, and density, and were used to develop detailed and quantifiable restoration objectives at the San Xavier site (see figure 2.2). In the largest river of the Iberian Peninsula, the Ebro River in northeastern Spain, the natural reserves of Alfaro and La Alfranca were established in river sections where natural meandering channel patterns and native plants have been maintained in the face of widespread flow regulation and flood control actions. The plant composition in these natural reserves was used as reference to develop vegetation stream restoration objectives for thirty-three restoration projects outside of these natural reserves (González et al. 2017b).

Two brief points conclude this tangential discussion on reference sites. First, however helpful identifying reference sites can be to developing restoration objectives, we caution to not throw all your eggs into the reference-site basket. It is always tempting to re-create the past, but streams follow complex trajectories that can make returning to a reference condition impossible (Dufour and Piégay 2009; Rohwer and Marris 2016). Much depends on the

**FIGURE 2.2** The riparian plant community along a protected portion of the Santa Cruz River through Tumacácori National Historical Park in southern Arizona (top, © Sonoran Institute 2020) was used as a reference site for developing plant species lists and densities for a riparian revegetation downstream on the Tohono O'odham Indian Reservation (bottom). Photo by Mark Briggs.

hydroecological condition of your stream, the sociopolitical and economic landscape that it passes through, and the extent that the drivers of hydroecological decline of your stream continue to persist. Second, there is a progressive shift in contemporary literature toward a more utilitarian view of restoration that maximizes ecosystem services and functions, implying a broadening of the reference-site concept to maximize processes and responses that not only focus on ecological conditions but also satisfy particular social or economic demands (e.g., improving aesthetics and access to river water for recreation and enhancing groundwater recharge for municipalities and agriculture) (Palmer et al. 2014b; Rohwer and Marris 2016). In summary, reference sites can provide information that is useful in developing certain aspects of your stream corridor restoration objective (e.g., types and quantities of plant species, per figure 2.2), but be careful how much weight you give them as you may not be able to duplicate reference-site conditions at your restoration site.

## Quality 3: Scientifically Defensible

Stream restoration objectives must be based on a scientifically sound understanding of the hydroecological condition of the stream corridor and its watershed (Rutherford et al. 2000; Laub et al. 2015; Wohl et al. 2015). Given stream conditions and trends, are restoration objectives achievable, and will proposed restoration tactics be effective toward achieving them? If the restoration objective is to reintroduce a native fish species into a stream where it has been extirpated, practitioners should conduct a data-based analysis of both the physiological requirements of the fish species that is being reintroduced and the current condition and trends of the stream environment, particularly those factors key to the survival and fecundity of the fish (such as base flow, seasonality of flooding, presence of suitable habitat, and water quality). Chapter 3 covers many of these topics.

## Quality 4: Economically Defensible

Practitioners should defend costs by demonstrating benefits. The likelihood of a restoration program being accepted and supported is enhanced if its direct value to society can be shown. For example, studies using hedonic analyses (an economic method used to determine the relative importance of variables that influence the price of a good or service) have shown that river restoration can increase the value of housing properties in depolluted urban streams of China (Chen 2017) or as a result of dam removal in Maine (Lewis et al. 2008) and Wisconsin (Provencher et al. 2008). Information comparing restoration costs with benefits distinguishes projects that may be expensive but of limited ecological or societal usefulness from those of high potential if funding is accessible. Cost-benefit analysis provides added incentive for funders or community and political leaders to be supportive of a restoration effort, particularly when resources are scarce (Acuña et al. 2013; Dave and Mittelstet 2017). Cost-benefit analysis is most useful when used to compare alternative strategies or potential scenarios. If one scenario indicates that positive results can be anticipated at low cost, such an analysis can be a huge selling point for moving forward with that scenario. On the other hand, if only minor improvements can be envisioned, you and your team may want to consider a different scenario that could involve scaling back objectives or using resources in a different setting that might produce more restoration bang for the money spent (see figure 2.3). Finally, investigate how society (or part of a society) may be willing to pay for restoration based on specific interests.

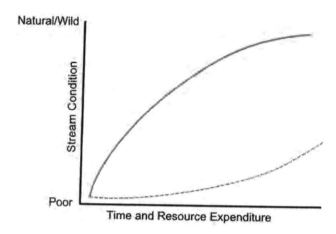

**FIGURE 2.3** Idealized graph showing two cost-benefit scenarios for restoration: one (the dashed line) is less desirable with little improvement in stream condition over time and with increasing resource expenditure; the other (the solid line) is more desirable, showing significant improvement with minor investment of time and resources. All else being equal, we want to be working on restoration efforts that reflect the solid line.

For example, local residents may be willing to pay increased county sales taxes for differing amounts of riparian restoration (Holmes et al. 2004), and anglers have been willing to pay for restoration to improve fishing opportunities (Robbins and Lewis 2008).

## Quality 5: Flexible

Restoration objectives need to specify expected accomplishments and a timetable for them, but ranges should also be included that acknowledge the inherently dynamic and unpredictable nature of stream environments that can produce variable results (Choi 2007; Nilsson et al. 2016). These ranges should be tight. Stating that a restoration objective will be achieved within five to one hundred years is certainly not appropriate. But, giving a range of a couple years (e.g., "our objective is to establish 1,000 cottonwoods and willows over a period of three to five years") is acceptable, providing temporal wiggle room to practitioners to conduct restoration actions when they are most likely to succeed. The importance of flexible timetables needs to be heard by those funding stream restoration efforts. The bottom line: funding and energy could be wasted if stream practitioners feel pressure to forge ahead with implementation when conditions are far from ideal (e.g., planting native obligate riparian plants during a particularly dry, hot year). The further in the future restoration objectives are projected, the less is our ability to predict how they might be affected by factors beyond project control. Consequently, ten-year objectives should build in greater flexibility than five-year objectives. For example, long-term monitoring in restored riparian forests in the Middle Rio Grande showed that low initial densities of cottonwood, a presumably negative result when considered in the short-term, led to cottonwoods of greater height that were better able to dominate the canopy cover (Bhattacharjee et al. 2009). Similarly, the positive effects on riparian vegetation of a multiyear restoration program in the lower Tarim River—a desert river of northwest China—were not evident until after a decade of yearly water releases (Zhang et al. 2013).

# PRIORITIZING RESTORATION LOCATION AND OBJECTIVES

As noted earlier in this chapter, practitioners may be planning their stream restoration response by considering multiple restoration objectives (e.g., reestablishing native fish and riparian habitat) and/or the locations where those objectives may best be achieved. If this is the case, there may be a need to prioritize when available resources are not sufficient to tackle all the objectives or implement restoration throughout the watershed all at once (see Case Study 2.2 below). The great benefit of prioritizing is that it directs actions to where maximum benefits are most likely to be achieved.

Stream restoration cannot be thoughtfully prioritized without sound restoration goals and objectives. Broad and effective prioritization also requires the development of appropriate criteria and indicators that are operational and appropriate in a broad range of spatial analysis and mapping contexts (Aguiar et al. 2011; Orsi et al. 2011; Macfarlane et al. 2017b). For stream restoration efforts whose central goal is to enhance biodiversity or protect key species, spatial prioritization of stream restoration response at the basin scale can be difficult. This can be due to the dendritic (branching) network structure of rivers, the unidirectional flow, patches of species possibly being distributed like pearls on a chain along the branches of the river networks, or a paucity of biodiversity data and information at appropriate scales (Dymond et al. 2008; Thomas 2014; Bourgeois et al. 2016).

A variety of decision-support systems have been used over the years to prioritize river-restoration actions (Llewellyn et al. 1995; Verdonschot and Nijboer 2002; Roni et al. 2002; González et al. 2018; Rohr et al. 2018). Such systems can be complex (Mobrand et al. 1997; Peters and Marmorek 2001; Game et al. 2018) or can be simple scoring systems that compare the relative merits of various projects (WPN 1999). In either case, the fundamental objective is to assemble, organize, and weigh information considered important to setting priorities (Cipollini et al. 2005). Simple scoring techniques, which offer the advantages of transparency and flexibility, are commonly used to rank restoration projects, most often in the form of a score sheet that addresses common evaluation criteria for stream and watershed restoration projects (e.g., the number of species that will benefit, project cost, and educational value). A disadvantage of such systems is that there may be a high degree of subjectivity in developing the system and completing a scorecard. On the other hand, complex scientific models can be difficult to implement because of extensive data and computation requirements (Comín et al. 2014, 2018). No matter how robust a prioritization approach may be, it has to be used in order to be effective. Methodical simplicity is therefore a key element in a successful prioritization tool (Peacock et al. 2012).

A challenge for prioritizing stream restoration projects with multiple objectives (like the Murray-Darling Case Study 2.2) is that different objectives may result in identification of different priority sites, establishment of different species, and selection of different restoration methods. Approaches must therefore be able to account for multiple objectives and enable their potential implications to be explored (Lamb et al. 2005; Orsi et al. 2011). Stream resto-

# CASE STUDY 2.2
Restoring Native Fish in the Murray-Darling River, Australia: One Restoration Goal Supported by Multiple Restoration Objectives

In 2004 in Australia, the Murray-Darling Basin Initiative—a cooperative arrangement between several state governments (New South Wales, Victoria, South Australia, and Queensland), the Australian Capital Territory, the Commonwealth of Australia, and a Community Advisory Committee—developed a fifty-year goal to rehabilitate native fish to 60 percent of estimated pre-European levels (Murray-Darling Basin Commission 2004) (see map in Case Study 6.6). This comprehensive goal is supported by thirteen objectives that include the following:

- Repair and protect key aquatic and riparian habitats deemed important for native fish species (target: a 20 percent increase in habitat-quality values);
- Revive links among terrestrial, wetland, and riverine ecosystems (target: functional processes and floodplain links established for 80 percent of remaining wetland habitats through improved flow management);
- Improve water quality (target: water quality to meet 90 percent of native-fish requirements);
- Provide adequate passage for native fish throughout the basin (target: reinstate 50 percent of migratory pathways and 50 percent of habitat for native species currently affected by structural barriers); and
- Augment community and partner ownership and support of the native-fish strategy (target: 90 percent of identified key partners actively engaged and 50 percent of the general community is aware and supportive of the strategy).

Lindsay White with a Murray Cod, one of numerous native fish species that are being protected under the Murray-Darling Basin Initiative.

ration responses can be prioritized at a variety of spatial scales, from potential restoration sites (at the reach scale or smaller) within a watershed or subwatershed, for tributary watersheds within a larger basin, or to entire watersheds at regional scales.

Below, we summarize criteria and considerations that can help practitioners to both prioritize stream-restoration response spatially and prioritize multiple restoration objectives. They are divided into two general groups: non-hydroecological (which include programmatic, socioeconomic, land-use, and organizational considerations) and biotic/hydroecologic. Not all will be appropriate to your situation. Take the ones to heart that are and skip the ones that are not.

### Non-hydroecological Considerations

- *Restoration goal*—The restoration goal that your team has established serves as the primary filter for prioritizing restoration objectives (for projects with multiple objectives) and the location of potential restoration sites. All else being equal, restoration sites and objectives that are most compatible with the goal should have the highest priority.

- *Mission*—Some objectives may be compatible with the restoration goal but peripheral to the mission of your organization. If the mission is incompatible, reassessment of objectives is needed.

- *Available tactics*—Lack of political will, community support, or funding may limit management options or restoration tactics that can achieve restoration objectives and their use at different restoration sites. Practitioners should also think about near-term and long-term strategies, conducting relatively easy-to-do actions in the near term that can serve as a foundation for addressing more difficult issues in a longer time frame. For example, low cost and cost-efficient restoration techniques and tactics, such as planting trees, may be strategically used to engage local communities through volunteering programs and generate a more favorable sociopolitical context to launch in the future more ambitious actions that today are not socially accepted, such as removing or reallocating levees, dikes, and flood defenses (González et al. 2018).

- *Minimizing or eliminating causes of deterioration*—High priority should be given to objectives and project locations where the impacts of a perturbation that is causing deteriorating stream conditions can be eliminated or reduced. Remember that stream corridor ecosystems are resilient even after years of impact and decline. Eliminating or reducing a central driver of stream decline will likely bring back some of the hydrological and ecological processes that are key to species recovery and ecosystem restoration. The *restoration of processes* is what needs to be embraced by the restoration community (Powers et al. 2018). If causes of stream decline can be directly addressed, the reestablishment of key processes may allow the stream to rebound dramatically and possibly without further intervention (see figure 2.4).

- *Citizen support and participation*—Heightened citizen or community support and participation can greatly increase the attractiveness of a project and thus raise its priority, regardless of whether the site is badly impaired (see Case Study 6.2, which focuses on the win-win between stream restoration and ranchers in Sonora, Mexico). Conversely, community disinterest or protest may render a project a low priority even if other considerations rank it much higher. Related factors that may affect prioritization include the level of political support and interest of funders in a particular location.

- *Ecosystem services*—Restoration of ecosystem services focuses on the human use or desire for often utilitarian objectives that specifically address such human needs as groundwater recharge

**FIGURE 2.4** Two photos of Cienega Creek Preserve near Tucson, Arizona (top taken in 1988, bottom in 2020) show recovery of native riparian plants following the abatement of stressors after Pima County acquired land in 1986 to create the Cienega Creek Natural Preserve (BLM 2003). Restoration objectives that aim to eliminate stressors as well as to provide public services (in this case, the Cienega Preserve offers a variety of recreation opportunities to the public) should be given high priority. Photos by David Scalero, Pima County Flood Control.

to improve community water access, improved water quality, and recreation attributes. Restoration projects that aim to enhance biodiversity/native species as well as ecosystem services that are important to the public can help foster citizen interest and participation, political will, and funding potential (Dufour and Piégay 2009; Palmer et al. 2014a; Comin et al. 2018) (see figure 2.4).

- *Sequencing of activities*—Some actions must be built on prior work. For example, upland restoration to improve water quality, reduce soil erosion, and/or increase streamflow may be necessary before bottomland activities can be started.
- *Severity of threat*—Streams often face multiple threats, but restoration objectives can be prioritized by the severity of the threats posed. Owing to the negative effects that stream impoundment causes on flow regimes and ecosystem processes, preventing a dam from being constructed or razing a dam (if it has become unsafe or obsolete) can be a high priority (Fischer et al. 2006; Bellmore et al. 2017; Foley et al. 2017).
- *Land-use change*—A low priority should be given to streams that are passing through landscapes where dramatic land-use changes (e.g., urbanization, construction of dams, channelization) are taking place that may significantly alter sediment and water runoff or directly impact the stream corridor itself.
- *Cost-benefit ratio*—Per previous conversation on this topic, knowing the benefit that can be realized relative to cost is critical to gaining project support. Projects that can provide substantial benefit for little cost will be much more attractive than the reverse and, all else being equal, should probably be given a higher priority. For example, Dave and Mittelstet (2017) determined the cost-efficiency of six streambank stabilization techniques using a case study of a river in Nebraska.
- *Location and access*—Projects should focus in areas where social or legal constraints to stream restoration are at a minimum. Isolated areas or sites with limited access are more difficult and probably more expensive to restore than sites that offer easier access. Particularly at the start of restoration efforts along a given stream, the easier sites should be given priority and can prove the restoration work's success and value.
- *Follow-up potential*—If follow-up after completion of a restoration objective is unlikely, practitioners should reconsider whether the project should be initiated. Always keep in mind the long-term nature of restoration (see chapter 8). Depending on where your stream is located, it may have taken hundreds of years and billions of dollars for humankind to deteriorate conditions (e.g., via construction of water diversion, channelization projects and dams, contamination from factories). In such instances, one restoration project will not be enough to bring back a stream. However, your restoration objective could be to serve as one in a string of responses that together will stem the stream's decline and improve conditions at a broader scale. Avoid "one and done" scenarios.

## Hydroecological Considerations

- *Protection versus restoration*—Rather than focusing on impaired portions of a basin, priority should be given to parts of the basin that still maintain a high degree of intact natural processes that allow passive self-restoration (Roni et al. 2012). In brief, restoration actions will likely be more successful, less costly, and foster a greater return to wild or natural conditions when conducted in locations where stream biophysical and chemical conditions have not

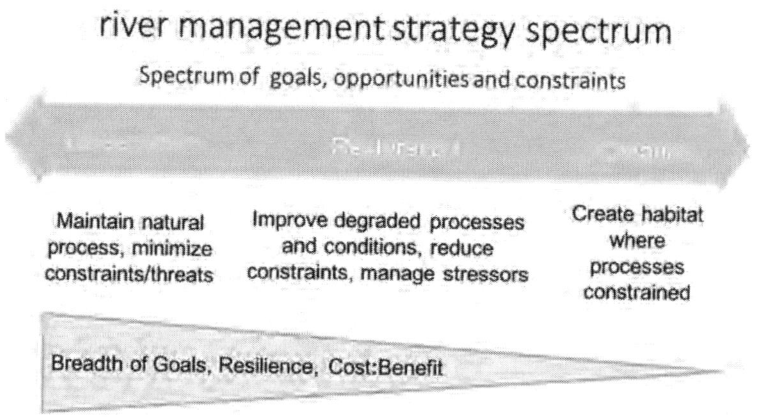

## river management strategy spectrum

Spectrum of goals, opportunities and constraints

Maintain natural process, minimize constraints/threats

Improve degraded processes and conditions, reduce constraints, manage stressors

Create habitat where processes constrained

Breadth of Goals, Resilience, Cost:Benefit

**FIGURE 2.5** Whether stream restoration goals emphasize conservation, restoration or the creation of a completely different ecosystem type (as compared to what existed prior to human interference) will depend largely on stream conditions, persistence of human impacts, and a variety of socioeconomic and political factors (after Gillilan 2005 and Skidmore et al. 2011).

been significantly affected by human perturbations (Gillilan et al. 2005; Beechie et al. 2010; Skidmore et al. 2011) (see figure 2.5). That is, there will be greater return on investment. An excellent example of prioritizing restoration response based on identification of relatively intact versus heavily perturbed parts of a basin is being carried out in the Colorado River Basin in the western United States (Skidmore 2017). Some recent approaches use land-cover classifications and a variety of biophysical indexes to compare and assess how riparian vegetation has departed from historic conditions, providing conservationists and restoration planners a tool to identify stream corridors and riparian areas that remain in relatively good condition (Macfarlane et al. 2017a).

- *Maintain keystone and foundation species, and/or functional diversity*—Keystone species are those that have a disproportionate effect on ecosystem processes; examples are large predators, beavers, and seed dispersers (Power et al. 1996; Galindo-Gonzalez et al. 2000; Soulé et al. 2005). Unlike keystone species, foundation species usually occupy lower trophic levels, are locally abundant and regionally common, and (as their key role), provide a stable habitat structure and fundamental ecosystem processes such as primary productivity (Ellison et al. 2005); an example is cottonwoods (*Populus* spp.). Stream corridors critical to maintaining or reestablishing keystone and foundational species may be a higher priority for restoration than others. Functional diversity is the number of sets of species showing either similar responses to the environment or similar effects on major ecosystem processes (Gitay and Noble 1997). The term often refers to the diversity of species' niches or functions, and to how this diversity responds to environmental stressors (Mayfield et al. 2010). A channel with high functional diversity could be a high priority for restoration, particularly if immediate threats to it are apparent. There are multiple examples in the literature that show that taxonomic and functional diversity do not always respond equally to restoration efforts (Fournier et al. 2015; Sullivan and Manning 2017; England and Wilkes 2018).

- *Control invasive species*—The presence of nonnative, invasive species may make a restoration intervention a high priority (for instance, when there is an opportunity to quickly stop the species' spread when it is initially introduced) or a low priority (as when an invasive species is already extensively established, making intervention prohibitively expensive). Factors that practitioners should consider include understanding why a particular invasive species is present and has become a problem in the first place. Will eradication be a never-ending battle? Other factors are the effectiveness and potential harm of control measures. For example, is an herbicide being used to control the species? What level of maintenance will be needed following removal or control? (See "Invasive Plant" section of chapter 6.)

- *Cross-linking*—Ecosystem cross-linking refers to persistent or recurring processes or attributes that connect different ecosystems in some manner (after Lamberti et al. 2010). Parts of the basin where there is a high degree of cross-linking with tributaries or other aquatic systems (such as lakes, ponds, and springs) should be given priority for restoration efforts over reaches where little cross-linking exists given the benefits that cross-linking has for habitat connectivity and supporting metapopulations.* Such nodes in stream networks should become barrier-free to allow the dispersal and free exchange of species,† enabling them to function as refuges from harmful disturbance events in single river branches (Thomas 2014). In stream systems, nodes of particular importance are where tributaries flow into a main stem, fostering morphological changes and increased physical heterogeneity (Rice et al. 2001; Knispel and Castella 2003), which can lead to an increase in the diversity of macroinvertebrates or fish (Gorman 1986; Knispel and Castella 2003; Fernandes et al. 2004; Kiffney et al. 2006; Pracheil et al. 2009).

- *Connection to headwater streams*—Compared to headwater streams, where the potential to increase biodiversity may be limited, higher recovery potential can be expected for reaches farther downstream that are connected to a large number of headwater streams (Thomas 2014). Proceed with caution, however, as the farther downstream you go, the more likely that there will also be increased significant, cumulative impacts from all the perturbations that have occurred upstream.

- *Refuge or biological hotspots*—Priority should be given to protecting or restoring biological hotspots and species pools where desired species can broaden out and reestablish in or along nearby reaches. Then, priority should be given to restoration actions that are adjacent to existing species pools (Diebel et al. 2010; Sundermann et al. 2011). This approach is often coined the "refuge approach," and is rooted in the idea of "protecting the best first" and expanding restoration outward from protected sites (Frissell and Bayles 1996; Ziemer 1997).

- *Restoration of entire stream branches*—Aim at identifying and restoring entire stream branches that have near-natural morphology and intact ecosystem processes (Bond and Lake 2003), rather than carrying out a diluted restoration response across an entire stream network.

We are now armed with the information we need to put together a stream restoration team, develop stream restoration goals and objectives, and prioritize where in the watershed restoration may be most effective. One brief point before moving on. Always keep in mind that good

---

* Metapopulations are groups of populations of the same species that are separated by space.

† A note of caution that is contrary to this point: fish biologists often point to the importance of maintaining barriers to prevent the spread of nonnative, invasive species into areas where native fish populations persist.

news is always welcome. When success occurs—even if successes are small steps—take the time to publicly highlight them. Sharing news about restoration success fosters goodwill and can increase public support for longer-term efforts. There will inevitably be bumps along the restoration road, but there will be pleasant surprises as well. When the positives occur, make sure to take the time to share them with your partners and the general public.

Why stress this point before even a shovel of soil has been turned on your restoration effort? Because forming your restoration team and restoration goals and objectives (as preliminary as they may be at this point in the process) is an important benchmark; one that may be worthy of highlighting. Something as simple as an announcement with a few photographs in a local paper can go a long way to garnering attention and support for your restoration effort that could have long-term payoffs.

> *The Spring Creek restoration team consisting of practitioners, community citizens, managers, and agency personnel announces the initiation of an effort to bring back Spring Creek. Specific objectives that we hope to accomplish include improving water quality, enhancing recreation opportunities, and bringing back native fish. We seek your support in this exciting project. Please contact us at. . . .*

Look for ways to highlight your stream restoration effort throughout the restoration process. Strategies for highlighting your restoration project in the public eye is the central theme of the final chapter of the guidebook (chapter 8). Now on to the next stop of the stream restoration journey—evaluating the biophysical conditions of the stream corridor and its watershed.

# REFERENCES

Acuña, V., J. R. Díez, L. Flores, M. Meleason, and A. Elosegi. 2013. "Does It Make Economic Sense to Restore Rivers for Their Ecosystem Services?" *Journal of Applied Ecology* 50 (4): 988–97.

Aguiar, F. C., M. R. Fernandes, and M. T. Ferreira. 2011. "Riparian Vegetation Metrics as Tools for Guiding Ecological Restoration in Riverscapes." *Knowledge and Management of Aquatic Ecosystems* 402:21.

Beechie, T. J., D. A. Sear, J. D. Olden, G. R. Pess, J. M. Buffington, H. Moir, P. Roni, and M. M. Pollock. 2010. "Process-Based Principles for Restoring River Ecosystems." *BioScience* 60 (3): 209–22.

Bellmore, J. R., J. J. Duda, L. S. Craig, S. L. Greene, C. E. Torgersen, M. J. Collins, and K. Vittum. 2017. "Status and Trends of Dam Removal Research in the United States." Wiley Interdisciplinary Reviews. *Water* 4 (2). https:/doi.org/10.1002/wat2.1164.

Bhattacharjee, J., J. P. Taylor Jr., L. M. Smith, and D. A. Haukos. 2009. "Seedling Competition Between Native Cottonwoods and Exotic Saltcedar: Implications for Restoration." *Biological Invasions* 11 (8): 1777–87.

BLM. 2003. *Approved Las Cienegas Resource Management Plan and Record of Decision.* U.S. Department of the Interior Bureau of Land Management, Arizona State Office, Tucson Field Office. https://www.cakex.org/sites/default/files/documents/LCROD-WEB.pdf.

Bohn, B. A., and J. L. Kershner. 2002. "Establishing Aquatic Restoration Priorities Using a Watershed Approach." *Journal of Environmental Management* 64 (4): 355–63.

Bond, N. R., and P. S. Lake. 2003. "Local Habitat Restoration in Streams: Constraints on the Effectiveness of Restoration for Stream Biota." *Ecological Management and Restoration* 4:193–98.

Bourgeois, B., E. González, A. Vanasse, I. Aubin, and M. Poulin. 2016. "Spatial Processes Structuring Riparian Plant Communities in Agroecosystems: Implications for Restoration." *Ecological Society of America* 26 (7): 2103–15. http://doi.org/10.1890/15-1368.1.

Briggs, M. K. 1996. *Riparian Ecosystem Recovery in Arid Lands: Strategies and References.* Tucson: University of Arizona Press.

Brooks, J. E., S. P. Platania, and D. L. Propst. 1991. *Effects of Pecos River Reservoir Operation on the Distribution and Status of Pecos Bluntnose Shiner (*Notropis simus pecosensis*): Preliminary Findings.* Report to U.S. Fish and Wildlife Service and U.S. Bureau of Reclamation, Albuquerque, New Mexico.

Cairns, J. 2000. "Setting Ecological Restoration Goals for Technical Feasibility and Scientific Validity." *Ecological Engineering* 15:171–80.

Chen, W. Y. 2017. "Environmental Externalities of Urban River Pollution and Restoration: A Hedonic Analysis in Guangzhou (China)." *Landscape and Urban Planning* 157:170–79.

Choi, Y. D. 2007. "Restoration Ecology to the Future: A Call for a New Paradigm." *Restoration Ecology* 15 (2): 351–53.

Cipollini, K. A., A. L. Maruyama, and C. L. Zimmerman. 2005. "Planning for Restoration: A Decision Analysis Approach to Restoration." *Restoration Ecology* 13:460–70.

Comín, F. A., B. Miranda, R. Sorando, M. Felipe-Lucia, J. J. Jiménez, and E. Navarro. 2018. "Prioritizing Sites for Ecological Restoration Based on Ecosystem Services." *Journal of Applied Ecology* 55 (3): 1–9. http://doi.org/10.1111/1365-2664.13061.

Comín, F. A., R. Sorando, N. Darwiche-Criado, M. García, and A. Masip. 2014. "A Protocol to Prioritize Wetland Restoration and Creation for Water Quality Improvement in Agricultural Watersheds." *Ecological Engineering* 66:10–18.

Dave, N., and A. R. Mittelstet. 2017. "Quantifying Effectiveness of Streambank Stabilization Practices on Cedar River, Nebraska." *Water* 9 (12): 930.

Diebel, M. W., J. T. Maxted, O. P. Jensen, and M. J. Vander Zanden. 2010. "A Spatial Auto-Correlative Model for Targeting Stream Restoration to Benefit Sensitive Nongame Fishes." *Canadian Journal of Fishes and Aquatic Sciences* 67:165–76.

Dudley, R. K., and S. P. Platania. 2007. "Flow Regulation and Fragmentation Imperil Pelagic-Spawning Riverine Fishes." *Ecological Society of America* 17 (7): 2074–86.

Dufour, S., and H. Piégay. 2009. "From the Myth of a Lost Paradise to Targeted River Restoration: Forget Natural References and Focus on Human Benefits." *River Research and Applications* 25:568–81.

Dufour, S., P. M. Rodríguez-González, and M. Laslier. 2019. "Tracing the Scientific Trajectory of Riparian Vegetation Studies: Main topics, Approaches and Needs in a Globally Changing World." *Science of the Total Environment* 653:1168–85. http://doi.org/10.1016/j.scitotenv.2018.10.383.

Dymond, J. R., A-G. E. Ausseil, and J. McC. Overton. 2008. "A Landscape Approach for Estimating the Conservation Value of Sites and Site-Based Projects, with Examples from New Zealand." *Ecological Economics* 66:275–81.

Ellison, A. M., M. S. Bank, B. D. Clinton, E. A. Colburn, K. Elliott, C. R. Ford, D. R. Foster, et al. 2005. "Loss of Foundation Species: Consequences for the Structure and Dynamics of Forested Ecosystems." *Frontiers in Ecology and the Environment* 3 (9): 479–86.

England, J., and M. A. Wilkes. 2018. "Does River Restoration Work? Taxonomic and Functional Trajectories at Two Restoration Schemes." *Science of the Total Environment* 618:961–70.

Falk, D., M. Palmer, J. Zedler, and R. Hobbs. 2006. *Foundations of Restoration Ecology*. Bibliovault OAI Repository at the University of Chicago Press.

Fernandes, C. C., J. Podos, and J. G. Lundberg. 2004. "Amazonian Ecology: Tributaries Enhance the Diversity of Electric Fishes." *Science* 305:1960–62.

Fischer, J. D., B. Lindenmayer, and A. D. Manning. 2006. "Biodiversity, Ecosystem Function, and Resilience: Ten Guiding Principles for Commodity Production Landscapes." *Frontiers in Ecology and the Environment* 4 (2): 80–86.

Foley, M. M., J. R. Bellmore, J. E. O'Connor, J. J. Duda, A. E. East, C. E. Grant, C. W. Anderson. 2017. "Dam Removal: Listening in." *Water Resources Research* 53 (7): 5229–46.

Fournier, B., F. Gillet, R. C. Le Bayon, E. A. Mitchell, and M. Moretti. 2015. "Functional Responses of Multitaxa Communities to Disturbance and Stress Gradients in a Restored Floodplain." *Journal of Applied Ecology* 52 (5): 1364–73.

Frissell, C. A., and D. Bayles. 1996. "Ecosystem Management and the Conservation of Aquatic Biodiversity and Ecological Integrity." *Water Resources Bulletin* 32:229–40.

Galindo-Gonzalez, J., S. Guevara, and V. J. Sosa. 2000. "Bat- and Bird-Generated Seed Rains at Isolated Trees in Pastures in a Tropical Rainforest." *Conservation Biology* 14:1693–703.

Game, E. T., L. L. Bremer, A. Calvache, P. H. Moreno, A. Vargas, B. Rivera, and L. M. Rodriguez. 2018. "Fuzzy Models to Inform Social and Environmental Indicator Selection for Conservation Impact Monitoring." *Conservation Letters* 11 (1): e12338.

Gillilan, S., K. Boyd, T. Hoitsma, and M. Kauffman. 2005. "Challenges in Developing and Implementing Ecological Standards for Geomorphic River Restoration Projects—a Practitioner's Response to Palmer and Others." *Journal of Applied Ecology* 42 (2): 223–27. http://dx.doi.org/10.1111/j.1365-2664.2005.01021.x.

Gitay, H., and I. R. Noble. 1997. "What Are Functional Types and How Should We Seek Them?" In *Plant Functional Types: Their Relevance to Ecosystem Properties and Global Change*, edited by T. M. Smith, H. H. Shugart, and F. I. Woodward, 3–19. Cambridge: Cambridge University Press.

González, E., V. Martínez-Fernández, P. B. Shafroth, A. A. Sher, A. L. Henry, V. Garófano-Gómez, and D. Corenblit. 2018. "Regeneration of Salicaceae Riparian Forests in the Northern Hemisphere: A New Framework and Management Tool." *Journal of Environmental Management* 218: 374–87. http://doi.org/10.1016/j.jenvman.2018.04.069.

González, E., M. R. Felipe-Lucia, B. Bourgeois, B. Boz, C. Nilsson, G. Palmer, and A. A. Sher. 2017a. "Integrative Conservation of Riparian Zones." *Biological Conservation* 211, part B: 20–29. http://doi.org/10.1016/j.biocon.2016.10.035.

González, E., A. Masip, E. Tabacchi, and M. Poulin. 2017b. "Strategies to Restore Floodplain Vegetation After Abandonment of Human Activities." *Restoration Ecology* 25:82–91.

González, E., A. A. Sher, E. Tabacchi, A. Masip, and M. Poulin. 2015. "Restoration of Riparian Vegetation: A Global Review of Implementation and Evaluation Approaches in the International, Peer-Reviewed Literature." *Journal of Environmental Management* 158:85–94.

Gorman, O. T. 1986. "Assemblage Organization of Stream Fishes—the Effect of Rivers on Adventitious Streams." *American Naturalist* 128:611–16.

Gumiero, B., J. Mant, T. Hein, J. Elso, and B. Boz. 2013. "Linking the Restoration of Rivers and Riparian Zones/Wetlands in Europe: Sharing Knowledge Through Case Studies." *Ecological Engineering* 56:36–50.

Gurnell, A. M., M. Rinaldi, B. Belletti, S. Bizzi, B. Blamauer, G. Braca, A. D. Bujise, et al. 2016. "A Multi-scale Hierarchical Framework for Developing Understanding of River Behaviour to Support River Management." *Aquatic Sciences* 78 (1): 1–16. http://doi.org/10.1007/s00027-015-0424-5.

Harrelson, C. C, C. L. Rawlins, and J. P. Potyondy. 1994. *Stream Channel Reference Sites: An Illustrated Guide to Field Technique: Gen. Tech. Rep. RM-245.* Fort Collins, Colo: Department of Agriculture, Forest Service, Rocky Mountain Forest and Range Experiment Station.

Harris, R. R. 1999. "Defining Reference Conditions for Restoration of Riparian Plant Communities: Examples from California, USA." *Environmental Management* 24 (1): 55–63.

Hatch, M. D., W. H. Baltosser, and C. G. Schmitt. 1985. "Life History and Ecology of the Bluntnose Shiner (*Notropis sinus pecosensis*) in the Pecos River of New Mexico." *Southwestern Naturalist* 30:555–62.

Hermoso, V., F. Pantus, J. Olley., S. Linke, J. Mugodo, and P. Lea. 2012. "Systematic Planning for River Rehabilitation: Integrating Multiple Ecological and Economic Objectives in Complex Decisions." *Freshwater Biology* 57 (1): 1–9. http://doi.org/10.1111/j.1365-2427.2011.02693.x.

Hinojosa-Huerta, O., M. Briggs, Y. Carrillo-Guerrero, E. P. Glenn, M. Lara-Flores, and M. Roman-Rodriguez. 2005. "Community-Based Restoration of Desert Wetlands: The Case of the Colorado River Delta." In *USDA Forest Service General Technical Report PSW-SGT-191*, 637–45. https://www.fs.usda.gov/rmrs/publications/.

Hoagstrom, C. W., J. E. Brooks, and S. R. Davenport. 2008. "Recent Habitat Association and Historical Decline of *Notropissimus pecosensis*." *River Research and Applications* 24:789–803.

Holmes, T. P., J. C. Bergstrom, E. Huszar, S. B. Kask, and F. Orr III. 2004. "Contingent Valuation, Net Marginal Benefits, and the Scale of Riparian Ecosystem Restoration." *Ecological Economics* 49 (1): 19–30.

Hughes, F. M. R., A. Colston, and J. O. Mountford. 2005. "Restoring Riparian Ecosystems: The Challenge of Accommodating Variability and Designing Restoration Trajectories." *Ecology and Society* 10 (1): 12. http://www.ecologyandsociety.org/vol10/iss1/art12/.

Jones, H. P., and O. J. Schmitz. 2009. "Rapid Recovery of Damaged Ecosystems." *PLoS ONE* 4 (5). https://doi.org/10.1371/journal.pone.0005653.

Jorda-Capdevila, D., and B. Rodríguez-Labajos. 2016. "Socioeconomic Value(s) of Restoring Environmental Flows: Systematic Review and Guidance for Assessment." *River Research and Applications* 33 (3): 305–20. http://doi.org/10.1002/rra.3074.

Kentula, M. 2000. "Perspectives on Setting Success Criteria for Wetland Restoration." *Ecological Engineering* 15:199–209.

Kiffney, P. M., C. M. Greene, J. E. Hall, and J. R. Davies. 2006. "Tributary Streams Create Spatial Discontinuities in Habitat, Biological Productivity, and Diversity in Mainstem Rivers." *Canadian Journal of Fisheries and Aquatic Sciences* 63:2518–30.

Kingsford, R. T., H. Dunn, D. Love, J. Nevill, J. Stein, and J. Tait. 2005. *Protecting Australia's Rivers, Wetlands and Estuaries of High Conservation Value.* Department of Environment and Heritage Australia. Product Number PR050823. www.lwa.gov.au/.

Knispel, S., and E. Castella. 2003. "Disruption of a Longitudinal Pattern in Environmental Factors and Benthic Fauna by a Glacial Tributary." *Freshwater Biology* 48:604–18.

Kondolf, G. M., S. Anderson, R. Lave, L. Pagano, A. Merenlender, and E. S. Bernhardt. 2007. "Two Decades of River Restoration in California: What Can We Learn?" *Restoration Ecology* 15 (3): 516–23.

Lamb, D., P. D. Erskine, and J. A. Parrotta. 2005. "Restoration of Degraded Tropical Forest Landscapes." *Science* 310:1628–32.

Lamberti, G. A., D. T. Chaloner, and A. E. Hershey. 2010. "Linkages Among Aquatic Ecosystems." *Journal of the North American Benthological Society* 29 (1): 245–63.

Laub, B. G., J. Jimenez, and P. Budy. 2015. "Application of Science-Based Restoration Planning to a Desert River System." *Journal of Environmental Management* 55 (4). http://doi.org/10.1007/s00267-015-0481-5.

Laub, B. G., and M. A. Palmer. 2009. "Restoration Ecology of Rivers." In *Encyclopedia of Inland Waters*, edited by G. E. Likins, 332–41. New York: Academic Press. https://doi.org/10.1016/B978-012370626-3.00247-7.

Lewis, L. Y., C. Bohlen, and S. Wilson. 2008. "Dams, Dam Removal, and River Restoration: A Hedonic Property Value Analysis." *Contemporary Economic Policy* 26 (2): 175–86.

Llewellyn, D. W., G. P. Schaffer, N. J. Craing, L. Creasman, D. Pashley, M. Swan, and C. Brown. 1995. "A Decision Support System for Prioritizing Restoration Sites on the Mississippi River Alluvial Plain." *Conservation Biology* 10:1446–55.

Macfarlane, W. W., J. T. Gilbert, M. L. Jensen, J. D. Gilbert, N. Hough-Snee, P. A. McHugh, J. M. Wheaton, and S. N. Bennett. 2017a. "Riparian Vegetation as an Indicator of Riparian Condition: Detecting Departures from Historic Condition Across the North American West." *Journal of Environmental Management* 202:447–60.

Macfarlane, W. W., C. M. Mcginty, B. G. Laub, and S. J. Gifford. 2017b. "High-Resolution Riparian Vegetation Mapping to Prioritize Conservation and Restoration in an Impaired Desert River." *Restoration Ecology* 25 (3): 333–41. http://doi.org/10.1111/rec.12425.

Mayfield, M. M., S. P. Bonser, J. W. Morgan, I. Aubin, S. McNamara, and P. A. Vesk. 2010. "What Does Species Richness Tell Us About Functional Trait Diversity? Predictions and Evidence for Responses of Species and Functional Trait Diversity to Land-Use Change." *Global Ecology and Biogeography* 19 (4): 423–31.

McDonald, T., C. D. Gann, J. Jonson, and K. W. Dixon. 2016. *International Standards for the Practice of Ecological Restoration—Including Principles and Key Concepts.* Washington, D. C.: Society for Ecological Restoration.

Metcalf, E. C., J. J. Mohr, L. Yung, P. Metcalf, and D. Craig. 2015. "The Role of Trust in Restoration Success: Public Engagement and Temporal and Spatial Scale in a Complex Social-Ecological System." *Restoration Ecology* 23 (3): 315–24.

Miller, J. R., and R. J. Hobbs. 2007. "Habitat Restoration—Do We Know What We Are Doing?" *Restoration Ecology* 15 (2): 382–90.

Mobrand, L. E., J. A. Lichatowich, L. C. Lestelle, and T. S. Vogel. 1997. "An Approach to Describing Ecosystem Performance 'Through the Eyes of Salmon.'" *Canadian Journal of Fisheries and Aquatic Sciences* 54:2964–73.

Morandi, B., H. Piégay, N. Lamouroux, and L. Vaudor. 2014. "How Is Success or Failure in River Restoration Projects Evaluated? Feedback from French Restoration Projects." *Journal of Environmental Management* 137:178–88.

Murray-Darling Basin Commission. 2004. *Native Fish Strategy for the Murray-Darling Basin: 2003–2013.* Canberra: Murray-Darling Basin Commission. https://www.mdba.gov.au/sites/default/files/pubs/NFS-for-MDB-2003-2013.pdf.

Nilsson, C., A. L. Aradottir, D. Hagen, G. Halldórsson, K. Høegh, R. J. Mitchell, K. Raulund-Rasmussen, K. Svavarsdóttir, A. Tolvanen, and S. D. Wilson. 2016. "Evaluating the Process of Ecological Restoration." *Ecology and Society* 21 (1): 41. http://dx.doi.org/10.5751/ES-08289-210141.

Orsi, F., D. Geneletti, and A. C. Newton. 2011. "Towards a Common Set of Criteria and Indicators to Identify Forest Restoration Priorities: An Expert Panel-Based Approach." *Ecological Indicators* 11:337–47.

Orth, K., J. W. Day, D. F. Boesch, E. J. Clairain, W. J. Mitsch, L. Shabman, C. Simenstas, B. Streever, C. Watson, J. Wells, and D. Whigham. "Lessons Learned: An Assessment of the Effectiveness of a National Technical Review Committee for Oversight of the Plan for the Restoration of the Mississippi Delta." *Ecological Engineering* 25:153–67.

Palmer, M. A., E. S. Bernhardt, J. D. Allan, P. S. Lake, G. Alexander, S. Brooks, J. Carr, S. Clayton, E. Sudduth, et al. 2005. "Standards for Ecologically Successful River Restoration." *Journal of Applied Ecology* 42:208–17.

Palmer, M. A., S. Filoso, and R. M. Fanelli. 2014. "From Ecosystems to Ecosystem Services: Stream Restoration as Ecological Engineering." *Ecological Engineering* 65:62–70.

Palmer, M. A., K. L. Hondula, and B. J. Koch. 2014. "Ecological Restoration of Streams and Rivers: Shifting Strategies and Shifting Goals." *Annual Review of Ecology, Evolution, and Systematics* 45:247–69.

Peacock, B. C., D. Hikuroa, and B. Morgan. 2012. "Watershed-Scale Prioritization of Habitat Restoration Sites for Non-point Source Pollution Management." *Ecological Engineering* 42:174–82.

Peters, C. N., and D. R. Marmorek. 2001. "Application of Decision Analysis to Evaluate Recovery Actions for Threatened Snake River Spring and Summer Chinook Salmon (*Oncorhynchus tshawytscha*)." *Canadian Journal of Fisheries and Aquatic Sciences* 58:2431–46.

Power, M. E., D. Tilman, and J. A. Estes. 1996. "Challenges in the Quest for Keystones." *BioScience* 46:609–20.

Powers, P. D., M. Helstab, S. L. Niezgoda. 2019. "A Process-Based Approach to Restoring Depositional River Valleys to Stage 0, an Anastomosing Channel Network." *River Research and Applications* 35 (1): 3–13. https://doi.org/10.1002/rra.3378.

Pracheil, B. M., M. A. Pegg, and G. E. Mestl. 2009. "Tributaries Influence Recruitment of Fish in Large Rivers." *Ecology of Freshwater Fish* 18:603–9.

Provencher, B., H. Sarakinos, and T. Meyer. 2008. "Does Small Dam Removal Affect Local Property Values? An Empirical Analysis." *Contemporary Economic Policy* 26 (2): 187–97.

Rice, S. P., M. T. Greenwood, and C. B. Joyce. 2001. "Tributaries, Sediment Sources, and the Longitudinal Organisation of Macroinvertebrate Fauna Along River Systems." *Canadian Journal of Fisheries and Aquatic Sciences* 58:824–40.

Robbins, J. L., and L. Y. Lewis. 2008. "Demolish It and They Will Come: Estimating the Economic Impacts of Restoring a Recreational Fishery." *JAWRA Journal of the American Water Resources Association* 44 (6): 1488–99.

Rohr, J. R., E. S. Bernhardt, M. W. Cadotte, and W. H. Clements. 2018. "The Ecology and Economics of Restoration: When, What, Where, and How to Restore Ecosystems." *Ecology and Society* 23 (2): 15. https://doi.org/10.5751/ES-09876-230215.

Rohwer, Y., and E. Marris. 2016. "Renaming Restoration: Conceptualizing and Justifying the Activity as a Restoration of Lost Moral Value Rather Than a Return to a Previous State." *Restoration Ecology* 24 (5): 674–79.

Roni, P., T. J. Beechie, R. E. Bilby, F. E. Leonetti, M. M. Pollock, and G. R. Pess. 2002. "A Review of Stream Restoration Techniques and a Hierarchical Strategy for Prioritizing Restoration in Pacific Northwest Watersheds." *North American Journal of Fisheries Management* 22:1–20.

Ruiz-Jaen, M. C., and T. M. Aide. 2005. "Restoration Success: How Is It Being Measured?" *Restoration Ecology* 13 (3): 569–77. https://doi.org/10.1111/j.1526-100X.2005.00072.x.

Rutherford, I. D., K. Jerie, and N. Marsh. 2000. *A Rehabilitation Manual for Australian Streams.* Vol 2. Land and Water Resources Research and Development Corporation, Department of Civil Engineering, Monash University, Clayton, Victoria. http://www.engr.colostate.edu/~bbledsoe/CIVE413/Rehabilitation_Manual_for_Australian_Streams_vol2.pdf.

Schroeder, R. L. 2006. "A System to Evaluate the Scientific Quality of Biological and Restoration Objectives Using National Wildlife Refuge Comprehensive Conservation Plans as a Case Study." *Journal for Nature Conservation* 14:200–206.

Seastedt, T. R., R. J. Hobbs, and K. N. Suding. 2008. "Management of Novel Ecosystems: Are Novel Approaches Required?" *Frontiers in Ecology and the Environment* 6 (10): 547–53.

Shafroth, P. B., V. B. Beauchamp, M. K. Briggs, K. Lair, M. L. Scott, and A. A. Sher. 2008. "Planning Riparian Restoration in the Context of Tamarix Control in Western North America." *Restoration Ecology* 16 (1): 97–112.

Skidmore, P. 2017. "Riparian Restoration in the Colorado River Basin." In *Case Studies of Riparian and Watershed Restoration in the Southwestern United States—Principles, Challenges, and Successes,* edited by B. E. Ralston and D. A. Sarr, 73–76. 2017 U.S. Geological Survey Open-File Report 2017–1091. https://pubs.usgs.gov/of/2017/1091/ofr20171091.pdf.

Skidmore, P. B., C. R. Thorne, B. L. Cluer, G. R. Pess, J. M. Castro, T. J. Beechie, and C. C. Shea. 2011. "Science Base and Tools for Evaluating Stream Engineering, Management, and Restoration Proposals." U.S. Department of Commerce, NOAA technical memo. NMFS-NWFSC-112.

Soulé, M. E., J. A. Estes, B. Miller, and D. L. Honnold. 2005. "Strongly Interacting Species, Conservation Policy, Management, and Ethics." *Bioscience* 55:168–76.

Sullivan, S. M. P., and D. W. Manning. 2017. "Seasonally Distinct Taxonomic and Functional Shifts in Macroinvertebrate Communities Following Dam Removal." PeerJ. April 6, 2017. https://peerj.com/articles/3189/.

Sundermann, A., S. Stoll, and P. Haase. 2011. "River Restoration Success Depends on the Species Pool of the Immediate Surroundings." *Ecological Applications* 21 (6): 1962–71.

Tear, T. H., P. Kareiva, P. L. Angermeier, P. Comer, B. Czech, and R. Kautz. 2005. "How Much Is Enough? The Recurrent Problem of Setting Measurable Objectives in Conservation." *BioScience* 55:835–49.

Thomas, G. 2014. "Improving Restoration Practice by Deriving Appropriate Techniques from Analysing the Spatial Organization of River Networks." *Limnologica* 45:50–60.

Thorpe, A. S., and A. G. Stanley. 2011. "Determining Appropriate Goals for Restoration of Imperiled Communities and Species." *Journal of Applied Ecology* 48:275–79.

U.S. Fish and Wildlife Service. 1992. *Pecos Bluntnose Shiner Recovery Plan.* U.S. Fish and Wildlife Service, Region 2, Albuquerque, New Mexico. https://www.fws.gov/southwest/es/Documents/R2ES/Pecos Bluntnose.pdf.

Van Riper III, C., K. L. Paxton, C. O'Brien, P. B. Shafroth, and L. J. McGrath. 2008. "Rethinking Avian Response to Tamarix on the Lower Colorado River: A Threshold Hypothesis." *Restoration Ecology* 16 (1): 155–67.

Verdonschot, P. F. M., and R. C. Nijboer. 2002. "Towards a Decision Support System for Stream Restoration in the Netherlands: An Overview of Restoration Projects and Future Needs." *Hydrobiologia* 478:131–48.

White, P. S., and J. L. Walker. 1997. "Approximating Nature's Variation: Selecting and Using Reference Information in Restoration Ecology." *Restoration Ecology* 5:338–49.

Williams, B. K. 2011. "Adaptive Management of Natural Resources—Framework and Issues." *Journal of Environmental Management* 92:1346–53.

Woelfle-Erskine, C., A. C. Wilcox, and J. N. Moore. 2012. "Combining Historical and Process Perspectives to Infer Ranges of Geomorphic Variability and Inform River Restoration in a Wandering Gravel-Bed River." *Earth Surface Processes and Landforms* 37:1302–12.

Wohl, E., S. N. Lane, and A. C. Wilcox. 2015. "The Science and Practice of River Restoration." *Water Resources Research* 51 (8): 5974–97. http://doi.org/10.1002/2014WR016874.

WPN (Watershed Professionals Network). 1999. *Oregon Watershed Assessment Manual*. Prepared for the Governor's Watershed Enhancement Board, Salem, Oregon. https://digital.osl.state.or.us/islandora/object/osl:18907.

Zhang, X., Y. Chen, W. Li, Y. Yu, and Z. Sun. 2013. "Restoration of the Lower Reaches of the Tarim River in China." *Regional Environmental Change* 13 (5): 1021–29.

Ziemer, R. R. 1997. "Temporal and Spatial Scales." In *Watershed Restoration: Principles and Practices*, edited by J. E. Williams, C. A. Wood, and M. P. Dombeck, 80–95. Bethesda, Md.: American Fisheries Society.

# 3

# Assessing the Hydrological and Physical Conditions of a Drainage Basin

*W. R. Osterkamp, Mark K. Briggs, David J. Dean, and Alfredo Rodriguez*

## INTRODUCTION

The focus of this chapter is to evaluate your stream, its corridor, and its drainage basin. The goal is to understand current physical and biological conditions and trends that are driving changes in those conditions. By having a robust scientific understanding of conditions and trends, stream practitioners will be able to develop realistic restoration objectives and identify the restoration tactics that will be effective to achieve them (Osterkamp and Toy 1997; Briggs and Osterkamp 2003; Shields et al. 2003; Corenbilt et al. 2007; Osterkamp et al. 2011). Gaining a firm grasp on conditions and trends requires a strategic yet thorough collection of scientific data and information on your watershed. Although stream restoration decisions should be backed by conclusive data, you can make progress by using the best available information, even if scientific uncertainty remains.

Evaluating the condition and trend of your stream and its watershed will require time and money. Keep in mind the ¼-½-¼ maxim for allocating stream restoration resources that we mentioned earlier: one-fourth for supporting planning work, such as evaluating stream corridor conditions and garnering public support; one-half for implementing restoration tactics; and one-fourth for monitoring and evaluation. The importance of evaluation to the success of your stream restoration program cannot be overemphasized. Even if your budget is tight, please do not discount or eliminate evaluation. Keep in mind that significant headway in addressing critical information needs can be made at low cost (for example, taking advantage of available natural resource data and information, field reconnaissance, and expert and local knowledge).

In this chapter, the evaluation of physical processes—in particular, the rate and movement (flow rate) of water, sediment, and the related nutrients and contaminants that are carried with it—is emphasized over biological and chemical attributes. Water and sediment constitute the physical inputs that determine the quality of the riparian zone, aquatic habitat, and the ecosystem services that the stream provides. Although biological and chemical processes are important, knowledge of how water and sediment are transported across the basin and how changes in water and sediment transport affect stream corridor conditions is fundamental to developing a restoration response. Even for stream corridors that are experiencing dramatic biological perturbations (e.g., large-scale increases in the extent and distribution of nonnative species), the roots of biological degradation are often deeply entrenched in the stream's altered physical environment. The altered physical environment needs to be understood.*

> It is a capital mistake to theorize in advance of the facts. Insensibly one begins to twist facts to suit theories instead of theories to suit facts.
>
> —SIR ARTHUR CONAN DOYLE IN
> *THE ADVENTURES OF SHERLOCK HOLMES*

Similar to what Sherlock Holmes does when piecing together clues of a crime scene, we are piecing together the necessary clues to understand what is driving the deterioration of stream corridor conditions that will allow us to make a solid case as to how best to respond. Specifically, the clues that we will be gathering will provide the information to address three questions:

- To what extent has your stream corridor changed physically, biologically, and chemically?
- When did changes begin and what are current trends?
- What are the main causes of the changes that have occurred?

Two sections are presented to guide the collection of data and information needed to answer the above questions:

1. "Using What Is Already Known and What You See to Characterize Your Watershed and Stream." This section emphasizes the use of readily available data and information, combined with field reconnaissance, to begin answering the above central questions.
2. "Addressing Critical Information Gaps." This section provides examples of how to identify and address information gaps with targeted analysis of available data as well as through new research. The introduction to this section includes questions that field practitioners should ask themselves when trying to ascertain what information gaps are in need of filling.

As you and your team evaluate the condition of your steam and its watershed, make sure to do the following:

---

* In contrast, the chapter of the guidebook on monitoring (chapter 7) emphasizes the measurement of biota to a much greater extent. This contrast is due to the differences between the goals of assessing stream corridor conditions versus monitoring. The goal of assessment (this chapter) centers on understanding what is going on with the system, why it is in the state that it's in, how and why it is changing, and future trends, all of which necessitate an emphasis on physical processes. In comparison, monitoring focuses on quantifying progress toward the realization of stream restoration objectives. Given that stream restoration objectives are often biologically orientated (e.g., establishing or protecting native riparian habitat, fish, birds, etc.), an essential monitoring activity has to focus on measuring the species that is/ are central to the restoration objective. If your restoration objective is to reestablish the Rio Grande silvery minnow, a key element of your monitoring program has to focus on the minnow.

*Learn from the Past.* Experiences are a central theme of the guidebook. Lessons gained from past evaluations of stream corridor conditions have much to tell us, and we encourage readers to profit from these efforts. You don't need to develop your evaluation approach from scratch. Consult past stream evaluation techniques to help understand how process and channel form may affect restoration (Schumm 1963; Kellerhals et al. 1976; Rosgen 1994; Montgomery and Buffington 1997; Brierley and Fryirs 2005). Review past stream corridor restoration projects that include strategies for evaluating stream corridor conditions (Briggs 1996; Goodwin et al. 1997; Roni et al. 2002; Beechie et al. 2010; Shafroth et al. 2013). Main points in this chapter complement many past evaluation approaches, many of which are cited to gain deeper insights into key aspects of the evaluation process.

*Avoid Laundry Lists of Scientific Needs.* Restoration practitioners need to attain the middle ground between proceeding with inadequate scientific inquiry or knowledge and wastefully compiling long lists of technical information that may be incidental to the success of the restoration project. Identify deficiencies in knowledge that pertain directly to your stream restoration objectives and develop an evaluation program that addresses those concerns. In this chapter, we provide a wide array of potential avenues to assess a watershed's condition. Do your best to pare down lists of scientific needs to the essential pieces that will be most cost-effective in your restoration activities.

*Do Your Background Work.* The more work you do to gather available data and background information on your stream and its watershed, the more effectively you can identify needs and focus on collecting information in the field to address them.

*Get to the Field.* A drainage basin cannot be understood from a desk, and conducting fieldwork is one of the rewards of restoration work. Visit your stream often, and have a plan of scientific inquiry based on specific questions related to your restoration objectives.

*Do Not Go It Alone! Scientists Should Be Directly Involved.* As mentioned as part of developing your restoration team (chapter 2), engaging stream scientists is critical to guiding the collection, analysis, synthesis, and transfer of data and information toward the development of realistic restoration goals. In addition to scientists, help is available from a variety of sectors (e.g., managers within government agencies)—there is no need to go it alone.

*Critical Thinking Is Basic to Stream Restoration.* Critical thinking is essential to combining and making sense of physical, biological, chemical, and social information regarding your watershed and its stream. This chapter does not provide checklists or scoresheets for formulating objectives and tactics. What is essential to the restoration process is a disciplined process that encourages readers to conceptualize, organize, synthesize, evaluate, and apply information gathered from observation, past efforts, experience, and reasoning to guide restoration action.

# USING WHAT IS ALREADY KNOWN AND WHAT YOU SEE TO CHARACTERIZE YOUR WATERSHED AND STREAM

In this section, we (1) provide background on key concepts (such as watersheds, water and sediment movement, and channel morphology), (2) review the main determinants of water and sediment movement in the watershed (from basic descriptors of the watershed itself to climate and land use), and (3) summarize readily available background information and mapping resources so you can begin answering the central questions posed in the introduction. In general, this section focuses on existing tools and resources and your own observations to characterize your watershed and to identify important gaps that may need attention before beginning restoration activities. The next section will help you fill these gaps by analyzing existing data and by using specific techniques or specialized expertise.

The introductory actions described in this section can be categorized under the general banner of "getting to know your watershed." There are numerous and diverse resources that can help in this process. Some are introductory, others are more advanced. Type "getting to know your watershed" into your search engine and you will be inundated with many such examples. Selected examples include:

- "¿Que es una cuenca?"—Centro de Información de Agua, MX: An interactive website that aims to improve the management of water in Mexico. https://agua.org.mx/que-es-una-cuenca/.
- "How to Conduct a Watershed Survey"—U.S. Environmental Protection Agency (EPA). https://archive.epa.gov/water/archive/web/html/vms31.html.
- "Getting to Know Your Local Watershed: A Guide for Watershed Partnerships"—Conservation Technology Information Center, U.S. https://www.ctic.org/.
- *Get to Know Your Watershed . . .*—Bathurst Sustainable Development, AU. Although focused on the Middle River and Carter Brook Watersheds in New Brunswick, this gem of a document provides background information on what a watershed is, why getting to know your watershed is important, water sources, and impacts on watershed health. http://www.bathurstsustainable development.com/userfiles/file/ETF%20Watershed%20Booklet%20EN%20electronic%20 version%20Feb2010.pdf.
- "Explore Your Watershed in Google Earth"—EarthLabs. https://serc.carleton.edu/eslabs /drought/2b.html.
- "Discover the Basin"—Murray-Darling Basin Authority, AU. Although focused on the Murray-Darling Basin, this website provides an example of how to convey important information about a watershed to the general public. https://www.mdba.gov.au/discover-basin.

## Watershed and Stream Basics

Before proceeding, we define several terms and provide some technical background relevant to water and sediment movement.

### General Terms and Concepts

"Watershed" is a good place to start. The watershed (or drainage basin, or catchment) defines the area that provides water and sediment to a river channel. Sediment entrainment (the process by which surface sediment is incorporated into the flow of air, water, or ice), transport, and storage are largely controlled by five variables: climate, geology and soils, topography, vegetation, and land use. Geology, soils, and topography are generally stable over long periods of time, whereas climate, vegetation, and land use can change substantially over short periods of time (decades or even a few years). Knowing how and why these variables have changed in your stream corridor is important. Human activities commonly alter water and sediment regimes (movements) within watersheds and impact their physical, biological, and chemical characteristics. Consequently, baseline characteristics (both basic and in-place attributes) of landscape processes and their interactions within a watershed must be identified before deviations can be evaluated.

"Streamflow" is the discharge that occurs in a natural channel (see glossary); it is the water remaining after losses of precipitation or snowmelt to evaporation or sublimation and after available groundwater has satisfied the needs of vegetation and replenishment of soil moisture. Streamflow is a principle variable of bottomland ecosystems and is a strong determinant of channel size. The habitat function of streams, as well as aquatic- and riparian-ecosystem health, is affected by flow velocity, flow depth, floodplain inundation, sediment fluxes (defined below), and shallow groundwater movement. Streamflow, therefore, is a basic control of channel morphology. Interactions of organisms, particulate matter, energy, and dissolved substances in the stream channel, floodplain, and hyporheic zone (the area of sediment adjacent to and beneath an alluvial stream channel, through which groundwater moves) are dependent on streamflow (Stanford et al. 1996; Poff et al. 1997; Ward 1989). Therefore, gathering data to describe past, recent, and current trends in streamflow is vital to an effective stream restoration program.

Streamflow is derived from a combination of surface runoff, soil water, and shallow groundwater, all of which are dependent on precipitation and temperature. Climate, geology, topography, soils, and vegetation affect the pathways by which precipitation reaches the stream channel and, ultimately, determines the stream's flow regime, which is described by five main components: discharge magnitude, frequency, duration of flows, predictability, and flashiness. *Discharge*, as a hydrological term of streamflow, is the movement of water downstream; water discharge is given in volume per unit time (see glossary). *Frequency* describes how often a flow greater or equal to a given magnitude occurs over a specific time period (e.g., a one hundred–year flow event describes a discharge that occurs on average once every one hundred years). *Duration* is the period of time associated with a specific flow condition (e.g., average number of days during the course of the year that a flow of specified discharge occurs). *Predictability* is the regularity of occurrence of different flow events (often specified high and low flows) deemed to be particularly important to stream biota or habitat (e.g., a high flow of sufficient discharge and duration to evacuate and move stored sediment downstream). And *flashiness* describes how quickly flow changes from one magnitude to another (Chow et al. 1988; Poff et al. 1997; Gordon et al. 2004).

An important characteristic of your stream is whether its flow is perennial, intermittent, or ephemeral. Perennial streams flow continuously, even during dry periods, and are primarily effluent (i.e., the channel bed is below the level of groundwater saturation and receives water from adjacent alluvium). Intermittent streams flow when they receive water from springs and surface runoff, and, depending on the season, can be effluent or influent (the channel bed is above the level of saturation and surface water flows into the underlying and adjacent alluvial deposits to become groundwater). Ephemeral streams are always influent and flow only during and immediately after rain or as a result of snowmelt. Continuity of flow has direct bearing on channel and floodplain form, and sediment erosion and deposition, as is discussed later in this section.

Sediment movement, erosion, and deposition are critical processes within your drainage basin and stream channel. A description of sediment movement along a stream typically begins by determining whether the stream is dominated by alluvium or bedrock (more simply, is the stream channel alluvial or nonalluvial?). Alluvial channels are formed in sediment, whereas bedrock channels are carved into bedrock and are typically in headwater areas. Headwater stream channels often have pools that store sediment during normal discharges, and flush the sediment during higher discharges. Nonalluvial streams will not be discussed in detail here; instead, we focus on alluvial stream channels, where the channel and floodplain are composed of sediment.

Sediment-transport rate is the rate at which a dry weight of sediment passes a section of a stream channel in a given time. The total amount of sediment transported, or total sediment flux, is the total quantity of sediment that is transported during a given time (often referred to as "sediment load"), as measured by dry weight or by volume. The sediment load of a stream is transported downstream in three ways: as bed load, suspended load, or dissolved load. *Bed load* refers to the portion of the sediment load that is transported along the bed by sliding, rolling, or hopping, and has a large impact on channel width. *Suspended load* is particulate sediment that is carried in the body of the flow and has a great impact on forming channel bank and floodplains. And *dissolved load* is material that is chemically carried in the water. The ratio between suspended load and bed load is a prime determinant of channel morphology—streams with a large bed load, for example, generally being wide or even braided.

To develop sound restoration objectives, it is important to determine whether sediment transport, and the amount of sediment stored within a river reach, has been stable, increased, or diminished. If the amount of sediment transported into a reach is more than the amount of sediment that exits a river reach, or vice versa, bottomland and stream instability results. *Erosion* refers to the entrainment and transport of sediment from the channel bed, banks, or floodplain; *deposition* refers to the settling out of sediment (sediment being deposited) on the channel bed, banks, or floodplain (increases in topographic elevation associated with deposition is called *aggradation*). The processes of erosion and deposition create the size and shape of the stream channel and floodplain.

If changes in the discharge of water and the transport of sediment are substantial, channel morphology change is inevitable (Schumm 1969, 1973). If sediment inflow exceeds outflow, sediment deposition may cause channel aggradation, channel narrowing, and the formation of bars. If there is adequate space within the alluvial valley for the river to change flowpaths, rivers can become more sinuous, with a resulting decrease in channel gradient. If the alluvial valley is narrow, and the river does not have the space to become more sinuous, then increases in channel gradient and channel straightening may occur. In some cases, the potential for avulsion (a rapid change in the course or position of a stream channel to bypass a meander and thereby shorten the channel's length and increase its gradient) will be increased, or the river may develop multiple threads (anabranches). Conversely, if sediment outflow exceeds inflow, channel-bed incision, channel-bed armoring, and channel widening can occur. These properties are interdependent and, during a given period of time, if one process changes, change becomes inevitable in the other processes.

## Channel Morphology

The morphology of an alluvial-stream channel is described by its range of widths, depths, and gradient through a specified reach of the channel. These metrics of morphology are determined by the fluxes of water and sediment that pass through the reach during a period of time, and they therefore determine the sediment properties of the channel bed and banks. Channel biota, especially vegetation, in turn are adjusted to that particular set of channel characteristics. Techniques of channel-morphology investigations and the collection of bed and bank sediment samples are discussed in Hedman and Osterkamp 1982.

Restoration practitioners need to pay attention to channel morphology because changes to channel morphology will affect the aquatic and riparian biota resources. For example, wide, shallow, sand channels provide the dynamic habitat critical for many desert fish species (e.g., pelagic minnows). Such habitat may be lost when increased delivery of fine sediment contributes to channel narrowing and more stable bed conditions (Hoagstrom et al. 2008). Near-channel vegetation can be substantially altered following changes in the stream's bed level (Hupp 1992; Bravard et al. 1999) and width (Johnson 1994; Friedman et al. 1996; Loheide and Booth 2011). Riverside residents and infrastructure may also be affected, and increased frequencies in flooding and flood damage may occur if aggradation and reduced channel capacity lower the ability of a channel to convey floods (Brookes and Gregory 1988; Dean and Schmidt 2011). All of the above underscore the importance of understanding water and sediment movements in your watershed and how changes in them can affect the condition of your stream corridor.

Streams in semiarid regions (areas receiving 250 to 500 mm of average annual precipitation) generally have high flow variability and large supplies of sediment for transport (Osterkamp and Friedman 2000). Boom-bust hydrology is typical, with periods of little or no flow interspersed with occasional, sometimes extreme floods that lead to episodic changes in the channels (Tooth 2000; Dean and Schmidt 2011; Dean and Schmidt 2013). Since arid-land streams often have steep flood-frequency curves (see glossary) and channels with sandy banks, the impacts of large floods may persist for long periods, with channel forms showing the effects of the last major event (Slatyer and Mabbutt 1964; Mabbutt 1977; Wolman and Gerson 1978). Channel narrowing and floodplain reconstruction* often follow the widening and scouring effects of floods, due to newly established vegetation on bars, banks, and islands (Osterkamp and Costa 1987; Friedman et al. 2005; Dean and Schmidt 2011; Manners et al. 2014; Räpple et al., 2017).

This set of processes is often complicated by local changes in water and sediment fluxes caused by humans (e.g., river impoundment, channelization, deforestation, agriculture, grazing, urbanization) and global changes (like global climate change). During the last sixty years, these changes have prompted substantial interest and study of the role of humans in altering stream channels (Warner 1984; Gregory 2006; Renwick and Andereck 2006; Manners et al. 2014).

The extent to which a channel adjusts to changes in water and sediment fluxes can be assessed by comparing its current cross sections, bed configuration, pattern, and gradient to past conditions. Considering channel adjustments over periods of hundreds or thousands of years provides the advantage of placing human-induced channel adjustment in the context of long-term natural changes. That said, such long-term studies can be expensive and

---

\* Floodplain reconstruction is the deposition of bottomland sediment as floodplain alluvium and the reestablishment of vegetation on the freshly deposited sediment, following an erosive flood.

**FIGURE 3.1** Photopoint at the confluence of Tornillo Creek with the RGB in Big Bend in 1936 ([a] top) and 2006 ([b] bottom) (Bennett 2010) underscores how changes in river hydrology can alter channel morphology and riparian biota. Along the RGB in Big Bend, a relatively unchanged sediment supply accompanied by highly diminished streamflow (as a result of upstream water diversions and impoundment) have resulted in channel narrowing and significant increase in riparian biomass, particularly owing to a dramatic increase in the extent and distribution of the invasive giant cane (*Arundo donax*). Note the dramatic reduction in the width of the RGB channel between 1936 and 2006.

time-consuming and are often beyond the means of marginally funded stream corridor restoration projects. Although practitioners try to look as far back as possible, great insights can be gleaned from assessing channel changes following European settlement, or even by comparing channel morphology before and after significant land-use change (e.g., dam construction, urbanization, overgrazing) by use of aerial photographs and/or on-the-ground surveying (see Case Studies 7.7 and 7.8).

Given the emphasis we are placing on sediment and water transport, and the impacts that changes in water and sediment transport can have on channel geometry, a summary of channel geometry—as described by pattern, gradient, bed form, and cross-section profile—is in order. We will be referring to many of these terms when we discuss mapping and field reconnaissance, as well as some of the methodologies discussed in the next section.

## Channel Pattern

Assessing changes in channel pattern following disturbance helps to interpret altered water and sediment fluxes. Channel pattern is typically classed as straight, braided, or meandering (Leopold and Wolman 1957). In many semiarid areas, channel patterns also include anastomosing (describing a wide, braided stream channel that divides and reunites into a network of shallow subchannels) and anabranching (diverging from the main channel and then rejoining downstream), and the pattern may vary along a channel reach. Channel patterns are controlled by local changes in sediment supply, valley width and slope, type and density of vegetation, and the sizes of sediment comprising the channel bed and banks. Heede (1980) suggested that the channel profile (the line described by the lowest points in a stream channel as it decreases downstream) is one of the major determinants of channel pattern. Meandering channels generally have low gradients of less than 1.5 percent (Osterkamp 1978; Rosgen 1988), whereas braided channels tend to have a steep gradient relative to discharge and bed sediment size and type (Wolman and Leopold 1957) and generally have abundant sediment sources.

## Channel Gradient and Longitudinal Profile

The gradient of a stream channel is the elevation change between two channel sites divided by the horizontal distance, measured along the thalweg (or channel centerline), between the sites. Stream gradient, with units of length over length (such as a fall in elevation of one meter over a channel length of one thousand meters), is nondimensional (Osterkamp 2008). Variations in gradient define the stream's longitudinal profile. Whereas site characteristics such as channel morphology, bed-sediment sizes, and density and distribution of plant species typically vary over short time periods owing to floods and lengthy low-flow periods, channel gradient is stable for years to a few decades if contributions of water and sediment are relatively steady. Channel gradient therefore reflects the integrated water/sediment inputs to a reach and is diagnostic of the longer-term conditions of the reach (Schumm 1977; Osterkamp 1978). The longitudinal profile of many alluvial channels is concave over the length of the basin, with a decrease in gradient from the steep upper reaches to the lower meandering reaches. Concave profiles are often pronounced in areas of high rainfall and may be absent in arid regions (Schumm 1977). Many stream channels in the deserts of the American Southwest, however, cross zones of rock that are relatively resistant to erosion or areas where sediment is abundant, which often results in a steeper gradient than would be expected based on the location within the watershed.

Gradient exerts basic controls on the biotic composition along a channel. Assessing the degree to which gradient varies provides insights into stream corridor biotic changes and may suggest which restoration objectives are viable. If the objective is to reestablish bottomland characteristics similar to those prior to significant changes or modification, the predisturbance channel gradient must be identified and reestablished. Methods for measuring channel gradient and the longitudinal profile are well documented (see, for example, Harrelson et al. 1994; Gordon et al. 2004).

## Bed Form

Bed form describes the topography and composition of a channel bed and includes sand and gravel bars, steps, and pool-riffle sequences. These features are related to the flow regime and sediment supply, and influence a channel's cross-section shape, resistance to flow, and mode of sediment transport. In absence of a sufficient sediment supply, or when streamflow increases, an armored bed of coarse channel sediment may accumulate. Typically, progressive downstream changes in bed forms include pools and riffles in the upper reaches of perennial and intermittent streams to sand-dominated bed forms in downstream reaches.

*Bars.* Bars are in-channel bed forms, typically consisting of sand- and gravel-sized sediment. They are deposited when high flows recede and are exposed during low flow; the upper surfaces of bars of perennial streams are typically equivalent to a stage of about 40 percent flow duration (Osterkamp 2008). Bars interactively influence the geometry of channel flows through changes in their size, height, or position. The type of bar (point, alternating, longitudinal, or tributary) that occurs at a site is an indicator of flow and channel conditions (Heede 1980).

*Pool-Riffle Sequences.* This term refers to sequences of one or more pairs of pools and riffles along the channel of an alluvial stream. The normal low water velocities in pools and higher velocities at riffles are reversed during floods, causing scour and removal of accumulated sediment from the pools and deposition of bed sediment on riffles (Keller 1971; Osterkamp 2008). Pool-riffle sequences are particularly important to ecologists given their importance in nutrient cycling, as areas of high productivity and biodiversity. Riffles support benthic invertebrates (those that live in or on the bottom sediments of rivers and streams) and are a vital food source for many species of fish. Although pool-riffle sequences are typically associated with gravel bed streams, well-defined sequences have been identified in boulder-dominated channels and ephemeral sand-bed streams. Because they are features produced by sediment sorting, they do not form in channels that transport and deposit uniform sizes of sediment.

*Cross-section Form.* This term describes the size and shape of a channel in its cross-section profile at a point or in a short reach of channel. It is a basic representation of channel morphology, describing channel width, bank height, and sizes and shapes of bars. An extended bottomland cross section includes floodplain and alluvial-terrace banks and surfaces. We will discuss monitoring channel cross-section change over time in chapter 7 (Case Study 7.7).

## Determinants of Water and Sediment Movement

One of the first exercises that practitioners need to do as part of the evaluation process is to identify, describe, or map watershed descriptors that have a strong bearing on how water and sediment move in the watershed. *A basic physical watershed description should include water-*

*shed size and shape, relief, stream patterns, and main tributaries.* Other critical descriptors pertain to climate, geology and soils, topography, vegetation, and land and water use. Why they are important to water and sediment movement is summarized below.

## Climate

Climate describes the mean and variability in temperature, surface pressure, wind (its movement and direction), humidity (amount of water vapor), and other elemental meteorological measurements of a particular area as measured over the most recent thirty-year period (IPCC 2001). Climate determines water availability, the magnitude and timing of water and sediment discharges (the flow at a given instant expressed as volume per unit of time), the frequency and severity of floods and drought, and the distributions of flora and fauna. The seasonality of precipitation, form (rain or snow), and average, as well as variability in both temperature and precipitation are especially pertinent.

Exceptional variability of rainfall in the southwestern U.S., northern Mexico, many parts of Australia, and similar dryland regions of the world produces similarly dramatic variability in streamflow, often characterized by short-duration flooding followed by extended periods of little or no flow. These episodes of intense periods of precipitation can produce marked changes in vegetation patterns and channel morphology. Thus practitioners should consider average annual and monthly precipitation as well as other statistics that offer insights into the natural variations of discharges that their streams will experience that, in turn, will help in formulating realistic restoration objectives. Some important questions are: What is the average annual precipitation? How does precipitation vary intra- and interannually? Are long-term records available to characterize the frequency, intensity, and duration of past droughts? Also important are averages and variability of the first and last frosts that could impact restoration activities, the duration and seasonality of high-intensity storms, and the distribution of rain and snow in time and physical space.

## Geology and Soils

As with climate, geology affects the entire watershed. Geology—the nature and history of the earth as demonstrated by its rocks and minerals—defines topography, watershed gradients (the rate of elevation changes per unit lengths of stream channels), and the erodibility of rocks, which, in turn, determine the size and quantity of the sediment that may be transported in the stream. Soils are an integration of rock types (surface and near surface) that reflect the combined effects of climate, land use, biology, and groundwater. A basic description of watershed geology should include the types and distributions of exposed rocks, faults, and other fractures, and stratigraphic relations specific to the watershed site. Knowledge of the soil types and thicknesses in the watershed and along the stream corridor can provide useful information regarding sediment deposition rates, groundwater chemistry, and organic material that may be contributed to the river channel.

## Topography

Topography, as applied here, describes the surficial expression of an area; it includes relief and position of features. Over long periods (i.e., thousands of years), topography is largely determined by climate and geology. During shorter periods, topography affects land-surface processes and vegetation patterns. The energy that drives channel processes comes from drainage-basin structure and the valley gradient, which are products of long-term geological processes* (Schumm 1977). Unlike channel gradient, the valley gradient is imposed on the stream corridor by long-term geomorphic processes in the drainage basin. The river, therefore, must adjust to the valley gradient and depends on the movement of water and sediment from higher in the basin.

## Vegetation

Vegetation plays an important role in erosion and deposition of sediment within the stream corridor. Plants stabilize hillslopes and promote the infiltration of precipitation. Bottomland vegetation provides roughness to floodplain and near-channel surfaces, thereby reducing flow velocities, promoting deposition, and helping to limit erosion. Likewise, the characteristics of landform shape, aspect (orientation or compass direction that a landform or surface faces), and proximity to river channels largely determine the type and density of vegetation. To understand the influence of vegetation on sediment movement and how landforms, runoff, and vegetation influence each other, knowledge of all three characteristics is imperative (see Osterkamp et al. 2011).

## Land and Water Use

Land use generally refers to the ways in which land is altered by human occupation or endeavors; understanding how land use has changed in your watershed is crucial to the evaluation process and provides a basis for identifying the stresses that drive stream corridor deterioration. Changes in land and water use can impact soil processes and water movement throughout the drainage system, potentially making stream channels unstable by altering erosion and sediment deposition. Depending on watershed size, the causes of ecological deterioration can be far above the bottomland. Almost always, reversal of the land and water uses that prompted disequilibrium must be the goal.

Both upland and bottomland land and water uses should be described and mapped. Upland activities may affect how water and sediment move through a watershed, potentially increasing or decreasing the amount of sediment delivered to a stream channel or degrading water quality. Examples of upland activities that may have a pronounced impact on water and sediment movement include urbanization, deforestation, catastrophic wildfire, mining, and overgrazing by livestock. In addition, practitioners should also be cognizant of upland "inactions" that may result in dramatic changes in how water and sediment move through a watershed. A good example of this includes fire suppression that results in a buildup of forest fuels, an increased potential for catastrophic fire, and an increased likelihood for erosion and downslope movement of water and sediment. Such a scenario is described in Case Study 8.3, which led to the establishment of the Rio Grande Water Fund, a diverse public/private partnership that focuses on upland watershed health.

Land uses that occur within the bottomland environment are important to document due to their more direct impact on stream corridor conditions. (Bottomlands are the parts of the

---

* These processes are commonly fluvial (associated with rivers and streams and the deposits and landforms created by them), glacial (from the movement of glaciers), and/or tectonic (associated with the structure of the earth's crust and the large-scale processes that take place within it).

watershed that are formed by the transport and deposition of sediment by streamflow.) Foremost in this category of land use is stream impoundment (e.g., dams), but other bottomland activities should be noted as well, including road and bridge construction, water diversions, agricultural activities, and bank and channel modifications such as the construction of levees and channelization (channel straightening), all of which can have dramatic impacts on the flow of water and sediment (Heede 1980; Renwick et al. 2005; Renwick and Rakovan 2010; Rakovan and Renwick 2011).

Make sure to take note of groundwater pumping. Pumping wells can be located within the bottomland environment or outside it; depending on well depth, the number of pumping wells, pumping rate, and aquifer characteristics, they can have a profound impact on the riparian water table with pronounced impacts on streamflow, channel morphology, biota, and riverside citizens. Although well fields—multiple pumping wells concentrated in a small area—should certainly be highlighted, keep in mind that scattered individual pumping wells can also have an accumulative and significant impact on aquifer conditions. A good example of how groundwater pumping was identified, addressed, and monitored as part of a stream restoration response is offered in Case Study 7.4.

If the deterioration of water quality is a significant issue, make sure to identify potential sources of water contamination as part of your exploration of land and water uses in your watershed. As with pumping wells, sources of contamination may occur both within or outside the bottomland environment. Sources of water contamination are typically described as point source or nonpoint source. "Point source" is a single, identifiable source of pollution, like a pipe, canal, or a drain associated with sewage treatment, industry, and construction sites. "Nonpoint-source pollution" (also called "diffuse" pollution) refers to contaminant inputs that occur over a wide area and may not be easily attributed to a single source. Land-use change as affected by agriculture, urbanization, forestry, and grazing, among many other examples, may produce nonpoint-source pollution, while industry, sewage treatment (facilities, lines, and septic), feedlots, and irrigated agriculture are often associated with point-source pollution.

Given its profound impact on sediment and water transport along a stream, let's spend a few more minutes on stream impoundment. Stream impoundment is a structure, typically a dam, that alters the natural flow of water and stores water for agriculture, power generation, and flood protection. Other impoundment structures, such as floodgates and dikes, manage or prevent water flow onto specific lands. Stream impoundments fundamentally change the way water and sediment is conveyed along a stream corridor. To offer an example from one study, the differences in streamflow between impounded and unimpounded reaches included an average 67 percent reduction in annual peak discharges (i.e., the highest flows or discharges of a water/sediment mixture that occur during flooding), a 60 percent decrease in the ratio of annual maximum to mean (average) flows, and a 64 percent decrease in the range of daily discharges (Graf 2006).

## Doing Your Homework: Gathering Readily Available Information and Mapping

Our focus here turns to doing background homework (or office work), which we describe as identifying, gathering, and synthesizing information and resources that are often readily available on the conditions and trends of your stream corridor and its watershed. The home-work activities described below should be done in tandem with field reconnaissance; they are key elements in developing sound restoration objectives and tactics.

## Sources of Information

A trove of information on watershed and stream corridor conditions is available for many watersheds in the form of published reports, watershed and climate summaries, maps, hydro-logical data, and unpublished documents and records. Much can be gained quickly via inter-net searches, preliminary phone calls, and email inquiries. Other mapping resources are also available—including topographical, soil, and geological maps; aerial photographs; Google Earth images and tools; and other remote sensing tools—that will allow you to identify a variety of watershed attributes important to your stream restoration project. Poring over avail-able mapping resources with your restoration team will not only further educate you about your watershed, but will be valuable for planning efficient and effective field reconnaissance. Gathering relevant information and mapping your watershed should continue throughout the restoration planning process, but try to be as thorough as possible from the start. The stronger the foundation, the greater will be your confidence in making a host of decisions on your restoration program.

The focus of your effort to gather available information and mapping resources should be on those watershed attributes presented above that have a strong bearing on sediment and water movement and conditions in your stream corridor and its watershed. Before we sum-marize resources specific to these attributes, we encourage practitioners to begin broadly by developing a complete list of the federal and state agencies, universities, NGOs, and consulting firms that work in your watershed. (Federal and state agencies that house natural resource information in Australia, Mexico, and the United States are summarized in appendix A.) For those outside this region, the list may still be of value by way of providing examples of the types of agencies that might have relevant or globally pertinent information (e.g., climatic studies with findings beyond the watershed scale). Of special value are reports summarizing biophysical conditions, land-use change, topographical and geological maps, satellite images, and aerial and ground photographs. Some reports and data sets of historical conditions may be unpublished or not electronically available, such as original inventories and assessments archived in local or regional offices.

Universities and research centers are especially important resources. Consult professors associated with studies in your watershed to help identify available natural resource data and information. Such a step may unearth valuable information, and may initiate the involvement of professors, researchers, and students in your project. In addition to the federal and state agencies summarized in this appendix, please also see appendix B, which summarizes selected universities and affiliations of northern Mexico, southwestern U.S., and Australia that are focused on changing climate conditions.

Be sure to include state governments and local water-management agencies and organi-zations in your search. City and county planning offices and flood-control districts also pro-vide natural resource services and maintain environmental and conservation departments. In addition, keep an eye out for NGOs that may be working in your watershed. Particularly those involved in environmental conservation efforts may already have gathered much of the information you are seeking. A great example of this is ongoing work in the Colorado River Delta, Mexico, where two NGOs (the Sonoran Institute and Pronatura) initiated a robust analysis of hydrological data to understand the streamflow conditions needed to bring back the Delta's riparian, wetland, and estuary ecosystems (Zamora-Arroyo et al. 2001; Glenn et al. 2017).

As you pursue all the above, look out for references to weather stations, stream gauges, and other instrumentation that may be collecting natural resource data in your watershed. If you are lucky, such instrumentation will be in your watershed and could provide long-term data on the parameters of interest.

Two specific points relating to available resources deserve attention before getting into specific "homework" activities: First, make sure to take advantage of available remote sensing data and information. Remote sensing provides a broad view of landscapes using a variety of technologies that gather information on natural resource conditions by measuring electromagnetic radiation reflected from Earth's surfaces. Kites, carrier pigeons, balloons, airplanes, drones, and space vehicles have all been used as platforms to gather remote sensing data (Wilkie and Finn 1996). We don't recommend carrier pigeons, as they don't always follow flightplans of interest. Today, practitioners can take advantage of an array of image sources and analysis techniques to understand how natural resource conditions are changing. "Change detection" tools—tools that identify changes in images or other data of the same piece of ground taken at different times—that use remote sensing data are very useful to understanding both short-term (e.g., seasonal, like flood inundation or snow cover) and long-term changes (e.g., decadal changes in vegetation cover due to deforestation or overgrazing) (Kennedy et al. 2009). Over the past several decades, remote sensing technology has shown increased utility for informing the state of, and pressures on, natural resource conditions at landscape, regional, ecosystem, continental, and global spatial scales (Roughgarden et al. 1991; Turner et al. 2003; Duro et al. 2007; Gillespie et al. 2008; Horning et al. 2010; Pettorelli et al. 2014).

With an increase in remote sensing imagery, short-term changes in natural resources can be documented. For example, a comparison of available imagery (potentially from multiple sources) may allow the nearly annual assessment of channel position, pattern and bed forms. Aerial photos and orthophotos can be acquired from land-management and development agencies, libraries, historical societies, the Earth Resources Observation and Science data center (edcwww.cr.usgs.gov and http://earthexplorer.usgs.gov/), the National Archives and Records Service (www.nara.gov), Google Earth, and the Aerial Photo Field Office in Salt Lake City, Utah (see appendix A). Acquisition of remote sensing data is also covered in chapter 7 as an important consideration for monitoring the results of stream restoration.

Second, we need to acknowledge that this chapter is an introduction on how to go about evaluating watershed and stream corridor conditions. For those of you who want to dive deeper (and we wholeheartedly encourage you to do so!), please consider the following excellent references that provide much greater detail on many of the topics that are highlighted in this chapter related to evaluating watershed and stream corridor conditions:

- *Methods in Stream Ecology*, 2nd edition (Hauer and Lamberti 2007)—This comprehensive reference written for stream ecologists covers all aspects of the watershed, including physical processes, stream biota, ecosystem processes, ecosystem quality, and much more.

- *Applied Hydrology* (Chow et al. 2010)—If you are looking for a handbook covering all aspects of hydrological processes and analysis, this book is for you.
- *Fluvial Forms and Processes: A New Perspective* (Knighton 1998)—As the name implies, this is an excellent reference on describing, mapping, and measuring fluvial processes.
- *Stream Hydrology: An Introduction for Ecologists*, 2nd edition (Gordon et al. 2004)—An excellent overview of methods to collect and analyze data on streamflow.
- *Geomorphic Analysis of River Systems: An Approach to Reading the Landscape* (Fryirs and Brierley 2013)—Focuses on field-based strategies to interpret rivers across a range of environmental and climatic settings.
- *Handbook of Applied Hydrology* (Singh 2017)—Covers data collection methods, analysis, and modeling.
- *Physical Hydrology, 3rd Edition* (Dingman 2015)—An excellent resource for addressing scientific and management questions about water resources.

## Describing and Mapping Attributes of Your Watershed

A review of available background information and mapping tools and resources pertinent to your watershed should focus on the attributes summarized below. By describing and (where appropriate) recording these observations on a map, you can develop a comprehensive image of the watershed and stream corridor conditions and how they are changing.

### Basic Physical Watershed Descriptors

Information on the size, shape, relief, topography, and other general physical attributes of your watershed is available in published and unpublished documents. For example, articles on many aspects of watersheds and streams begin with a concise description of the watershed where the study took place. Many of the sources mentioned above may have this information as well.

If documented information on your watershed is not available, basic watershed descriptors can be easily measured from topographical maps and digital elevation models. Manual measurements of watershed area are summarized in a variety of texts, including several of those cited above. Google Earth features can also be used to calculate most watershed descriptors. EarthLabs offers excellent guidance on this topic at https://serc.carleton.edu/eslabs/drought/2b.html. In addition to all the above, geospatial-analysis programs such as ESRI ArcMap have features that can calculate many watershed parameters.

### Climate

Climate data, including daily, monthly, and annual observations of temperature, precipitation, and other variables, for Australia, Mexico, and the United States can be accessed from a variety of sources. With climate change and the development of effective climate adaptation strategies becoming a priority concern in many parts of the world, there has been a proliferation of climate informational and service organizations. (See chapter 4 on climate adaptation as well as appendix B for a list of these sources of information and data on climate).

### Topography, Geology, and Soils

Before heading to the field, obtain recent topographic, geologic, and soil maps of your area. Topographic, geologic, and soil information about your watershed can also be gleaned from a variety of published and unpublished reports that may be available through some of the sources described above, and such information can be helpful to stream practitioners in a variety of ways.

In particular, describe geological features that control river morphology and that provide sources of sediment; include exposed bedrock, steep cliffs with disaggregate rock at the base, springs, waterfalls, and deposited sediment at tributary mouths. Topographical maps can help identify reaches of a stream or channel for rehabilitation. For example, if practitioners are aiming to reestablish native riparian vegetation, they may need to avoid narrow valley-confined reaches in lieu of wider, alluvial reaches that offer protection from flood scour (the erosion, or scouring action, that occurs during a flood in an alluvial stream channel). Cross-sectional diagrams of the watershed geology can help distinguish patterns of groundwater movement and thus identify the channel reaches where riparian vegetation has access to shallow groundwater. A basic understanding of the rock types found on geological maps also helps identify abundant sediment source areas, such as large deposits of alluvium, or thick beds of weathered sandstone or shale.

In Australia, topographical, soil, and geological maps can be found in a variety of sources. Geoscience Australia produces topographical maps for the entire country at varying scales as well as a variety of thematic maps that include information on land tenure and use, railways, and more (http://www.ga.gov.au/scientific-topics/national-location-information/topographic -maps-data/printed-topographic-maps). States and territories of Australia have departments and resources that can provide additional coverage. For example, the Department of Environment and Water of South Australia has completed a mapping program that covers soil and land mapping information for all of South Australia (https://www.environment.sa.gov .au/Knowledge_Bank/Information_data/soil-and-land/mapping-soil-and-land). New South Wales Office of Environment and Heritage uses a Google Maps–based information system called eSpade that allows users to access a wealth of soil and land information from all across New South Wales (https://www.environment.nsw.gov.au/news/introducing-espade). And the state of Victoria Energy and Earth Resources features an "earth resources online store" where users can access a variety of geophysical data packages (many at no cost) (http://earthresources .efirst.com.au/categories.asp?cID=13).

In Mexico, Instituto Nacional de Estadistica, Geografia e Informacion (INEGI) is the best place to look for all kinds of data and information on Mexico, including a variety of maps (http://montanismo.org/2013/mapas-topograficos-de-mexico-gratuitos/).

For the United States, a good place to start is the National Geospatial Program of the U.S. Geological Survey (USGS) (https://www.usgs.gov/core-science-systems/national-geospatial -program/us-topo-maps-america?qt-science_support_page_related_con=0#qt-science_support _page_related_con).

Online resources such as Google Earth and topographic-map.com can provide global coverage, with the level of detail dependent on a variety of factors.

## Vegetation

Vegetation and related land-cover maps for large areas can be generated using global and regional databases such as MODIS, which covers spatial resolutions of one kilometer (Friedl et al. 2002), and for variable scales using GLC2000 (Bortholomé and Belward 2005). At the drainage-basin scale, vegetation and land-cover data and maps are generally available through state and county governments and universities. Regardless of scale and area covered, maps generated at different times for a specific area may provide information on vegetation, land-cover, and land-use changes during recent decades.

In the United States, the National Land Cover Database 2001 (https://www.mrlc.gov/data /statistics/national-land-cover-database-2001-nlcd2001-statistics) is a land-cover classification scheme for the United States and Puerto Rico at a spatial resolution of thirty meters. Although the spatial resolution of this dataset is coarse, it can provide good general information on the type and coverage of vegetation and land use. For Mexico, the Instituto Nacional de Estadistica, Geografia e Informacion offers geographic and statistical data on a range of environmental themes that include land cover, sociodemographics, and economics (http://www .inegi.org.mx/) (see appendix A). The Australian National Earth Observation Group provides spatial information on environmental and community safety issues, including earth observations and satellite imagery, geographic information, land cover, marine and coastal conditions, and minerals (http://www.ga.gov.au/about) (see appendix A).

In reference to riparian vegetation, Congalton et al. (2002) offer a cost-effective approach to classify riparian vegetation using aerial photography. Nagler et al. (2004) show that a normalized difference vegetation index measured from low-level aerial photography is useful for determining vegetation cover and estimating radiative properties of canopies in the Colorado River Delta of northwest Mexico.

## Water and Sediment Movement

Regarding sources of data and information specific to water and sediment movement, there are a variety of sources that should be explored, most being governmental agencies that compile and publish data from routine monitoring or as parts of a network (e.g., weather stations, streamflow gauges, water-quality monitoring). These compilations generally characterize sediment discharge in a regional manner as areal averages that provide little interpretation of small-scale inputs; their application to a specific channel reach therefore may be limited. That noted, weather stations and streamflow gauges are becoming more common. Initial questions you will want to answer include: Are there weather stations and streamflow gauges in my watershed? How many and where are they located? What is the period of record (period of data collection) associated with them and what information can be developed from the collected data on long-term averages, variability, and trends? We'll dive deeper into the analysis of streamflow data later in this chapter (see "When Streamflow Data Are Available").

The Australian Government Bureau of Meteorology houses a variety of data and information on water, including current conditions, reservoir storage, forecasts, regulations, and more (http://www.bom.gov.au/waterdata/). In Mexico, CONAGUA collects, analyzes and manages surface and groundwater data for the entire country (https://datos.gob.mx/busca/organization /conagua). In addition, every year, CONAGUA publishes an overview of water statistics for the country in any of three magazines: (1) Estadisticas del Agua en Mexico, edicion 2018 [CONAGUA 2018]; (2) Atlas del agua en México; and (3) Numeragua (2 and 3 can be found at http://sina.conagua.gob.mx/sina/index.php?publicaciones=1).

A robust dataset for U.S. streams is the USGS summary of instantaneous discharge and sediment transport data (http://waterdata.usgs.gov/nwis/). Additional USGS sediment transport data are also provided for a number of rivers in western U.S. at https://www.gcmrc.gov

/discharge_qw_sediment/. Also, check "Science in Your Watershed" (https://water.usgs.gov /wsc/databases.html), a USGS-managed website that provides links to many databases and other websites with a range of natural resource information (including many that focus on water) in a variety of regions of the United States. Other important compilations include "The Incidence and Severity of Sediment Contamination in Surface Waters of the United States National Sediment Quality Survey Database: 1980–1999" of the EPA (https://archive .epa.gov/water/archive/polwaste/web/pdf/nsqs2ed-complete-2.pdf), and listings of sediment-deposition surveys in reservoirs by the U.S. Department of Interior's (USDOI's) Bureau of Reclamation (https://www.usbr.gov/tsc/techreferences/reservoir.html). Sediment-deposition rates in reservoirs are catalogued in numerous state, local, and drainage-basin publications such as those of Renwick and colleagues (2005) and Renwick and Andereck (2006).

A *sediment budget* is an accounting of the sediment entering and leaving a specified area or stream reach. The relative balance or imbalance of those inputs and exports constitutes the change in sediment storage within that reach. Understanding the manner, extent, and timing of channel change helps identify the causes of the changes (e.g., land use that has altered water and sediment inputs), how the changes may be affecting stream biota and near-channel landowners, and how the sediment budget has been affected. All are crucial for developing achievable rehabilitation objectives.

Identifying sediment sources (areas of origin) and sinks (areas of deposition) that could affect the target reach of your stream is an important step to understanding your watershed's sediment budget. Zones of erodible rocks (e.g., weakly cemented sandstone or weathered granite) or human stresses such as gravel mines and large-scale agriculture may release copious amounts of sediment to the stream network. Soil loss from agricultural fields has been modeled extensively and described by the U.S. Department of Agriculture's (USDA's) Agricultural Research Service (see, for example, Wischmeier and Smith 1978; Foster et al. 2000). A tributary that transmits sediment to the main channel can cause abrupt changes to the geomorphology of a stream and its sediment-transport characteristics. Analysis of aerial photographs to identify sediment sources and sinks is reviewed in Brierley and Fryirs (2005).

## Channel Morphology

Mapping and measuring channel morphology over successive time periods can be done at the watershed or reach scale. Maps, aerial photographs, ground photography, repeated field surveys, interbasin comparisons, and field reconnaissance provide a means to assess channel change. Examples of such studies for streams in arid and semiarid climates include Friedman et al. 1996; Friedman et al. 1998; Allred and Schmidt 1999; Gaeuman et al. 2005; and Dean and Schmidt 2011.

Valley and channel gradient, channel pattern, channel width, bed configuration, channel cross section, and even pool/riffle sequences often can be determined using standard maps and remote sensing techniques. Examples of resources that practitioners can use to assess channel change are summarized below. Examples are from both arid and nonarid environments because the relevant methodologies can be employed in either.

- *Maps*—Quantifying channel change via analysis and comparison of historical hydrological maps are described for the Danube River near Vienna by Baart and colleagues (2013).

- *Interbasin Comparison*—Interbasin comparisons of channel morphologies in undeveloped and developed basins can provide insights into preimpact conditions (see Pizzuto et al. 2000).
- *Ground Photography*—Techniques of comparing historical ground photographs to assess channel change and changes in near-channel vegetation are offered by Dean and Schmidt (2011), and an extensive library of historical ground photographs in the southwestern United States and northwest Mexico is housed at the USGS office in Tucson, Arizona (Webb et al. 2007; Shroder 2013).
- *Aerial Photography*—Aerial photography has been vital in comparing changes in channel boundaries of the lower Rio Grande (Dean and Schmidt 2011); bank erosion and channel migration in the Bellinger River Valley, New South Wales, and the Latrobe River, Victoria, Australia (Reinfelds et al. 1995); bed load and sediment storage along the Waimakariri River, New Zealand (Carson and Griffiths 1989), and the Kowai River, New Zealand (Hoey 1994). Remote sensing was used to map channel morphology and habitat along Soda Butte Creek, Yellowstone National Park (Legleiter et al. 2004). Methods to assess geomorphological change using satellite and aerial photography are described by Shroder (2013). There is also a practical handbook on how to use aerial photos and maps to interpret channel position (Lagasse et al. 2004).

## Water Quality

Unlike the other attributes in this summary, water quality does not directly affect water and sediment movement, yet understanding water quality along your stream can be central to developing restoration objectives and effective tactics when deterioration of water quality is an issue. If impaired water quality is a concern for your stream, water-quality data will need to be collected. The degree of ecological impairment will motivate the extent of data collection. An expansive water-quality program enables an assessment of trends and the identity of sites and sources of contamination. Reduced water quality between monitored sites may help identify where contaminates enter the channel and possibly the source of contamination.

In Australia, there are several sources that describe water-quality sampling protocols and summaries of water-quality conditions and trends:

- Assessment (with results provided as grades) of the state and trends of inland water quality of Australia (2016, 2011): https://soe.environment.gov.au/assessment-summary/inland-water/state-and-trends-water-quality.
- State or Territory water agencies: For example, the Department of Water and Environmental Regulation, Government of Western Australia: http://www.water.wa.gov.au/water-topics/water-quality; the Department of the Environment, Queensland Government: https://environment.des.qld.gov.au/water/monitoring/assessment/water_quality_data.html; the Environmental Protection Agency, Government of Western Australia: https://www.epa.sa.gov.au/environmental_info/water_quality.
- Federal Initiatives: for example, Water Quality Australia: http://www.waterquality.gov.au/introduction.

In Mexico, water management is conducted (a) at the federal level by CONAGUA, (b) at the state level by state water commissions Comisiones Estatales del Agua—CEAs (or similar state agencies), and (c) by basin authorities and councils (see appendix A).

In the United States, the USGS is the lead agency for sampling protocols and the collection of water-quality data (Wilde 2011). Standard water-quality methods of the USGS are peer reviewed, current, and published in the National Field Manual for the Collection of Water-Quality Data (http://pubs.water.usgs.gov/twri9A/). The USGS also manages the National Water-Quality Assessment Program, which reports water-quality conditions and trends and describes how natural features and human activities affect those conditions (http://water.usgs .gov/nawqa/). Each western state of the United States has its own department or division charged with water-quality monitoring and assessment (see appendix A).

There may also be institutions or agencies unique to rivers that form or cross international borders. For example, along the U.S.-Mexico border, the International Boundary and Water Commission, U.S. Section, Texas Clean Rivers Program, in partnership with the Texas Commission on Environmental Quality, collects water-quality data from the Texas portion of the Rio Grande Basin along the border downstream of El Paso/Juarez (http://www.tceq.texas.gov /waterquality/clean-rivers).

## Land and Water Use

It may not be necessary to describe and map all the land and water uses in your watershed. Priority should be to describe those deemed most likely to be causing hydroecological damage to your stream corridor. Practitioners should describe and map land and water use by the following conditions:

- *Type, intensity and/or size*—How you portray these characteristics will depend on the type of land use. For example, for grazing you would describe the area of the watershed affected and intensity (number of cattle per hectare). For dams, you would describe the size and capacity.
- *Location*—Is the land or water use you are describing located in the stream corridor and/or outside the stream corridor (i.e., upland)?
- *Timing*—The date(s) when changes in land or water use occurred or ceased, and whether they continued to be made. For example, dates of completion for major dams provide clear before-and-after delineations for hydrological analysis (e.g., average streamflow before and after dam construction).
- *Trends*—Are the land-use changes that practitioners are seeing increasing, stable, or declining (e.g., changes in watershed area that is being urbanized)? It is important to note that some land-use changes occur over many decades and thus make it difficult to discern a clear before-and-after picture of how they may be affecting stream corridor conditions (Leopold et al. 2005).

Types of upland land and water use that should be described and mapped include livestock grazing, deforestation, agriculture, urbanization, road networks, mining operations, and well-field construction and operation (these land and water uses may also be occurring in the bottomland environment). For bottomlands, look for dams, well fields, flood control actions (channelization, levees), water diversion canals, bridges and other road crossings, and sand and gravel mining operations.

If dams are present in your stream corridor, it is important to determine the degree to which your stream is regulated and how the natural flow regime has been altered. The level of impairment on impounded streams is a function of the size of the dam, its reservoir capacity, the proximity of the restoration site to the dam, and tributaries. Topographical maps, aerial photographs, and inputs from river-management agencies can provide information on dam construction, location, and capacity.

Australia has 501 large dams, defined as those having a wall height greater than 15 meters, a reservoir capacity of more than 1 million cubic meters ($m^3$), or the ability to store a flood discharge of more than 2,000 cubic meters per second ($m^3s^{-1}$) (Australian Water Association 2007). A source of information is the Australian National Committee on Large Dams (http://www.ancold.org.au/?page_id=3469), an apolitical organization that disseminates knowledge, develops capability, and provides guidance on dam engineering and management (including environmental considerations). In Mexico, CONAGUA is the source for information on the design, construction, and management of dams. Dam characteristics in Mexico can be found in http://sina.conagua.gob.mx/sina/.

Most large dams in the western United States are managed by Reclamation, although some are managed by local governments. The USACE is commissioned to inventory major dams in the United States (USACE 2010a). Roughly 4,000 dams have been constructed in Mexico; nearly 700 are considered large and so fall under the management directive of CONAGUA.

To understand the extent that your stream's flow has been altered by impoundment and other human interferences, Indicators of Hydrologic Adjustment (IHA) can be employed. IHA methods originated from research on river ecology and restoration of flow along impounded rivers, and are often applied to gauged streams (Poff et al. 1997; Richter et al. 1998). See chapter 5 for a review of both IHA and similar methodologies as well as for an overall approach to develop an environmental flow program to improve stream corridor conditions along impounded streams.

## Embarking on Field Reconnaissance

With data and information on the above topics gathered and digested from readily available sources and mapping, we are now better prepared to go to the field for targeted reconnaissance in a manner that effectively informs our stream restoration goals, objectives, and tactics. By "targeted," what we mean is to use your newly acquired knowledge of your stream and its watershed, as well as your restoration goals and objectives, as a lens to identify priority sites in your watershed that are essential to be seen firsthand in the field. That is, if you and your team cannot visit your entire watershed, what sites or parts of your watershed should be a priority?

"Hold the horses!" you may be thinking. "You told us that the central aim of conducting an evaluation is to inform our stream restoration objectives. How can we use restoration objectives as a lens to narrow mapping and field reconnaissance if they have yet to be developed?" A fair question, indeed. In response, we assume at this point that even though your stream restoration objectives may not be fully developed (i.e., may not have all the qualities of a robust restoration objective summarized in chapter 2), you and your team can probably state at least vaguely what your stream restoration objectives are. For example, "Our aim is . . . to enhance riparian habitat" or "to reintroduce native fish" or "to improve water quality and/or aesthetics." The point is to use the current state of your stream restoration objectives as a guide to identify field reconnaissance priorities and knowledge gaps that may need to be addressed via additional data collection and analysis (data collection and analysis is the focus of the next section). If your stream restoration objective centers on reintroducing native fish, examining pool-riffle sequences may be more of a priority than if your restoration objective concerns restoration of riparian habitat. If improving water quality is a big part of your restoration objective, seeing potential sources of water contamination firsthand will be a priority. Your restoration objective can also help you to prioritize the timing of field reconnaissance. For example, particularly in arid regions, you may want to see the streamflow conditions along your stream during the height of the dry season. We encourage you to spend as much time in the field as possible, while understanding the need to prioritize.

### Who Should You Bring with You?

Given the focus on streamflow, going to the field with a hydrologist is essential. Ideally, the hydrologist who accompanies you has undertaken (or is at least aware of) the data and some of the analysis that you have already conducted to address the questions posed as part of the "Water and Sediment Movement" section, above. Also invite other relevant experts: if your stream restoration objective is oriented around fish, inviting a fish biologist will be important; if the focus is to achieve a desired channel morphology, bring a geomorphologist; if recreation objectives are central, visit the stream with boaters (plus, they'll bring cold beer!), and so forth. Don't forget to include local community members who likely know your stream better than anyone. Ranchers, farmers, local water managers, and other riverside citizens can tell you about recent flow conditions and provide such valuable insights as when the last flood was, the effects of that flood, and when key pools or backwater areas go dry. Such information can be invaluable for understanding how well current streamflow conditions will fit with your stream corridor restoration objectives.

### Where and When Should You Go?

Endeavor to go to the field with your technical team to see as much of the stream system and its watershed as possible. Depending on the spatial scale of your restoration effort, your field visits may focus predominately on a particular site or parts of your stream that experts consider important to realizing objectives. For the establishment of an experimental population of Rio Grande silvery minnow in the RGB, fish biologists and hydrologists evaluated pool-riffle sequences and backwater areas, and walked or floated along significant parts of the river to get a sense of overall habitat and streamflow conditions during critical times of the year (see Case Study 6.9). If the low-flow period during the hot, dry season has been identified as a critical part of the hydrograph, a priority will be to visit the stream as many times as possible during that season in order to provide your experts insight as to how streamflow conditions during that period manifest themselves on the ground and how well those conditions support your stream restoration objectives. Visiting your stream with your experts during multiple years will improve the likelihood of observing streamflow conditions during different weather patterns (e.g., during times of below-average, average, and above-average rainfall).

### What Information Should Be Gathered Before Going to the Field?

First, to the extent possible, gather the information summarized in the subsection "Doing Your Homework." You do not need to have 100 percent of your "homework" completed and in-hand before going to the field. The two actions—gathering background information and field reconnaissance—go hand in hand and inform one another. But the more informed you are before heading to the field, the more productive your field visits will be.

Second, as mentioned in the section on water and sediment movement, if streamflow data are available, conduct some basic hydrological analysis. Jumping ahead a bit, several basic yet informative hydrological analyses ideas are offered in the next section; show this section of the chapter to the hydrologist and other members of your technical team to identify which initial analyses may be most useful to conduct. Additional thoughts on the types of basic hydrological analyses that can be fruitful to conduct prior to going to the field are offered in chapter 4 (in the context of understanding consequences of climate change) and chapter 5 (in the context of understanding potential need for environmental flow).

Third, download or have a sound understanding of the most recent streamflow data (flow discharge for the day you go to the field as well as average flow during the current critical period). For example, if you and your team visit your stream during the dry season (for much of the southwestern United States and northern Mexico, the hot, dry season begins in April), make sure to log in (if possible) to see what the discharge is on the day you go to the field. In addition, download daily data for the year to the date you are in the field. If you are in the field on June 15, downloading streamflow data from April to June 15 will give you a sense of flow conditions for the current hot, dry period. Per comment above on conducting targeted basic analysis of available streamflow data, conducting an analysis of the available streamflow record for the same April–June 15 time period will allow you to compare quantitatively (as well as see firsthand when you are in the field) how current flow conditions measure up to the long-term record. For example, are streamflow conditions on the day you visit the stream and the current season's average above or below the long-term average?

Fourth, go to the field well-versed in how temperature and rainfall conditions for the month of your field visit stack up against long-term averages. Are you visiting the stream during a time of above-average precipitation and relatively cool temperatures or during a drought? Combining such insights with an understanding of how recent streamflow compares to long-term average (per above paragraph) can provide great insight on how streamflow conditions (as well as other bottomland biophysical conditions) measure up and inform your stream restoration objectives. For example, if your restoration objective centers on establishing native fish along a stream whose flow becomes a trickle during the hot, dry premonsoon months, going to the field during those months makes a lot of sense. If streamflow and precipitation for the month of your field visit are below long-term averages (and temperatures are above average), yet pools and backwater parts of the stream environment (e.g., secondary channels—see "Oxbows and Secondary Channels" below) that provide critical habitat for native fish still contain ample amounts of water, such may be one positive indication that the stream environment is suitable to your native fish species. On the other hand, if current streamflow data and precipitation for the month of your site visit are above average (and temperatures are below), yet many pools and backwater areas are dry, such may be an indication that the stream environment is not suitable for establishing native fish (i.e., if many of the pools that your fish species depends on have gone dry during a period of above-average precipitation, they'll certainly be dry when streamflow and precipitation are average or below average).

## What Equipment Should You Bring?

Your experts will certainly know what constitutes typical field gear that will allow them to make the most of each site visit. A hydrologist may bring basic survey equipment to measure channel width, channel slope, or depth of key pools, as well as a pygmy meter or portable weir to measure flow along key parts of the stream. The fish biologist may bring sampling gear to get a sense of your stream's macroinvertebrate population. If you have river guides with you, they'll bring refreshing beverages, which are always important for postfieldwork wrap-up conversations. In addition, bring a portable water-quality meter with you if water-quality conditions are an important aspect of your stream restoration objective (e.g., for restoration objectives focused on native fish, does water quality deteriorate past important thresholds during low-flow or no-flow periods?).

### What to Look for in the Field Regarding Streamflow

*High-Water Marks.* Identifying and recording high-water marks and lines of a recent flood, as shown by debris and sediment on canyon walls, large trees, and bridge abutments, can be helpful in estimating the frequency and magnitude of inundation in a bottomland. If you can link a high-water line in a reach of your stream to one that is near human settlement, interviewing locals may allow you to identify the flow date. This information may help you estimate discharge and recurrence intervals if stream gauges are available along your stream or in nearby watersheds (see the subsection "When Streamflow Data Are Available"). Other approaches to gain insights into the flow regime of ungauged streams are also available (see the subsection "When Streamflow Data Are Not Available"). If you can correlate discharge estimates to multiple high-water marks via repeat visits or to several lines during one visit, a rough stage-discharge relation can be constructed to meet restoration objectives.

*Presence of Springs or Seeps.* Springs and seeps can strongly impact discharge and water availability, particularly during low-flow periods. Noting the locations, number, and discharges of springs is a step toward understanding their significance. Spring discharges may contribute significantly to base flow during dry, premonsoon months, making their protection a conservation priority. Evidence of spring discharge that should be noted includes: (a) dense growth of riparian plants on floodplains or channel surfaces, (b) change in water temperature as indicated by vapors at the surface or melting of snow and ice, and (c) changes in color and odor where seeps or springs enter a stream, which may indicate precipitation of metal oxides, hydroxides, or salts as carbonates, sulfates, or halides. Similar changes can occur as a result of pollution from a sewer or drainage pipe, so be careful to identify the source.

### What to Look for in the Field Regarding Channel Morphology

Protocols to document channel morphology change by field reconnaissance are provided by Fitzpatrick et al. (1998). Geomorphic assessment and rapid assessment protocols have been developed by most federal and state agencies involved in river restoration and are very helpful in guiding field reconnaissance. Brierley and Fryirs (2005) published a standard handbook for describing channel conditions. References such as Harrelson et al. (1994) offer guidance on developing site maps and basic survey methods.

Geomorphic features to take note of in the field are the following:

- *Oxbows and secondary channels*—Oxbows, which are normal features of mature bottomlands, are parts of abandoned meander loops of channel reaches that remain on a floodplain or alluvial terrace. Secondary channels are channel segments that are split from a dominant alluvial stream channel and convey water and sediment during high-flow events. Both of these features can offer conditions of morphological complexity, streamflow, and water availability that can be conducive to aquatic and riparian restoration objectives.

- *Floodplain depressions*—These natural pits or depressions that occur on a floodplain surface may be caused by flood scour, tree toppling, or near-surface groundwater dynamics. Depending on the stream's physical environment, floodplain depressions can retain water and provide refuge for a variety of aquatic and riparian species.
- *Outer and inner banks of meander bends*—Where an outer bank of a channel bend or meander loop is steeply sloping and unvegetated, the implication is active meander migration and channel instability; this area should be avoided regarding riparian revegetation efforts because anything planted there is likely to be scoured out. Conversely, the inside of meander bends are often areas of active aggradation and stability and may offer suitable stability (as well as water availability) for riparian revegetation efforts.
- *Sand and gravel bars*—Channel bars of coarse sand or gravel are the result of sorting processes by streamflow. If bars are absent, it may be related to channel stability or the lack of sand and gravel for stream transport.
- *Eroding banks*—In alluvial channels, you will inevitably find an eroded bank. As indicated above in discussion on channel meanders, this is a normal part of an active and healthy stream environment. However, if eroding banks are found along a significant part of a stream reach, this may suggest excessive streamflow relative to the sediment available for transport and indicate that the stream channel is becoming wider and possibly deeper. Practitioners may want to avoid implementing on-the-ground restoration actions along such reaches, and adopt a wait and see approach.
- *Alluvial-valley width*—This is the width of a stream bottomland that is underlain by alluvium of the floodplain and terrace surfaces. In general, wide alluvial valleys relative to stream size indicate long-term sediment storage and stability.
- *Channel and floodplain widths*—Channel width is an indicator of stream discharge characteristics, and floodplain width indicates whether large or small amounts of sediment are stored within the valley. In addition, a wider floodplain surface often provides morphological complexity conducive to both aquatic (at higher-flow discharges) and riparian restoration objectives that can be lacking in narrow configurations.
- *Sites of channel constraint*—These include places of natural constriction, as by bedrock, and artificial constriction, as by artificial flood control structures. Since these areas are constricted relative to reaches up- and downstream, they are often characterized by steep stage-discharge curves (i.e., a relatively dramatic rise in stage per increase in discharge) that lead to channel instability and frequent bottomland disturbance. Such parts of a stream may not be suitable for on-the-ground restoration actions.
- *Gradient*—The gradient of an alluvial-channel reach is the rate of thalweg-elevation decline between two sites along the reach and therefore is an indicator of the energy conditions of the stream.
- *Sizes of bed sediment*—Bed sediment size is an indicator of both the size of sediment supplied to the stream channel and the energy conditions of the stream. Riffles composed of coarse, loose gravel and cobbles indicates high-energy conditions and the ability of floods to move the coarse sediment. Bed sediment composed of sand suggests a large sediment supply of fine sediment. Sand is generally mobile during most high flows. Harrelson et al. (1994) offer an easy approach to characterize bed and bank material.

Repeated field observations can indicate whether a channel is sediment-rich or sediment-deficient. These observations may justify more detailed investigations to quantify sediment movement (see the next section).

Signs of sediment surplus include the following:

- abundant channel bars that lack mature vegetation
- evidence of lateral channel migration, such as bars inside of bends or cutbanks on the outside
- gravel "plugs" at tributary junctions that grow with time and are relatively stationary, causing backwater upstream and riffles downstream (see figure 3.2a)
- evidence of progressive channel narrowing

Signs of sediment deficit include the following:

- a channel bed with coarse sediment of gravel and cobbles
- evidence of channel incision, such as steep channel banks lacking vegetation, root flares of riparian vegetation exposed by erosion of bank sediment (see figure 3.2b), or evacuation of sediment from bridge pylons (see figure 3.2c)
- lack of channel bars and floodplains

Repeat surveys of channel profiles and cross sections allow an appraisal of temporal change in width, mean depth, wetted perimeter, hydraulic radius, and width-to-depth ratio. Cross-section surveys are basic to investigations such as hydraulic modeling. Channel form can vary tremendously at the reach scale and therefore, the measurement of many cross sections may be necessary to permit hydraulic modeling. Knighton (1998), Gordon et al. (2004), and Harrelson et al. (1994) summarize protocols for surveying longitudinal profiles and channel cross sections. We will broach this topic again in chapter 7.

## What to Look for in the Field Regarding Stream Biota

Evaluating biota (e.g., the extent and distribution of riparian vegetation communities) and relating biology to hydrology and land use can provide great insight toward understanding why conditions changed. For example, changes in the extent and distribution of a riparian plant community can be noted from many of the sources of information noted previously, as well as by mapping the riparian community via a comparison of aerial photographs from different time periods. The cause of change to the riparian community could be pieced together by looking at streamflow records and indicators of high-flow occurrences in the field as well as by identifying human interferences in maps, aerial images, or Google Earth. Relating spatial and temporal changes of bottomland biota to groundwater pumping (when, where, how much) may yield links between groundwater extractions and loss of native plants and animals.

OBLIGATE RIPARIAN AND AQUATIC PLANTS

The presence of *Populus* spp. (cottonwoods), *Salix* spp. (willows), *Fraxinus veluntina* (velvet ash), *Platinus wrightii* (Arizona sycamore), *Juglans major* (Arizona walnut), *Eucalyptus camaldulensis* (river red gum) and herbaceous aquatic genera such as *Eleocharis, Scirpus, Cyperus,*

A

B

C

**FIGURE 3.2** Photographs depicting sediment surplus and deficit along selected streams: (a) gravel plug at confluence of Tornillo Creek (entering from the left side of the photo) and Rio Grande/Bravo in Big Bend National Park (David Dean, USGS); (b) exposed root flare of a sycamore tree (*Platanus occidentalis*) at Warm Springs Run, Berkeley Springs, W.Va. (Edward Schenk, U. S. Geological Survey); (c) sediment evacuation from bridge pylon foundations along the Río San Pedro, near Meoqui, Chihuahua (Alfredo Rodriguez, WWF).

*Typha*, and *Carex* provides evidence of near-surface permanent water (see Stromberg et al. [2006], who developed a model that evaluates nine riparian vegetation traits determined to be sensitive to water availability).

If woody riparian plant communities are present, note their composition and age classes. Although conclusions need to be drawn cautiously, the likelihood that conditions of water availability and channel stability are conducive to the establishment and persistence of obligate riparian trees (species that can only survive the hot, dry season if their roots can access saturated soils) is strengthened when both young and mature obligate riparian trees are found. On the other hand, a single-age class of mature cottonwoods may indicate that formerly shallow groundwater has declined to levels too deep for the survival of seedlings or saplings with developing shallow root systems.

AQUATIC BIOTA

*Presence/Absence.* In arid regions, stream practitioners work in situations where evaluating surface water permanence to inform the development of realistic stream restoration objectives can be a priority. Whether a pool of water identified in the spring season persists through the hot, dry summer is often valuable information. Although the presence of fish in an isolated part of a stream does not definitively answer this question, it does indicate high potential for perennial surface water, with verification dependent on further assessment. Regardless, if fish are present along a stream that practitioners feel may be on the borderline between perennial and intermittent, such is important to note.

*Fish Populations.* In addition to presence/absence information, practitioners should make note of the specific species present and collect samples, which will provide practitioners with data and is highly encouraged. See chapter 7 for discussion of fish-sampling techniques.

*Benthic Invertebrates.* Benthic invertebrates—organisms that live in or on the bottom sediments of rivers, streams, and lakes—consume algae, biofilms, and organic matter, and are important links in the food chain, including to fish. They are often sampled as part of an aquatic inventory and monitoring program because they are diverse, generally sedentary, and responsive to environmental alterations. As such, they are good indicators of ecosystem productivity and health. Jones (2011) reviews procedures associated with site selection, sampling, sample preservation, and identification and listing invertebrates. Cuffney et al. (1993) summarize protocols for collecting benthic invertebrates for the USGS water-quality assessment program; included are recommendations organized by substrate or bed-sediment type. Hotzel and Croome (1999) present procedures for sampling, preserving, identifying, and enumerating phytoplankton in Australian surface waters, and recommended approaches for quality control and data storage.

## What to Look for in the Field Regarding Water Quality

If deterioration of water quality is an issue for your stream, field reconnaissance activities should focus on collecting clues as to what might be driving deterioration and pinpointing locations along your stream where sources of contamination are entering or where deterioration of water quality is most severe. Actions to consider while in the field that are related to water quality include the following:

- Note with GPS unit or on a map, or simply describe, potential point and nonpoint sources of contamination.
- Note abrupt changes in water color, odor, foam on water, and algae.
- Collect your own chemical data! Obtain a portable water quality meter and take a few measurements along your stream. For example, if you are walking along 5 miles of your stream, take a measurement every half mile, or if you located a source of contamination, concentrate measurements above and downstream of the source. A good portable water quality meter can be purchased for less than US$1,000 and can measure temperature, pH, salinity-compensated dissolved oxygen measurements, and conductivity. Meters in the range of US$2,000 can also measure ammonium, ammonia, chloride, and nitrate, and provide GPS coordinates with each reading.

To dive deeper into water quality sampling methods, consider the following excellent resources:

- Government of Western Australia's field-sampling guidelines (2009);
- Tasmania Department of Primary Industries Water and Environment's community water-quality sampling protocols (2004);
- Programa Nacional de Reservas de Agua, Mexico (Lanza Espino 2014); and
- USGS's National Field Manual for the Collection of Water-Quality Data (2018).

Also, consider collecting your own aquatic biota data! We talked briefly about macroinvertebrates above, but remember that macroinvertebrates (which generally need identification by an aquatic biologist) are excellent indicators of water quality. By knowing which macroinvertebrates are highly sensitive or somewhat sensitive to water-quality changes, aquatic ecologists can determine if a stream's water quality is increasing or decreasing. For example, if water quality is becoming an issue for your stream, such macroinvertebrates as gilled snails and mayfly nymphs will be absent, while more tolerant species, such as leeches and pouch snails, will be present. You will need to have skills and training to identify macroinvertebrates, especially at lower taxa. As with water-quality sampling, there are numerous excellent references available on all aspects of sampling, preserving, identifying, and interpreting macroinvertebrate diversity: New South Wales Office of Environment and Heritage n.d.; Cuffney et al. 1993; Bug Lab—BLM/USU National Aquatic Monitoring Center (https://www.usu.edu/bug lab/MonitoringResources/MonitoringProtocols/#item=26), and CONAGUA 2014 are but a few examples.

# ADDRESSING CRITICAL INFORMATION GAPS VIA ANALYSIS OF EXISTING DATA AND CONDUCTING OTHER INVESTIGATIONS

In this section, we turn our attention to addressing critical information gaps that you have identified as part of your homework in the previous section through analysis of existing data as well as by conducting your own investigations (i.e., collecting your own data). As with the previous section, emphasis is on hydrology (versus biology and chemistry). Whereas case studies are provided to underscore key points in other chapters, here we provide examples to help readers walk through the main steps of selected analyses.

The information gaps that you will have identified reflect some level of scientific uncertainty on stream and watershed conditions. Determining whether these gaps are sufficiently wide to halt forward progress with restoration is a common challenge that practitioners face. There is no green/red light scorecard. What we are aiming for regarding understanding watershed and stream corridor conditions is a middle ground between doing no evaluation (and implementing restoration tactics with little understanding of watershed and stream conditions) and conducting never-ending research. Keep in mind that the stream restoration response is itself a scientific experiment whereby a restoration action is implemented, monitoring data are collected and interpreted, and the restoration approach/tactics are modified accordingly and then (hopefully) implemented again, with the hope that restoration tactics will be more effective and realize a greater restoration footprint.

The challenge at this point in developing your stream corridor restoration response is to identify the information gaps that are essential to address before moving forward. Identifying which gaps are essential and which are not is often not clear-cut and can spark quite a debate among team members. Gaps that should be addressed before moving forward are associated with an inadequate understanding of the following:

a. The causes of ecological decline. Successful stream corridor restoration hinges on understanding the causes behind the deterioration of your stream. If such is not understood, do not move forward. More investigation is needed.

b. How your stream corridor has responded to external stresses, particularly stresses that are having a dramatic input on water and sediment transport. For example, how is diminished sediment supply from river impoundment causing a stream to incise its bed? How are unaltered sediment inputs combined with diminished flow fostering aggradational processes that are burying high-quality aquatic and riparian habitat?

c. The level of ecological recovery that the stream corridor might experience if main stresses are reduced or eliminated. If you are able to reduce or eliminate a main stressor (remove cattle, reduce groundwater pumping, get rid of a dam), understanding the extent that the stream ecosystem might bounce back without further intervention is an important piece of information. Stream ecosystems are highly resilient and can come back strongly after main stressors have been addressed. If strong natural recovery is likely after a stressor has been eliminated or reduced, all you may need to do is sit back and watch natural recovery processes occur without conducting on-the-ground actions.

d. The potential that direct (on-the-ground) restoration actions will be successful. More ecological information will need to be collected and assessed if there is only a vague notion of the likelihood that restoration actions will achieve their objectives (e.g., level of certainty that the native fish you are reintroducing or native riparian species you are planting will establish and persist in the long term).

In this section, we review several methodologies that are commonly conducted to address information gaps related to sediment and water transport. The methodologies that are summarized run the gamut from analytical methods employed when streamflow data are available for your watershed, as well as when they are not available, to developing stage-discharge relations and measuring losses and gains to streamflow.

Before getting into specifics, we want to caution that this section is perhaps the most technical of the guidebook. Parts may be daunting, particularly if you lack experience in natural resources or data analysis. First and foremost, the selected investigations outlined here are provided as *potential* investigations that practitioners could pursue to address critical gaps in knowledge. We are not at all implying that they are a prerequisite to moving forward with a stream restoration program. Second, remember that the purpose of this section is *not* to transform a lay person into a hydrologist, but to provide technical guidance. If you find parts of this chapter rough going, please note the basic concepts from those parts and skip to the next section. Please don't close the book! Skim through the information as a means to get the wheels turning as to the investigations you may need to conduct to move thoughtfully forward. Third, be assured that help is available. Keep your restoration team in mind (see chapter 2) and acknowledge the importance of seeking assistance from someone well-versed in hydrology. There may be faculty at a nearby university who would be interested in involving their students, or land or water managers who would jump at the opportunity to help you evaluate conditions and trends in the watershed in which they work. Don't go it alone. Also, take advantage of the background literature and other sources of information that are cited in this chapter and don't hesitate to reach out to authors of the literature of particular interest.

## When Streamflow Data Are Available

If your stream is gauged, you will have access to flow data and a variety of analysis opportunities. Knowing where gauges are, who manages them, the types of data collected, periods of record, frequencies and times of data collection, and quality of data are all pertinent.

Numerous methods to quantify and describe the stream's seasonal pattern of flow over the course of a year can be applied if flow data are available. Selected types of flow measurements and statistics that describe hydrological conditions and trends are listed and explained below. We only summarize them, as all are well-documented (Gordon et al. 2004; Gore 2007; Chow et al. 2010). It is also important to note that a variety of factors related to the homework you did in the previous section can help immensely to guide the type of analysis you conduct. For example, knowledge of when dramatic changes in land or water use occurred can help you hone analyses. If a large dam was constructed in your watershed in 1930, focusing analysis of streamflow data on years prior to 1930 and the years after will allow you to quantify impacts of impoundment on streamflow. A long-term record of precipitation and temperature can provide the means to focus the analysis of streamflow data on drought and nondrought years. Some or all of the analyses below may already have been conducted by the agency that is managing the gauge. If not, they can be easily computed using established methods reviewed in the literature cited in this paragraph.

### Continuity of Flow

Access to a long-term flow record will allow you to describe flow during years of interest as *perennial, intermittent,* or *ephemeral.* If discharge data are above zero for the entirety of a typical year, your stream is perennial. If flow goes to zero only during the height of the dry season, your stream may be intermittent. If streamflow is zero during much of the dry season, your stream is intermittent or ephemeral. If you have long-term precipitation data for your watershed, you could also compare continuity of flow along your stream for years of above-average precipitation and years of drought.

## Average Flow

The average flow for a period, usually calculated annually, is the average of all flow measurements of that period. Comparing average flows over the period of record is useful for identifying trends. If analysis reveals a trend or significant change in average annual discharge, further analyses may suggest the main causes of change, with past climate and land-use change being top candidates.

## Stream Hydrograph

A stream hydrograph depicts discharge in terms of stage, water depth, or runoff volume over time, and indicates streamflow patterns of a year, season, or single runoff event. Knowing the mean annual discharge for a period of record can help you discern trends (gradual or abrupt) in discharge, and suggests the time and cause of change. In chapter 5, we recommend developing an annual hydrograph as a helpful first step to understanding how streamflow typically changes during the course of the year. Flow characteristics that can be generated include a ranking of mean annual flows, discharge per unit area, and single-event hydrographs.

## Flow Duration

Flow duration refers to the percentage of time that a specified discharge (flow rate) is equaled or exceeded (Osterkamp 2008). Flow-duration curves are useful where aquatic and terrestrial habitats are threatened by lack of water and project managers are assessing environmental flow requirements. The high ends of curves indicate the periods or duration that channel banks and floodplain surfaces are inundated, which could be important for restoration practitioners looking for suitable surfaces for planting native riparian plants. The low ends can indicate the time a specific backwater area is under water (potentially identifying suitable fish habitat). The flows analyzed are for a specified duration (e.g., one day, ten days, six months, one year), and are available from established gauged sites (see appendix A for federal and state sources).

If discharge data are adequate to generate a flow-duration curve and sediment-concentration data are available for much of the same period, a sediment-rating curve, or sediment-transport curve (Glysson 1987) can be constructed. A sediment-rating curve is a line that defines the relation between discharge and the concentration of sediment in transport; it indicates mean variation in sediment concentration with respect to water discharge for the period of data collection (Osterkamp 2008). The availability of a flow-duration curve and a sediment-rating curve permits the development of concentration duration, or sediment duration; a concentration-duration curve indicates the percentage of time that various concentrations of suspended sediment were equaled or exceeded (Osterkamp 1977). Analyzing specific portions of such a curve allows estimates of how much sediment passes a site relative to the longer-term ranges of water and sediment discharges. Such estimates can indicate whether sediment fluxes are in balance, which is an important measure of the health of the hydrological system.

## Intra-Annual Variation

Intra-annual variation describes the seasonal flow variability of a stream. For semiarid streams, this usually includes snowmelt (which causes moderate spring and early-summer discharges), summer monsoonal storms that yield floods, and very low flows at other times. Understanding seasonal-flow variation is essential to developing a rehabilitation program that recognizes the needs of native vegetation and bottomland fauna. Analysis of monthly flows quantifies seasonality of changes. From gauge records, peak flow (the maximum discharge for a specified period) can be identified. Knowing the magnitudes, timing, and durations of annual peak flows, and whether they changed during a time interval, may indicate the cause and extent of impairment along a stream reach.

## Flow Stage at Key Locations

Stage is the height of a water surface at a reference point that is either naturally or artificially occurring, such as one associated with a streamflow gauge. For streamflow gauges, a stage-discharge curve, or rating curve, is developed via repeated measurements at various stages, producing a graph showing the relation between the gauge height and the amount of water flowing in a channel (expressed as volume per unit of time) (see "Developing Stage-Discharge Relations" below). Regarding naturally occurring reference points, practitioners may want to know the discharge needed to inundate a specific channel morphological surface of interest. If discharge can be measured or estimated for a specific natural stage of interest, then practitioners can estimate the frequency of the discharge of interest based on the long-term streamflow record (see "recurrence interval," below). For example, if restoration actions involve planting obligate riparian trees on a particular floodplain surface, knowing how regularly that surface experiences flow is valuable. If flow occurs frequently (e.g., a couple times a year), newly planted seedlings may be vulnerable to scour. If it occurs once in a blue moon, the surface may be too dry. Restoration objectives will determine whether ballpark recurrence interval of discharges at specific stages would suffice (e.g., every year versus once in a blue moon) or whether more accuracy is needed.

## Length of No-Flow Periods

The lengths of no-flow periods can be an important metric for developing realistic stream restoration objectives and tactics in arid regions. Natural no-flow periods (such as months prior to the onset of the monsoon season) can be modified by climate change, groundwater extractions, impoundments, and flow diversions, with potential adverse consequences to native aquatic and riparian biota, and riverside citizens. Correlating changes in no-flow periods with knowledge of land-use change that alters flow (e.g. dams, diversions, groundwater pumping) is important to demonstrating causation as well as determining what may or may not be a realistic restoration objective (e.g., if no-flow periods are regularly long enough to dry out key pools for native fish, then reintroduction of native fish may not be a viable option).

## Recurrence Interval

The recurrence interval of a hydrological event is the estimated average time, in years, that a specified flow rate for a flood is expected to occur. A flow rate, or discharge, with a ten-year recurrence interval has a probability of one in ten of occurring each year. Questions to consider include: Have statistical recurrence intervals associated with defined high flow rates changed over time (e.g., before and after dam construction, before and after such upland modifications as deforestation or urbanization, or as a result of climatic change)? What are the distributions of low and high flows and how have they changed?

Methods for calculating return periods of selected discharges are well-documented. The USACE Hydrologic Engineering Center (HEC) maintains the Statistical Software Package (HEC-SSP) that computes flood-flow frequency curves, as described in FEMA's Interagency Advisory Committee on Water Data (1981). HEC-SSP can be downloaded for free and is linked directly to the USGS database. In addition, the USGS maintains the National Streamflow Statistics program (http://water.usgs.gov/software/NSS/), which already has these statistics calculated.

IHA programs often are applied to gauged streams to quantify the alterations that have occurred along impounded rivers, using statistics for flow parameters (Poff et al. 1997; Richter et al. 1998). IHA programs are among a variety of methods that stream practitioners can apply to improve stream corridor conditions. Effective restoration may require policy initiatives, such as modifying releases from impoundments, to achieve environmental flows to benefit native ecosystems (see chapter 5).

## Flow Modeling

Flow modeling provides a means to expand relations between stage and inundation of bottomland surfaces. The USACE River Analysis System (HEC-RAS) and other one-dimensional (1-D)* flow models compute water-surface elevation and a downstream velocity vector for designated discharges (USACE 2002). The primary inputs to 1-D models are a reliable stage-discharge relation at the downstream end and a series of cross sections upstream. (HEC-RAS can be downloaded at https://www.hec.usace.army.mil/software/hec-ras/).

Various studies have combined 1-D hydraulic-flow models with data and information from channel cross sections, aerial photographs, and Landsat images. Selected examples of tools and resources that might address information gaps associated with stream corridor restoration, categorized by objective or focus, are provided below. If these or similar analyses are of interest, we encourage you to use them to help you dive deeper.

- Channel-morphology and sediment-dynamics modeling of water-surface elevation, stream power, and channel-morphological change produced by a high-magnitude flow event (Big Bend reach of the RGB, U.S. and Mexico [Dean and Schmidt 2013]);
- Estimation of riparian tree establishment based on flood evaluation (Bill Williams River, Colorado River Basin [Shafroth et al. 2010]);
- Assessment of sediment dynamics along pool-riffle sequences in a mountain stream (N. F. Cache La Poudre River, Colorado [Rathbun and Wohl 2003]);
- Mapping flood-prone reaches along ungauged channels (Hali River Basin, Kerman Province, southeastern Iran [Sarhadi et al. 2012]);

---

* One-dimensional modeling assumes that the vast majority of the water is moving in only one dimension—upstream to downstream. In comparison, two-dimensional computer models provide a better understanding of flow that moves in more than one direction.

- Assessment of the sensitivity of 1-D hydraulic models to different types of river-morphology data to develop flood-forecast mapping applications at regional scales (Saleh et al. 2012);
- Aquatic-habitat simulation and assessment of ecological effects of physical-habitat alteration by changed river management (Zwalm River, northern Belgium [Mouton et al. 2007]);
- Assessment of fish habitat based on synthetic hydraulic and water-quality parameters (Arno and Serchio Rivers, Tuscany Region, Italy [Marsili-Libelli et al. 2013]).

Modeled simulations of streamflow can be an effective way to evaluate how freshwater eco-systems respond to disturbed conditions, identify appropriate sites for restoration actions, and evaluate system response to different scenarios such as changes in land use or climate change (McKinney et al. 1999; Merenlender and Matella 2013). Integrated physical and ecological modeling methods, however, are fraught with challenges relating to data limitations and high uncertainty in our understanding of the relation between land use and physical and ecological responses (Nilsson et al. 2003). An important question is whether the additional cost of modeling is justified, particularly if less expensive approaches, such as relying on expert opinion, are available that would provide the precision required to address pertinent information gaps.

## Basic Analysis of Streamflow Data from Multiple Gauges

Having one streamflow gauge with a long-term record on your stream will provide great insight into characterizing your stream's flow regime and how it has changed over time (at least, at the point where flow data are collected). Consider yourself super lucky if your stream has multiple gauges that, in addition to providing temporal insights on streamflow variability, will allow you to understand how your stream's flow regime changes spatially. This example describes basic analyses of streamflow data collected from several gauges along the RGB in the Big Bend region of Texas, Chihuahua, and Coahuila, prior to and after construction of major dams throughout the watershed. The RGB drains the southern Rocky Mountains of Colorado and New Mexico and forms the United States–Mexico boundary between Ciudad Juarez, Mexico, and Brownsville, Texas. Near Presidio, Texas, and Ojinaga, Chihuahua, the RGB is joined by the Río Conchos, its second largest tributary, which originates in the Sierra Madre Occidental of southern Chihuahua (see figure 3.3).

From 1915 through the 1960s, several large dams were constructed on the upper Rio Grande in New Mexico and on the Río Conchos in Chihuahua that have impacted the natural flow regime of the river along the international border through the Big Bend region. For example, dam operations and irrigation diversions in New Mexico and the El Paso–Juarez Valley have eliminated annual snowmelt floods of the upper Rio Grande. Simple analyses using streamflow records illustrate the magnitude of change. These analyses can be performed for any gauge with long-term records.

Once available streamflow data are downloaded, the first main step is to organize the data by either calendar year or water year. The definition of water year (sometimes called discharge year or flow year) varies. As defined by the USGS, a water year is October 1 through September 30. In Mexico, CONAGUA typically uses a calendar year, closing water balances for all reservoirs on December 31 of each year to calculate water distribution for the next farming year. In Australia, the water year is July 1 to June 30 (as defined by the Bureau of Meteorology, Australian Government). The differences in data organization can be for political and/or climatic reasons. For example, for watersheds that receive significant amounts of snow in late fall,

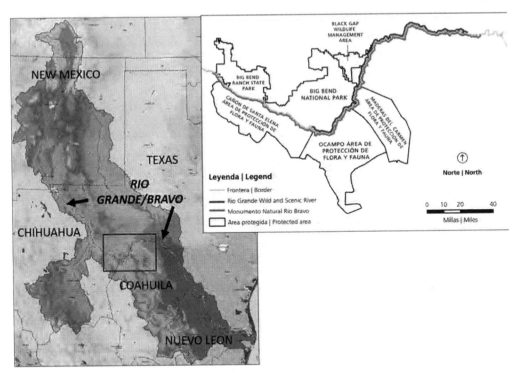

**FIGURE 3.3**  Regional map of the Rio Grande Basin (left) with a detailed map of the RGB in the Big Bend binational region with protected areas in Mexico and the United States depicted. Right map by Marie Landis, Big Bend National Park.

which often does not manifest itself as runoff until the spring of the following year, organizing streamflow data by water year (as defined by USGS) does not split snowmelt-driven discharge between two different years. In contrast, in watersheds that experience late summer–early fall moisture, it may be better to organize streamflow data by calendar year. For example, in the southern portion of the RGB, a significant amount of precipitation occurs during the monsoon season, which can extend into October, and thus, the monsoon season could be potentially split into separate years (if data are organized October 1–September 30). The main point: organize streamflow data as best suited to your geographic location.

The next step is to calculate the median and average of the mean-daily discharges for each day of the year for the period of interest. This was the approach used for streamflow data collected at gauges located upstream and downstream of RGB's confluence with the Río Conchos for time periods prior to large-scale dam construction as well as during an extended drought in the 1990s and 2000s. Table 3.1 shows how mean-daily discharges were organized by each day of the year, and the averages and medians of those data were calculated for the period of 1901 to 1913, which was the period of record before major dam construction. Once that is done, computation of the mean-annual flow for each year (the average of all mean-daily discharges), the annual flow volume (mean-annual flow times the number of days in the year times 86,400 seconds per day), and the maximum and minimum mean-daily flow can be easily accomplished (see table 3.2).

Graphing results of the above analysis of data collected from two streamflow gauges will allow the identification of both spatial and temporal changes in the flow regime. For example,

**TABLE 3.1** Computing average daily flow and median daily flow records

| | Mean daily discharges | | | | | | | | | | Average. daily flow (1901–13) | | Median daily flow (1901–13) | |
| | 1901 | | 1902 | | 1903 | | ... | | 1913 | | | | | |
| Day of the month | m³s⁻¹ | ft³s⁻¹ | m³s⁻¹ | ft³s⁻¹ | m³s⁻¹ | ft³s⁻¹ | m³s⁻¹ | ft³s⁻¹ | m³s⁻¹ | ft³s⁻¹ | m³s⁻¹ | ft³s⁻¹ | m³s⁻¹ | ft³s⁻¹ |
|---|---|---|---|---|---|---|---|---|---|---|---|---|---|---|
| 1-Jan | 0.16 | 5.7 | 0.20 | 7.1 | 0.30 | 10.3 | ... | ... | 0.59 | 21.2 | 0.84 | 29.87 | 0.59 | 21.2 |
| 2-Jan | 0.16 | 5.7 | 0.19 | 6.8 | 0.30 | 10.6 | ... | ... | 0.59 | 21 | 0.79 | 28.07 | 0.59 | 21 |
| 3-Jan | 0.16 | 5.7 | 0.18 | 6.5 | 0.30 | 10.6 | ... | ... | 0.54 | 19.3 | 0.75 | 26.9 | 0.65 | 23.1 |
| 4-Jan | 0.16 | 5.7 | 0.17 | 6.2 | 0.25 | 8.9 | ... | ... | 0.53 | 19 | 0.75 | 26.61 | 0.64 | 22.7 |
| 5-Jan | 0.15 | 5.4 | 0.18 | 6.5 | 0.25 | 8.9 | ... | ... | 0.53 | 19 | 0.71 | 25.18 | 0.61 | 21.8 |
| ... | ... | ... | ... | ... | ... | ... | ... | ... | ... | ... | ... | ... | ... | ... |
| ... | ... | ... | ... | ... | ... | ... | ... | ... | ... | ... | ... | ... | ... | ... |
| 23-Dec | 0.21 | 7.4 | 0.33 | 11.9 | 0.18 | 6.5 | ... | ... | 0.17 | 6 | 0.21 | 7.36 | 1.38 | 49.3 |
| 24-Dec | 0.23 | 8.2 | 0.32 | 11.3 | 0.18 | 6.5 | ... | ... | 0.19 | 6.7 | 0.23 | 8.21 | 1.36 | 48.7 |
| 25-Dec | 0.23 | 8.2 | 0.32 | 11.5 | 0.18 | 6.5 | ... | ... | 0.20 | 7.2 | 0.23 | 8.21 | 1.36 | 48.7 |
| 26-Dec | 0.22 | 7.9 | 0.32 | 11.6 | 0.17 | 6 | ... | ... | 0.20 | 7.1 | 0.22 | 7.93 | 1.18 | 42.2 |
| 27-Dec | 0.23 | 8.2 | 0.33 | 11.8 | 0.16 | 5.8 | ... | ... | 0.19 | 6.8 | 0.23 | 8.21 | 1.17 | 41.9 |
| 28-Dec | 0.29 | 10.5 | 0.33 | 11.8 | 0.16 | 5.7 | ... | ... | 0.19 | 6.7 | 0.29 | 10.5 | 1.31 | 46.7 |
| 29-Dec | 0.22 | 7.9 | 0.32 | 11.6 | 0.16 | 5.8 | ... | ... | 0.19 | 6.8 | 0.22 | 7.93 | 1.31 | 46.7 |
| 30-Dec | 0.22 | 7.9 | 0.32 | 11.5 | 0.17 | 6 | ... | ... | 0.19 | 6.9 | 0.22 | 7.93 | 1.21 | 43.3 |
| 31-Dec | 0.21 | 7.4 | 0.30 | 10.8 | 0.17 | 6.1 | ... | ... | 0.19 | 6.9 | 0.21 | 7.36 | 1.13 | 40.2 |

An example of computing average daily flow and median daily flow from average daily flow records as they are typically presented by the entity that manages the streamflow gauge (though, in the United States, discharges are usually presented only in Imperial [or English] units, whereas SI [or metric] units are used everywhere else in the world). In this example, data are from USGS stream gauge 08374200 of the RGB below Rio Conchos near Presidio, Texas, and focus is on analyzing the years of record prior to major dam construction along the RGB (in the RGB, the first major dam was constructed in 1915). The last two columns on right show annual daily average and median for the period of record from 1901 to 1913 and are used to develop the graphs presented in figure 3.4.

**TABLE 3.2** Yearly computation of selected streamflow parameters

| Selected parameters computed for the 1901–13 period of record, RGB | 1901 | | 1902 | | 1903 | | 1913 | |
|---|---|---|---|---|---|---|---|---|
| | $m^3s^{-1}$ | $ft^3s^{-1}$ | $m^3s^{-1}$ | $ft^3s^{-1}$ | $m^3s^{-1}$ | $ft^3s^{-1}$ | $m^3s^{-1}$ | $ft^3s^{-1}$ |
| Total flow volume ($m^3$) | 6.97E+08 | 2.47E+10 | 1.89E+09 | 6.69E+10 | 1.21E+09 | 4.28E+10 | 8.04E+08 | 2.85E+10 |
| Mean annual flow | 22.11 | 782.92 | 59.93 | 2122.12 | 38.5 | 1363.29 | 25.5 | 902.96 |
| Median annual flow | 11.3 | 400.13 | 9.06 | 320.81 | 24.4 | 864.00 | 15.7 | 555.94 |
| Peak mean daily Q | 161 | 5701.01 | 1270 | 44970.70 | 229 | 8108.89 | 278 | 9843.98 |
| Minimum mean daily Q | 0.57 | 20.18 | 0.28 | 9.91 | 2.97 | 105.17 | 1.7 | 60.20 |

Using data collected from the IBWC streamflow gauge at 08374200 (see table 3.1), the yearly computation of selected streamflow parameters was conducted for the years 1901–13; the years 1901, 1902, 1903, and 1913 are shown as examples. Computation of the selected annual parameters in this table was used to develop the graphs presented in figure 3.4. Q is discharge; E is exponential or scientific notation (e.g., 6.97E+08 = 6.97 x 10⁸).

*Note:* Total flow volume is in cubic meters and cubic feet, while values in all other rows are in $m^3s^{-1}$ and $ft^3s^{-1}$.

graphing median and mean-daily discharges above and below the RGB's confluence with the Río Conchos (a major tributary that drains significant parts of the state of Chihuahua) prior to 1915 and for the time period between 1992 and 2010 not only shows the impact of the Río Conchos's inputs on the RGB but also shows the dramatic differences that construction of major dams had on RGB. As compared to the 1992–2010 period, flows along the RGB prior to major dam construction were not only much more variable, but also reveal a two-season high-flow period: one in the monsoon season driven by high flows from the Río Conchos and another of lesser average magnitude high flows during the spring driven by snowmelt from the northern branch of the watershed (above the Río Conchos) (see figure 3.4). In comparison, analyses of flow during the 1992–2010 period show a lack of spring floods and little increase during the monsoon (see figure 3.4). Calculating both the mean and median is valuable given the sensitivity of mean-daily discharge analysis to extreme flows (e.g., a rare large flood in 2008 is not revealed when median data are considered) (see figure 3.4).

Among other considerations, such an analysis can help practitioners identify priority information gaps that need to be addressed. For example, although annual hydrographs (like those depicted in figure 3.4) can reveal general changes to streamflow, practitioners will often need to know more specifics as to the extent and timing of alteration. Analysis of total annual-flow volume and mean annual flow for an entire period of record can reveal cycles of high and low flows (see table 3.2). For example, along the RGB, large flow volumes occurred from 1900 until the mid-1940s and were followed by widespread drought and low flows in the 1950s and early 1960s. Flow volumes increased in the 1960s, peaked around 1990, and then declined again in response to drought (see figure 3.5). Analysis of annual flow peaks show similar trends to the trends in flow volumes and mean annual flow. High-magnitude peak flows occurred from 1900 to the mid-1940s, but after the 1940s, peak flows were clearly lower, with large floods occurring every ten to twenty years (see figure 3.5c) and with less frequency. Although flows were high during the 1980s, few large floods occurred. Large floods occurred in 1978, 1990, and 1991, but peak flows were generally low to moderate. Thus the high-flow volumes and mean annual flows were caused by large base flows and intermediate floods, not by large floods. If

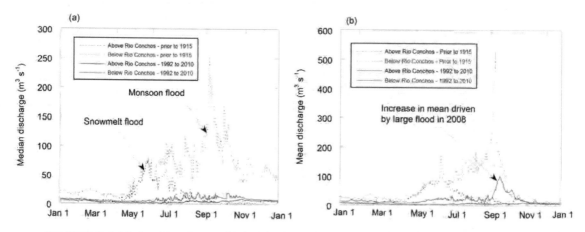

**FIGURE 3.4** Median (a) and mean (b) daily hydrographs compiled from gauge records above and below the Río Conchos prior to 1915 and for 2002 through 2010 (Dean and Schmidt 2011).

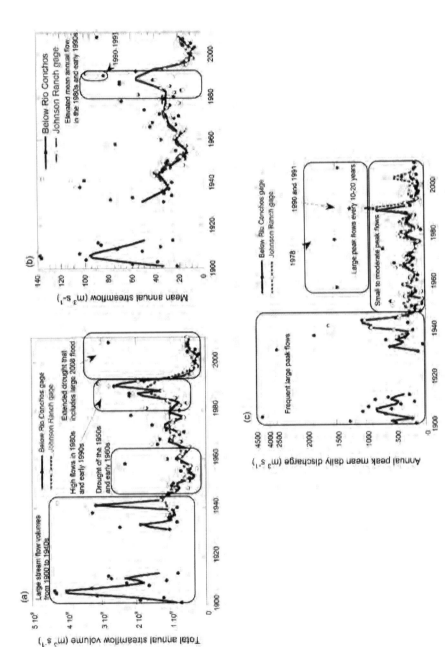

**FIGURE 3.5** Time series of (a) total annual flow volumes, (b) mean annual flow, and (c) annual flow peaks for the RGB below the Río Conchos and at Johnson Ranch. Data points are shown for each year. In (a), a five-year running average is shown for the gauge below the Río Conchos, and a ten-year running average is shown for the Johnson Ranch gauge. The differences between the two running averages illustrates how adjusting the time-span over which the running average is calculated can highlight the nuanced trends (five-year) and general trends (ten-year) of a time series. In (b), ten-year running averages are shown, and in (c), five-year running averages are shown. Also note that the ten-year running average in (a) is essentially the same as the ten-year running averages in (b), because total annual flow volume and mean annual flow volume are essentially the same metrics, just accumulated over different time periods (e.g., flows summed over the entire year in [a], and averages of each mean daily flow over the entire year [b]).

practitioners know the discharge thresholds needed for significant geomorphic change to occur, analysis of annual flow peaks can provide insight as to how the frequency of such important flow events have changed during the period of record (see figure 3.5c).

For the analysis of streamflow data along the RGB, analysis of data collected from two or more gauges (in this case, data collected at the Río Conchos gauge and the Johnson Ranch gauge, which lies 140 km downstream of the Río Conchos gauge) reveals occasional divergence owing to main-stem floods caused by tributary inputs between the two gauges. When flood peaks at the two gauges are similar, the Río Conchos was typically the source of the flood. When flood peaks at the gauges differed, the floods were probably derived from tributaries. This difference is important because ephemeral-stream tributaries can deliver large amounts of sediment to the trunk channel and can change the channel and floodplain morphology (Dean and Schmidt 2011). The main point for practitioners is the value of analyzing flow data collected from gauges in different locations.

## When Streamflow Data Are Not Available

Many stream restoration projects will not have the luxury of long-term flow records. If streamflow data are not available, options include the following: (a) using models to estimate runoff from precipitation measurements with the application of curve numbers, (b) estimating flow characteristics from measurements of channel morphology and bed and bank sediment, (c) using regression equations to estimate flood discharges of specified recurrence intervals from regional curves that relate documented flood magnitudes to basin characteristics, and (d) estimating streamflow using the slope-area method.

### Estimating Runoff from Rainfall

A standard way to estimate runoff is through the use of models that involve a curve number (CN), which is a parameter used to predict the amount of precipitation that will infiltrate into the soil versus the amount that will directly run off (Tedela et al. 2008; Hawkins et al. 2009). The CN method was developed by the USDA Natural Resources Conservation Service from analyses of runoff in small (generally less than 100 km²) watersheds.* It is based on soil group, land use, and hydrological condition (USDA 1986). A high CN indicates high runoff and low infiltration, whereas a low CN indicates low runoff and high infiltration.

Several watershed-runoff and nonpoint-source-pollution models use a CN to determine runoff. They include AGNPS (Young et al. 1987), EPIC (Williams 1995), SWAT (Arnold et al. 1996), and WMS (https://www.aquaveo.com/software/wms-watershed-modeling-system -introduction). The ArcCN-Runoff tool, an extension of ESRI GIS software, both computes the CN and calculates runoff or infiltration for a rain event, thereby reducing computational time significantly (Zhan and Huang 2004). Sediment and nutrient transport can then be calculated based on computed runoff.

If you are interested in using a CN to estimate runoff, the models cited above can be explored in tandem with current literature that has information pertinent to your watershed.

---

* The runoff equation is $Q = (P - 0.2S)^2 / (P + 0.8S)$, in which runoff, $Q$, rainfall, $P$, and infiltration, $S$, are in millimeters (mm); $S$ is computed using a curve number, CN: $S = (25,400 / CN) - 254$.

A simple rational model, for example, yields estimates of rainfall and snowmelt runoff using daily precipitation and mean daily temperatures (e.g., Lindsey et al. 1982; Chow 2010). One empirical rainfall-runoff model developed for hillslopes in a semiarid region indicated that catchment-surface characteristics are critical in yielding sufficient runoff during low rainfall years (van Wesemael et al. 1998).

Although the use of CNs is useful to estimate runoff in watersheds that lack instrumentation, we emphasize that runoff values generated using a CN should be applied cautiously, as the method often yields large errors when results are compared with measured runoff volumes (Hawkins et al. 2009). Curve numbers were designed primarily for agricultural and urban watersheds that have tight soils and significant runoff following rainfall, so larger errors should be expected in forested watersheds (Hawkins et al. 2009; USACE 2010b). In support of the curve-number research, Boughton (2006) developed ratios of calculated to actual runoff to determine the ranges of error from the different periods of data used for calibration.

## Estimating Streamflow Characteristics from Measurements of Channel Morphology and Sediment

If discharge data on your ungauged stream are not available, selected flow metrics can be estimated fairly rapidly and inexpensively by measuring and describing channel morphology. Hedman and Osterkamp (1982) developed equations for predicting annual runoff from channel-geometry measurements based on the particle sizes of the material on the channel perimeter. Particle sizes are determined by collecting three composite samples from the perimeter of the active channel (Hedman and Osterkamp 1982). The number of samples that make up each composite sample varies with channel size. One composite sample should consist of material collected at equal intervals across the channel bed, and the other two should consist of samples taken at equal intervals up each bank to the active-channel reference point. The appropriate equation for the stream of interest can be selected based on flow frequency, taking into account the effect of channel material and runoff characteristics (see table 3.3).

Hedman and Osterkamp (1982) also provide equations for determining flood-frequency discharges in stream channels of the western United States. The equations are for floods with recurrence intervals of two through one hundred years and also are based on measurements of channel morphology and sediment characteristics.

## Estimating Streamflow Characteristics Using Regression Equations

The USGS has developed hundreds of regional regression equations for flow-characteristic determination in ungauged basins. In the southwestern United States, these equations apply mainly to floods of specified recurrence intervals and can be accessed at the USGS National Flood Frequency Program website (http://water.usgs.gov/software/NFF/). Regression equations for flood recurrence intervals in the southwestern United States can also be found in Thomas et al. (1997); table 3.4 lists their regional equations for flows of specified return periods in southern Arizona. These equations can, with limitations, be applied elsewhere in the semiarid regions of the southern United States and northern Mexico. Note that the standard errors of these predictions are high. In addition, there is an online tool to estimate flood frequencies for ungauged catchments in Australia: https://rffe.arr-software.org.

TABLE 3.3  Equations for determining mean annual runoff for streams in the western United States

| Flow frequency | Areas of similar regional-runoff characteristics | Percentage of time having discharge | Channel-material characteristics[a] | Equation[b] | Standard error of estimate (percent) |
|---|---|---|---|---|---|
| Perennial | Alpine | >80 | Silt-clay and armored | $Q_A = 0.74 \times 10^6 \, W^{1.88}$ | 28 |
| Intermittent | Plains north of 39° N latitude | 10 to 80 | Silt-clay and armored | $Q_A = 0.42 \times 10^6 \, W^{1.80}$ | 50[c] |
| | | | sand | $Q_A = 0.35 \times 10^6 \, W^{1.65}$ | |
| | Plains south of 39° N latitude | 10 to 80 | Silt-clay and armored | $Q_A = 0.18 \times 10^6 \, W^{1.65}$ | 50[c] |
| | | | sand | $Q_A = 0.16 \times 10^6 \, W^{1.55}$ | |
| Ephemeral | Northern and southern plains and intermountain areas | 6 to 9 | Silt-clay and armored | $Q_A = 78 \times 10^3 \, W^{1.55}$ | 40[d] |
| | | 2 to 5 | sand | $Q_A = 73 \times 10^3 \, W^{1.50}$ | 40[d] |
| | | | | $Q_A = 29 \times 10^6 \, W^{1.50}$ | |
| | | | | $Q_A = 26 \times 10^3 \, W^{1.40}$ | |
| | Deserts of southwest | <1 | Silt-clay and armored | $Q_A = 390 \, W^{1.75}$ | 75 |
| | | | sand | $Q_A = 260 \, W^{1.40}$ | 75[c] |

*Note*: Modified from Hedman and Osterkamp 1982.

[a] *Silt-clay* channels are bed material $d_{50}$ less than 0.1 mm or bed material $d_{50}$ equal to or less than 5.0 mm and bank silt-clay content equal to or greater than 70 percent, *Sand* channels are bed material $d_{50} = 0.1$ to 5.0 mm and bank silt-clay content less than 70 percent, and *Armored* channels are bed material $d_{50}$ greater than 5.0 mm.

[b] Active-channel width, W, in meters; discharge, $Q_A$, mean annual runoff, in square meters.

[c] Approximate standard error of estimate of the basic regression equation.

[d] Standard error of estimate determined by graphic analysis.

**TABLE 3.4**  Generalized least-squares regression equations for estimating regional flood-frequency relations for southern Arizona

| Recurrence interval, in years | Equation | Average standard error of prediction, in percent | Equivalent years of record |
|---|---|---|---|
| 2 | $Q_2 = 1.15 \times 10^{(6.38-4.29\,A-0.06)}$ | 57 | 2.0 |
| 5 | $Q_5 = 2.06 \times 10^{(5.78-3.31\,A-0.08)}$ | 40 | 6.2 |
| 10 | $Q_{10} = 3.00 \times 10^{(5.68-3.02\,A-0.09)}$ | 37 | 11.1 |
| 25 | $Q_{25} = 4.48 \times 10^{(5.64-2.78\,A-0.10)}$ | 39 | 15.0 |
| 50 | $Q_{50} = 5.63 \times 10^{(5.57-2.59\,A-0.11)}$ | 43 | 15.9 |
| 100 | $Q_{100} = 7.29 \times 10^{(5.52-2.42\,A-0.12)}$ | 48 | 16.1 |

*Note*: Modified from Thomas et al. 1997, table 17.

Peak discharges, $Q_2$ through $Q_{100}$, are in $m^3s^{-1}$, and areas, A, are in square kilometers ($km^2$); coefficients 1.15 through 7.29 are conversions to metric units from English equivalents.

In the western United States, regression equations for flow duration are available for Utah, Nevada, New Mexico, and Texas and can be found at the USGS National Streamflow Statistics Program website (http://water.usgs.gov/osw/programs/nss/pubs.html).

Recently, equations based on statistical application of the rational approach (discharge estimated from basin size and characteristics) have been proposed for much of Montana (Sando and Chase 2017). The technique is offered as being applicable to studies of fish populations. The model relies on random forest-regression analysis and does not include considerations of process. The average mean absolute percent error was reported to be about 73 percent, and thus the technique should be used cautiously until further testing has been conducted.

*Estimating Streamflow Using the Slope-Area Method*

The slope-area method permits the estimation of a specific flood discharge using only survey equipment and an estimation of channel roughness (Benson and Dalrymple 1967). It can also be used to estimate a one- to two-year return-interval flow event for perennial streams ("bankfull discharge"), but this method should be used cautiously due to the challenges associated with estimating water-surface slope (Smith et al. 2010).

The slope-area employs the Manning equation or other flow-resistance equations for open-channel uniform flow to relate discharge to cross-section geometry, roughness, and channel gradient (slope). The Manning equation is:

$$Q = (1.49/n)A(R^{2/3})(S^{1/2}), \text{ English units}$$

$$Q = (1.0/n)A(R^{2/3})(S^{1/2}), \text{ Metric units}$$

where $Q$ is discharge ($ft^3s^{-1}$, English units, or $m^3s^{-1}$, for metric), $n$ is the Manning's roughness coefficient, $A$ is the cross-section area, $R$ is the hydraulic radius, and $S$ is the slope. Manning's $n$ can be estimated using tables (Chow 2009) or pictures of channel conditions (Barnes 1967).

Manning's n can also be back-calculated if the discharge, area, hydraulic radius, and slope are known (Chow 2009). Care should be taken when calculating or estimating Manning's *n*, because *n* will vary with discharge on any specific stream. As discharge rises, *n* generally declines because there is a smaller percentage of the flow that is in contact with the roughness elements of the bed and banks. Based on changing roughness values associated with changing discharge and the inability to back-calculate *n* if the discharge is unknown, the slope-area method should only be considered an approximation technique. Calculation methods are provided in numerous hydrology texts (e.g., Linsley et al. 1982; Chow 2009). For channels that have complex geometry, a hydraulic model such as HEC-RAS can help determine accurately the normal depth.

## Developing Stage-Discharge Relations

A stage-discharge relation correlates stage (water-surface elevation at a specific channel site) with the discharge at that site; stage-discharge relations can quantify how frequently various bottomland surfaces are inundated (Osterkamp 2008). The shape of stage-discharge curves and channel cross sections are closely related. A stage-discharge relation is a function of channel shape, size, gradient, and roughness. Changes in the curve indicate that channel adjustment is occurring or has occurred. Developing stage-discharge curves requires multiple discharge measurements at varying water-surface elevations. Once a stage-discharge record is quantified, practitioners can review the streamflow record to determine how frequently flows inundate the different surfaces adjacent to the channel. The upper part of a stage-discharge relation is defined by floods that inundate the floodplain, and thus the dramatic increase in the wetted width as the floodplain gets inundated is the reason the slope of the relation becomes much less steep.

For stream reaches of interest, you may need to develop your own stage-discharge relation, which will give you discharges for channel surfaces of interest (e.g., discharges needed to inundate primary and secondary floodplain surfaces and terraces) and, as a subsequent step if a long-term flow record is available, to estimate return intervals for discharges of interest (e.g., how often on average is the primary floodplain where we are planning to plant one thousand cottonwood poles inundated?). Other uses may include identifying the inundation potential of specific features in transporting sediment, prescribing environmental flows that will achieve specific flow depths and shear stresses on the channel bed, or flushing fine sediment from side channels.

Relations between stage and inundation of channel and bottomland surfaces can be expanded using the HEC-RAS 1-D flow model (USACE 2010c). HEC-RAS and other 1-D flow models calculate the water-surface elevation and a downstream velocity vector for discharges of interest (2- and 3-D models also calculate the cross-stream and vertical velocity components). Primary inputs to 1-D models are a reliable stage-discharge relation at the downstream end, and a series of cross sections through the channel reach for which the model calculations are run. HEC-RAS uses the stage-discharge relation (or a specified gradient) as a starting point for model calculations, and then iteratively solves step-backwater calculations for consecutive upstream cross sections to estimate water-surface profiles for given discharges. Thus, at each cross section, the water-surface elevation and downstream velocity can be calculated for a selected discharge. A benefit of 1-D models is that stage-discharge relations can be computed for portions of the channel where stage-discharge relations have not been developed. For additional information on constructing and calibrating 1-D models, see the *HEC-RAS Technical Reference Manual* (USACE 2010c).

Below is an example of HEC-RAS output for a reach of the RGB in Big Bend National Park, Texas (see figure 3.6). The map shows the sites of measured cross sections of the reach; flow is left to right. The cross-section plot for one cross section, indicated by the red circle on the map, depicts computed stages for discharges of 1, 112, 315, and 525 $m^3s^1$. The stage-discharge relation (water-surface elevation against river discharge) was developed using field measurements collected at USGS stream gauge 08374550, RGB, near Castolon, Texas (cross-section 1).

Developing a stage-discharge relation requires measuring discharge at numerous water-surface elevations and is the means by which discharge is computed at a gauging station. Stage-discharge data produce "at-a-station" relations because they apply to the cross section where the rating was developed. The accuracy of a relation varies with the stability of hydraulic controls such as bedrock in the channel, riffles, or structures; accuracy, therefore, requires establishing a cross section in a stable location. Accordingly, most USGS stream gauges are near bridges or natural controls that provide stable cross-sectional geometry, and the shape of the stage-discharge curve is a direct result of the channel cross section.

By measuring several stages and discharges directly, a curve is fit to the data, allowing discharge to be estimated for any water depth. The procedure is simple and safe if the channel can be crossed at all flow levels. It becomes troublesome is the flow is too deep to wade safely, in which case measurements may need to be taken from bridges or boats.

Follow the following steps to develop an accurate rating curve empirically:

1. Select a reach with a morphology that allows access to the highest possible discharges (e.g., a reach characterized with a relatively wide and shallow channel morphology). If high flows cannot be measured, uncertainty for discharges greater than those defining the rating curve will be large.
2. Select sites whose channel morphology is not likely to change significantly over time. This is particularly important if the rating curve is to be used for an extended period. Bedrock-dominated reaches are preferable. Bridge abutments, riffles, weirs, and bedrock ledges also provide stable conditions for developing a rating curve.
3. Choose a relatively straight reach with simple cross sections where flow within a cross section is relatively uniform and bed and water-surface slopes are similar and reasonably parallel.
4. Use a current meter (old school) or acoustic doppler current profiler (newer school) to measure the discharge of various flow stages. Take discharge measurements at three or more stages, from low to high flows.
5. At the site where measurements are taking place, install a staff gauge that is marked at appropriate increments into the channel bed (or on a bridge abutment or bedrock outcrop) to allow the water surface to be easily recorded and correlated to the rating curve.

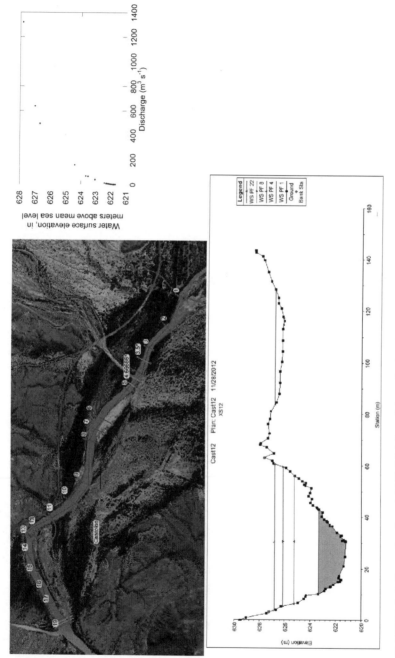

**FIGURE 3.6** Example of partial HEC-RAS model results for the RGB in Big Bend National Park: map showing modeled reach and sites of measured cross sections, the cross-section (red circle of map) plot for stages (water-surface elevations) at discharges of 1 (WS PF 1), 112 (WS PF 4), 315 (WS PF 8), and 525 m³s⁻¹ (WS PF 22), and the stage-discharge relation determined from field measurements collected at USGS stream gauge 08374550, RGB, near Castolon, TX (cross-section 1).

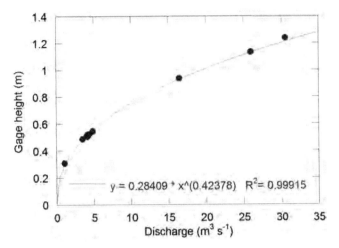

**FIGURE 3.7** Example of a stage-discharge relation with a power-fit curve.

6. If feasible, use pressure transducers that measure flow depth to construct detailed records of stage (see Case Study 7.6). If discharge is available from a nearby gauge, an extensive stage-discharge record can be developed with the stage record from the transducer. After a relation is developed, frequent stage measurements (e.g., fifteen-minute intervals) obtained from pressure transducers can provide the information to build a near-continuous discharge record.

7. Plot the water-surface elevation or stage (the independent variable) on the y-axis and the discharge (the dependent variable) on the x-axis. Plotting points for stage (y axis) and discharge (x axis) on log-log graph paper should yield a relatively straight-line approximation of the stage-discharge relation.

8. To construct an accurate rating curve, fit the rating relation using $Q = a(h - z)b$, in which $h$ represents gauge height or surface water elevation, $z$ is the gauge height at "zero flow," and $a$ and $b$ are regression constants. In some instances, a power-fit curve like the one below may be inadequate, and in those instances, other curve-fitting techniques may be applied. Regression techniques are described in many standard texts on statistics.

Three excellent resources for measuring streamflow are: the Bureau of Meteorology, Australia 'Water Information' (http://www.bom.gov.au/water/standards/documents/WISBF_GL _100_04-2013.pdf), SEMARNAT, Mexico 'Manual para la Medición del Agua:Producción, Operación y Consumo' (https://files.conagua.gob.mx/conagua/mapas/SGAPDS-1-15-Libro9 .pdf), and the U.S. Geologic Survey's 'How Streamflow is Measured' (https://www.usgs.gov /special-topic/water-science-school/science/how-streamflow-measured?qt-science_center_ objects=0#qt-science_center_objects), Finally, see a review of methods for monitoring streamflow (Dobriyal et al. 2017).

## Measuring Losses and Gains to Streamflow

Quantifying water requirements of aquatic and riparian species to maintain desired ecosystem conditions and meet the objectives of a rehabilitation program is often an essential step to restoration. A particularly important part of the process is quantifying flow needs at a particular site (or sites) downstream of where flows are measured or released (in the case of impounded streams). Significant losses or gains in flow between the point of release and the targeted restoration sites can be caused or affected by the presence of bottomland springs, distance from point of release, time of year (changes in evapotranspiration rates), transmission losses, elevation of the bottomland zone of saturation, and a variety of other factors.

These losses and gains need to be quantified to provide restoration practitioners sound estimates of flow characteristics along a targeted reach. Practitioners can assess how much water is lost or gained at sites along a channel reach by conducting seepage runs (which consist of making flow measurements along them at approximately the same time) and seepage studies (which more broadly entail the use of streamflow gauges or discharge estimates to quantify inflows and outflows) (Riggs 1972). A reduction in flow rate between two sites represents a loss of streamflow by seepage through the channel bed. Increases in flow rate indicate streamflow gains by contributions of groundwater from seeps and springs.

Specific questions that seepage studies and runs can answer are the following: (1) Is the stream or channel gaining or losing flow? and (ii) What are the changes in gains or losses relative to varying discharge rates?

Measuring surface water gains and losses (seepage runs) are relatively straightforward and consist of the collection of discharge measurements at different locations downstream along a stream reach. Seepage-run methods are discussed in detail by Cox et al. (2002), and field techniques are described by Harrelson et al. (1994). These investigations can be conducted during base flow, when discharge is relatively steady without runoff contributions, or during high flows, if discharge is stable. Stream gauges can add flow data at sites within the reach of interest. Gains and losses can be computed using the following equation of Raines et al. (2012):

$$G = Qd - Qu - I + D - R + E + ET$$

in which:

G is the streamflow gain or loss (groundwater inflow or outflow)
Qd is measured streamflow at the downstream site
Qu is measured streamflow at the upstream site
I is measured tributary inflow
D is measured outflows (such as diversions)
R is return flows to the reach
E is evaporation
ET is evapotranspiration

## Estimating Groundwater / Surface Water Exchange

Along a stream reach that has been identified for restoration, understanding groundwater–surface water interactions can help practitioners determine site suitability and the likelihood of and possible strategies to ensure restoration success. As a flood wave passes the target reach, groundwater levels in adjacent alluvium will typically rise and then recede. The speed at which groundwater levels decline is especially important to understand during periods when surface water is scarce, helping practitioners to identify areas likely to be underlain by saturated alluvium or moist soils that are potential sources of water for aquatic and riparian plants and animals. An analysis of this sort can eliminate potential restoration sites that lack persistent soil moisture. Assessing groundwater–surface water interactions is also important to understanding hyporheic (essentially, the saturated sediments surrounding the open channel) processes to the ecology of fish, macroinvertebrates, and other organisms in stream ecosystems (e.g., Stanford and Ward 1993; Boulton et al. 1998; Wondzell 2011).

Groundwater–surface water interactions can be investigated by collecting groundwater data from observation wells or piezometers. Piezometers are pipes with an open point at one end that are driven into the streambed or adjacent bottomland surfaces; water enters the pipe only at the depth to which it has been driven. Data collected by piezometers, especially if equipped with pressure transducers and thermometers, can be used in a variety of ways to estimate the hydraulic conductivity of channel alluvium and alluvium underlying surrounding surfaces, and to estimate vertical and horizontal flow rates and hydraulic gradients into or out of the stream (Malard et al. 2001; Baxter et al. 2003). Such estimates can provide stream restoration practitioners the data and information to answer such specific questions as the following:

- How does the availability of shallow groundwater change at different discharge rates and times of year?
- How far do discharge losses move laterally into adjacent alluvium?
- What is the depth to groundwater in bottomland areas and how do the depths change with discharge and seasonality?
- How long are groundwater levels elevated after floods?

Piezometer installation involves inserting several (three to five) piezometers at equidistant intervals perpendicular to the orientation of the stream channel. The number needed will vary with width of the active floodplain environment and morphological complexity. The length of the piezometer should be greater than the expected range of water table levels. That is, piezometer length should be greater than what you expect the lowest elevation of shallow groundwater will be. Installation is commonly done with augers or by pounding a metal pipe into the soil with a fence post driver. There are numerous detailed and excellent descriptions of constructing and installing piezometers (e.g., McLean 2011; Cunningham and Schalk 2011). Please see a detailed and excellent description of installing minipiezometers in remote locations (Baxter et al. 2003).

Once installed, you need to ensure that piezometers are communicating with shallow groundwater, and the top should be capped to prevent dust, leaves, insects, and other debris from entering. A variety of methods can be used to measure water elevations inside the piezometer (e.g., chalk line, demarcated pole, pressure transducer, bubbler). For more information on installation and interpretation of data collected from piezometers, refer to Gordon et al. (2004) and Hauer and Lamberti (2007).

Although outside our dryland focus, an excellent resource on groundwater–surface water interactions can be found at the Environment Agency of England and Wales, which has organized the Hyporheic Network, where the *Hyporheic Handbook* and other resources can be found (https://assets.publishing.service.gov.uk/government/uploads/system/uploads /attachment_data/file/291621/scho1009brdx-e-e.pdf). Reports of many other studies are available that use similar strategies (e.g., Boulton et al. 1998; Simonds and Sinclair 2002; Boulton et al. 2010).

# REFERENCES

Allred, T. M., and J. C. Schmidt. 1999. "Channel Narrowing by Vertical Accretion Along the Green River Near Green River, Utah." *Geological Society of America Bulletin* 111 (12): 1757–72.

Arnold, J. G., P. M. Allen, M. Volk, J. R. Williams, and D. D. Bosch. 2010. "Assessment of Different Representations of Spatial Variability." *Transactions of the American Society of Agricultural and Biolgical Engineers* 53 (5): 1433–43. https://naldc.nal.usda.gov/download/46705/PDF.

Australian Water Association. 2007. *Water in Australia: Facts and Figures, Myths and Ideas.* Sydney: Australian Water Association, National Library of Australia.

Baart, I., S. Hohensinner, I. Zsuffa, and T. Hein. 2013. "Supporting Analysis of Floodplain Restoration Options by Historical Analysis." *Environmental Science and Policy* 34:92–102.

Barnes, H. H., Jr. 1967. "Roughness Characteristics of Natural Channels." U.S. Geological Survey Water-Supply Paper 1849.

Baxter, C., F. R. Hauer, and W. W. Woessner. 2003. "Measuring Groundwater–Stream Water Exchange: New Technique for Installing Minipiezometers and Estimating Hydraulic Conductivity." *Transactions of the American Fisheries Society* 132:493–502.

Beechie, T. J., D. A. Sear, J. D. Olden, G. R. Pess, J. M. Buffington, J. Moir, P. Roni, and M. M. Pollock. 2010. "Process-Based Principles for Restoring River Ecosystems." *BioScience* 60 (3): 209–22.

Bennett, J. B. 2010. Resource Management on the Rio Grande Big Bend Ecosystem, National Park Service. Annual Meeting of the Rio Grande Basin Initiative, May 17–20, 2010, at Sul Ross State University, Alpine, Texas.

Benson, M. A., and T. Dalrymple. 1967. "General Field and Office Procedures for Indirect Measurements." In *U.S. Geological Survey Techniques of Water-Resources Investigations*, bk. 3. http://pubs .usgs.gov/twri/twri3-a1/.

Bonar, S. A., W. A. Hubert, and D. W. Willis. eds. 2009. *Standard Methods for Sampling North American Freshwater Fishes.* Bethesda, Md.: American Fisheries Society.

Bortholomé, E., and A. S. Belward. 2005. GLC 2000: "A New Approach to Global Land Cover Mapping from Earth Observation Data." *International Journal of Remote Sensing* 26 (9): 1959–77.

Boughton, W. C. 2006. "Effect of Data Length on Rainfall-Runoff Modeling." *Environmental Modeling and Software* 22:406–13.

Boulton, A. J., S. Findlay, P. Marmonier, E. H. Stanley, and H. M. Valett. 1998. "Significance of the Hyporheic Zone in Streams and Rivers." *Annual Review of Ecology and Systematics* 29:59–81.

Boulton, A., T. Kasahara, T. Datry, and M. Mutz. 2010. "Ecology and Management of the Hyporheic Zone: Stream-Groundwater Interactions of Running Waters and Their Floodplains." *Journal of the North American Benthological Society* 29 (1): 26–40.

Bravard, J. P., G. M. Kondolf, and H. Piégay. 1999. "Environmental and Societal Effects of Channel Incision and Remedial Strategies." In *Incised River Channels*, edited by S. E. Darby and A. Simon, 303–41. New York: John Wiley.

Brierley, G. J., and K. A. Fryirs. 2005. *Geomorphology and River Management: Applications of the River Styles Framework*. Malden, Mass.: Blackwell.

Briggs, M. K. 1996. *Riparian Ecosystem Recovery in Arid Lands*. Tucson: University of Arizona Press.

Briggs, M. K., and W. R. Osterkamp. 2003a. "Developing Recovery Plans for Riparian Ecosystems." *Southwest Hydrology* 2 (2): 18–19.

Briggs, M. K., and W. R. Osterkamp. 2003b. "Important Concepts for Riparian Recovery." *Southwest Hydrology* 2 (2): 26.

Brookes, A., and K. J. Gregory. 1988. "Channelization, River Engineering, and Geomorphology." In *Geomorphology in Environmental Planning*, edited by J. M. Hooke, 145–68. Chichester, UK: Wiley.

Carson, M. A., and G. A. Griffiths. 1989. "Gravel Transport in the Braided Waimakariri River: Mechanisms, Measurements, and Predictions." *Journal of Hydrology* 109:201–20.

Chow, V. T., D. R. Maidment, and L. W. Mays. 2010. *Applied Hydrology*. New York: McGraw-Hill.

Chow, V. T. 2009. *Open-Channel Hydrology*. Caldwell, N.J.: Blackburn Press.

CONAGUA. 2018. *Estadísticas del agua en México, edición 2018*. México, D.F.: Comisión Nacional del Agua.

CONAGUA. 2014. *Guia para la colecta, manejo y las observaciones de campo para bioindicadores de la calidad del agua*. Mexico, D.F.: Comisión Nacional del Agua. http://biblioteca.semarnat.gob.mx/janium/Documentos/Ciga/Libros2014/229011.pdf.

Congalton, R. G., K. Birch, R. Jones, and J. Schriever. 2002. "Evaluating Remotely Sensed Techniques for Mapping Riparian Vegetation." *Computers and Electronics in Agriculture* 37:113–26.

Corenbilt, D., E. Tabacchi, J. Steiger, and A. M. Gurnell. 2007. "Reciprocal Interactions and Adjustments Between Fluvial Landforms and Vegetation Dynamics in River Corridors: A Review of Complementary Approaches." *Earth-Science Reviews* 84:56–86.

Cox, M., D. Rosenberry, G. Su, T. Conlon, K. Lee, and J. Constantz. 2002. "Comparison of Methods to Estimate Streambed Seepage Rates." *In Proceedings of the AWRA Summer Specialty Conference, Ground Water / Surface Water Interactions*, 519–23. American Water Resources Association (https://www.awra.org/AWRA/Publications/Members/Publications/Publications.aspx?hkey=4711ebd8-2326-47b2-945b-127b8fef5b8a).

Cuffney, T. F., M. E. Gurtz, and M. R. Meador. 1993. *Methods for Collecting Benthic Invertebrate Samples as Part of the National Water-Quality Assessment Program*. U.S. Geological Survey Open-File Report 93–406.

Cunningham, W. L., and C. W. Schalk. 2011. *Groundwater Technical Procedures of the U.S. Geological Survey: U.S. Geological Survey Techniques and Methods 1–A1*. https://pubs.usgs.gov/tm/1a1/.

Dean, D. J., and J. C. Schmidt. 2011. "The Role of Feedback Mechanisms in Historic Channel Changes of the Lower Rio Grande in the Big Bend Region." *Geomorphology* 126:333–49.

Dean, D. J., and J. C. Schmidt. 2013. "The Geomorphic Effectiveness of a Large Flood on the Rio Grande in the Big Bend Region: Insights on Geomorphic Controls and Post-flood Geomorphic Response." *Geomorphology* 201:183–98.

Dingman, S.L. 2015. *Physical Hydrology*. 3rd ed. Long Grove, Ill.: Waveland Press.

Dobriyal, P., R. Badola, C. Tuboi, S. A. Hussain. 2017. "A Review of Methods for Monitoring Streamflow for Sustainable Water Resource Management." *Applied Water Science* 7:2617–28.

Duro, D., N. C. Coops, M. A. Wulder, and T. Han. 2007. "Development of a Large Area Biodiversity Monitoring Driven by Remote Sensing." *Progress in Physical Geography* 31:235–60.

Fitzpatrick, F. A., I. R. Waite, P. J. D'Arconte, M. R. Meador, M. A. Maupin, and M. E. Gurtz. 1998. *Revised Methods for Characterizing Stream Habitat in the National Water-Quality Assessment Program.* U.S. Geological Survey Water-Resources Investigations Report 98–4052. https://pubs.er.usgs.gov/.

Foster, G. R., D. C. Yoder, D. K. McCool, G. A. Weesies, T. J. Toy, and L. E. Wagner. 2000. "Improvements in Science." In *RUSLE2: Paper No. 00–2147*, 1–19. St. Joseph, Mich.: ASAE.

Friedl, M. A., D. K. McIver, J. C. F. Hodges, X. Y. Zang, D. Muchoney, A. H. Strahler, C. E. Woodcock, S. Gopal, A. Schneider, A. Cooper, A. Bacini, F. Gao, and C. Schaaf. 2002. "Global Land Cover Mapping from MODIS: Algorithms and Early Results." *Remote Sensing of Environment* 82:287–302.

Friedman, J. M., W. R. Osterkamp, and W. M. Lewis. 1996. "Channel Narrowing and Vegetation Development Following a Great Plains Flood." *Ecology* 77 (7): 2167–81.

Friedman, J. M., W. R. Osterkamp, M. L. Scott, and G. T. Auble. 1998. "Downstream Effects of Dams on Channel Geometry and Bottomland Vegetation: Regional Patterns in the Great Plains." *Wetlands* 18 (4): 619–33.

Friedman, J. M., K. R. Vincent, and P. B. Shafroth. 2005. "Dating Floodplain Sediments Using Tree-Ring Response to Burial." *Earth Surface Processes and Landforms* 30 (9): 1077–91.

Fryirs, K. A., and G. J. Brierley. 2013. *Geomorphic Analysis of River Systems: An Approach to Reading the Landscape.* West Sussex, UK: John Wiley. www.wiley.com/go/fryirs/riversystems.

Gaeuman, D., J. C. Schmidt, and P. R. Wilcock. 2005. "Complex Channel Responses to Changes in Stream Flow and Sediment Supply on the Lower Duchesne River, Utah." *Geomorphology* 64 (nos. 3–4): 185–206.

Gillespie, T. W., G. M. Foody, D. Rocchini, A. P. Giorgi, and S. Saatchi. 2008. "Measuring and Modelling Biodiversity from Space." *Progress in Physical Geography* 32:203–21.

Glenn, E., K. Flessa, E. Kendy, P. B. Shafroth, J. Ramírez-Hernández, M. Gomez-Sapiens, P. L. Nagler, eds. 2017. "Environmental Flows for the Colorado River Delta: Results of an Experimental Pulse Release from the US to Mexico." *Ecological Engineering* 106 (part B):629–808.

Glysson, G. D. 1987. *Sediment-Transport Curves.* USGS Open-File Report 87–218.

Goodwin, C. N., C. P. Hawkins, and J. L. Kershner. 1997. "Riparian Restoration in the Western United States: Overview and Perspective." *Restoration Ecology* 5 (4S): 4–14.

Gordon, N. D., T. A. McMahon, B. L. Finlayson, C. J. Gippel, and R. J. Nathan. 2004. *Stream Hydrology: An Introduction for Ecologists.* 2nd ed. West Sussex, UK: John Wiley.

Gore, J. A. 2007. "Discharge Measurements and Streamflow Analysis." In *Methods in Stream Ecology.* 2nd ed. Edited by F. R. Hauer and G. A. Lamberti. New York: Elsevier.

Government of Western Australia, Department of Water. 2009. *Field Sampling Guidelines: A Guideline for Field Sampling for Surface Water Quality Monitoring Programs.* Perth, Western Australia: Department of Water. https://www.water.wa.gov.au/__data/assets/pdf_file/0020/2936/87154.pdf.

Graf, W. L. 2006. "Downstream Hydrologic and Geomorphic Effects of Large Dams on American Rivers." *Geomorphology* 79:336–60.

Gregory, K. J. 2006. "The Human Role in Changing River Channels." *Geomorphology* 79:172–91.

Harrelson, C. C., C. L. Rawlins, and J. P. Potyondy. 1994. *Stream Channel Reference Sites: An Illustrated Guide to Field Technique.* General Technical Report RM-245. Fort Collins, Colo.: U.S. Department of Agriculture, Forest Service, Rocky Mountain Forest and Range Experiment Station.

Hauer, F. R., and G. A. Lamberti. 2007. *Methods in Stream Ecology*. 2nd ed. Elsevier.

Hawkins, R. H., T. J. Ward, D. E. Woodward, and J. A. Van Mullem, eds. 2009. *Curve Number Hydrology*. Reston, Va.: American Society of Civil Engineers.

Hedman, E. R., and W. R. Osterkamp. 1982. "Streamflow Characteristics Related to Channel Geometry of Streams in Western United States," U.S. Geological Survey Water-Supply Paper 2193. https://pubs.er.usgs.gov/.

Heede, B. M. 1980. *Stream Dynamics: An Overview for Land Managers*. USDA Forest Service Technical Report RM-72. https://www.fs.usda.gov/rmrs/publications/series/general-technical-reports.

Hoagstrom, C. W., J. E. Brooks, and S. R. Davenport. 2008. "Spatiotemporal Population Trends of *Notropis primus pecosensis* in Relation to Habitat Conditions and the Annual Flow Regime of the Pecos River, 1992–2005." *Copeia* 1:5–15.

Hobbs, R. J., and D.A. Norton. 1996. "Towards a Conceptual Framework for Restoration Ecology." *Restoration Ecology* 4 (2): 93–110.

Hoey, T. B. 1994. "Patterns of Sediment Storage in the Kowai River, Torlesserange, New Zealand." *Journal of Hydrology* (New Zealand) 32:1–15

Horning, N., J. A., Robinson, E. J. Sterling, W. Turner, S. Spector. 2010. *Remote Sensing for Ecology and Conservation*. New York: Oxford University Press.

Hotzel, G., and R. Croome. 1999. *A Phytoplankton Methods Manual for Australian Freshwaters*. Wobonga, Victoria: Department of Environmental Management and Ecology, La Trobe University. http://lwa.gov.au/products/pr990300.

Hupp, C. R. 1992. "Riparian Vegetation Recovery Patterns Following Stream Channelization: A Geomorphic Perspective." *Ecology* 73 (4): 1209–26.

IPCC. 2001. *Climate Change 2001: Synthesis Report*. "A Contribution of Working Groups I, II, and III to the Third Assessment Report of the Intergovernmental Panel on Climate Change." Cambridge: Cambridge University Press.

Interagency Advisory Committee on Water Data. 1981. *Guidelines for Determining Flood Flow Frequency*. Bulletin #17B, Hydrology Subcommittee. U.S. Geological Survey, Office of Water Data Coordination. https://pubs.er.usgs.gov/.

Johnson, W. C. 1994. "Woodland Expansion in the Platte River, Nebraska: Patterns and Causes." *Ecological Monographs* 64 (1): 45–84.

Jones, N. E. 2011. *Benthic Sampling in Natural and Regulated Rivers: Sampling Methodologies for Ontario's Flowing Waters*. Peterborough, Ont.: Aquatic Research and Development Section, Ontario Ministry of Natural Resources.

Keller, E. A. 1971. "Areal Sorting of Bed-Load Material: The Hypothesis of Velocity Reversal." *Geological Society of America Bulletin* 82 (3): 753–56.

Kellerhals, R., M. Church, and D. I. Bray. 1976. "Classification and Analysis of River Processes." *Journal of Hydraulic Engineering*:813–29.

Kennedy, R. E., P. A. Townsend, J. E. Gross, W. B. Cohen, P. Bostad, Y. Q. Wand, and P. Adams. 2009. "Remote Sensing Change Detection Tools for Natural Resource Managers: Understanding Concepts and Tradeoffs in the Design of Landscape Monitoring Projects." *Remote Sensing Environment* 113 (7): 1382–96.

Knighton, D. 1998. *Fluvial Forms and Processes: A New Perspective*. Oxford: Oxford University Press.

Lagasse, P. F., W. J. Spitz, and L. W. Zevenbergen. 2004. *Handbook for Predicting Stream Meander Migration*. National Cooperative Highway Research Program Report 533. Washington, D.C.: Transportation Research Board.

Lanza Espino, G. 2014. *Programa Nacional de Reservas de Agua: Protocol para el muestreo de calidad del agua en rios endorréicos y exorréicos, y en humedales para la aplicacion de la Norma de Caudal Ecologico.* http://www.ibiologia.unam.mx/aguas/Protocolo%20calidad%20de%20aguai.pdf.

Legleiter, C. J., D. A. Roberts, W. A. Marcus, and M. A. Fonstad. 2004. "Passive Optical Remote Sensing of River Channel Morphology and In-Stream Habitat: Physical Basis and Feasibility." *Remote Sensing of Environment* 9:493–510.

Leopold, L., R. Huppman, and A. Miller. 2005. "Geomorphic Effects of Urbanization in Forty-One Years of Observation." *Proceedings of the American Philosophical Society* 149 (3): 349–71.

Leopold, L., and M. Wolman. 1957. "River Channel Patterns: Braided Meandering and Straight." USGS Professional Paper 282-B. https://pubs.er.usgs.gov/publication/pp282B.

Linsley, R. K., J. L. Paulhus, and M. A. Kohler. 1982. *Hydrology for Engineers.* New York: McGraw-Hill.

Loheide, S. P., and E. G. Booth. 2011. "Effects of Changing Channel Morphology on Vegetation, Groundwater, and Soil Moisture Regimes in Groundwater-Dependent Ecosystems." *Geomorphology* 126 (3): 364–76.

Mabbutt, J. A. 1977. *Desert Landforms.* Canberra: Australian National University Press.

Malard, F., A. Mangin, V. Uehlinger, and J. V. Ward. 2001. "Thermal Heterogeneity in the Hyporheic Zone of a Glacial Floodplain." *Canadian Journal of Fishedries and Aquatic Sciences* 58 (7): 1319–35.

Manners, R. B., J. C. Schmidt, and M. L. Scott. 2014. "Mechanisms of Vegetation-Induced Channel Narrowing of an Unregulated Canyon River: Results from a Natural Field-Scale Experiment." *Geomorphology* 211:100–15.

Marsili-Libelli, S., E. Giusti, and A. Nocita. 2013. "A New Instream Flow Assessment Method on Fuzzy Habitat Suitability and Large Scale Modelling." *Environmental Modelling and Software* 41:27–38.

McKinney, D. C., X. Cai, M. W. Rosegrant, C. Ringler, and C. A. Scott. 1999. *Modeling Water Resources Management at the Basin Level: Review and Future Directions.* IWMI Books, Reports H024075. Colombo, Sri Lanka: International Water Management Institute.

McLean, S. 2011. *Standards/Guidelines for Installation and Management of Testwells and Piezometers.* Department of Primary Industries Sustainable Landscapes, Mallee, Melbourne, Victoria. http://vro.agriculture.vic.gov.au/dpi/vro/vrosite.nsf/pages/vrohome.

Mercado-Silva, N., J. D. Lyons, G. Sagado-Maldonado, and M. Medina Nava. 2002. "Validation of a Fish-Based Index of Biotic Integrity for Streams and Rivers of Central Mexico." *Reviews in Fish Biology and Fisheries* 12 (2–3): 179–91.

Merenlender, A. M., and M. K. Matella. 2013. "Maintaining and Restoring Hydrologic Habitat Connectivity in Mediterranean Streams: An Integrated Modeling Framework." *Hydrobiologia* 719 (1): 509–25.

Montgomery, D. R., and J. M. Buffington 1997. "Channel-Reach Morphology in Mountain Drainage Basins." *Geological Society of America Bulletin* 109:596–611.

Mouton, A. M., M. Schneider, J. Depestele, P. L. M. Goethals, and N. De Pauw. 2007. "Fish Habitat Modeling as a Tool for River Management." *Ecological Engineering* 29:305–15.

Nagler, P. L., E. P. Glenn, T. L. Thompson, and A. Huerte. 2004. "Leaf Area Index and Normalized Difference Vegetation Index as Predictors of Canopy Characteristics and Light Interception by Riparian Species on the Lower Colorado River." *Agricultural and Forest Meteorology* 125:1–17.

New South Wales Government Office of Environment and Heritage. n.d. *River Macroinvertebrate Sampling Manual for Volunteers.* https://www.environment.nsw.gov.au/-/media/OEH/Corporate-Site/Documents/Research/Citizen-science/river-macroinvertebrate-sampling-manual-150075.pdf.

Nilsson, C., J. E. Pizzuto, G. E. Moglen, M. A. Palmer, E. H. Stanley, N. E. Bockstael, and L. C. Thompson. 2003. "Ecological Forecasting and the Urbanization of Stream Ecosystems: Challenges for Economists, Hydrologists, Geomorphologists, and Ecologists." *Ecosystems* 6:659–74.

Osterkamp, W. R. 1977. *Fluvial Sediment in the Arkansas River Basin, Kansas.* Kansas Water Resources Board Bulletin 19.

Osterkamp, W. R. 1978. "Gradient, Discharge, and Particle-Size Relations of Alluvial Channels of Kansas, with Observations on Braiding." *American Journal of Science* 278:1253–68.

Osterkamp, W. R. 2008. *Annotated Definitions of Selected Geomorphic Terms and Related Terms of Hydrology, Sedimentology, Soil Science, Climatology, and Ecology.* U.S. Geological Survey Open-File Report 2008–1217. https://pubs.er.usgs.gov/.

Osterkamp, W. R., and J. E. Costa. 1987. "Changes Accompanying an Extraordinary Flood on a Sand-Bed Stream." In *Catastrophic Flooding,* edited by L. Mayer and D. Nash, 201–24. Boston: Allen and Unwin.

Osterkamp, W. R., and J. M. Friedman. 2000. "The Disparity Between Extreme Rainfall Events and Rare Floods—with Emphasis on the Semiarid American West." *Hydrological Processes* 14 (16–17): 2817–29.

Osterkamp, W. R., C. R. Hupp, and M. Stoffel. 2011. "The Interactions Between Vegetation and Erosion: New Directions for Research at the Interface of Ecology and Geomorphology." *Earth Surface Processes and Landforms* 37 (1): 23–36. https://doi.org/10.1002/esp.2173.

Osterkamp, W. R., and T. J. Toy. 1997. "Geomorphic Considerations for Erosion Prediction." *Environmental Geology* 29 (nos. 3–4): 152–57.

Pettorelli, N., W. F. Laurance, T. G. O'Brien, M. Wegmann, H. Nagendra, and W. Turner. 2014. "Satellite Remote Sensing for Applied Ecologists: Opportunities and Challenges." *Journal of Applied Ecology* 51:839–48.

Pizzuto, J. E., W. C. Hession, and M. McBride. 2000. "Comparing Gravel-Bed Rivers in Paired Urban and Rural Catchments of Southeastern Pennsylvania." *Geology* 28 (1):79–82.

Poff, N. L., J. D. Allan, M. B. Bain, J. R. Karr, K. L. Prestegaard, B. D. Richter, R. E. Sparks, and J. C. Stromberg. 1997. "The Natural Flow Regime." *Bioscience* 47 (11): 769–84.

Raines, T. H., M. J. Turco, P. J. Connor, and J. B. Bennett. 2012. *Streamflow Gains and Losses and Selected Water-Quality Observations in Five Subreaches of the Rio Grande / Río Bravo del Norte from near Presidio to Langtry, Texas, Big Bend Area, United States and Mexico, 2006.* U.S. Geological Survey Scientific Investigations Report 2012–5125. https://doi.org/10.3133/sir20125125.

Rakovan, M. T., and W. H. Renwick. 2011. "The Role of Sediment Supply in Channel Instability and Stream Restoration." *Journal of Soil and Water Conservation* 66 (1): 40–50.

Räpple, B., H. Piégay, J. C. Stella, and D. Mercier. 2017. "What Drives Riparian Vegetation Encroachment in Braided River Channels at Patch to Reach Scales? Insights from Annual Airborne Surveys (Dröme River, SE France, 2005–2011)." *Ecohydrology* 10 (8). https://doi.org/10.1002/eco.1886.

Rathburn, S., and E. Wohl. 2003. "Predicting Fine Sediment Dynamics Along a Pool-Riffle Mountain Channel." *Geomorphology* 55 (1–4): 111–24.

Reinfelds, I., I. Rutherfurd, and P. Bishop. 1995. "History and Effects of Channelisation Along the Latrobe River, Victoria, Australia." *Australian Geographical Studies* 33 (1): 60–76.

Renwick, W. H., and Z. D. Andereck. 2006. "Reservoir Sedimentation Trends in Ohio, USA: Sediment Delivery and Response to Land-Use Change." In *Sediment Dynamics and the Hydromorphology of Fluvial Systems*, 341–47. Proceedings of a symposium held in Dundee, UK, July 2006. International Association of Hydrological Sciences Publication.

Renwick, W. H., and M. T. Rakovan. 2010. "Sediment Supply Limitation and Stream Restoration." *Journal of Soil and Water Conservation* 65 (3): 67A. http://doi.org/10.2489/jswc.65.3.67A.

Renwick, W. H., S. V. Smith, J. D. Bartley, and R. W. Buddmeier. 2005. "The Role of Impoundments in the Sediment Budget of the Conterminous United States." *Geomorphology* 71:99–111. https://pubs.er.usgs.gov/publication/70028277.

Richter, B. D., J. V. Baumgartner, D. P. Braun, and J. Powell. 1998. "A Spatial Assessment of Hydrologic Alteration Within a River Network." *Regulated Rivers: Research and Management* 14:329–40.

Riggs, H. C. 1972. *Low-Flow Investigations. U.S. Geological Survey Techniques of Water-Resources Investigations*. United States Government Printing Office, Washington, D.C. https://scholar.google.com/scholar?q=low+flow+investigations+us+geological+survey+techniques+of+water-resources&hl=en&as_sdt=0&as_vis=1&oi=scholart.

Roni, P., T. J. Beechie, R. E. Bilby, F. E. Leonetti, M. M. Pollack, and R. G. Pess. 2002. "A Review of Stream Restoration Techniques and a Hierarchical Strategy for Prioritizing Restoration in Pacific Northwest Watersheds." *North American Journal of Fisheries* 22 (1): 1–20.

Rosgen, D. L. 1988. "Conversion of a Braided River Pattern to Meandering: A Landmark Restoration Project." Paper presented at the California Riparian Systems Conference, Sept. 22–24, Davis, Calif.

Rosgen, D. L. 1994. "A Classification of Natural Rivers." *Catena* 22:169–99.

Roughgarden, J., S. W. Running, and P. A. Matson. 1991. "What Does Remote Sensing Do for Ecology?" *Ecology* 72, 1918–22.

Saleh, F., A. Ducharne, N. Flipo, L. Oudin, and E. Ledoux. 2012. "Impact of River Bed Morphology on Discharge and Water Levels Simulated by a 1D Saint-Venant Hydraulic Model at Regional Scale." *Journal of Hydrology* 476:169–77.

Sando, R., and K. J. Chase. 2017. *Estimating Current and Future Streamflow Characteristics at Ungagged Sites, Central and Eastern Montana, with Application to Evaluating Effects of Climate Change on Fish Populations*. U.S. Geological Survey Scientific Investigations Report 2017–5002. http://doi.org/10.3133/sir20175002.

Sarhadi, A., S. Soltani, and R. Modarres. 2012. "Probabilistic Flood Inundation Mapping of Ungauged Rivers: Linking GIS Techniques and Frequency Analysis." *Journal of Hydrology* 458–459:68–86.

Schumm, S. A. 1963. *A Tentative Classification of Alluvial Rivers*. U.S. Geological Survey Circular 477. https://pubs.er.usgs.gov/.

Schumm, S. A. 1969. "River Metamorphosis." *Proceedings of the American Society of Civil Engineers, Journal of the Hydraulics Division* 95:255–73.

Schumm, S. A. 1973. "Geomorphic Thresholds and Complex Response of Drainage Systems." In *Fluvial Geomorphology*, edited by M. Morisawa, 299–310. Binghamton: Publications in Geomorphology, State University of New York.

Schumm, S. A. 1977. *The Fluvial System*. New York: John Wiley.

Schumm, S. A. 1984. "Patterns of Alluvial Rivers." *Annual Review of Earth and Planetary Sciences* 13:5–27.

Shafroth, P. B., D. M. Merritt, M. K. Briggs, V. B. Beauchamp, K. Lair, M. L. Scott, and A. A. Sher. 2013. "Riparian Restoration in the Context of Tamarix Control in the Western United States." In *Tamarix:*

*A Case Study of Ecological Change in the American West*, edited by A. A. Sher and M. Quickly, 404–25. New York: Oxford University Press.

Shafroth, P. B., A. C. Wilcox, D. A. Lytle, J. T. Hickey, D. C. Andersen, V. B. Beauchamp, and A. Warne. 2010. "Ecosystem Effects of Environmental Flows: Modeling and Experimental Floods in a Dryland River." *Freshwater Biology* 55 (1): 68–85.

Shields, F. D., C. M. Cooper Jr., S. S. Knight, and M. T. Moore. 2003. "Stream Corridor Restoration Research: A Long and Winding Road." *Ecological Engineering* 20 (5): 441–54.

Shroder, J. F. 2013. "Ground, Aerial, and Satellite Photography for Geomorphology and Geomorphic Change." In *Treatise on Geomorphology*. Vol. 3, *Remote Sensing and GIScience in Geomorphology*, edited by J. F. Shroder and M. P. Bishop, 25–42. San Diego, Calif.: Academic Press.

Simonds, F. W., and K. A. Sinclair. 2002. *Surface Water–Ground Water Interactions Along the Lower Dungeness River and Vertical Hydraulic Conductivity of Streambed Sediments, Clallam County, Washington, September 1999–July 2001*. U.S. Geological Survey Water-Resources Investigations Report 02–4161. https://doi.org/10.3133/wri024161.

Singh, V. P. 2017. *Handbook of Applied Hydrology*. 2nd ed. New York: McGraw Hill Education.

Slatyer, R. O., and J. A. Mabbutt. 1964. "Hydrology of Arid and Semi-arid Regions." In *Handbook of Applied Hydrology*, edited by V. T. Chow, 24–1 to 24–46. New York: McGraw-Hill.

Smith, C. F., J. T. Cordova, and S. M. Wiele. 2010. *The Continuous Slope-Area Method for Computing Event Hydrographs*. U.S. Geological Survey Scientific Investigations Report 2010–5241. https://pubs.usgs.gov/sir/2010/5241/sir2010-5241.pdf.

Stanford, J. A., and J. V. Ward. 1993. "An Ecosystem Perspective of Alluvial Rivers: Connectivity and the Hyporheic Corridor." *Journal of the North American Benthological Society* 12 (1): 48–60.

Stanford, J. A., J. V. Ward, W. J. Liss, C. A. Frissell, R. N. Williams, J. A. Lichatowich, and C. C. Coutant. 1996. "A General Protocol for Restoration of Regulated Rivers." *Regulated Rivers: Research and Management* 12:391–413.

Statzner, B., J. A. Gore, and V. H. Resh. 1988. "Hydraulic Stream Ecology: Observed Patterns and Potential Applications." *Journal of the North American Benthological Society* 7:307–60.

Stromberg, J. C., S. L. Lite, T. J. Rychener, L. R. Levick, M. D. Dixon, and J. M. Watts. 2006. "Status of the Riparian Ecosystem in the Upper San Pedro River, Arizona: Application of an Assessment Model." *Environmental Monitoring and Assessment* 115:145–73.

Tasmania, Department of Primary Industries Water and Environment. 2004. *Community Water Quality Sampling Protocols and Standards: Reference Resources of State, Australia, and International Standards for Water Quality Monitoring*. https://dpipwe.tas.gov.au/Documents/Community_WQ_Sampling_-Protocols.pdf?details=true.

Tedela, N., S. McCutcheon, T. Rasmussen, and W. Tollner. 2008. *Evaluation and Improvements of the Curve Number Method of Hydrological Analysis on Selected Forested Watersheds of Georgia*. Georgia Water Resources Institute. https://ce.gatech.edu/category/georgia-water-resources-institute.

Texas Commission on Environmental Quality. 2011. *Texas Surface Water Quality Monitoring and Assessment Strategy, FY 2012–2017*. Austin, Tex.: Surface Water Quality Monitoring Program, Water Quality Planning Division.

Thomas, B. E., H. W. Hjalmarson, and S. D. Waltemeyer. 1997. "Methods for Estimating Magnitude and Frequency of Floods in the Southwestern United States." U.S. Geological Survey Water-Supply Paper 2433.

Tooth, S. 2000. "Process, Form, and Change in Dryland Rivers: A Review of Recent Research." *Earth-Science Reviews* 51:67–107.

Turner, W., S. Spector, N. Gardiner, M. Fladeland, E. Sterling, and M. Steininger. 2003. "Remote Sensing for Biodiversity Science and Conservation." *Trends in Ecology and Evolution* 18:306–14.

USACE. 2002. *HEC-RAS River Analysis System*. Davis, Calif.: Hydrologic Engineering Center.

USACE. 2010a. *U.S. Army Corps of Engineers National Inventory of Dams*. https://nid.usace.army.mil.

USACE. 2010b. *Hydrologic Engineering Center Hydrologic Modeling System HEC-HMS, Technical Reference Manual*. https://usace.contentdm.oclc.org/.

USACE. 2010c. *Hydrologic Engineering Center River Analysis System HEC-RAS, Hydraulic Reference Manual*. https://usace.contentdm.oclc.org/.

USDA Soil Conservation Service. 1986. *Urban Hydrology for Small Watersheds*. Technical Release No. 20, rev. ed. Technical Release 55 (TR-55). Washington, D.C.: Natural Resources Conservation Service. https://www.nrcs.usda.gov/Internet/FSE_DOCUMENTS/stelprdb1044171.pdf.

U.S. Geological Survey. 2018. *National Field Manual for the Collection of Water-Quality Data* (ver. 1.1, June 2018). U.S. Geological Survey Techniques and Methods. https://doi.org/10.3133/tm9A0.

Utah State University Extension. "Marcoinvertebrate Sampling." http://extension.usu.edu/utahwaterwatch/monitoring/field-instructions/macroinvertebratesampling/.

van Wesemael, B. J. Poesen, A. S. Benet, L. C. Barrionuevo, and J. Puigdef Ábregas. 1998. "Collection and Storage of Runoff from Hillslopes in a Semi-arid Environment: Geomorphic and Hydrologic Aspects of the Aljibe System in Almeria Province, Spain." *Journal of Arid Environments* 40 (1): 1–14.

Ward, J. V. 1989. "Riverine-Wetland Interactions." In *Freshwater Wetlands and Wildlife*, edited by R. R. Sharitz and J. W. Gibbons, 385–400. DOE Symposium Series No. 61. Oak Ridge, Tenn.

Warner, R. F. 1984. "Man's Impacts on Australian Drainage Systems." *Australian Geographer* 16:133–41.

Webb, R. H., D. E. Boyer, R. M. Turner, and S. H. Bullock. 2007. *The Desert Laboratory Repeat Photography Collection: An Invaluable Archive Documenting Landscape Change*. U.S. Geological Survey Fact Sheet 2007–3046. https://pubs.er.usgs.gov.

Wilde, F. D. 2011. *Water-Quality Sampling by the U.S. Geological Survey: Standard Protocols and Procedures*. U.S. Geological Survey Fact Sheet 2010–3121. http://pubs.usgs.gov/fs/2010/3121.

Wilkie, D. S., and J. T. Finn. 1996. *Remote Sensing Imagery for Natural Resources Monitoring: A Guide for First-Time Users*. New York: Columbia University Press.

Williams, J. R. 1995. "The EPIC Model." In *Computer Models of Watershed Hydrology*, edited by V. P. Singh, 909–1000. Highlands Ranch, Colo.: Water Resources Publications.

Winward, A. H. 2000. *Monitoring the Vegetation Resources in Riparian Areas*. USDA General Technical Report RMRS-GTR-47. https://www.ars.usda.gov/.

Wischmeier, W. H., and D. D. Smith. 1978. *Predicting Rainfall Erosion Losses: A Guide to Conservation Planning*. USDA, Agriculture Handbook No. 537. https://www.ars.usda.gov/.

Wolman, M. G., and R. Gerson. 1978. "Relative Scales of Time and Effectiveness of Climate in Watershed Geomorphology." *Earth Surface Processes and Landforms* 3:189–208.

Wolman, M. G., and L. B. Leopold. 1957. "River Flood Plains: Some Observations on Their Formations." U.S. Geological Survey Professional Paper 282-B.

Wondzell, S. M. 2011. "The Role of the Hyporheic Zone Across Stream Networks." *Hydrologic Processes* 25 (22): 3525–32.

World Meteorological Organization (WMO). 2010. *Manual on Stream Gauging.* 2 vols. WMO No. 1044. Geneva, Switzerland: World Meteorological Association. https://www.wmo.int/pages/prog/hwrp /publications/stream_gauging/1044_Vol_I_en.pdf; http://www.wmo.int/pages/prog/hwrp/publications /stream_gauging/1044_Vol_II_en.pdf.

Young, D. F., and J. N. Carleton. 2006. "Implementation of a Probabilistic Curve Number Method in the PRZM Runoff Model." *Environmental Modeling and Software* 21:1172–79.

Zamora-Arroyo, F., P. L. Nagler, M. K. Briggs, D. Radtke, H. Rodriquez, J. Garcia, C. Valdes, A. Huete, and E. P. Glenn. 2001. "Regeneration of Native Trees in Response to Flood Releases from the United States into the Delta of the Colorado River, Mexico." *Journal of Arid Environments* 49:49–64.

Zhan, X., and M. L. Huang. 2004. "ArcCN-Runoff: An ArcGIS Tool for Generating Curve Number and Runoff Maps." *Environmental Modeling and Software* 19:875–79.

# Adapting Your Stream Restoration Project to Climate Change

Sarah R. LeRoy, Megan Friggens, Gregg Garfin, Rebeca González Villela,
Sue Harvison, Katharine Hayhoe, Sharon J. Lite, Martín José Montero Martínez,
John Nielsen-Gammon, Jeff Renfrow, David Rissik, Julio Sergio Santana Sepúlveda,
Bart (A.J.) Wickel, and Mark K. Briggs

## INTRODUCTION

In this chapter, we focus on climate change issues and strategies that stream practitioners can use to improve the likelihood of restoration success in the context of a rapidly changing climate. In developing a restoration response for your target stream, you may already be dealing with a variety of human-related impacts that are complicating the design and implementation of your stream corridor restoration program. If that is the case, the idea of adding a global issue such as climate change into the mix may send you running for the hills, or at least skipping to the next chapter.

Before doing either, consider this: the impacts of climate change could affect the conditions of your target stream's watershed to the point of threatening the success of your entire stream restoration effort. Although there are ideas and strategies in this chapter that you may not be able to undertake on your own (e.g., conducting your own analysis of climate data), please keep in mind the long-term, iterative nature of stream restoration. Topics related to the nexus of stream restoration and climate change that seem far afield today may be completely relevant and necessary to tackle tomorrow. Also, keep in mind that other entities may be collecting and analyzing climate data in your watershed or your region. If so, an understanding of your watershed's current climate and both past and projected future trends may be a phone call or Google search away.

Before we go any further, let's define a few terms. First, what is the difference between "weather" and "climate"? "Everybody talks about the weather, but nobody does anything about it," quipped Mark Twain. When someone complains that a particular day is too hot, too cold, too rainy, too dry, they are referring to the weather. *Weather* is the state of the atmosphere at a particular place and time as described by heat, dryness, sunshine, wind, and rain, among other descriptors. *Climate* is the prevailing weather conditions for a given area over a long period (e.g., the average precipitation over a thirty-year period for Ciudad Chihuahua in the summer). *Climate change* is a long-term trend in the statistics of weather over periods ranging from decades to thousands of years. It may manifest as a change in average weather conditions (including long-term temperature and precipitation), as a change in the frequency or magnitude of extreme weather events, or both. In the past, climate has changed naturally over long periods of time as a result of changes in energy from the sun, the Earth's orbit, natural cycles, and long-term geological processes. Today, climate is changing at an unusually rapid rate, primarily as a result of human activities (Wuebbles et al. 2017a; IPCC 2013) and is impacting key biophysical and chemical processes that underpin the functioning of stream corridor conditions.

*Resilience* is another term frequently heard in the context of discussing climate change and is defined as "the capacity of ecosystems to persist and absorb change and disturbance, while maintaining key relationships among important system variables or populations" (Holling 1973). Stream corridor restoration can be an appropriate response to climate change, particularly if the focus is to enhance ecosystem resilience. Riverine ecosystems (benthic, aquatic, and riparian) are naturally resilient and capable of rapidly bouncing back following high-energy disturbances such as flooding. These disturbances are essential to the ecosystem's long-term viability. River restoration efforts that seek to enhance this natural resilience by reducing vulnerability and increasing the adaptive capacity of riverine ecosystems and species will likely be successful under a wide range of climatic conditions (Frankhauser et al. 1999; Ingram 2009). However—and this is a big "however"—in order to be successful, stream restoration objectives and tactics themselves need to be adapted to a rapidly changing climate.

In the context of restoration, climate change presents an entirely new challenge for practitioners: managing for the unexpected, including conditions that may lie beyond those experienced in the historical record at that location (Matthews and Wickel 2009; Ingram 2009; Cook et al. 2012). Not only is the climate changing at a pace and an amount never before experienced by human civilization, but even if human carbon emissions are stopped today, some impacts of climate change would remain irreversible for centuries to millennia to come. Uncertainty about the future is not a new problem, but climate change significantly increases that uncertainty because of the difficulty in predicting the magnitude and details of changes, and because future changes depend upon human choices that are extremely difficult, if not impossible, to predict (Hayhoe et al. 2017).

A central reason to include the topic of climate change as part of planning your stream corridor restoration response is that understanding climate change and its potential impacts on your watershed and stream corridor will help you to develop restoration objectives and tactics that will work in the future, not just in the past. Returning stream ecosystems to a previous state (e.g., before human settlement) may be a noble cause, yet completely unrealistic if climate conditions in the near future will no longer support past species and processes. As major natural disturbances such as wildfires, cyclones, floods, and droughts become more frequent, widespread, and intense in many areas (Cochrane and Barber 2009; Pittock 2009; IPCC 2012), stationarity—the idea that natural systems fluctuate within an unchanging range of variability—has become an outdated concept (Milly et al. 2008).

If human management of ecosystems and resources is based on stationarity at a time when climate is changing rapidly, this assumption will make the ecosystems as well as the human communities we are trying to protect even more vulnerable to changing climate conditions.

Instead, what communities and practitioners need are climate adaptation plans that integrate climate change information into development and conservation (IPCC 2014; Bierbaum et al. 2013). How far in the future your temporal planning window needs to extend will depend on the objectives of your own restoration project. That said, it's a safe bet that whether your objective is to bring back native fish, riparian plant communities, or other aquatic communities, or to bring benefits to citizens of riverside towns, the desire is to produce results that last well into the future.

The bottom line when it comes to stream restoration and climate change is that restoration projects that are climate-adapted are more likely to be successful and viable in the long term than those that are not (Perry et al. 2015). As such, the central aims of this chapter are to help stream practitioners gauge the vulnerability of their stream restoration project to climate change and provide strategies to improve the ability of the project to provide resilience to climate change. Is your stream restoration project adapted to current climate conditions and trends? And, if not, what can be done about it? Those are the two central questions we address here.

To address these questions, we need two prerequisites in hand: a sound and realistic restoration objective and a solid understanding of current streamflow conditions and trends. Realistic and detailed stream restoration objectives are the foundation of any stream restoration project. Without them, there is no way to assess the potential that climate change may have for impacting the success of your project. A sound understanding of current streamflow characteristics is essential for understanding what is happening to your stream's benthic, aquatic, and riparian communities; what is driving their ecological deterioration; and what can realistically be done to bring them back. Therefore, understanding potential impacts of climate change on streamflow is key to gauging the vulnerability of your stream restoration project to climate change.

The first section of this chapter reviews climate change projections and impacts, discusses where practitioners can find sources of climate data and projections for their area of interest, and describes strategies for applying large-scale climatic information to the subbasin scale. The focus is how you can access and hone data and information on climate change to the watershed you are working in. The second section focuses on assessing the vulnerability of your stream corridor restoration project to climate change and summarizes measures and practices to make your project climate-adapted.

## UNDERSTANDING OUR CHANGING CLIMATE

Why does climate change matter when developing a stream restoration project? Among the many ramifications of rapid climate change are its impacts on your stream's flow regime, with some research emphasizing the negative consequences of warming and drying trends in low- and mid-latitude streams of West Africa, southern Europe, south and east Asia, eastern Australia, western Canada and the United States, and northern South America (Dai 2013; Koirala et al. 2014; Jiménez Cisneros et al. 2014). The bottom line is that climate change will affect flood and drought intensities and frequencies, altering stream sediment and water fluxes, channel morphology, and water quality and temperature, among other parameters (Baron et al. 2002; Sabater and Tockner 2010), all of which may impede or even prevent the realization of your stream corridor restoration objectives (what realistically can be realized) and the effectiveness of the tactics that are employed to achieve them (Perry et al. 2015).

We begin this section with a brief overview of climate change, then summarize sources of climate data and information that practitioners can take advantage of, then conclude this section with a review of methods for honing climate data and information to relevant spatial scales. The central aim is to provide a foundation of references on climate change. Some of the background you may already know, while some background and strategies may not be appropriate for your situation. Regardless, take in those that make sense to you and skip (and maybe come back to later on) parts that do not. But please do not close the chapter if you find yourself getting bogged down. In the next section, we assess the vulnerability of your stream restoration project to climate change and provide strategies for strengthening how well your restoration project is adapted to a rapidly changing climate.

## Climate Change Basics

### *Why Is Climate Changing Today?*

The climate has always changed in the past, due to natural factors. Over months to decades, natural cycles such as the El Niño–Southern Oscillation affect the exchange of heat between the ocean and atmosphere, making one part of the world cooler and another warmer as they redistribute heat and moisture through the earth's climate system. A large volcanic eruption can cool the climate for months to years; over hundreds of thousands to millions of years erosion and weathering also play a role in altering climate (Fahey et al. 2017). As energy from the sun increases or decreases over decades to eons, the earth's climate responds by warming and cooling. And over tens of thousands of years, natural cycles in the Earth's orbit and its tilt affect where and when sunlight falls on the earth and drive the ice ages, or glacial maxima, and the warm interglacial periods—such as the one that Earth is currently experiencing.

Today, according to natural factors, climate should be cooling slightly (Bindoff et al. 2013). Instead, the Earth is warming, and doing so at a rate that, according to the Fourth U.S. National Climate Assessment (NCA4), is unprecedented in the history of human civilization; global temperatures "appear to have risen at a more rapid rate during the last 3 decades [from 1986 to 2015], than any similar period possibly over the past 2,000 years or longer" (Wuebbles et al. 2017a, 2017b).

Why is the earth warming? Based on observational evidence, there is no convincing alternative explanation for recent warming trends outside of human activities, which include fossil-fuel combustion, deforestation, agriculture, waste disposal, and other activities that produce greenhouse (heat-trapping) gases such as carbon dioxide and methane (IPCC 2013; Wuebbles et al. 2017a). Like a blanket that is too thick would cause us to overheat by trapping too much of our body heat, increased levels of greenhouse gases in the atmosphere trap more of the earth's heat than would otherwise escape to space, increasing the average temperature of the planet.

Two-thirds of the warming from human emissions of greenhouse gases is due to carbon dioxide ($CO_2$), emitted mainly from the burning of coal, gas, and oil. Just over 20 percent comes from methane ($CH_4$), another important greenhouse gas, with stronger heat-trapping properties than $CO_2$, but a shorter atmospheric lifetime and smaller concentrations. In addition to the human-caused emissions of $CH_4$ from the extraction and transport of coal and natural gas, and livestock and agricultural production, there are many natural sources of $CH_4$, such as decaying organic material in wetlands and thawing permafrost, and many of these sources are increasing as the result of a warming planet. The remaining 12 percent comes from a number of different gases, including nitrous oxide ($N_2O$) from fertilizer and chlorofluorocarbons from industrial processes. Water vapor is also a powerful greenhouse gas but in the lower atmosphere, the amount of water vapor is controlled primarily by temperature. Thus as increases in other greenhouse gases warm the lower atmosphere, water vapor increases in response, further amplifying the warming effect.

For more information on climate change, including responses to many frequently asked questions, such as "How can we predict what the climate will be like in 100 years if we can't even predict the weather next week?" and "What is (and is not) debated among climate scientists about climate change?," we strongly recommend the Third U.S. National Climate Assessment's (NCA3) appendix 3, "Climate Science Supplement" (Walsh et al. 2014) and the Fourth U.S. National Climate Assessment's (NCA4) appendix 5, "Frequently Asked Questions" (Dzaugis et al. 2018). For more detailed information on the latest climate science, interested readers are directed to the first volume of the NCA4 (USGCRP 2017) and chapter 2 of the second volume (Hayhoe et al. 2018). Highlights from the NCA3 report, and the NCA4 volume 2 Report-in-Brief, are also available in Spanish (https://www.ccass.arizona.edu/resources /spanish-language), as are other resources, including selected reports of the Fifth Assessment of the IIPCC and the American Association for the Advancement of Science's What We Know and How We Respond reports (https://whatweknow.aaas.org/wp-content/uploads/2014/07 /whatweknow_report_espanol_v4.pdf and https://howwerespond.aaas.org/document/ada -reportspanish/).

## How Is Climate Change Affecting Temperature?

Global average near-surface temperature anomalies are the most common measure of the magnitude of climate change. Annual average temperatures have increased by about 1.0°C (1.8°F) from 1901 to 2016, and recent decades have been the warmest in the past 1,500 years in the Northern Hemisphere (IPCC 2013; Wuebbles et al. 2017a, 2017b; Vose et al. 2017). Although changes at the local scale may be greater or less than the global average, similar changes are seen across most large regions of the world, including the contiguous United States, where annual average temperatures have increased by 0.7° to 1.0°C (1.2° to 1.8°F) from 1895 to 2016. In Mexico, the 1990s and 2000s were the warmest decades of the last 110 years (Redmond and Abatzoglou 2014). Australia has experienced an increase in temperature of about 1°C (1.8°F) since the beginning of the twentieth century (CSIRO and Bureau of Meteorology 2015).

Since the mid-1960s, the frequency, intensity, and duration of heat waves have increased, and over the last two decades in the United States, the number of record high temperatures has far exceeded the number of record low temperatures, a trend also seen on a global scale (CSIRO and Bureau of Meteorology 2015; Vose et al. 2017). In Australia, temperature records have been broken on a regular basis over the past 15 years; for example, 2013 was the warmest year on record, very warm months have increased fivefold, and the hottest day on record for the country was set in January 2013, then again in December 2019 (CSIRO and Bureau of Meteorology 2015).

Some amount of future climate change is already inevitable due to past emissions; but the larger part of future change "will depend primarily on the amount of greenhouse (heat-trapping) gases emitted globally and on the remaining uncertainty in the sensitivity of Earth's climate to those emissions," according to the NCA4 (Wuebbles et al. 2017b). For that reason, climate projections are typically developed for a range of possible futures, corresponding to higher or lower future greenhouse gas emissions, or even (more recently) for a range of global mean temperature changes, from 1°C to 3°C or more. A full description of these scenarios is provided in Hayhoe et al. (2017).

Global annual average temperatures are projected to rise 0.6° to 2.2°C (1° to 4°F) under a lower scenario, and 3.3° to 5.6°C (6° to 10°F) under a higher scenario (see figure 4.1) (Hayhoe et al. 2017). The number of hot days and nights are projected to increase, the number of cold days and nights are projected to decrease, and heat waves will become more frequent and last longer (Meehl and Tebaldi 2004; IPCC 2013).

Across North America, relatively greater increases are projected for more northern regions such as Alaska and the northern half of Canada, but annual average temperatures in the southwestern United States and northern Mexico are still projected to increase by about 2.8° to 5.6°C (5° to 10°F) by 2100 under a higher scenario (Cayan et al. 2013; Vose et al. 2017). That may not sound like much, but it means that by the 2050s, average summertime temperatures in the southwestern United States and northern Mexico will be consistently warmer than what has been recorded in the twentieth and early twenty-first centuries, and what would be considered an extremely hot summer today will be a normal occurrence in the future (McRoberts and Nielsen-Gammon 2010). In Australia, for example, where the summer average temperature (as distinct from the summer average high temperature) in southwest Queensland is already 29°C (84°F)—it may rise to over 30°C (86°F) by as early as 2030, and to over 33°C (91°F) by 2070. Figure 4.2 shows projected annual average temperatures for the end of the century for North America and Australia. Table 4.1, at the end of section 1a, lists important trends (such as the increasing trend in temperature) in current and future climate, and trends in climate change impacts.

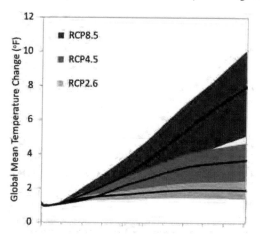

**FIGURE 4.1** Global mean temperature change (in °F) projected for 2000–2100 relative to 1900–1960 under different scenarios, as simulated by CMIP5 models. Figure by K. Hayhoe, adapted from Hayhoe et al. 2017.

## How Is Climate Change Affecting Precipitation?

Observed changes in average and seasonal precipitation are more region-specific than temperature. In the United States, annual precipitation over the last century has decreased in the Southeast and much of the western half of the country (especially in the Southwest), and increased in the Midwest, Northeast, and Plains regions (Easterling et al. 2017).

## Projected Changes in Annual Average Temperature
### Late 21st Century

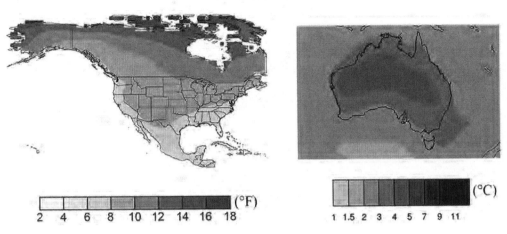

**FIGURE 4.2** Graph showing projected changes in annual average temperatures for North America (left, in °F) and Australia (right, in °C) under a higher future scenario (RCP 8.5). Projections are for 2070–2099, compared to 1976–2005. The figure for North America is adapted from figure 6.7 of Vose et al. 2017. The figure for Australia was created using the full set of CMIP5 models from the Royal Netherlands Meteorological Institute (KNMI) Climate Explorer website (http://climexp.knmi.nl/).

Important changes are also happening at timescales finer than annual. In the North American monsoon (NAM) core region, including the southwestern United States and northern Mexico, for example, no distinct *annual* precipitation trends have been seen over the last half of the twentieth century, but this doesn't mean that no changes have occurred. Rather, rain events have become more intense but less frequent, and the length of the monsoon season has decreased (Englehart and Douglas 2006; Anderson et al. 2010; Arriaga-Ramirez and Cavazos 2010; Luong et al. 2017). There has also been an increase in extreme precipitation events associated with land-falling hurricanes (Cavazos et al. 2008).

Precipitation patterns in Australia also exhibit regional differences. Northern and central Australia have experienced a significant increase in wet season (October–April) rainfall since the early 1970s, while the southern part of the country has experienced a 10 to 20 percent reduction in rainfall during the cool season (April–September) (CSIRO and Bureau of Meteorology 2015). In southeast Australia, there has been a 15 percent reduction in late fall and early winter Southern Hemisphere rainfall since the mid-1990s, and in the Southwest the decline has been as much as 40 percent during winter since around 1970 (CSIRO and Bureau of Meteorology 2015).

Air can hold more water as its temperature increases, at a rate of 6 to 7 percent more per degree Celsius of warming, according to the Clausius-Clapeyron relationship (Easterling et al. 2017). This means that heavy rain events should increase in frequency and intensity as temperatures warm from climate change, and in fact, this is being documented across the globe. A study focusing on Australia found that heavy rain events are increasing two to three times faster than the Clausius-Clapeyron relationship indicates (Guerreiro et al. 2018), and at the global scale, heavy rain events have increased in frequency by more than 20 percent relative to preindustrial conditions, a direct result of warming (Fischer and Knutti 2015). Heavy rain events have increased across the United States since the 1950s, most notably in the Northeast and Midwest (Easterling et al. 2000, 2017). In Australia, the fraction of the country receiving a high proportion of rainfall from extreme rain events has increased as well since the 1970s (CSIRO and Bureau of Meteorology 2015).

In the future, the location and timing of annual and seasonal precipitation is expected to continue to shift. Precipitation projections vary substantially by region and season. In North America, for example, precipitation is projected to increase by 20 to 30 percent under a higher scenario (RCP 8.5) by the end of the century across the northern half of the continent, especially in winter and spring, but decrease by 20 to 30 percent across Mexico and the southwestern United States during the same seasons (Hayhoe et al. 2018). In summer and fall, on the other hand, precipitation is expected to change by no more than around 10 percent (either way) across most of North America (Easterling et al. 2017). Even though precipitation is not projected to change as much in the summer months for the southwestern United States and northern Mexico, consistently warmer summertime temperatures projected for the future means even higher evapotranspiration rates during a time of year when much of the precipitation falls in the region. Therefore, water availability in the region's many reservoirs will depend more on cool-season precipitation, which is normally relatively low and, as stated above, is projected to decrease even further (McRoberts and Nielsen-Gammon 2010; Easterling et al. 2017). In Australia, end-of-century rainfall projections show a large amount of variability and possibilities of a drier or wetter climate, depending on the region.

In a warmer world, climate scientists are confident that many regions will experience more intense downpours and longer and/or more intense droughts. Heavy precipitation events are expected to continue to become more frequent and more intense across nearly the entire world. In general, the subtropics are expected to become drier as the climate warms, with more consecutive dry days (Held and Soden 2006), with increased evaporation across most land surface areas except the Amazon and decreases in soil moisture across most land areas except for a few locations, such as southeastern South America and central Asia (Collins et al. 2013) (see figure 12.26 in Collins et al. 2013, which illustrates the above points on anticipated changes in the frequency of intense periods of precipitation and number of dry days. Also see figure 12.23, which illustrates the projected change in annual mean soil moisture in the top 10 cm [~4 inches] of soil, for 2081–2100 relative to 1986–2005. Both figures can be found at https://www.ipcc.ch/report/ar5/wg1/long-term-climate-change-projections-commitments -and-irreversibility/).

NAM precipitation accounts for at least half of the annual precipitation in many parts of the southwestern United States and northern Mexico. Compared to the precipitation of winter storms, NAM precipitation is difficult to simulate, because it involves a subtle balance among wind, heat, moisture, and thunderstorms on spatial scales that are too small to be simulated directly by global climate models. Even though the models capture broad-scale features associated with the NAM seasonal cycle (Liang et al. 2008; Gutzler et al. 2009), it has been difficult to simulate many important related phenomena (Castro et al. 2007; Lin et al. 2008; Cerezo-Mota et al. 2011). Over the coming century, model simulations generally project reduced precipitation in the core zone of the NAM and the rest of Mexico (IPCC 2013), but this signal is not particularly consistent across models, even under the highest scenario (Cook and Seager 2013). Thus confidence is low in projections of monsoon precipitation changes (IPCC 2013). Scientists are working on improving the accuracy of regional-scale projections, and, for the NAM region, recent research suggests a slight decrease in daily average monsoon precipitation, but an increase in the amount of precipitation delivered by the most extreme thunderstorms (Cavazos and Arriaga-Ramirez 2012: Luong et al. 2017).

## What Are the Impacts of Climate Change on Stream Hydrology?
### Heavy Precipitation and Flooding

As discussed above, heavy precipitation events are increasing in frequency and intensity across most of the globe, and this trend is expected to continue in the future because a warmer atmosphere holds more moisture (IPCC 2013). More intense rainfall combined with little change or even a decrease in average precipitation implies more runoff in small basins, interspersed with longer dry periods. In a study of the Río Conchos Basin in Mexico, Rivas Acosta and Montero Martínez (2013) found that even though annual precipitation (and the subsequent runoff) may decrease, the frequency and magnitude of intense storms could increase. However, recent research in Australia suggests that soil moisture plays a notable role in future changes to flooding from more intense storms (Wasko and Nathan 2019). For floods that occur once a year, flood magnitudes have been decreasing in Australia in areas where soil moisture has also been decreasing, despite increasing rainfall extremes. For rarer extreme rainfall events, however, flood magnitude is likely to increase even with decreased soil moisture.

In the southwestern United States, rain-on-snow events, occurring when late winter or early spring moisture falls as rain rather than snow, can lead to catastrophic flooding. In contrast, floods from snowmelt in late spring and early summer are expected to decrease in frequency and intensity because of lowered snowpack. At the same time, rapid melting of snowpack would increase the susceptibility of these streams to extreme drying later in the year, as streamflows peak earlier in the year than normal (Stewart et al. 2005; Karl et al. 2009; Georgakakos et al. 2014).

The uncertainty in projections of NAM precipitation described above translates directly into a lack of confidence regarding summer flooding from monsoon storms in the NAM region. However, recent research suggests that the most extreme monsoon storms are becoming even more intense in the southwestern United States and northern Mexico, increasing the flood risk (Luong et al. 2017). Additionally, atmospheric rivers—elongated, moisture-rich atmospheric features that are the "primary drivers of flood damages in the western United States" (Corringham et al. 2019)—are projected to increase in frequency and intensity as a result of increasing evaporation from the oceans and higher atmospheric water vapor from warmer temperatures (Hagos et al. 2016; Kossin et al. 2017). Changes in atmospheric rivers and their impacts is an area of active research (e.g., Guirguis et al. 2018; Aguilera et al. 2019; Huning et al. 2019).

### Drought

Although the influence of climate change on drought is complex, recent droughts have reached record intensity in some regions of the United States (Wehner et al. 2017). Record-setting droughts have been sustained and exacerbated in the southwestern United States by a combination of warmer temperatures, decreased snowpack, and sustained ridges of high pressure that deflect storms, resulting in reduced soil moisture and increased evapotranspiration (Cayan et al. 2010; IPCC 2012; Ault et al. 2016). In this region, where drought is endemic and occurs as part of the climatic variability, future droughts may be more severe, even if precipitation stays the same, given that these droughts will take place in a warmer environment, resulting in higher evaporation rates (Garfin et al. 2013; Seager et al. 2015; Wehner et al. 2017). There is already evidence that human-induced trends in temperature, relative humidity, and precipitation pushed the most recent drought in southwestern North America from an otherwise moderate drought to one comparable to the worst megadroughts in the past 1200 years (Park Williams et al. 2020). Another study estimates streamflow reductions of 20 to 35 percent by the mid to end of the twenty-first century in the Colorado River Basin of the United States just from temperature increases and resulting higher evaporation (Udall and Overpeck 2017). In the south-central United States, warming temperatures will likely exacerbate the risk of a semipermanent high-pressure system over the region in the summer, leading to stronger, more prolonged summer drought (Ryu and Hayhoe 2017).

Drought is a common phenomenon in California's climate, which is historically variable. Although natural precipitation variability sparked the 2011–17 California drought (Seager et al. 2015), the drought's severity was exacerbated by human-induced temperature increases. One study estimated that warmer temperatures amplified the drought by 8 to 27 percent in 2012–14 (Williams et al. 2015). There is a clear connection between years that are hot *and* dry, and the risk of such years increases as the world warms (Diffenbaugh et al. 2015).

Scientific confidence in the magnitude and sometimes even the direction of projected changes in precipitation and drought vary from region to region. For the southwestern and south-central United States and northwest Mexico, however, there is relatively high confidence in projections of longer, more frequent, and more severe droughts. For Mexico, Prieto-González and colleagues (2011) concluded that projected drought events will surpass the length, magnitude, and frequency of those of the second half of the twentieth century. Projections are similar for Australia, where the time spent in drought is projected to increase, along with the occurrence of more frequent and longer extreme droughts and fewer moderate to severe droughts (CSIRO and Bureau of Meteorology 2015).

## Runoff, Streamflow, and Impacts to Species

Changes in temperature and precipitation—including changes in annual and seasonal averages, as well as changes in intensity and variability—affect soil moisture, runoff, streamflow, and water availability in a watershed. Although studies often focus on changes in the central tendency or average of these metrics, the extremes are important to riparian system function (see, for example, Leigh et al. 2015) and should be incorporated into projects that use climate projections in future planning.

In general, scientists are confident in the projected impacts of warmer temperatures on riparian areas. Increases in average river temperatures will affect the distribution of freshwater species, and higher evaporation rates will likely contribute to increased drying of wetlands and waterholes in ephemeral rivers (Pratchett et al. 2010). In Australia, where some regions are projected to be drier, this could mean that some permanent habitats will become intermittent (Kennard et al. 2010).

Changes in temperature can alter the seasonal timing of surface flows, as can droughts and changes in average annual precipitation. This is particularly true in the snow-dominated watersheds of California and the southern part of the southwestern United States, as the amount of precipitation that falls as rain versus snow increases with temperature, and snow-melt shifts to earlier in the year (Knowles et al. 2006; Berghuijs et al. 2014; Dettinger et al. 2015; Solander et al. 2017; Harpold and Brooks 2018). (High altitude buffers some watersheds in Utah and Colorado from this transition.) Such shifts can create difficulties for time-sensitive water allocations or management plans.

A central take away is that the impacts of climate change cascade throughout the ecosystem. Changes in temperature and precipitation affect streamflow processes that, in turn, impact biota. For example, warmer water temperatures, increased salinization, and low dissolved oxygen, such as are projected for the Rio Grande Basin (Llewellyn and Vaddey 2013), may increase the likelihood that nonnative aquatic species will invade (Rahel and Olden 2008; Bond et al. 2008). In southeastern Australia, a globally invasive fish species, the eastern mosquitofish (*Gambusia holbrooki*) has been shown to be more resistant to high water temperatures and hypoxia than a native fish, the locally endangered southern pygmy perch (*Nannoperca australis*) (Stoffels et al. 2017).

Longer, more severe droughts could decrease runoff and eliminate ephemeral lakes and spring pools, as well as drive perennial streams toward intermittent flow, and intermittent streams to ephemeral (Grimm et al. 1998). Droughts are also detrimental for riparian seedling survival, especially cottonwood and willow (*Populus* and *Salix* spp.) seedlings, in contrast to tamarisk (*Tamarix* spp.) seedlings, which are more adaptable to low-flow conditions (Garssen et al. 2014). During a time of markedly reduced surface flows (2002–2007) in the Colorado River Delta in Mexico, overall vegetation cover was unchanged, but there was a reduction in native tree cover (mostly *Populus fremontii* and *Salix gooddingii*) and an increase in shrub cover (mostly *Tamarix* spp.) (Hinojosa-Huerta et al. 2013).

Decreased streamflow can also have a number of indirect effects on aquatic ecosystems and water supply, such as reducing groundwater recharge (Lettenmaier et al. 2008; Niraula et al. 2017). One study estimates a 10 to 20 percent reduction in total recharge across aquifers in the southwestern United States (Meixner et al. 2016). Decreased streamflow can also reduce stream capacity to buffer pollutants (Hurd and Coonrod 2007), triggering a cascade of effects throughout the riparian food web (Janetos et al. 2008). For example, in southeastern Australia, severe drought from 1997 to 2009 reduced streamflow by more than 50 percent in some areas, which in turn substantially reduced the prevalence of many flow-dependent macroinvertebrates (Thomson et al. 2012). Of particular priority is to conduct the necessary studies to understand the consequences of climate change on watersheds in water-deficient regions that are characterized by water supplies that are already overallocated (see Case Study 4.1 about the Río Conchos in Chihuahua, Mexico).

Climate change can also affect streamflow through interactions of vegetation, wildfire, and hydrology. Across the western United States, the area burned by wildfires has more than doubled since the 1980s relative to what would be expected without human-induced climate change (Wehner et al. 2017), and coniferous forests across western North America are experiencing widespread mortality as a result of drought, insect outbreaks, and wildfire associated with climate change (Breshears et al. 2005; Adams et al. 2009; van Mantgem et al. 2009;

# CASE STUDY 4.1
## Analysis of Trends in Climate and Streamflow to Inform Water Management Decisions in the Río Conchos, Chihuahua, Mexico

Rebeca González Villela, Martín José Montero Martínez,
and Julio Sergio Santana Sepúlveda

## Introduction

The Río Conchos originates in the pine and oak forests of the eastern slope of the Sierra Madre Occidental (known regionally as the Sierra Tarahumara) and crosses the wide plains of the Chihuahuan Desert before joining the RGB at Ojinaga, Chihuahua. The Río Conchos Basin covers 68,606 km$^2$ (26,489 mi$^2$), which is roughly 8 percent of the entire surface area of the Rio Grande Basin and about 26 percent of the state of Chihuahua (Gutiérrez and Carreón 2004; INEGI 2013). The Río Conchos and its tributaries supply water to three large irrigation districts (the largest being Irrigation District 005, in Delicias, which has 12,503 members and covers 90,589 ha), thirty-seven municipalities of the state of Chihuahua, and three municipalities in the state of Durango (Rocha 2005). Overall, over 90 percent of its water is used for irrigated agriculture. In addition to meeting water demands for municipalities and irrigated agriculture, the Río Conchos contributes flow in compliance with the 1944 Water Treaty between Mexico and the United States and sustains riparian and freshwater ecosystems that are noted for their richness in biodiversity and high rate of endemism.

Given that the Río Conchos' waters are already overallocated and that the basin is in an arid region where the future climate is trending hotter and drier, it is critical to predict trends and changes in flow in order to sustainably manage its water resources (Jones and Wigley 2010). To do this effectively, the impacts of climate change, socioeconomic and demographic impacts on water demand (which includes water needed for agriculture, municipalities, recreation and electric power), and the water needed to support native riverine flora and fauna, all need to be understood. This case study summarizes how we quantified the availability of water in the Río Conchos by analyzing the quantity and frequency of precipitation; estimating the quantity, frequency, and magnitude of the flow needed to maintain native species (i.e., environmental flows); and assessing flow alteration by climate change and human impacts.

## Methods

To assess trends and variation in precipitation and temperature, a database created by Centro de Investigación Científica y de Educación Superior de Ensenada, Baja California (CICESE) (http://clicom-mex.cicese.mx/malla) was used. This database compiles the climate information of more than five thousand stations stored in the CLICOM (CLImatological COMputing) system. Basic quality analysis was conducted on all raw data to remove: (1) negative precipitation values, (2) maximum temperatures greater than or equal to the minimum temperature of the same day, and (3) the same temperature and precipitation values for ten or more consecutive times (except for zero precipitation). The remaining data for the Río Conchos were interpolated to a regular mesh Synographic Mapping System (Symap) (Shepard 1984), which uses weighted averages of all the records to produce a daily database for precipitation and maximum and minimum surface temperatures (Zhu and Lettenmier 2007; Muñoz-Arriola et al. 2009).

To analyze trends and variation in flow, we analyzed streamflow data collected at La Boquilla Dam and at two sites below it (see map in this case study). Flow data were analyzed using the IHA V7

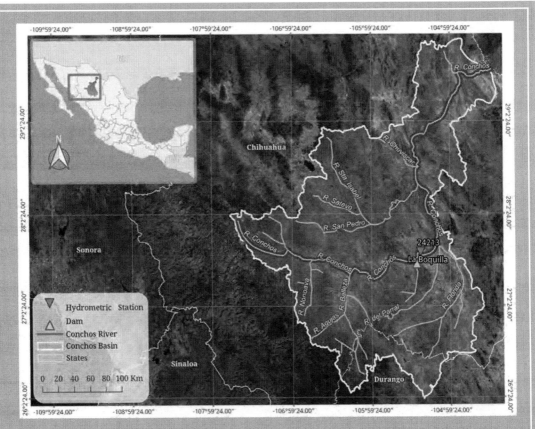

Inset map and satellite image showing location of Río Conchos watershed in Chihuahua, México (adapted from González-Villela et al. 2017). The confluence of the Río Conchos with the Rio Grande/Bravo, which forms part of the international border between Mexico and the United States, is at the upper right of image.

program (The Nature Conservancy 2006), focusing in particular on periods of extreme low flow, moderate low flow, high flow, pulses of high flow, and small and large floods of two and ten years of return, respectively. For each indicator, parametric and nonparametric analyses were performed that estimated the means or medians and deviations for the years considered. This range of variability was used to assess how the Río Conchos' streamflow regime has changed and how those changes are likely impacting native riverine flora and fauna. See González-Villela et al. (2017) and González-Villela et al. (2018) for a full discussion of methods.

## Study Results

Several results from this study highlight how impacts of climate change on the Río Conchos' flow regime may be exacerbating impacts from such human interferences as dam construction, deforestation, and overgrazing. For example:

1. In general, and not surprisingly, years that had the greatest number of days with zero precipitation coincided with drought periods that regularly affect this region. During these periods, low flow and reduced reservoir volumes were the norm, forcing populations to rely more and more on pumping

*(continued)*

from basin aquifers that are already overexploited. Owing to population growth and industrialization—two realities linked to the integration of the economies of Mexico and the United States via the North American Free Trade Agreement—water demand for both industrial and domestic uses have increased dramatically, placing even greater pressure on the Río Conchos's water supplies.

2. Deforestation, overgrazing, and other upper watershed activities have reduced infiltration, accelerated soil erosion rates, and increased sediment input into streams, which have in turn affected the Río Conchos's streamflow, channel morphology, and water quality (Gutiérrez and Carreón 2004; Rocha 2005).

3. Analyses of historical data indicate a decrease in average monthly minimum precipitation (the average monthly minimum precipitation for the entire record that was analyzed) and in the number of storms, but also an increase in storms that produce significant rainfall in all months of the year except January and August. Taken together, these results seem to indicate changing patterns of humidity that are making the weather more extreme (González-Villela et al. 2017).

4. Dam construction has homogenized the magnitude, frequency, periodicity and duration of the Río Conchos's flow. As compared to the river's natural flow regime, the current flow regime shows little variability throughout the year. The combination of the impacts of dam construction and climate change have precipitated changes in the composition, structure, and function of the river and the riparian ecosystem—as well as changes in water quality (the water is more brackish, especially during low flow periods)—and the decline of native aquatic fauna. Native fish have been particularly impacted, especially species that prefer clear, cold, and well-oxygenated water as well as those that depend on sand, gravel, and rock substrate for food and nesting; much of this habitat is being buried by sediment.

## Future Priorities

This study underscored some of the water-related challenges facing water-deficient watersheds worldwide. Based on its results, we put forward the following priorities as potential next steps:

- In general, studies such as this one need to be conducted in watersheds throughout Mexico—particularly in those where overallocation of water resources is already an issue—in order to better guide water-management decisions toward sustainable levels.

- Addressing the deterioration of upper watershed conditions is an important climate-adapted response. Any action that slows runoff and increases water retention will be positive for both people and wildlife. We want the watershed to be more like a sponge rather than a sluice.

- For the Río Conchos, additional and constant monitoring of water and sediment flow is needed. Collection of these data is critical for evaluating how future management decisions meet stated objectives.

- To compensate for some of the impacts of dams, other human interferences, and climate change on native riverine flora and fauna, improving water management as well as securing environmental flows needs to be a priority (environmental flow is the focus of the next chapter). Such a priority can also be a win-win for both people and native species, but requires close collaboration among community leaders, water managers, researchers, and policymakers (King and Brown 2006).

Allen et al. 2010). Analysis of these interactions suggests that sediment yields will increase in response to climate change (Goode et al. 2012).

### Sea-Level Rise and Tropical Storms

Stream restoration efforts in coastal areas or just upstream from estuaries may also be affected by sea-level rise and changes in tropical storms, from impacts such as saltwater intrusion, high storm surges, and coastal retreat. The mean sea level is rising as the oceans warm and expand, and land-based ice sheets and glaciers melt (Church et al. 2013). Since 1900, the global mean sea level has increased by 18 to 20 cm (7 to 8 in.), but the rate has doubled since the satellite record began in the 1990s (Nerem et al. 2018). Sea level is expected to increase by at least 30 to 130 cm (1.0 to 4.3 ft.) by 2100; given recent findings regarding the mechanisms that control the Antarctic ice sheet melt, however, NCA4 concludes that a rise of 2.4 m (8 ft.) by 2100 cannot be ruled out (Sweet et al. 2017).

At the local scale, global changes in sea level are modified by changes in ocean circulation and by local uplift (upward elevation of the Earth's surface) and subsidence (sudden or gradual sinking of the Earth's surface). In a few places in Alaska, for example, relative sea level is actually dropping as the land is rising faster than the sea level. Across much of the Gulf of Mexico, however, relative sea level is much higher than the global average, as the land is sinking. Projections for the rise in sea level on Mexican coasts by the end of the century are consistent with the global average, at 0.24 to 1.24 m (0.8 to 4.0 ft.) (Slangen et al. 2014; Zavala-Hidalgo and Ochoa-de la Torre 2015). Projected rates of sea-level rise are even higher for parts of Australia (Slangen et al. 2014), with huge potential impacts (see Case Study 4.2) to a variety of coastal wetlands that provide habitat for hundreds of species of fish and migratory shorebirds. For example, as mangrove ecosystems—a tidal swamp ecosystem of tropical coasts, deltas, estuaries, lagoons, and islands—are inundated more often and for longer periods of time, they could potentially retreat landward, dramatically impacting vulnerable species (Pratchett et al. 2011). In northern Australia, one study suggests a large loss of freshwater floodplains (around 65 percent) by 2100 due to saltwater inundation (Bayliss et al. 2016).

Finally, coastal ecosystems are also at risk due to observed and projected changes in tropical storms, including those known as hurricanes, cyclones, or typhoons, depending on their geographic location. Named storms are not increasing in frequency but are becoming stronger, bigger, slower, and intensifying faster, with heavier and more intense rainfall as the world warms (Kossin et al. 2017). This implies an increased risk of both inland flooding as well as stronger storm surges. Future changes to tropical cyclone frequency and intensity is an active area of research, but more and more studies (e.g., Knutson et al. 2015; Kossin et al. 2017; Bacmeister et al. 2018) agree that, while little change is expected in the overall frequency of tropical cyclones, the frequency of intense category 4 and 5 storms will significantly increase, as will precipitation rates. All of these factors combined indicate a high potential for flooding of coastal areas in the future.

## Sources of Climate Data, Predictions, and Projections

Most readers will be familiar with the concept of weather services, which offer forecasts of weather conditions (including both averages and extremes) over timescales of hours to a week or more. In contrast, climate services focus on changes over weeks to centuries (National

# CASE STUDY 4.2

Protecting the Kooragang Wetlands from Rising Sea Levels: A Holistic
Climate-Adapted Response in New South Wales, Australia

David Rissik

The Kooragang Wetland complex is situated near Newcastle on the central New South Wales (NSW) coastline (see map in this case study). This wetland complex is part of the Hunter Estuary, which lies above a busy port and is part of a well-developed catchment that experiences a range of different pressures, many of which will be affected by climate change (Rogers 2016).

The Kooragang Wetlands were originally a saltwater/estuarine system with vast extents of salt-marsh and mangroves. By the 1980s a system of constructed levees, floodgates, and channels had reduced the influx of saltwater into the system as well as enabled the wetland to be used for farming (Svoboda 2017). The deterioration of this wetland complex, as well as recognition of the importance of protecting this saltwater system for bird habitat, resulted in the Kooragang wetlands being listed as a Ramsar site in the mid-1980s.

As part of a holistic, climate-adapted response, rehabilitation activities are taking place to improve wetland conditions, and an Estuary Management Plan (EMP) has been developed. In addition, new legislation in NSW has prepared the way for Coastal Management Programs (CMPs) that provide

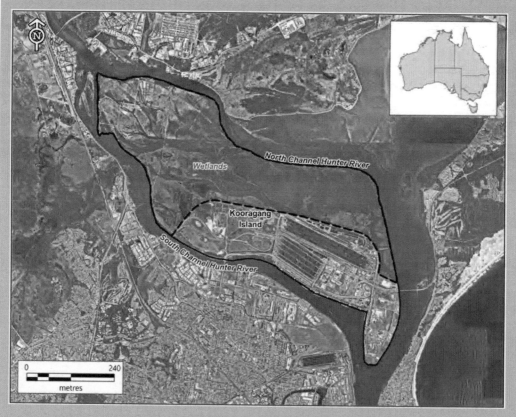

Map showing location of Kooragang Wetlands on Australia's southeast coast with detailed satellite image of Kooragang Wetland complex.

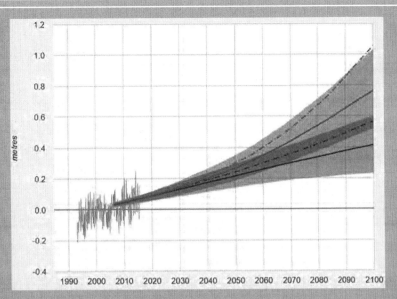

Sea level rise projections for the Kooragang area. The slide shows projections for RCP8.5 (high emission scenario) and for RCP2.6 (very low emission scenario). The graph also shows observed sea level data collected using satellites. Data are from Coastadapt.com.au

opportunities to further protect estuary habitat with improved management actions that account for climate change.

With a grant provided as part of the EMP, the Kooragang Wetland Rehabilitation Project (KWRP) was launched in 1993 to help compensate for the loss of fisheries and other wildlife habitat at suitable sites in the estuary. Central objectives of the project include

- protecting threatened/protected species and communities,
- enhancing estuarine habitat for flora and fauna,
- increasing the frequency of the flushing of tidal creeks, and
- restoring creek ecosystems.

Sea-level rise projections (see graph in this case study), along with detailed Light Detection and Ranging (LiDAR) data, have been used to determine the extent of tidal changes over time. Sea-level rise as a result of climate change and the impacts of sea-level rise on groundwater conditions poses several challenges to the rehabilitation effort. Rising sea levels and saltwater intrusion further inland could affect surface water and the quality of coastal groundwater aquifers as well as the saltmarsh ecosystem and the ecosystem services it provides. Such impacts may be exacerbated if existing coastal barriers prevent inland migration of the saltmarsh as sea levels continue to rise. To address this challenge, one of the rehabilitation activities is to open floodgates and remove coastal barriers.

In addition to sea-level rise, climate projections for this region show increases in temperatures and longer and more severe droughts (with corollary increases in evaporation) that are punctuated by periods of intense rainfall and flooding. These forecasts have raised the likelihood of bushfires and accelerated soil erosion, which could compromise saltmarsh accretion.

*(continued)*

To address these issues, cooperation with local landowners is essential to both improve management and to acquire targeted lands for maintaining corridors and biodiversity (Rogers et al. 2013). The co-funded EMP and CMP will assist local governments to work with all relevant stakeholders to use new and existing knowledge to support them in developing robust iterative, coastal climate adaptation plans that will sustain this critical wetland ecosystem for people and wildlife far into the future.

### Key Take Home Points

This case study illustrates that coastal ecosystems, like the Kooragang Wetlands, and human communities are going to be impacted by climate change–driven sea-level rise. Here are two key takeaway points:

- Conducting the studies needed to understand the extent and consequences of sea-level rise is essential to developing management and rehabilitation plans that will be effective in the face of a rapidly changing climate (i.e., management and rehabilitation plans that are climate-adapted) (e.g., Finlayson et al. 2017);
- For the Kooragang Wetlands and other important ecosystems of this region, public-private partnerships are key to developing sustainable and effective climate-adapted responses that will be win-win for people and the environment. But, per above point, in order to be effective, such responses have to be based on a sound understanding of the challenges these systems are facing.

Research Council 2001). Here we use *climate services* as an umbrella term that refers to any group involved in gathering, analyzing, and distributing climatic data, projections, or information to the public. Climate services can help restoration teams in a variety of ways, including assisting with the collection, analysis, and interpretation of climate data and securing additional resources to address gaps in knowledge.

The type of climate data offered by many climate service providers include historical collections at weather stations of hourly, daily, and monthly temperature, precipitation, and more. Some datasets have synthesized this weather station data into a grid (a spatial area varying in size), providing information for a larger area or entire country. Climate services can also provide *seasonal predictions* that describe upcoming seasons as warmer or cooler or wetter or drier than average. Depending on the region and the climate service provider, practitioners may be able to obtain long-term climate projections that extend through the end of the century.

Land- and water-management areas may be small geographically, but they are linked in complex ways to many other natural and human ecosystems. Consequently, river practitioners require a combination of narrowly constrained and site-specific information that climate service providers can assist with, as well as regional-scale information on how upland ecosystems interact with climate (for example, the impacts of climate change on mountain headwater areas that in turn affect the streamflow of a targeted stream reach) (Perry et al. 2012).

Stream practitioners can be challenged by the fact that climate data, predictions, and projections may be made available in different formats and at very different levels of accessibility

**TABLE 4.1** Trends in current and future climate indicators and impacts relevant to streamflow, for the southwestern United States, northern Mexico, and Australia

| Parameter | Trend | Confidence | Sources |
|---|---|---|---|
| Annual average temperature | Increasing | High | Cayan et al. 2013; Vose et al. 2017; CSIRO and Bureau of Meteorology 2015 |
| Heat waves | Increasing in frequency, duration, and intensity | High | Vose et al. 2017; CSIRO and Bureau of Meteorology 2015 |
| Winter/spring precipitation for southwestern U.S./Mexico | Decreasing | Medium to high | Easterling et al. 2017 |
| Summer/fall precipitation for southwestern U.S./Mexico | Little to no change in amount, increase in storm intensity | Low | Easterling et al. 2017; Luong et al. 2017 |
| Winter/spring precipitation for Australia | Decreasing for southern Australia | High | CSIRO and Bureau of Meteorology 2015 |
| Summer/fall precipitation for Australia | Decreasing for southwestern Victoria in fall, and western Tasmania in summer | Medium | CSIRO and Bureau of Meteorology 2015 |
| Heavy precipitation events | Increasing in frequency and intensity | Medium to high | Easterling et al. 2000, 2017; CSIRO and Bureau of Meteorology 2015 |
| Soil moisture | Decreasing | Medium to high | Collins et al. 2013, Wehner et al. 2017 |
| Drought severity | Increasing in many areas | High | Garfin et al. 2013; Prieto-González et al. 2011; CSIRO and Bureau of Meteorology 2015 |
| Streamflow | Decreasing (Southwest U.S. and northern Mexico); Increasing (most of Australia) | Low to medium | Garfin et al. 2013; CSIRO and Bureau of Meteorology 2015 |
| Runoff | Flows peak earlier in the year in snowmelt-driven regimes | High | Garfin et al. 2013; CSIRO and Bureau of Meteorology 2015 |
| Invasive species | Thriving | High | Rahel and Olden 2008; Bond et al. 2008; Stoffels et al. 2017 |
| Groundwater recharge | Decreasing | Medium to high | Meixner et al. 2016 |
| Sea-level rise | Increasing | High | Church et al. 2013; Sweet et al. 2017; Dangendorf et al. 2019 |
| Tropical storms | Strengthening and intensifying faster, becoming bigger and slower, with more intense rainfall | High | Kossin et al. 2017; Bacmeister et al. 2018; Knutson et al. 2015 |

for the watershed of interest. The different datasets have their own specific sets of caveats and guidance regarding data quality, methods for addressing missing datapoints, fidelity of recording instruments, uncertainty associated with projections, and applicability across space and time. Climate-service data managers and analysts can help practitioners by conducting basic data-quality analyses (per the Río Conchos, Case Study 4.1) and by standardizing data in an appropriate, user-friendly format.

Climate service providers can also help garner additional resources to conduct applied science research or experiments on topics of mutual interest. For example, in the United States, the Western Regional Climate Center's Program for Climate, Ecosystem and Fire Applications (http://cefa.dri.edu/) generates fire and smoke forecasts and climate-fire analysis products. The USGS's Climate Adaptation Science Centers (CASCs) provide guidance to resource managers on using the latest climate projections and offer trainings, webinars, podcasts, fact sheets, and other information that may be useful to resource management practitioners and the public with whom they work.

In Australia, Mexico, and the United States, a variety of regional and national services can provide stream restoration practitioners some or all of the above-mentioned resources. Appendix B identifies selected climate services from these three countries and describes the help and types of services they offer and how to find them. Although the geographic focus is on these three countries, the information summarized in Appendix B may still be useful outside these regions. For example, some agencies, such as CONAGUA, provide climate services outside national boundaries (e.g., for the U.S.-Mexico border region). U.S. climate service providers include the CASCs, National Oceanic and Atmospheric Administration's (NOAA's) Regional Integrated Sciences and Assessments (RISAs), state climatologists, and various other regional centers hosted by universities, nonprofits, and state and federal government agencies. These organizations produce information and products inspired by and sometimes in direct collaboration with the needs of end users. Some of these entities have similar missions and provide similar services but differ in terms of the end-user communities with which they work, the degree to which they partner with end users, and the scope of their activities (e.g., data collection, climate education, and training).

In addition to the organizations listed in appendix B, local universities are often a trusted source of climatic information, and may host useful tools, such as Droughtview from the University of Arizona (https://droughtview.arizona.edu/), a list of visualization tools from the Applied Climate Science Lab at the University of Idaho (http://climate.northwestknowledge .net/gallery_vis.php), and a Digital Climate Atlas for Mexico from Universidad Nacional Autónoma de México (http://uniatmos.atmosfera.unam.mx/ACDM/servmapas). Another valuable university resource is the researchers themselves, who are often willing to collaborate with practitioners on issues of mutual interest.

## Strategies for Applying Large-Scale Climate Information to the Subbasin Scale

Observed climate trends, seasonal predictions, and long-term projections need to be applied at the scale of the watershed where a stream restoration project is taking place in order to be most useful and relevant. The challenge is that climate information is often provided at regional or broader spatial scales, which, in the case of climate projections, often requires downscaling from a regional to a subbasin scale. In addition, stream practitioners may want to do additional analyses, such as incorporating historical climate data and future projections into a hydrological model to estimate how future temperature and precipitation changes may impact evaporation, runoff, streamflow, and other variables of interest.

However, there is no need to immediately dive into expensive and data-intensive modeling. Simple and cheap strategies are available that can provide insight into what climate change means for your area of interest and a sufficient knowledge base for climate adaptation, while identifying information gaps that may be a priority to address at a later date when resources are available (as in Case Study 4.4). Regardless of the strategy used, well-developed restoration objectives will help determine how much climate information is necessary to achieve your goals.

This section reviews several strategies that allow practitioners to better understand climate change and its impacts at subbasin scales. Selecting the most appropriate method will depend on the climate-related questions that you feel you need to address before moving forward with your stream restoration project, your available resources, and the level of support you have from climate service organizations. The aim is to select a strategy that will provide the information needed to adjust your restoration objectives and tactics in a manner that increases likelihood for long-term success in a changing climate. Depending on what your investigation reveals, you may need to make no changes, make minor changes (for example, the quantity or type of species to plant for a riparian revegetation project), or make significant changes, such as refocusing restoration efforts to different drainages or dramatically changing your stream restoration objectives (e.g., from reestablishing obligate riparian plants to nonobligate riparian plants).

## Using Observations to Characterize Current Conditions and Past Trends

Comparing current precipitation and temperature patterns to historical trends is central to understanding how climate change is already impacting your watershed. The homework you did in chapter 3 to investigate sources of natural resource data (including climate data) that may be available for your watershed is important here. If you are lucky, your watershed will have weather stations with long-term records of temperature and precipitation (and sometimes other variables, including humidity, wind speed and direction, solar radiation, and cloudiness). Not surprisingly, the longer the period of record, the more likely you will be able to develop a detailed and nuanced picture of past and current climate conditions, including trends in variability and long-term averages.

Finding weather stations in your watershed that are managed by federal agencies can be relatively simple. Global daily weather station data from the Global Historical Climatology Network (GHCN) are available free of charge from the U.S. National Centers for Environmental Information (NCEI) via their "Climate Data Online" tool (Peterson and Vose 1997). Agencies in Australia and Mexico that can provide this information are the Bureau of Meteorology (BOM) and Servicio Meteorológico Nacional (SMN), respectively. In addition to providing climatic data, these same agencies (as well as other agencies listed in appendix B), local academic institutions, and climate service providers may also be available to help practitioners fine-tune their climate-related questions and conduct necessary analysis.

Standard datasets such as GHCN undergo substantial testing and quality control before publication, but many local or regional datasets may have useful data without quality control. As you and your restoration team accumulate additional resources and expand the footprint of your restoration response, analyzing climate data that are available in your watershed may become more of a realistic priority. If this is something that you take on, be wary of numerous errors or artificial changes that can creep into these data sources. These errors are usually not large in an absolute sense and may not be worrisome if you are simply trying to characterize the climate of a region. However, if you are attempting a more sophisticated analysis (e.g., characterizing the rate of climate change in a region, or the return period of a rare event such as a one-in-ten-year rain event), systematic errors of one degree or a few millimeters in the period of record can make the difference between inferring a significant trend and no trend.

Systematic errors typically occur due to changes in the observing location, type of instrument used, time of observation, the vegetation beneath and around the instrument, and the surrounding human and natural environment. In a data time series, changes caused by these and other factors are called *inhomogeneities*. When you obtain data, be sure to find out whether they are raw or have been "homogenized" or "adjusted" as the GHCN data have been (see the data-quality analysis conducted as part of the Río Conchos Case Study 4.1 as an example). Such corrections are automated, and while they generally work well, if you have any doubts you may consider reviewing the station histories and utilizing only those stations that have consistent long-term records.

To analyze temperature records, choose stations that are located in environments that are similar to your study area. Urban stations are affected by heat islands caused by the urban infrastructure, so they will not provide good proxy data unless you are restoring an urban stream. Despite their name, weather stations are not always stationary. Beware of those that have been moved up or down in altitude, from inside to outside a town, or have had a town develop around them. Such changes can introduce step-changes in the time series and lead to false interpretations of the actual temperature change.

Similar types of errors can be introduced for precipitation records if a station's location has changed relative to major topographical features, especially from being shifted between the windward and leeward sides of a mountain or changing elevation. For example, as part of McRoberts and Nielsen-Gammon's (2010) investigation of droughts in the Big Bend region of western Texas, they noted that the weather station in the town of Alpine moved at least ten times in its 110-year history, sometimes located on the valley floor and sometimes on the slope of a hill, creating artificial jumps in precipitation amounts each time. In contrast, the nearby COOP station at Mount Locke moved once in 75 years, providing a more stable record of temperature and precipitation. However, since this station is located on a mountaintop, it provides suitable data for local montane ecosystems but not lowland desert ecosystems. Homogenization of data attempts to minimize the impacts of these sorts of changes on the climate record.

If quantitative data are not available for your watershed, it is possible to get an idea of historical change from photographic records or through discussions with the local community. Gridded observational datasets often merge multiple sources of observational data (as from stations, satellites, and aircraft) into a single dataset, using statistical interpolation techniques (e.g., Higgins et al. 1997; Feng et al. 2010). They allow the user to obtain meteorological data for their location, even if weather stations are not nearby. The resolution of these datasets can range from a few miles or kilometers per side (for a national dataset, such as Livneh et al. 2013) to a relatively coarse five degrees per side (for an international dataset such as the Global Precipitation Climatology Project [GPCP]). They can be daily or monthly (e.g., Adler et al. 2003; Mitchell and Jones 2005).

Another commonly used gridded observational dataset is the PRISM (Parameter-elevation Regression on Independent Slopes Model) dataset produced by Oregon State University (http://prism.oregonstate.edu). PRISM uses weather station observations with at least twenty years of data and accounts for variations in weather and climate due to complex terrain, rain shadows, elevation, and aspect (Daly et al. 1994). PRISM data is monthly since 1895, and daily since 1981.

Reanalysis is a method that incorporates all available observational data every six to twelve hours into a weather model, producing continuous time series extending from the present time back to the mid-1900s or even further, depending on the dataset. The results, which are interpolations between available data, represent the state of the atmosphere and the earth's surface at each time step and on a gridded scale. Some commonly used gridded reanalysis datasets include the North American Regional Reanalysis (Mesinger et al. 2006), NCEP/NCAR Reanalysis 1 (Kalnay et al. 1996), and ERA-interim (Dee et al. 2011). Reanalyses.org provides access to and allows for comparison of different reanalysis datasets. Reanalyses typically do not use precipitation observations but instead simulate precipitation from the analyzed weather conditions.

As good as these gridded datasets are at producing data with a consistent spatial and temporal resolution and filling gaps where observations are lacking, they are only as good as the observations that inform them and are only accurate to the scale of multiple grid cells, which makes it challenging to apply them to small spatial areas. Changes in observation type (for example, satellite vs. aircraft), observation time period, or location can affect their accuracy; these limitations should be carefully considered when using gridded data (Daly 2006; Dee et al. 2016). Despite these challenges, these datasets can be extremely useful for identifying past patterns and current trends for climate change adaptation planning. For example, Merritt and Poff (2010) utilized PRISM precipitation data to inform their study of tamarisk distribution in riparian corridors in the western United States. The authors measured recruitment and abundance at sixty-four sites along thirteen perennial rivers, and the PRISM gridded dataset allowed them to easily look at past precipitation over these large areas, as opposed to choosing specific weather station sites which only provide data for a single location. In another example (González et al. 2017), researchers used PRISM temperature and precipitation datasets to assess success of tamarisk control treatments along rivers in six U.S. states.

If data and observations are lacking for your watershed, gridded databases can help to identify nearby basins with similar climates where more extensive local observations might be available. Those proxy observations may not capture the same weather information, but may reflect similar weather and climate statistics and long-term trends.

## Using Paleoclimate Data to Investigate a Broader Range of Climate Variability

Instrumental records, though useful for describing recent climate, only span a small fraction of Earth's history. Analyses of prehistoric or paleoclimate data can give practitioners a perspective on climatic variability that can provide important context to future climate forecasts. For example, in the early 2000s, the U.S. Southwest experienced one of the worst droughts in recorded history. One might infer from this information that the intensity, frequency, and duration of future droughts will be similar to these conditions. However, reconstructions of past climate showed that longer, more severe "megadroughts" have occurred over the past 2,000 years (e.g., Woodhouse and Overpeck 1998; Cook et al. 2004; Routson et al. 2011). Although not good news for the southwestern U.S., these paleoclimate reconstructions provide a more realistic range of possible future conditions that practitioners can use to develop climate-resilient restoration projects.

We are not implying that you need to become directly involved in paleoclimatic studies, but that you should look for past and ongoing paleoclimatic studies that can more deeply inform your understanding of your watershed's climate. Obtaining paleoclimate data from tree-ring analysis (dendroclimatology) is a common practice throughout midlatitude areas. Such analyses offer annual, and sometimes seasonal, reconstructions of past climate conditions. For example, Cleveland and colleagues (2011) used tree-ring analysis to determine the Palmer Drought Severity Index (PDSI) in western and central Texas in the past. The PDSI is a commonly used drought index that combines the effects of precipitation (water gain) and temperature (evaporative water loss); it is discussed later in this chapter. They found that the drought of the 1950s, which most Texas water planners use as a worst-case scenario, was neither as long nor as severe as other droughts that occurred in the region over the past five hundred years. This finding is consistent with other climate reconstructions that have been made in recent years for the southwestern United States.

Tree-ring analysis can even provide the means to reconstruct paleo streamflow, particularly related to annual or seasonal flow of large rivers with perennial flows (e.g., Gray et al. 2011; Margolis et al. 2011; Allen et al. 2015). A well-known example is for the Colorado River in the southwestern United States, where tree-ring streamflow reconstructions have helped identify the troubling fact that the time period in the early twentieth century when the river water was formally allocated was one of the wettest periods in the past five centuries (Woodhouse et al. 2006). Scientists have also used tree-ring analysis to reconstruct fire histories (e.g., O'Donnell et al. 2010; Falk et al. 2011).

Paleoclimate data can also come from pollen and dust deposits, stalactites and stalagmites in caves, sediment cores, and even packrat middens (e.g., Thompson and Anderson 2000; McCarthy and Head 2001; Metcalfe et al. 2010). These records usually have a coarser temporal resolution than do tree rings but still provide valuable evidence of ecosystem change related to climate and other factors.

There are abundant online resources for paleo data, including the NOAA Paleoclimatology Program (https://www.ncdc.noaa.gov/data-access/paleoclimatology-data/datasets), which houses reconstructions of climate, streamflow, and fire for the entire globe (it is best to stick to reconstructions, as raw datasets can be difficult to interpret). The NOAA's Paleoclimatology Program also features datasets of drought indices, most notably the PDSI, including a North American Drought Atlas, Mexican Drought Atlas, and Eastern Australia and New Zealand Drought Atlas (https://www.ncdc.noaa.gov/data-access/paleoclimatology-data/datasets/tree -ring/drought-variability). TreeFlow (http://www.treeflow.info/), produced by the Western

Water Assessment (WWA) and Climate Assessment for the Southwest (CLIMAS) RISAs, is an especially relevant resource for riparian scientists and practitioners in the southwestern U.S. and northeastern Mexico (for more information, see appendix B under RISA).

## Using Climate Model Projections to Quantify Future Trends

Two of the central climate-related questions that you as a practitioner will want to answer are: (1) What will the climate of your watershed be like in the future? and (2) Will future climate conditions be conducive to supporting key hydroecological thresholds* that are important for the long-term viability of your stream restoration objective (e.g., the survival and fecundity of the native fish [or plant] being reintroduced into your stream), or will climate change be so rapid and extreme to put in question the validity of the entire stream restoration program? Getting at these questions requires projections of future climate.

As stated earlier in this chapter, the primary drivers of climate change today are human emissions of heat-trapping gases from fossil fuel use, deforestation, and agriculture. For that reason, future climate projections begin with scenarios of how human activities, based on the energy and land-use choices made in coming decades, may affect the climate. These scenarios typically cover a broad range of possibilities, from a future where emissions continue to grow, to one where net emissions decrease and even become negative† before the end of the century. Over the next few decades, which scenario the world follows makes little difference to projected changes in climate at regional-to-local scales, but by the midcentury and beyond, changes can be very different depending on which scenario is used. Hayhoe and colleagues (2017) provide a comprehensive and up-to-date discussion of the different types of scenarios scientists use to develop climate projections, including the Special Report on Emission Scenarios, Representative Concentration Pathways (RCPs), and Global Mean Temperature scenarios. The first two types of scenarios are used as input for global climate model (GCM) simulations to calculate the impact of a given future scenario on global- and regional-scale climate; the last type is calculated from the results of these simulations. Since no likelihood can be assigned to any given scenario (the likelihood being determined by human actions and choices which are nearly impossible to predict with certainty), it is recommended that practitioners consider at a minimum the changes resulting from both a higher and a lower scenario for projects with time horizons longer than two or three decades. A lower future will quantify the changes you'll need to prepare for even if the world significantly reduces its emissions, consistent with targets such as the Paris Agreement; a higher future will quantify the changes that will occur if little to no action is taken, as well as the benefits that would result to your system from reducing emissions.

---

* If you are not familiar with hydroecologic thresholds, background on importance of identifying key hydroecologic thresholds is presented in chapter 5 (see the first section: "Assessing the Need for an Environmental Flow Program").

†Negative emissions will occur when humans take up more carbon than they produce. This can be accomplished through traditional practices including reforestation, carbon farming, and managed grazing as well as through modern technology that can capture carbon directly from the air and sequester it or turn it into a range of products, from stone to fuel.

GCMs are complex physical models that simulate the most important processes at work in the climate system, including in the atmosphere, the ocean, the land surface, and the cryosphere (ice and snow). They are used to study climate in the recent and distant past, to simulate conditions on other planets, and—when driven by the future scenarios described above—to quantify projected changes as a result of human activities (Hayhoe et al. 2017). GCMs divide the planet into three-dimensional grids which typically range from 25 to 500 km (around 15 to 310 miles) per side. Within each grid cell, the models use physics and chemistry to compute the dynamics of the atmosphere, land, and ocean, including such parameters as changes in sea ice, large-scale heat and water transport, and the effects of aerosols (small particles such as dust and soot) on precipitation.

GCMs are typically used to simulate past and future changes at large spatial scales that can be used to resolve regional trends in average annual and seasonal temperature and precipitation as well as extreme events such as heat waves, droughts, and atmospheric rivers. Regional, national, and global maps based on GCM simulations are available through assessments such as those undertaken by the U.S. National Climate Assessment and the IPCC. For many applications, simply knowing the direction of the projected trend and its approximate magnitude is sufficient (for example, "runoff is projected to increase by 10 percent and heavy precipitation by 60 percent") and the general information provided by regional, national, and international assessments is helpful.

The challenge for many stream restoration efforts is that GCMs fail to resolve many small-scale geographic and atmospheric features, such as the valleys and peaks in mountainous areas and convective rainfall systems, that affect climate and streamflow at the watershed scale. To address this challenge, a process known as "downscaling" is commonly used to translate higher-resolution information.

There are two approaches commonly used to generate high-resolution climate projections by downscaling GCM output to a given location or region from GCM output: (1) dynamical downscaling with regional climate models (RCMs), and (2) empirical statistical downscaling models (ESDMs) that use a statistical model to bias-correct GCM output based on long-term observations.

RCMs cover a limited rectangular area that typically encompass a region or a country. They take GCM output at the boundaries of their area and combine it with much more localized information on topography and physical processes to produce climate information at a spatial resolution typically ranging from 10 to 50 km (6 to 30 mi.) (Hayhoe et al. 2017). These physical models are data-intensive and computationally complex but their output is becoming more readily available in numerous parts of the world. For example, RCM output for a limited number of future scenarios and GCMs is available for North America from the North American CORDEX project, and more simulations are expected to be added in the future (NA-CORDEX, https://na-cordex.org/). For Australia, projections from both dynamically and statistically downscaled GCM output are available on the Climate Change in Australia website (https://www.climatechangeinaustralia.gov.au/en/).

ESDMs use statistics to build relationships between fine-grain local observations and coarser, more regional historical GCM simulations. They combine these relationships with GCM projections of the future to remove the bias or offset from GCM simulations relative to observations and produce site-specific climate projections at the spatial resolution of the original observations (Wilby et al. 2004; Hayhoe et al. 2017). A comprehensive description of ESDMs and a list of available datasets for the United States is provided in Kotamarthi et al. (2016, 2020).

Many datasets of both RCM and ESDM projections can be downloaded online for free. No single best method can be recommended for every analysis. Both dynamical and statistical downscaling have unique strengths and limitations, and understanding these limitations is essential to an accurate interpretation of any resulting information that is used (see the comprehensive discussion in Kotamarthi et al. 2016, 2020; see also Ekström et al. 2015 and Gutmann et al. 2014). The best approach for a given purpose depends on the resources, familiarity with the data and methods used, whether the dataset provides information at the spatial and temporal scales of interest, and the specific focus and questions of each individual application, including which climate variables are needed for your analysis (Mote et al. 2011; Ekström et al. 2015).

There are many resources that can be of assistance in choosing a downscaling method (Kotamarthi et al. 2016; see also USAID 2014; Ekström et al. 2015). Partnering with a climate scientist, statistician, hydrologist, or ecologist who has experience or expertise with high-resolution climate projections can help you identify the most appropriate method or dataset to use and keep you up-to-date with developments in climate models (USAID 2014; Harris et al. 2014). At minimal or sometimes no cost, many climate-service providers, such as those described in the previous section of this chapter, can help you identify the most suitable datasets of observations or high-resolution projections. If they are unable to provide this service, they can often connect you with others who can. If larger computational endeavors are necessary, they may offer to collaborate on a grant proposal that will provide the necessary supportive funding.

### Accounting for Uncertainty in Future Projections

Regardless of which method you use to understand what climate change means for your stream's watershed, keep in mind that there are several well-known uncertainties when projecting future climate (for a more detailed discussion, see Hayhoe et al. 2017). These include (1) *natural or internal variability* of the climate system and future changes in this variability; (2) scenario uncertainty that stems from human behavior and the decisions society will make that affect emissions, which become increasingly important after the midcentury, and more so for average and extreme temperature and extreme precipitation as compared to average and seasonal precipitation; and (3) scientific uncertainty, which relates to the ability of scientists to understand the climate system's response to human activities.

The climate experts and resources that you involve in your stream restoration project will help you to account for these uncertainties. Some questions that practitioners may have regarding the future climate of their target watershed can be answered using a specific scenario. For example, will a certain stream-restoration plan achieve its goals under the amount of change expected under a higher future scenario? Or, assuming the world is able to limit warming below the 2°C goal established by the Paris Agreement (at the 2015 United Nations Framework Convention on Climate Change), will your restoration plan still function as designed?

A few guidelines that you and your climate experts should take note of to account for the uncertainty inherent in future climate projections:

1. For projects with planning horizons shorter than twenty to thirty years, natural variability is the most important source of uncertainty; using different scenarios won't have a big effect. Using multiple GCMs and even multiple simulations from those GCMs is recommended, if possible. For projects with planning horizons longer than twenty to thirty years, however, it is important to use a higher and a lower future scenario to encompass the range of uncertainty in human choices and their effect on the amount and the rate of future climate change.

2. Given that the largest sources of uncertainty in regional precipitation projections are natural variability and climate model differences (Hawkins and Sutton 2011), we encourage the use of multiple climate models (not just one) to encompass most of the range of scientific uncertainty. When selecting which climate model to use, give priority to models from CMIP version 5 or later (Taylor et al. 2012; CMIP6 was released in 2020), as well as those with a long development history that are well documented in the literature and are currently on their third, fourth, or even fifth version (e.g., CCSM, CNRM, CSIRO, GFDL, Had, or MIROC). While attempts to identify a "better" or "best" GCM are generally discouraged,* one relatively minor refinement is to consider a model-weighting scheme, such as that used in NCA4, that accounts for similarities between different climate models as well as their basic performance (Sanderson et al. 2017).

3. Always group and average future simulations across climate timescales of twenty to thirty years. This period represents climate, while variation within this period provides a range to quantify uncertainty due to natural variability in the climate system and scientific uncertainty. Consider the range or standard deviation, not just the multimodel mean, and focus on restoration solutions that are robust across that range (restoration solutions that can cope with both the highs and lows of the range) (Harris et al. 2014; Nover et al. 2016).

4. All analyses (e.g., calculating seasonal precipitation or counting the number of dry days, extreme precipitation events, or heat waves) should be done on *individual* GCM simulations first. Then, you can calculate the mean and the range in projected values across all the GCMs used: this will give you insight into the uncertainty due to both natural variability and scientific uncertainty. But never average across different scenarios, as they represent different pictures of the future that are inconsistent with each other.

## Using and Developing Climatic Analogs

With a clearer understanding of what the future climate may look like for your watershed, the next step is to understand the consequences of climate on your watershed's natural resource conditions. Climatic analogs can be useful tools in this regard. Climatic analogs compare climatically similar sites across time (temporal analogs) or space (spatial analogs). Temporal analogs allow you to look at past climate events in your region or watershed of interest (such

---

* Since GCMs generally perform differently for different regions, for different variables, and even over different time scales (such as simulating seasonal variability versus long-term change), simple attempts to identify one or more GCMs as "best" for a specific watershed generally will not improve the quality of your future projections and may even decrease them.

as historic floods or drought) that are similar to those predicted to occur in the future and see how they affected natural resource conditions. Spatial analogs identify regions with biophysical characteristics similar to the watershed where your restoration site is located and that are currently experiencing climatic conditions similar to those forecast for your restoration area.

### Temporal Climate Analogs

Both instrumental and paleoclimate data can be used to identify temporal climate analogs. Analyzing how natural resource conditions were affected during the temporal analog periods that are identified provides a strong basis for adapting stream restoration objectives and tactics in ways that improve likelihood for long-term success. One of the inherent challenges in using the temporal analog approach is assuming the past is a good representation of the future. The complexity and unknowns associated with the cascading effects of climate on a diversity of natural resource parameters can produce erroneous conclusions. Nonetheless, as long as temporal analogs can be identified for your watershed, and as long as information on natural resource conditions during the analog periods is available, this method can be a particularly useful method to better adapt your stream restoration project to a rapidly changing climate.

In the Big Bend region of the Chihuahuan Desert, McRoberts and Nielsen-Gammon (2010) identified recent drought analogs using weather-station data. They looked for years that had both low precipitation totals and above-normal temperatures. For winter season analogs, they found the years 2000 and 2006 had precipitation and temperatures comparable to what is predicted to be the average in the near future, and for summer season the analogs were found to be 1934 and 1998. Each of these analogs occurred when temperatures were about 1.7°C (3°F) above normal, representative of projections for the next few decades. However, they found no analogs that characterized droughts during an unusually warm near-future year, or for droughts occurring in the middle to late twenty-first century, when temperatures are projected to be much warmer than 1.7°C above normal. For the south-central United States, the spatial extent and magnitude of the number of days per year over 100° F in 2011 was very similar to what is projected to be the average summer by midcentury (Karl et al. 2009). Thus, in terms of extreme heat, that year can be a good analog for future average conditions.

For analogs of a more distant future climate, or of "worst-case-scenario" droughts, it is perhaps more useful to use paleo analogs or spatial analogs (see *Spatial Climate Analogs* below) (McRoberts and Nielsen-Gammon 2010). As discussed previously, paleoclimatic analyses expand the climate window to include prehistoric droughts that may have been longer and more severe than droughts that took place during the period of record. This can make them good analogs for future droughts. For example, the mid-twelfth-century drought that occurred in the southwestern United States was anomalously warm, severe, and long, and occurred over a very large area. As such, it may serve as a conservative analog for future droughts more severe than any recorded over the last century (Woodhouse et al. 2009).

Most paleo-drought studies reconstruct the PDSI, a commonly used drought index that combines the effects of precipitation (water gain) and temperature (evaporative water loss). The fidelity of the analog will depend on how well each of those components matches future conditions individually. A drought driven by precipitation will have different characteristics from a drought driven by temperatures (Udall and Overpeck 2017). Important parameters

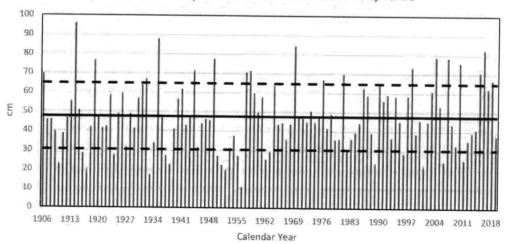

**FIGURE 4.3** Total annual precipitation at Del Rio, Texas, from the years 1906 to 2019, with one standard deviation (dotted lines) around the mean (dashed line) for the period of record. Access to long-term precipitation records can help identify temporal analogs for climate conditions forecast for your region. In this example, if your stream is located in the Del Rio Region, understanding how stream hydroecological conditions were affected by the drought years 1950 to 1956 can provide insight as to how stream conditions may respond to the drier, hotter conditions that are forecasted for the future and what might constitute realistic restoration objectives for your stream given those future conditions.

such as average streamflow, peak streamflow, timing of peak streamflow, and water temperature will react differently to changes in temperature and changes in precipitation.

The analog period for an extreme precipitation event can be as short as a single day or week. Ideally, however, the analog period for a drought should be long enough to provide insight into the environmental response (at least several years). In this sense, the longer the analog period, the better. The amount of natural resource information and data that are available will depend on the watershed and analog period selected. In general, the further back in time the analog period is, the more difficult it will be to find data and information. If your stream has a streamflow gauge, and if the gauge record includes the selected analog period, this would provide great insight as to how streamflow was impacted during the analog period. Did streamflow continue during the hot, dry months of the analog period or did it disappear completely? Such insights can be crucial to understanding how future climate conditions may impact your restoration objectives. If historical aerial photography is available for the analog period, practitioners may be able to glean insights as to how natural (and human) resources were affected both during and immediately after the analog period. Historical photographs can be used to assess riparian plant communities, presence of streamflow and pools, channel conditions, and so on. If the analog is not too far in the past, there may be residents who lived through the period and can provide insights as to how the climate during the analog period impacted both them and natural resources.

Whether the source of information is a streamflow gauge, photographs, people, or a combination thereof, the aim is for you to better understand how your stream environment was affected by a past event and, based on those insights, how your restoration objective may be impacted in the future by similar climatic conditions. For example, in northern Chihuahua and northern Coahuila in Mexico, and western Texas in the United States, the severe drought of the 1950s is a good analog to understand how both human and natural systems responded to conditions that were dramatically drier and hotter than normal. As this drought took place during recorded history, insights into ecosystem response can be gained from many of the sources mentioned above (see figure 4.3).

## Spatial Climate Analogs

The advantage of spatial analogs is that they allow all manner of direct study into current biophysical conditions and trends, as well as development of insights and even experiments regarding what management actions might be most effective in reducing the vulnerability of key natural resources to the impacts of climate.

The challenge of this approach is selecting a watershed that has biophysical characteristics that are sufficiently similar to your target watershed. Climate characterizations tend to include indicators of temperature, precipitation, or a combination of both. Commonly used climate variables include mean annual temperature, mean temperature of the coldest month, growing degree days (the number of days above a minimum temperature that allow for a particular plant species' development) above 5°C (41°F), annual precipitation, and annual water deficit (see, for example, Ohlemüller et al. 2006; Bergmann et al. 2010). When finding analogs for urban areas, indicators can include an index of aridity, heating degree days (a measurement that quantifies the demand for energy needed to heat a building), and cooling degree days (a measurement that quantifies the demand for energy needed to cool a building) (e.g., Kopf et al. 2008). Using summer heat index, for example, Frumhoff et al. (2007) identified analogs based on how hot the summers would feel to inhabitants of the northeastern United States. For example, they found that by midcentury, the average Massachusetts summer would resemble that of New Jersey today under a lower future scenario, and that of Maryland under a higher future. By the end of the century, summers in the greater New York City area could resemble those of Virginia today under a lower future scenario, and South Carolina under a higher one.

The high-resolution PRISM climate analyses for the United States (http://prism.oregonstate .edu/) produced by Oregon State University can be very useful for identifying spatial analogs. After determining what the future climate will look like in your watershed, you can use PRISM tools (such as the comparison tool) to identify areas that are currently experiencing those climate conditions. For Australia, the Climate Change in Australia website contains an "Analogues Explorer" tool (https://www.climatechangeinaustralia.gov.au/en/climate-projections /climate-analogues/about-analogues/). The tool uses annual average rainfall and maximum temperature to match the projected future climate of a region with a region that is currently experiencing those conditions (CSIRO and Bureau of Meteorology 2015).

Given the inherent similarity in biophysical and chemical conditions and processes between different locations in the same watershed, selecting spatial analog sites within the same watershed is often better than sites in different watersheds. This can potentially be accomplished by simply a change in elevation from your target site. Generally, as elevation decreases, temperatures increase and precipitation decreases. Thus current conditions at a lower altitude site may be similar to what is projected for a higher altitude site (e.g., Hallegatte et al. 2007) in a warming climate. The Southwest Experimental Garden Array (http://sega.nau.edu/) in the southwestern United States is an experimental field site, spanning a large elevation gradient, that uses the spatial-analog strategy for quantifying the ecological impacts of climate change. By transplanting species downslope at sites that are an effective analog for the future, they study plant response to climate warming. Depending on location, each degree Celsius of warming can be equivalent to two hundred meters of elevation change. Lower-altitude drainages can thus serve as proxies for future higher-altitude drainages.

## ASSESSING VULNERABILITY AND CLIMATE ADAPTATION

### Assessing Vulnerability

Climate change adaptation is deeply tied to the concept of vulnerability—the extent to which a species, ecosystem, or area is likely to be harmed as a result of climate change and associated stresses (Brooke 2008). The main purpose of assessing the vulnerability of species, human communities, or different management actions to climate change is to establish measures that will reduce the likelihood of harm while also promoting actions that will bring the most benefit (Chambers 1989). In its simplest form, a climate change vulnerability assessment examines the sensitivity of a current decision or set of decisions to the projected climate change impacts (Klein et al. 2001). Once you understand the vulnerability of your system to climate change, you are in a position to identify appropriate solutions.

In the context of stream restoration, the vulnerability question revolves around the extent that stream restoration objectives and the restoration tactics to achieve them may be compromised by the impacts of climate change. How likely is it that climate change will prevent your restoration objectives from being achieved? While there is no algorithm that will generate a definitive answer to this question, much insight can be gained by looking at the potential impacts to the stream's flow regime.

Streamflow is derived from a combination of surface runoff, soil water, and shallow groundwater, all of which are driven by precipitation and temperature. Climate, geology, topography, soils, and vegetation all affect the pathways by which precipitation reaches the stream channel and, ultimately, the stream's flow regime, which is described by five main components: discharge magnitude, frequency, duration, predictability, and flashiness (Poff et al. 1997). *Discharge* is simply the volume of water moving past a fixed location per unit of time (e.g., cubic meters per second, cubic meters per month, or cubic meters per year). *Frequency* describes how often a flow greater or equal to a given magnitude occurs over a specific time period (e.g., a one hundred–year flow event describes a discharge that occurs on average once every one hundred years). *Duration* is the period of time associated with a specific flow condition (e.g., the average number of days during the course of the year that a flow of specified discharge occurs). *Predictability* is the regularity of occurrence of different flow events, often specified high and low flows deemed to be particularly important to stream biota or habitat (e.g., a high flow of sufficient discharge and duration to evacuate and move stored sediment downstream). And *flashiness* describes how quickly flow changes from one magnitude to another (Chow et al. 1988; Gordon et al. 1992; Poff et al. 1997; see also chapter 3 for greater detail).

Climate change can impact all five flow regime components, which, in turn, can affect such primary regulators of stream ecosystem integrity as water quality, energy sources, physical habitat, and biotic interactions (Poff et al. 1997; Poff et al. 2010). Given that restoration practitioners are working on streams whose ecological integrity is already compromised, the impacts of climate change can compound deteriorating trends in a stream's flow regime that may be difficult or impossible to disentangle from other drivers of change. For example, streamflow that is already diminished from its natural state because of municipal or agricultural flow diversions or impoundment may be compromised further by future droughts that occur more frequently, are of longer duration, and are more intense. In such a future climate, a perennial stream may transition from perennial to intermittent, or even to ephemeral.

Such dramatic changes in stream hydrology often carry with them ecological and societal tipping points that can dramatically affect stream biota and the ecosystem services that streams provide to humans. In addition, such changes can affect the likelihood of realizing restoration objectives (distinguishing what is realistic and what is not) and the restoration tactics best suited to achieve them. A restoration objective that aims to reestablish native fish will be much harder to achieve if climate change impacts compound already occurring trends that are fostering reduced flow. If reduction in flow is forecast to be significant, stream practitioners may need to consider a variety of climate-adapted options that include changing restoration strategies (such as developing an environmental flow program to protect streamflow from human interferences), modifying restoration objectives (perhaps focusing more on fish species that are adapted to surviving the hot, dry season in isolated pools or dropping fish reintroduction altogether in favor of a different ecological focus), or moving the location of their restoration project areas where climate impacts will not be so severely felt.

Regardless, understanding how climate change may impact your stream's flow regime is predicated on a sound understanding of current conditions and trends of your stream and its watershed.

## Strategies for Assessing Vulnerability

In the context of climate change, vulnerability has three components: sensitivity, exposure, and adaptive capacity (Dawson et al. 2011; Glick et al. 2011). *Sensitivity* refers to innate characteristics of an organism or ecosystem (e.g., tolerance to temperature changes) that predispose it to being more or less susceptible to climate change. *Exposure* refers to the amount of change either in climate or in climate-driven factors that a species or system will experience. *Adaptive capacity* is the ability of a species or system to reduce the impacts or take advantage of climate change. Documenting aspects of all three of these vulnerability components informs adaptation planning by providing valuable insights about which species or systems will be affected most significantly by climate change (Glick et al. 2011; Lawler et al. 2013).

Strategies for assessing the effects of climate change on streamflow vary widely. On one end of the spectrum are sophisticated and relatively expensive approaches based on a range of

climate projections, which enable experts to model changes in a variety of streamflow parameters (see Case Study 4.3). On the other extreme are less sophisticated, qualitative approaches
that rely on expert opinion and field observations (see Case Study 4.4). Both approaches can
provide considerable insight into understanding the impacts of climate change on streamflow.

In Australia, standard approaches have been developed for assessing changes in streamflow. For example, the Australian Rainfall and Runoff guide (arr.ga.gov.au) provides a standard
method to link rainfall intensity and temperature. Given a projected temperature change, it is
possible to calculate the rainfall intensity increase. This is useful for estimating the change in
flood frequency. Similarly, there are standard rainfall sequences that can be used to estimate
changes in daily and annual flows (State of Victoria Department of Environment, Land, Water
and Planning 2016).

The most time-consuming and costly steps in developing hydrological models capable of
forecasting potential changes to a stream's flow regime is often the collection and organization of hydrological data and the calibration of the model. Many years of data collection by
streamflow gauges are needed to adequately describe a stream's flow regime. The longer the
period of record, the stronger the foundation for modeling potential changes brought on
by climate change. For watersheds that lack or have limited streamflow records, a variety of
methods can be used to provide the required data, each with its own inherent cost and error
(see chapter 3).

Several recent studies have assessed the impact of climate change on streamflow and are
good examples of how to apply these methods to stream restoration projects. These studies
also suggest strategies for mitigating climate impacts and ways that the results can be presented to inform policy.

- Dey and Mishra (2017) provide a comprehensive review of methodologies and critical assumptions employed by various hydrological models that attempt to differentiate between climate
  and human impacts on streamflow.
- Jiang and colleagues (2007) compared the results of six hydrological models that simulated the
  hydrological impacts of climate change in the Dongjiang Basin in South China, the source of
  about 80 percent of Hong Kong's annual water supply. The hydrological models were run using
  historical climate data to simulate current water-balance components and were compared to
  scenarios that incorporated hypothetical climate change scenarios to simulate the impacts on
  hydrology.
- Pervez and Henebry (2015) used the Soil and Water Assessment Tool (SWAT) as part of their
  study to evaluate sensitivities and patterns in freshwater availability due to projected climate
  and land-use changes in the Brahmaputra Basin in south Asia. SWAT is well-documented
  in the literature (Ullrich and Volk 2009; Arnold et al. 2010; Sun and Ren 2013) and has been
  applied to evaluate a variety of parameters (e.g., erosion, sediment transport, pesticide transport, water management) as well as to assess the impacts of climate change (Jha et al. 2006).
- Serrat-Capdevila and colleagues (2007) used future average annual precipitation estimates and
  a realistic representation of spatial-recharge distribution to evaluate the long-term climate
  change effects on the water balance and impacts on both the aquifer and the riparian area of the
  San Pedro River—a transboundary watershed located in southeastern Arizona and northern
  Sonora, Mexico, that occupies roughly 9,450 square kilometers.

- Mantua et al. (2010) estimated summertime stream temperatures, seasonal low flows, and changes in peak and base flows in several watersheds in the state of Washington that provide key habitat for salmon. To accomplish this, they evaluated weekly summertime water temperatures and extreme daily high and low streamflow under multimodel composites for two future scenarios.
- A useful risk-assessment and decision-support tool for managing wetland and cave ecosystems that depend on groundwater was developed and tested in southwestern Australia, a global biodiversity hotspot and one of the earliest regions impacted by climate change. It was then modified to help manage similar ecosystems across Australia (Beatty et al. 2013; Chambers et al. 2013).

Although potentially worth pursuing for larger-scale, well-funded restoration projects, these modeling examples are likely outside the time and budget constraints of many. Understanding the value of undertaking such endeavors will be based on both the cost of conducting such modeling and the knowledge gained that allows practitioners the knowledge base to move confidently forward.

### A Rapid, Low-Cost Approach for Assessing Vulnerability

Regarding the above debate on pursuing high- or low-cost vulnerability assessment strategies for your stream, it needs to be emphasized that it does not have to be either/or. You can begin with a low-cost approach that provides great insights and is followed by higher-cost approaches that target priority gaps in knowledge. Case Study 4.4 (on pages 146–47) provides a low-cost approach that narrows the assessment of vulnerability to climate change to those elements of the stream's flow regime deemed most critical to the realization of restoration objectives. This approach may not provide all the information you need to characterize the vulnerability of your stream restoration project to climate change, but it can constitute a sound first step.

The results of the first-step approach to assess vulnerability of stream restoration objectives at 3 Bar Ranch indicate that the objective is vulnerable to climate change. That is, it seems likely that future climate conditions will foster scenarios where shallow groundwater frequently drops below tipping points required for the survival of obligate riparian plants. This is an important piece of information that encourages Terlingua Creek practitioners to take a hard look at how restoration objectives and tactics could be improved so the stream is better adapted to future climate conditions. We will turn our attention to adaptive strategies in the section immediately below, "Strategies to Strengthen the Climate Adaptive Capacity of On-the-Ground Stream Restoration Actions."

Before concluding this section, we want to note that the low-cost approach that was conducted at the 3 Bar Ranch will not provide all the information that practitioners may need to understand the vulnerability of stream restoration objectives to climate change. Certainly, hydrological studies (e.g., using approaches summarized in chapter 3), a longer monitoring period, and hydrological modeling will provide greater certainty about current hydrological conditions and trends. Nonetheless, such an approach can provide important information in a timely manner that, among other considerations, will help to identify knowledge gaps that more costly assessments may be able to address. In the Terlingua Creek example, long-term monitoring of initial revegetation results and shallow groundwater elevations will provide

# CASE STUDY 4.3
Using Vulnerability Assessments to Strengthen Climate-Adaptive Conservation Response to Climate Change

Megan Friggens

Vulnerability assessments are designed to organize and communicate information and reduce the "analysis paralysis" that can accompany climate change adaptation planning. In this case study, three unique but interrelated vulnerability assessments were conducted for riparian and aquatic systems of the Middle Rio Grande in New Mexico. The fifth longest river in North America, the RGB has been substantially modified with profound impacts for the river's hydrology and natural riparian forest (or "bosque") habitats (USACE 2011).

In addition to supporting agricultural and domestic needs, the RGB supports diverse flora and fauna, including critical habitat for the endangered southwestern willow flycatcher (*Empidonax traillii extimus*) and silvery minnow (*Hybognathus amarus*) and the threatened yellow-billed cuckoo (*Coccyzus americanus*).

Climate change is likely to exacerbate the challenges the river already faces from water deficits, invasive species, and pollution. Assessments that provide guidance on reducing negative impacts from a changing climate are therefore a priority.

The first Rio Grande vulnerability assessment (RGVA) was conducted in 2012 along the Middle Rio Grande in New Mexico (see map in this case study) as a collaborative effort of the Rocky Mountain Research Station, U.S. Forest Service (USFS), U.S. Fish and Wildlife Service (USFWS), Department of Defense, and The Nature Conservancy. In this effort, the System for Assessing Vulnerability of Species (SAVS) tool was used to identify the riparian obligate vertebrate species of the Rio Grande that are most vulnerable to climate change impacts and, based on this information, identify effective climate-adapted management actions (Friggens et al. 2013). As part of the assessment, 117 riparian obligate terrestrial vertebrate species were ranked according to traits that predict population trends under climate change. The ratings were intended to identify priority species as well as the factors that make them vulnerable (such as habitat, physiology, phenology, species interactions) and so prioritize targets for future management actions. The graphic on the final page of this case study shows vulnerability scores for bird species. Regarding sensitive amphibian populations, potential actions identified to reduce impacts of drought and warming temperatures included establishing vegetation to shade the water, erecting fencing to reduce erosion from grazing, reducing nonclimate stressors, and managing exotic species. The Middle Rio Grande assessment also noted common issues among species and helped to identify "low-hanging fruit" and best-practice options, based on changed species behavior in response to climate change (such as altered migration and breeding behavior).

The second RGVA combined elements of the trait-based scores computed in SAVS with an environmental niche model (ENM) to provide a geographically defined assessment of climate change impacts along the Rio Grande (Friggens et al. 2014; Friggens and Finch 2015). ENMs are commonly used to measure climate-related changes to species' habitat availability. They correlate a species' presence with environmental conditions (including climate), then use the modeled relationship to identify new suitable habitats under various future conditions. ENMs are especially useful for identifying which species are likely to be impacted and where those impacts are likely to occur. In collaboration with the Southern Rockies Landscape Conservation Cooperative, Rocky Mountain Research Station developed species niche models for 12 obligate riparian species along the entire Rio Grande in New Mexico and

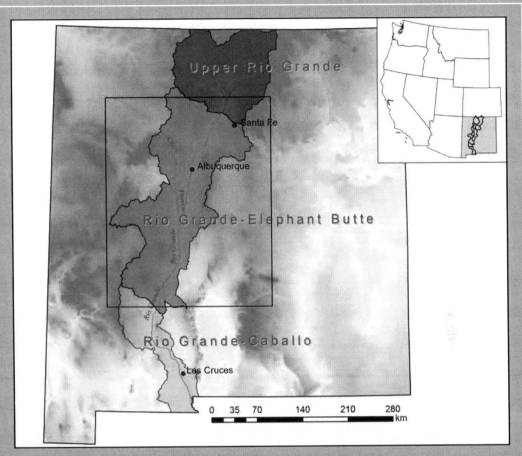

The Middle Rio Grande (indicated by the area within the square) comprises the area from the Cochiti Dam north of Albuquerque to the Elephant Butte Reservoir in the south.

The endangered southwest willow flycatcher. Vulnerability assessments for this species indicate it will be negatively impacted by climate-related changes in hydrology and wildfire. Its suitable habitat is projected to shift north. Photo from NRCS, USDA [https://www.nps.gov/grca/learn/nature /southwestern-willow-flycatcher.htm].

*(continued)*

projected future suitable habitat under three climate futures for two time periods (2060 and 2090). Results showed large shifts in suitable habitat for nearly all species by the year 2090 (see graphic below). For some, like the southwestern willow flycatcher (see photo in this case study), suitable habitats were expected to shift north and higher in elevation (Friggen and Finch 2015). Others, like the yellow-billed cuckoo, appear to lose habitat and are only expected to be found within a few refugia by the end of the century (Friggens and Finch 2015).

The third RGVA, initiated in 2015 in partnership with the Southern Rockies Landscape Conservation Cooperative, produced actionable science through a collaborative process wherein scientists,

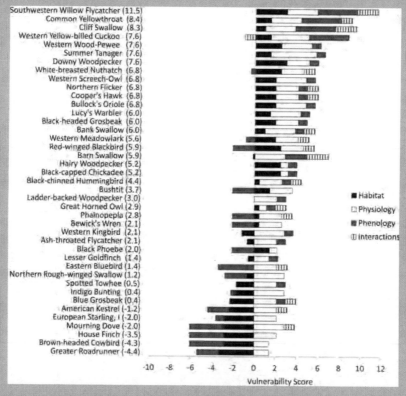

Climate change vulnerability for birds that breed within the Middle Rio Grande Bosque. Species are ordered least to most vulnerable (highest score) by overall vulnerability scores (−20 to +20) shown in parentheses. Colored bars represent scores for each of four primary mechanisms of vulnerability (−5 to +5). Overall vulnerability ranks identify both currently threatened species and species that may become increasingly threatened by climate-related changes in riparian habitat. Physiological sensitivity to drought and heat waves was commonly indicated (in physiology scores) across bird species. Resilience (as indicated by lowest, negative scores) was most commonly associated with expected increases in a species' habitat under climate change and for species that have flexible phenologies. (Asterisks indicate introduced species.)

managers, and stakeholders worked together to define the scope, measures, and methods used to produce and use science information (Beier et al. 2017). Through a series of workshops, webinars, and small group meetings, the Rocky Mountain Research Station developed a toolbox for landscape conservation design (described at https://www.fs.usda.gov/rmrs/science-spotlights/co-producing-landscape-scale-vulnerability-assessments) to assess aquatic systems based on spatially explicit data identified by local managers. This RGVA, which produced maps identifying subwatershed vulnerability, was conducted to inform future adaptive-management planning among a diverse group of stakeholders.

Reports and data products produced from all three RGVAs highlighted common goals among stakeholders by outlining high-priority areas within the landscape and specific issues of common interest. Efforts in other watersheds have further customized landscape planning and considered the social components of vulnerability (e.g., Furniss et al. 2013), an aspect not addressed in the RGVAs. It is important to remember that vulnerability assessments are only one step in a larger process of developing adaptive-management strategies.

## Lessons Learned

- Conducting vulnerability assessments can be challenging for two main reasons. First, it can be difficult to solicit vulnerability assessment targets from managers for a variety of reasons. In the Middle Rio Grande, we attempted to co-produce assessment targets and tailor them to potential end users, but the project was never able to get commitment for a final list of targets. Second, vulnerability assessments can require significant time and resources, making implementation difficult in many cases without additional funding.

- Vulnerability assessments are most effective when they are developed at a scale and scope appropriate to the system being targeted, and designed and implemented in close collaboration with restoration practitioners to meet their specific objectives and information needs (Beier et al. 2017).

- In the second RGVA, the greatest benefit of combining elements of the trait-based scores computed in SAVS with ENM was the capacity to identify the areas that were most likely to support species under a variety of future potential climate scenarios.

- The structure and application of a vulnerability assessment will differ depending on whether the objective is to identify priority species within an ecosystem, identify refugia under climate change, or assess the effectiveness of current management practices under future scenarios. If it is sufficiently flexible, it will allow a level of customization so that it can address a wide range of potential management questions, including developing proactive strategies to prevent additional species listings, identifying potential refugia and biodiversity hotspots, and nurturing potential partnerships between land owners and land management agencies (Sadoti et al. 2017; Wiles and Kalasz 2017).

- Successful utilization and application of information derived from a vulnerability assessment requires a formal plan to incorporate results into management planning. Discussion should consider how vulnerability assessment results will be presented to managers at the beginning of the planning process so that there is sufficient support for implementing the plan.

- Although the RGVAs were informative and have inspired additional studies of climate change impacts in the region, they are likely to remain underutilized within local management communities without some effort to support additional engagement between scientists and managers to identify specific actions for species and habitat conservation under climate change.

practitioners much greater insight regarding current water availability conditions along Terlingua Creek at the 3 Bar Ranch.

To sum up, the following elements are needed to conduct the vulnerability-assessment approach outlined above:

- A sound stream-restoration objective (chapter 2).
- A sound understanding of current streamflow conditions and trends (chapter 3). Of particular importance is knowledge about the physiological requirements of the species that are central to realizing the restoration objectives to changes in stream hydrology. In the Terlingua Creek case study, the focus was on the tolerance of obligate riparian trees to a decline in the elevation of the soil saturation zone. Depending on your stream restoration objectives (whether objectives concern obligate riparian plants, fish, mammals, or something else entirely), other tipping points could be related to the duration of baseflow, frequency of flood flows, temperature, water quality, or a combination of those and/or other parameters. Understanding these tipping points is central to assessing the vulnerability of your restoration objectives to climate change. Information (documented or not) on the sensitivity of the species that are central to your restoration objective to changing water and other environmental conditions is often available (e.g., depth to saturated soils tolerated by obligate riparian plants per Case Study 4.4). In addition, reach out to scientists and other experts who have studied the species that are central to your restoration objective (e.g., a riparian ecologist in the instance of Case Study 4.4) to help identify tipping points.
- Identification of the key parts of your stream's flow regime that are particularly important to the species central to the restoration objectives. This important part of the vulnerability assessment encourages practitioners to narrow focus on the hydrological thresholds that are considered of critical importance to the viability of the species central to their stream-restoration objectives.
- A good understanding of climate change projections for the target watershed, particularly regarding how changes in temperature and precipitation will affect the parts of your stream's flow regime that have been identified as critically important to realizing restoration objectives (per above). In the Terlingua Creek case study, practitioners were able to take advantage of recent climate studies in their region. In situations where recent climate studies have not been conducted for the area of interest, practitioners may need to pursue (with collaborating climatologists) one or more of the strategies described in the previous section (e.g., downscaling or using temporal and/or spatial analogs).

## Climate Adaptation

Climate adaptation describes practices and adjustments that enhance resilience and/or reduce vulnerability to changes in climate (Adger et al. 2007). Although adaptation technically mitigates impacts, climate change adaptation should not be confused with what is typically referred to as climate change mitigation, which concerns large-scale efforts (for example, at the state, national, or even global scale) to curb heat-trapping greenhouse-gas emissions.

People, societies, and even species and ecosystems have adjusted to and coped with climate variability throughout history (IPCC 2014). In populated areas, people have implemented engineered solutions, such as coastal surge protection and green infrastructure, as well as institutional solutions, such as early warning systems and building codes, to plan for and minimize impacts from weather extremes (Matthews et al. 2011). In riparian areas, many plant species are adapted to various disturbance regimes (such as fire or episodic flooding) and variations in seasonal and annual environmental conditions (Naiman and Décamps 1997; Seavy et al. 2009). Unfortunately, in the context of climate change, we cannot use the past as an example for the future, and actions previously used to adapt to climate variability may no longer be sufficient in many cases as change is occurring much more rapidly than in the past. Nevertheless, we can learn from the past and supplement our adaptation toolbox to prepare for a future that looks quite different from what we have previously experienced (Biagini 2014).

One of the great challenges for stream practitioners who seek to strengthen the climate-adaptive capacity of their stream restoration response is that there is no single solution or approach that will be effective (Bierbaum et al. 2013). One size does not fit all. Before we discuss climate adaptation as it specifically relates to stream restoration, remember that on-the-ground restoration (such as those efforts presented in chapter 6) is one of several potential climate adaptation options available. Barmuta and colleagues (2013) identified three other broad categories of adaptation options in stream environments, namely:

- *Water management options* that intervene in the water or flow regime. These can include changing water supply through managing dams and weirs and trading or managing water licenses. This is discussed further in chapter 5.
- *Catchment management options* that help to reduce pressures on the waterway and riparian zones through planning mechanisms, funding programs, and training practitioners. Catchment management can also help to ensure that a strategic high-level view of several projects is taken and any possible interactions between projects are understood.
- *Policy options*, where activities such as fishing in the waterways are restricted to reduce pressure on stock. These options can generally be achieved through influencing policy makers rather than by the restoration teams themselves.

Kareiva et al. (2008) is a good resource that outlines many adaptation options for climate-sensitive ecosystems and resources. Another excellent resource is the Collaborative Conservation and Adaptation Strategy Toolbox, or CCAST (https://usbr.maps.arcgis.com/apps /Cascade/index.html?appid=01245fcb9dec43938996e18b53f0f142), which is an online management toolbox that provides case studies on management actions, partnership and collaboration, monitoring, and adaptive management. Case study topics include fish and wildlife, water resources, restoration, landscape and watershed-scale management, and connectivity and corridors. There are also a variety of adaptation planning cycles that practitioners can follow and apply to their own restoration planning process (e.g., Stein et al. 2014; Rissik et al. 2014; Palutikof et al. 2018). Figure 4.4 shows an example of a framework for adaptation planning and implementation.

"Scenario planning" helps users develop plausible alternative scenarios of future conditions based on a variety of environmental, social, political, economic, or technical factors, which provide the opportunity to assess the potential effects of these different scenarios on your objectives (National Park Service 2013). These are not the scenarios of future global emis-

# CASE STUDY 44

Assessing the Vulnerability of Reestablishing Obligate Riparian Trees Along a Desert Stream to Climate Change: An Example from Terlingua Creek, West Texas

Jeff Renfrow, Sue Harvison, and Mark K. Briggs

The central objective of riparian revegetation efforts along Terlingua Creek, near the Mexico-U.S. border in the Chihuahuan Desert, is to reestablish obligate riparian trees where the native plants have been overgrazed by cattle and overharvested to support local mining activities. Mining no longer occurs in this region, and cattle have been removed from the site, so both of these impacts have been eliminated. The stream is intermittent and drains an area of about 2,600 sq. km (1,000 sq. mi.). The restoration objective for the project is to establish over two thousand cottonwood (*Populus fremontii*) and willow trees (*Salix* spp.) along a 3 km (about 2 mi.) reach of Terlingua Creek that passes through the 3 Bar Ranch—an eleven thousand–acre ranch surrounded by mountains and canyons—which lies just upstream of the creek's confluence with the RGB on the U.S.-Mexico border. The restoration project is a collaboration between the 3 Bar Ranch, the Texas Parks and Wildlife Department, local consultants, a nongovernmental organization, and several private foundations.

In desert stream systems, the boom-and-bust nature of precipitation (long dry and hot periods interrupted by short periods of intense precipitation and flooding) often makes riparian revegetation projects a challenge, with high losses of planted materials expected as a result of flood scour and/or desiccation. To assess the vulnerability of this project's restoration objective to climate change, this project focused on the hot, dry season (late spring to midsummer, the "bust" season, when water availability is lowest) with the central question being, Does the elevation of the zone of soil saturation at planting sites drop to levels that are likely to not permit the establishment of obligate riparian trees?

To answer this question, three main pieces of information were needed: (1) Tolerance thresholds of the trees being planted to decline in the elevation of the soil saturation zone. In essence, how far could the elevation of soil saturation drop before obligate riparian trees are compromised? What was the tipping point?; (2) The elevation change in the level of soil saturation at 3 Bar Ranch planting sites during the hot, dry season; and (3) Climate projections for the region, with particular attention on how future changes in precipitation and temperature could reduce water availability during the hot, dry season and make it even more difficult for obligate riparian species to survive this critical season in marginal areas (areas where water availability is already at threshold levels for obligate riparian plants).

Regarding tolerance thresholds of many obligate riparian trees, to survive the hot, dry season, obligate riparian trees like cottonwoods and willows need to have their roots in the saturated zone of the soil. Cottonwoods and willows do not have a tap root and are relatively shallow-rooted, with even mature root systems not extending much more than three meters (ten feet) below the surface of the soil (Stromberg et al. 1996). Seedlings succumb much more rapidly than mature trees to a drying of the root zone during this crucial time, given their comparatively lower biomass and less-developed root systems (Stromberg et al. 2007; Coble and Kolb 2012).

Concerning the elevation of the soil saturation zone at 3 Bar planting sites, it was necessary to monitor shallow groundwater levels in order to understand how far and how rapidly the soil saturated zone dropped during the hot, dry season. Unfortunately, the limited project budget did not permit multiple-year monitoring with piezometers (see chapter 6 for details on monitoring shallow groundwater). To accommodate budget constraints, a backhoe provided by the 3 Bar Ranch was used to excavate

Deep test holes were excavated in numerous potential revegetation sites at the 3 Bar Ranch to assess how far the elevation of the saturated soil zone dropped during the hot, dry season. Given budget limitations, this was a cheap strategy that allowed insight into determining how vulnerable the objective of our restoration project might be to climate change.

a series of test holes in several promising planting sites the year before planting was scheduled to begin. Test holes were excavated as deep as possible (most were dug to a depth of about 2.5 m), and 3 Bar personnel monitored depth to water in the test holes during the entire hot, dry season prior to planting (see photo above). During the course of the 2017 hot, dry season many of the test holes went completely dry at a depth of 2.5 m, only two remained moist at the bottom of the hole (i.e., moist soils but no pool of water at the bottom of the hole), and one had a pool of water at the bottom of the hole during the entire hot, dry season. 3 Bar personnel also noted that the flow in Terlingua Creek during May and June is typically rare, which means that shallow groundwater recharge during the hot, dry season as a result of streamflow occurrence cannot be counted on.

Regarding climate, climate records for this region showed that 2017—the year the test holes were monitored—was average with respect to temperature and precipitation. That is, 2017 was not in the midst of an overly wet period, nor were the years considered drought years. Finally, the restoration team reached out to climatologists from Texas A&M University, who had recently conducted an analysis of long-term climatic data from this region that indicated with a high degree of confidence that it will experience an increase in temperatures ranging from 1.6°C (3°F) to 2.8°C (5°F) over the next fifty years (McRoberts and Nielsen-Gammon 2010). Results also predict with an equally high degree of confidence that droughts will occur with increasing severity, frequency, and duration. Although predictions of precipitation do not carry the same level of confidence, the general consensus from climate experts at Texas A&M University is that future monsoon activity

(*continued*)

will be unpredictable and accompanied by an overall reduction in winter storms, which provide much of the annual moisture.

With the above information, the following conclusions were made regarding the vulnerability of planting obligate riparian trees along Terlingua Creek at the 3 Bar Ranch:

- *Sensitivity of obligate riparian trees to impacts of climate change on water availability was deemed high.* Obligate riparian trees will not survive if the elevation of the saturated soil zone falls below the root zone (Stromberg et al. 1996). The sensitivity of newly planted seedlings with developing root systems will be even higher.

- *Exposure to deleterious conditions will be greater in the future.* Results from monitoring the test holes indicate that the elevation of the zone of soil saturation dropped to 2.5 m below land surface in the majority of test holes that were monitored, which indicates that many of the planting sites at the 3 Bar Ranch were already at or close to tipping points regarding water availability for obligate riparian trees. Given that monitoring took place during an average year of precipitation and temperature, and that the future climate in the region is forecast to be hotter and drier with less reliable monsoon rains, it appears likely that the exposure of planted vegetation to conditions ill-suited for their long-term survival, particularly during the hot, dry season, will increase. For many ephemeral streams like Terlingua Creek, where the saturated soil zone is already at threshold depths for establishing obligate riparian trees, this means an even bleaker prognosis for obligate riparian seedlings and saplings.

- *Adaptive capacity of obligate riparian trees is low.* Unlike wildlife species and humans, plants cannot pick up roots (so to speak) and move to locations that offer conditions more amenable to their survival. In this sense, their adaptive capacity is low. However, there are planting strategies that we can use at 3 Bar to improve opportunities for success (see the "Climate Adaptation" discussion below).

sions (e.g., RCP 4.5 or 8.5) that we discuss in the previous section, but rather are plausible scenarios of future local conditions, created by the users, to assist in planning for an uncertain future. Scenario planning tools are available to guide practitioners in the selection of climate-adaptation strategies that may be most effective in situations of high uncertainty and complexity. Rowland et al. (2014) outlined the scenario-planning process and provided examples and other resources. In the context of stream restoration, scenario planning can be used to assess the impacts of climate change on a variety of natural resources and to identify and prioritize sites for restoration and conservation. A good example is the Rising Waters project described in Aldrich et al. (2009), wherein participants used scenario planning methodologies "to strengthen the preparedness and adaptive capacity of the Hudson River Estuary Watershed to meet the impacts of future climate change."

### Strategies to Strengthen the Climate-Adaptive Capacity of On-the-Ground Stream Restoration Actions

If the results of a vulnerability assessment indicate that your current stream restoration objectives will likely not be achieved by tactics that you have identified, there are three main options: alter restoration objectives, change restoration tactics, or stop your restoration effort altogether.

1. *Alter your restoration objectives.* As we learned in chapter 2, restoration objectives describe the time frame, location, and end result of your project. Each of these descriptors can be adjusted to improve the adaptive capacity of your stream restoration project to climate change.

    i. *Change the time frame.* Changing the start and/or end date of your restoration project sounds easy, but admittedly can be challenging to do, particularly in situations where your boss, funders, or other powers that be have tied restoration resources to a tight project implementation window. However, if you can orchestrate a more flexible time frame, you will have more time to fill information gaps, alter restoration tactics, involve additional players, and so on, all of which can have a profound positive impact on restoration outcomes. A decade from now, as you and your restoration team look at the native species you were able to establish or hear from the satisfied riverside citizens whom you worked with, no one is going to remember that the implementation of your project was delayed a couple of years.

    ii. *Change location.* Changing the location of your stream restoration project is certainly an option if the planned restoration site, reach, segment, or even basin where your

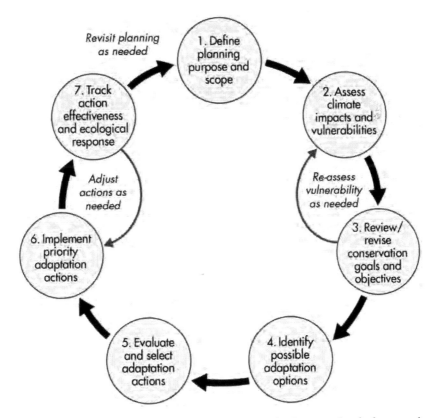

**FIGURE 4.4** The Climate-Smart Conservation Cycle: An example of a framework for adaptation planning and implementation (Stein et al. 2014).

stream restoration project is currently being planned is not conducive to realizing desired results. This option won't be available to you if your stream restoration project is tied to a specific site. Such is one of the great benefits of planning restoration at the scale of the watershed. The broader the geographic area being considered, the more options will likely be available to you to address climate vulnerability issues, including having the option to move a restoration project's location upstream (to a higher elevation) or to another tributary entirely where climate and other conditions are more conducive to success (e.g., Lukasiewicz et al. 2015). A good example of this approach is in Glacier National Park, Montana, where managers are translocating threatened bull trout to lakes where threats from nonnative fish and climate change impacts are reduced (Gray 2016).

iii. *Change desired outcomes.* If the desired outcome for your stream restoration project is vulnerable to climate change impacts, consider changing your goals and objectives. If establishing native biota is central to your stream restoration objective, switching to species that are better adapted to future conditions is an option. At Terlingua Creek (Case Study 4.4), changing the restoration outcome from an emphasis on establishing native obligate riparian trees to nonobligate riparian species that are more tolerant of drier conditions (e.g., *Celtis* spp., *Acacia* spp., *Prosopis* spp.) is a possibility. Establishing such species can still improve the quality of the bottomland riparian habitat (i.e., still meet the objectives of your stream restoration project) while also increasing the likelihood for long-term success, given that such species are better adapted than obligate riparian plants to the environmental conditions predicted for the future. We need to maintain a long-term outlook (at least a couple decades) when it comes to our restoration objectives. Just as the ice-hockey great Wayne Gretzky once quipped, "Skate to where the puck's gonna be, not to where it's been," we want the species that we are reestablishing as part of our restoration objectives to be adapted to the conditions that are gonna be.

2. *Alter your restoration tactics.* Depending on the objective of your stream restoration project, there are a variety of ways restoration tactics can be adjusted or changed completely to improve chances for success in a rapidly changing climate. Four selected approaches are highlighted below.

i. *Alter on-the-ground restoration tactics.* Changing the on-the-ground restoration tactics that are being used to implement restoration can improve likelihood for success in the long term. This could potentially involve a host of tactical changes depending on your stream restoration objectives. For example, at Terlingua Creek, practitioners used longer branches ("poles" or "pole plantings") in parts of the 3 Bar Ranch that allowed the species that were planted to reach the saturated zone of the soil profile at greater depth (Case Study 4.4). This is a cost-effective approach that allows obligate riparian plants to be established in environments where the saturated zone of the soil profile is declining (possibly due to climate change as well as other factors) (Gonzalez et al. 2018). For restoration efforts that focus on native fish reintroduction, placing more emphasis on reducing threats from nonnative fish (instead of or in addition to reintroducing natives) may be an option. Chapter 6 reviews lessons learned from a variety of stream restoration projects and describes options during implementation.

ii. *Change the emphasis from restoration to protection.* If stream restoration objectives are not likely to be achieved, another option is to place greater emphasis on protecting whatever critical native habitat continues to persist in your watershed rather than on restoring habitat that has already been damaged. As noted in chapter 2, it can be a good idea to debate over whether to protect or restore (or a combination of both). Objectives focused on protecting existing habitat may be more cost- and labor-effective (and successful) than those focused on restoring lost native habitat. In the context of climate change, protecting or creating refugia that allow key species to survive extreme climatic events should be part of this debate. For example, deep water holes, springs, or small reservoirs can serve as refugia for aquatic and riparian species during droughts, providing pockets of resilience that can be critical for the survival of certain species in the long-term.

The protection versus restoration debate, however, is predicated on operating within a geographic spatial scale that is broad enough to provide a variety of options on where to place your resources. A distinct benefit of operating at such a scale is having more latitude to select a range of protection and restoration priorities that are geared to the unique biophysical and sociopolitical characteristics of the landscape that is being considered. One essential priority (whether it is in the vein of restoration or protection) is to protect landscape connectivity, which is a commonly recommended adaptation approach for addressing climate threats to biodiversity (Heller and Zavaleta 2009). This approach can improve species' mobility and probability of survival by enabling them to follow their climate niche (Heller and Zavaleta 2009; Lindenmayer et al. 2010).

iii. *Protect streamflow.* Restoration tactics can be shifted from emphasizing on-the-ground restoration actions to an emphasis on protecting streamflow (via changes in management, policy, regulations, market forces, etc.). In dryland environments, it's all about streamflow. Your restoration objectives will be much more difficult to achieve if your stream's flow regime is dramatically changing as a result of human interferences and climate change impacts. Strategies to protect streamflow can include improving water management to address issues associated with water supply and demand, management of dams, diversions and weirs, and water trading, among other actions. This is the central theme of chapter 5.

In addition, policy initiatives that protect space for rivers so that they are able to adjust geomorphologically to changing flow and sediment regimes in the face of human development has to be a priority in the future. Streams are resilient and can adapt just fine over time if they have space to adjust within their corridors.

iv. *Genetics.* New technology and research platforms are making it possible for researchers to identify individual species and populations that will survive in the climates of the future and in the face of the myriad cascading effects of climate change. Such innovative combined restoration-research experiments could allow practitioners to select genotypes of desired species that will have a better chance of surviving future climate conditions (e.g., see Southwest Experimental Garden Array [SEGA] at https://sega.nau.edu).

v. *Address stressors.* The elimination or reduction of stress from human interferences should always be the priority. Eliminating, or at least reducing, the impacts of the primary stressor may be the single-most effective climate-adaptive action that practitioners can contemplate and will likely produce benefits regardless of what future climate becomes the reality.

3. *Push the pause button.* In addition to the strategies outlined above, don't hesitate to pause implementation of your stream corridor restoration project if the information you have gathered indicates that your restoration objectives and tactics are vulnerable to climate change impacts. Understand the limits to adaptation and adaptive capacity and consider that there will be times when you have to go back to the drawing board. Much can be gained through an honest discussion with all stakeholders. In the end, you may decide that halting a project makes the most sense, or you may decide to pursue one or more of the options discussed above.

By way of concluding this chapter, as well as foreshadowing the focus in chapter 7, we want to note the importance of monitoring and adaptive management. Only by monitoring the results of your stream restoration project, and combining those data with ongoing data collection on your watershed's climate and stream corridor conditions (maybe done by other entities) will you have the data (and ultimately the information) to evaluate how well you are realizing restoration objectives in the context of a rapidly changing climate. This will ultimately provide the basis for understanding how restoration objectives and tactics could be changed for greater effectiveness. Monitoring strengthens decision-making in the face of uncertainty (Skidmore et al. 2011) and allows you to gauge the effectiveness of the climate-adaptive measures you have implemented.

# REFERENCES

Adams, H. D., M. Guardiola-Claramonte, G. A. Barron-Gafford, J. Camilo Villegas, D. D. Breshears, C. B. Zou, P. A. Troch, and T. E. Huxman. 2009. "Temperature Sensitivity of Drought-Induced Tree Mortality Portends Increased Regional Die-Off Under Global-Change-Type Drought." *Proceedings of the National Academy of Sciences* 106:7063–66.

Adger, W. N., S. Agrawala, M. M. Q. Mirza, C. Conde, K. O'Brien, J. Pulhin, R. Pulwarty, B. Smit, and K. Takahashi. 2007. "Assessment of Adaptation Practices, Options, Constraints and Capacity: Climate Change 2007; Impacts, Adaptation and Vulnerability." Contribution of Working Group II to the *Fourth Assessment Report of the Intergovernmental Panel on Climate Change.* Edited by M. L. Parry et al., et al., 717–43. Cambridge: Cambridge University Press.

Adler, R. F., G. J. Huffman, A. Chang, R. Ferraro, P. Xie, J. Janowiak, B. Rudolf, U. Schneider, S. Curtis, D. Bolvin, A. Gruber, J. Susskind, P. Arkin, and E. Nelkin. 2003. "The Version-2 Global Precipitation Climatology Project (GPCP) Monthly Precipitation Analysis (1979–Present)." *J. Hydrometeor* 4:1147–67.

Aguilera, R., A. Gershunov, and T. Benmarhnia. 2019. "Atmospheric Rivers Impact California's Coastal Water Quality via Extreme Precipitation." *Science of the Total Environment* 671:488–94.

Aldrich, S., M. Dunkle, and J. Newcomb. 2009. *Rising Waters: Helping Hudson River Communities Adapt to Climate Change; Scenario Planning 2010—2030 Executive Summary.* The Nature Conservancy Eastern NY Chapter. https://www.nature.org/media/newyork/rw_070509_exec.pdf.

Allen, C. D., A. K. Macalady, H. Chenchouni, D. Bachelet, N. McDowell, M. Vennetier, T. Kitzberger, et al. 2010. "A Global Review of Drought and Heat-Induced Tree Mortality Reveals Emerging Climate Change Risks for Forests." *Forest Ecology and Management* 259:660–84.

Allen, K. J., S. C. Nichols, R. Evans, E. R. Cook, S. Allie, G. Carson, F. Ling, and P. J. Baker 2015. "Preliminary December–January Inflow and Streamflow Reconstructions from Tree Rings for Western Tasmania, Southeastern Australia." *Water Resources Research* 51:5487–503. http://doi.org/10.1002/2015WR017062.

Anderson, B. T., J. Wang, G. Salvucci, S. Gopal, and S. Islam. 2010. "Observed Trends in Summertime Precipitation over the Southwestern United States." *Journal of Climate* 23:1937–44. https://doi.org/10.1175/2009JCLI3317.1.

Arnold, J. G., R. Srinivasan, R. S. Muttiah, and J. R. Williams. 2007. "Large Area Hydrologic Modeling and Assessment Part 1: Model Development." *Journal of the American Water Resources Association* 34 (1): 73–89.

Arriaga-Ramírez, S., and T. Cavazos. 2010. "Regional Trends of Daily Precipitation Indices in Northwest Mexico and Southwest United States." *Journal of Geophysical Research: Atmospheres* 115(D14). https://doi.org/10.1029/2009JD013248.

Ault, T. R., J. S. Mankin, B. I. Cook, J. E. Smerdon. 2016. "Relative Impacts of Mitigation, Temperature, and Precipitation on 21st-Century Megadrought Risk in the American Southwest." *Science Advances* 2 (10). http://doi.org/10.1126/sciadv.1600873.

Bacmeister, J. T., K. A. Reed, C. Hannay, P. Lawrence, S. Bates, J. E. Truesdale, N. Rosenbloom, M. Levy. 2018. "Projected Changes in Tropical Cyclone Activity Under Future Warming Scenarios Using a High-Resolution Climate Model." *Climatic Change* 146:547–60.

Barmuta, L, P. Davies, A. Watson, M. Lacey, B. Graham, M. Read, S. Carter, and D. Warfe. 2013. *Joining the Dots: Integrating Climate and Hydrological Projections with Freshwater Ecosystem Values to Develop Adaptation Options for Conserving Freshwater Biodiversity.* National Climate Change Adaptation Research Facility, Griffith Univerity, Gold Coast, Australia. https://www.nccarf.edu.au/.

Baron, J. S., N. L. Poff, P. L. Angermeier, C. N. Dahm, P. H. Gleick, N. G. Hairston, R. B. Jackson, C. A. Johnston, B. D. Richter, and A. D. Steinman. 2002. "Meeting Ecological and Societal Needs for Freshwater." *Ecological Applications* 12:1247–60.

Bayliss, P., K. Saunders, L. X. C. Dutra, L. F. C. Melo, J. Hilton, M. Prakash, and F. Woolard. 2016. "Assessing Sea Level-Rise Risks to Coastal Floodplains in the Kakadu Region, Northern Australia, Using a Tidally Driven Hydrodynamic Model." *Marine and Freshwater Research* 69 (7):1064–78.

Beatty, S., D. Morgan, J. Keleher, A. Lymbery, P. Close, P. Speldewinde, T. Storer, and A. Kitsios. 2013. *Adapting to Climate Change: A Risk Assessment and Decision Making Framework for Managing Groundwater Dependent Ecosystems with Declining Water Levels.* National Climate Change Adaptation Research Facility, Griffith University Gold Coast, Australia. https://www.nccarf.edu.au/.

Beier, P., L. J. Hansen, L. Helbrecht, and D. Behar. 2017. A How-To Guide for Coproduction of Actionable Science. *Conservation Letters* 10 (3): 288–96.

Berghuijs, W. R., R. A. Woods, and M. Hrachowitz. 2014. "A Precipitation Shift from Snow Towards Rain Leads to a Decrease in Streamflow." *Nature Climate Change* 4:583–86. http://doi.org/10.1038/nclimate2246.

Bergmann, J., S. Pompe, R. Ohlemüller, M. Freiberg, S. Klotz, and I. Kühn. 2010. "The Iberian Peninsula as a Potential Source for the Plant Species Pool in Germany Under Projected Climate Change." *Plant Ecology* 207:191–201.

Biagini, B., R. Bierbaum, M. Stults, S. Dobardzic, and S. M. McNeeley. 2014. "A Typology of Adaptation Actions: A Global Look at Climate Adaptation Actions Financed Through the Global Environment Facility." *Global Environmental Change* 25:97–108.

Bierbaum, R., J. B. Smith, A. Lee, M. Blair, L. Carter, F. S. Chapin III, P. Fleming, S. Ruffo, L. Verduzco. 2013. "A Comprehensive Review of Climate Adaptation in the United States: More Than Before, but Less Than Needed." *Mitigation and Adaptation Strategies for Global Change* 18:361–406. http://doi .org/10.1007/s11027-012-9423-1.

Bindoff, N. L., P. A. Stott, K. M. AchutaRao, M. R. Allen, N. Gillett, D. Gutzler, K. Hansingo, et al. 2013. "Detection and Attribution of Climate Change: From Global to Regional." In *Climate Change 2013: The Physical Science Basis*, edited by T. F. Stocker et al. Cambridge: Cambridge University Press.

Bond, N. R., P. S. Lake, and A. H. Arthington. 2008. "The Impacts of Drought on Freshwater Ecosystems: An Australian Perspective." *Hydrobiologia* 600:3–16.

Breshears, D. D., N. S. Cobb, P. M. Rich, K. P. Price, C. D. Allen, R. G. Balice, W. H. Romme, J. H. Kastens, C. W. Meyer, et al. 2005. "Regional Vegetation Die-Off in Response to Global-Change-Type Drought." *Proceedings of the National Academy of Sciences of the USA* 102:15144–48.

Brooke, C. 2008. "Conservation and Adaptation to Climate Change." *Conservation Biology* 22:1471–76.

Castro, C. L., R. A. Pielke Sr., and J. O. Adegoke. 2007. "Investigation of the Summer Climate of the Contiguous United States and Mexico Using the Regional Atmospheric Modeling System (RAMS), Part 1: Model Climatology (1950–2002)." *Journal of Climate* 20:3844–65.

Cavazos, T., and S. Arriaga-Ramírez. 2012. "Downscaled Climate Change Scenarios for Baja California and the North American Monsoon During the 21st Century." *Journal of Climate* 25 (17): 5904–15. http://doi.org/10.1175/JCLI-D-11-00425.1.

Cavazos, T., C. Turrent, and D. P. Lettenmaier. 2008. "Extreme Precipitation Trends Associated with Tropical Cyclones in the Core of the North American Monsoon." *Geophysical Research Letters* 35 (21). https://doi.org/10.1029/2008GL035832.

Cayan, D. R., T. Das, D. W. Pierce, T. P. Barnett, M. Tyree, and A. Gershunov. 2010. "Future Dryness in the Southwest US and the Hydrology of the Early 21st Century Drought." *Proceedings of the National Academy of Sciences of the United States of America* 107:21271–76. http://doi.org/10.1073/pnas .0912391107.

Cayan, D. R., M. Tyree, K. E. Kunkel, C. Castro, A. Gershunov, J. Barsugli, A. J. Ray, P. Duffy, et al. 2013. "Future Climate: Projected Average." In *Assessment of Climate Change in the Southwest United States: A Report Prepared for the National Climate Assessment*, edited by G. Garfin et al., 101–25. Washing-ton, D.C.: Island Press.

Cerezo-Mota, R., M. Allen, and R. Jones. 2011. "Mechanisms Controlling Precipitation in the Northern Portion of the North American Monsoon." *Journal of Climate* 24:2771–83.

Chambers, J., G. Nugent, B. Sommer, P. Speldewinde, S. Neville, S. Beatty, S. Chilcott, P. Davies, et al. 2013. *Adapting to Climate Change: A Risk Assessment and Decision Making Framework for Managing Groundwater Dependent Ecosystems with Declining Water Levels Development and Case Studies.* National Climate Change Adaptation Research Facility, Griffith University, Gold Coast, Australia. https://www.nccarf.edu.au/.

Chambers, R. C. 1989. "Editorial Introduction: Vulnerability, Coping, and Policy." *IDS Bulletin* 20 (2): 1–7.

Chow, V. T., D. R. Maidment, and L. W. Mays. 1988. *Applied Hydrology.* New York: McGraw Hill.

Church, J. A., P. U. Clark, A. Cazenave, J. M. Gregory, S. Jevrejeva, A. Levermann, M. A. Merrifield, et al. 2013. "Sea Level Change." In *Climate Change 2013: The Physical Science Basis.* Contribution of

Working Group I to the Fifth Assessment Report of the Intergovernmental Panel on Climate Change. Edited by T. F. Stocker et al., 1137–216. Cambridge: Cambridge University Press.

Cleaveland, M. K., T. H. Votteler, D. K. Stahle, R. C. Casteel, and J. L. Banner. 2011. "Extended Chronology of Drought in South Central, Southeastern and West Texas." *Texas Water Journal* 2 (1): 54–96.

Coble, A. P., and T. E. Kolb. 2012. "Riparian Tree Growth Response to Drought and Altered Streamflow Along the Dolores River, Colorado." *Western Journal of Applied Forestry* 27:205–11.

Cochrane, M. A., and C. P. Barber. 2009. "Climate Change, Human Land Use and Future Fires in the Amazon." *Global Change Biology* 15 (3): 601–12.

Collins, M., R. Knutti, J. Arblaster, J.-L. Dufresne, T. Fichefet, P. Friedlingstein, X. Gao, et al. 2013. "Long-Term Climate Change: Projections, Commitments and Irreversibility." In *Climate Change 2013: The Physical Science Basis*. Contribution of Working Group I to the Fifth Assessment Report of the Intergovernmental Panel on Climate Change. Edited by T. F. Stocker et al., 1029–136. Cambridge: Cambridge University Press.

Cook, B. I., and R. Seager. 2013. "The Response of the North American Monsoon to Increased Greenhouse Gas Forcing." *Journal of Geophyical Research* 118 (4): 1690–99.

Cook, E., C. Woodhouse, C. Eakin, D. Meko, and D. Stahle. 2004. "Long-Term Aridity Changes in the Western United States." *Science* 306 (5698): 1015–18.

Cook, J., S. Freeman, E. Levine, and M. Hill. 2012. *Shifting Course: Climate Adaptation for Water Management Institutions*. Washington, D.C.: World Wildlife Fund.

Corringham, T. W., F. M. Ralph, A. Gershunov, D. R. Cayan, and C. A. Talbot. 2019. "Atmospheric Rivers Drive Flood Damages in the Western United States." *Science Advances* 5 (12). https://doi.org/10.1126/sciadv.aax4631.

CSIRO and Bureau of Meteorology. 2015. *Climate Change in Australia Information for Australia's Natural Resource Management Regions*. Technical Report. CSIRO and Bureau of Meteorology, Australia. https://www.csiro.au/en/Research/.

Dai, A. 2013. "Increasing Drought Under Global Warming in Observations and Models." *Nature Climate Change* 3:52–58.

Daly, C. 2006. "Guidelines for Assessing the Suitability of Spatial Climate Data Sets." *International Journal of Climatology* 26:707–21. https://doi.org/10.1002/joc.1322.

Daly, C., R. Neilson, D. Phillips. 1994. A Statistical-Topographic Model for Mapping Climatological Precipitation over Mountainous Terrain. *Journal of Applied Meteorology* 33:140–58.

Dangendorf, S., C. Hay, F. M. Calafat, M. Marcos, C. G. Piecuch, and J. Jensen. 2019. "Persistent Acceleration in Global Sea-Level Rise Since the 1960s." *Nature Climate Change* 9:705–10. https://doi.org/10.1038/s41558-019-0531-8.

Dawson, T. P., S. T. Jackson, J. I. House, I. C. Prentice, and G. M. Mace. 2011. "Beyond Predictions: Biodiversity Conservation in a Changing Climate." *Science* 332:53–58.

Dee, D. P., J. Fasullo, D. Shea, J. Walsh, and the National Center for Atmospheric Research Staff, eds. 2016. "The Climate Data Guide: Atmospheric Reanalysis: Overview & Comparison Tables." https://climatedataguide.ucar.edu/climate-data/atmospheric-reanalysis-overview-comparison-tables.

Dee, D. P., S. M. Uppala, A. J. Simmons, P. Berrisford, P. Poli, S. Kobayashi, U. Andrae, et al. 2011. "The ERA-Interim Reanalysis: Configuration and Performance of the Data Assimilation System." *Quarterly Journal of the Royal Meteorological Society* 137:553–97. https://doi.org/10.1002/qj.828.

Dettinger, M., B. Udall, and A. Georgakakos. 2015. "Western Water and Climate Change." *Ecological Applications* 25 (8): 2069–93.

Dey, P., and A. Mishra. 2017. "Separating the Impacts of Climate Change and Human Activities on Streamflow: A Review of Methodologies and Critical Assumptions." *Journal of Hydrology* 548:278–90.

Diffenbaugh, N. S., D. L. Swain, and D. Touma. 2015. "Anthropogenic Warming Has Increased Drought Risk in California." *Proceedings of the National Academy of Sciences* 112 (13): 3931–36. http://doi.org /10.1073/pnas.1422385112.

Dzaugis, M. P., D. R. Reidmiller, C. W. Avery, A. Crimmins, L. Dahlman, D. R. Easterling, R. Gaal, et al. 2018. "Frequently Asked Questions." In *Impacts, Risks, and Adaptation in the United States: Fourth National Climate Assessment*. Vol. 2. Edited by D. R. Reidmiller et al., 1444–515. Washington, D.C.: U.S. Global Change Research Program. https://doi.org.10.7930/NCA4.2018.AP5.

Easterling, D. R., K. E. Kunkel, J. R. Arnold, T. Knutson, A. N. LeGrande, L. R. Leung, R. S. Vose, D. E. Waliser, and M. F. Wehner. 2017. "Precipitation Change in the United States." In *Climate Science Special Report: Fourth National Climate Assessment*. Vol. 1. Edited by Maycock et al., 207–30. Washington, D.C.: U.S. Global Change Research Program. http://doi.org/10.7930/J0H993CC.

Easterling, D. R., G. A. Meehl, C. Parmesan, S. A. Changnon, T. R. Karl, and L. O. Mearns. 2000. "Climate Extremes: Observations, Modeling, and Impacts." *Science* 289 (5487): 2068–74.

Ekström, M., M. Grose, P. Whetton. 2015. "An Appraisal of Downscaling Methods Used in Climate Change Research." *WIREs Clim Change* 6:301–19. http://doi.org/10.1002/wcc.339.

Englehart, P. J., and A. V. Douglas. 2006. "Changing Behavior in the Diurnal Range of Surface Air Temperatures over Mexico." *Geophysical Research Letters* 32 (1). http://doi.org/10.1029/2004GL021139.

Fahey, D. W., S. J. Doherty, K. A. Hibbard, A. Romanou, and P. C. Taylor. 2017. "Physical Drivers of Climate Change." In *Climate Science Special Report: Fourth National Climate Assessment*. Vol. 1. Edited by D. J. Wuebbles et al., 73–113. Washington, D.C.: U.S. Global Change Research Program. http:// doi.org/10.7930/J0513WCR.

Falk, D. A., E. K. Heyerdahl, P. M. Brown, C. Farris, P. Z. Fulé, D. McKenzie, T. W. Swetnam, A. H. Taylor, and M. L. Van Horne. 2011. "Multi-scale Controls of Historical Forest-Fire Regimes: New Insights from Fire-Scar Networks." *Frontiers in Ecology and the Environment* 9:446–454. http://doi.org/10 .1890/100052.

Feng, J., J. Li, and Y. Li. 2010. "A Monsoon-Like Southwest Australian Circulation and Its Relation with Rainfall in Southwest Western Australia." *Journal of Climate* 23:1334–53. http://doi.org/10.1175/2009 JCLI2837.1.

Finlayson, C. M., S. J. Capon, D. Rissik, J. Pittock, G. Fisk, N. C. Davidson, K. A. Bodmin, et al. 2017. "Policy Considerations for Managing Wetlands Under a Changing Climate." *Australian Journal for Marine and Freshwater Research* 68:1803–15. https://doi.org/10.1071/MF16244.

Fischer, E. M., and R. Knutti. 2015. "Anthropogenic Contribution to Global Occurrence of Heavy-Precipitation and High-Temperature Extremes." *Nature Climate Change* 5:560–64.

Frankhauser, S., J. B. Smith, and R. S. J. Tol. 1999. "Weathering Climate Change: Some Simple Rules to Guide Adaptation Decisions." *Ecological Economics* 30:67–78.

Friggens M., and D. Finch. 2015. "Implications of Climate Change for Bird Conservation in the Southwestern U.S. Under Three Alternative Futures." *PLoS ONE* 10 (12) https://doi.org/10.1371/journal .pone.0144089.

Friggens, M., D. Finch, K. Bagne, S. Coe, and D. Hawksworth. 2013. *Vulnerability of Species to Climate Change in the Southwest: Terrestrial Species of the Middle Rio Grande*. USFS-RMRS-GTR-306. Fort Collins, Colo.: U.S. Department of Agriculture, Forest Service, Rocky Mountain Research Station.

Friggens, M., R. Loehman, L. Holsinger, and D. Finch. 2014. *Vulnerability of Riparian Obligate Species to the Interactive Effect of Fire, Climate and Hydrological Change.* Final Report for Interagency Agreement #13-IA-11221632–006. Albuquerque, N.M.: U.S. Department of Agriculture, Forest Service, Rocky Mountain Research Station. https://www.fs.fed.us/rmrs/publications/vulnerability-riparian-obligate-species-interactive-effect-fire-climate-and.

Frumhoff, P. C., J. J. McCarthy, J. M. Melillo, S. C. Moser, and D. J. Wuebbles. 2007. *Confronting Climate Change in the U.S. Northeast: Science, Impacts, and Solutions.* Synthesis report of the Northeast Climate Impacts Assessment (NECIA). Cambridge, Mass.: Union of Concerned Scientists (UCS).

Furniss, M. J., K. B. Roby, D. Cenderelli, J. Chatel, C. F. Clifton, A. Clingenpeel, P. E. Hays, and M. Weinhold. 2013. *Assessing the Vulnerability of Watersheds to Climate Change: Results of National Forest Watershed Vulnerability Pilot Assessments.* Gen. Tech. Rep. PNW-GTR-884. Portland, Ore.: U.S. Department of Agriculture, Forest Service, Pacific Northwest Research Station.

Garfin, G., A. Jardine, R. Merideth, M. Black, and S. LeRoy, eds. 2013. *Assessment of Climate Change in the Southwest United States.* A Report Prepared for the National Climate Assessment by the Southwest Climate Alliance. Washington, D.C.: Island Press.

Garssen, A. G., J. T. A. Verhoeven, and M. B. Soons. 2014. "Effects of Climate-Induced Increases in Summer Drought on Riparian Plant Species: A Meta-analysis." *Freshwater Biology* 59 (5):1052–63.

Georgakakos, A., P. Fleming, M. Dettinger, C. Peters-Lidard, T. C. Richmond, K. Reckhow, K. White, and D. Yates. 2014. "Water Resources." In *Climate Change Impacts in the United States: The Third National Climate Assessment,* edited by J. M. Melillo, T. C. Richmond, and G. W. Yohe, 69–112. N.p.: U.S. Global Change Research Program. http://doi.org/10.7930/J0G44N6T.

Glick, P., B. A. Stein, and N. A. Edelson. 2011. *Scanning the Conservation Horizon: A Guide to Climate Change Vulnerability Assessment.* Washington, D.C.: National Wildlife Federation.

González, E., V. Martínez-Fernández, P. B. Shafroth, A. A. Sher, A. L. Henry, V. Garófano-Gómez, and D. Corenblit. 2018. "Regeneration of Salicaceae Riparian Forests in the Northern Hemisphere: A New Framework and Management Tool." *Journal of Environmental Management* 218:374–87.

González, E., A. A. Sher, R. M. Anderson, R. F. Bay, D. W. Bean, G. J. Bissonnete, B. Bourgeois et al. 2017. "Vegetation Response to Invasive *Tamarix* Control in Southwestern U.S. Rivers: A Collaborative Study Including 416 Sites." *Ecological Applications* 27:1789–804. http://doi.org/10.1002/eap.1566.

González-Villela, R, M. J. Montero Martínez, and J. S. Santana Sepúlveda. 2017. "Repercusiones del cambio climático en el caudal ecológico del Río Conchos." In *La Cuenca del Río Conchos: Una mirada desde las ciencias cnte elcCambio climático,* edited by M. J. Montero Martínez and O. F. Ibáñez Hernández. Morelos, Mex.: Instituto Mexicano de Tecnología del Agua. https://www.imta.gob.mx/biblioteca/download/?key=230.

González-Villela, R., M. J. Montero-Martínez, and J. S. Santana Sepúlveda. 2018. "Effects of Climate Change on the Environmental Flows in the Conchos River (Chihuahua, Mexico)." *Ecohydrology and Hydrobiology* 18:431–40. http://doi.org/10.1016/j.ecohyd.2018.10.004.

Goode, J. R., C. H. Luce, and J. M. Buffington. 2012. "Enhanced Sediment Delivery in a Changing Climate in Semi-arid Mountain Basins: Implications for Water Resource Management and Aquatic Habitat in the Northern Rocky Mountains." *Geomorphology* 139–40:1–15.

Gordon, N. D., T. A. McMahon, and B. L. Finlayson. 1992. *Stream Hydrology: An Introduction for Ecologists.* West Sussex: John Wiley.

Gray, B. 2016. "Glacier National Park Reshuffles Native Trout." *High Country News*, May 5, 2016. https://www.hcn.org/articles/change-in-management-strategy-means-change-of-scenery-for-threatened-trout-species.

Gray, S. T., J. J. Lukas, and C. A. Woodhouse. 2011. "Millennial-Length Records of Streamflow from Three Major Upper Colorado River Tributaries." *Journal of the American Water Resources Association* 47 (4): 702–12. http://doi.org/10.1111/j.1752-1688.2011.00535.x.

Grimm, N. B., A. Chacón, C. N. Dahm, S. W. Hostetler, O. T. Lind, P. L. Starkweather, and W. W. Wurtsbaugh. 1998. "Sensitivity of Aquatic Ecosystems to Climatic and Anthropogenic Changes: The Basin and Range, American Southwest and Mexico." *Hydrological Processes* 11 (8): 1023–41.

Guerreiro, S. B., H. J. Fowler, R. Barbero, S. Westra, G. Lenderink, S. Blenkinsop, E. Lewis, and X. Li. 2018. "Detection of Continental-Scale Intensification of Hourly Rainfall Extremes." *Nature Climate Change* 8:803–7.

Guirguis, K., A. Gershunov, T. Shulgina, R. E. S. Clemesha, and F. M. Ralph. "Atmospheric Rivers Impacting Northern California and Their Modulation by a Variable Climate." *Climate Dynamics* 52:6569–83. http://doi.org/10.1007/s00382-018-4532-5.

Gutiérrez, M., and E. Carreón. 2004. "Salinidad en el bajo Río Conchos: Aportes y tendencias." *Terra Latinoamericana* 22 (4): 499–506.

Gutmann, E., T. Pruitt, M. Clark, L. Brekke, J. Arnold, D. Raff, and R. Rasmussen. 2014. "An Intercomparison of Statistical Downscaling Methods Used for Water Resource Assessments in the United States." *Water Resources Research* 50:7167–86. http://doi.org/10.1002/2014WR015559.

Gutzler, D. S., L. N. Long, J. Schemm, S. B. Roy, M. Bosilovich, J. C. Collier, M. Kanamitsu, et al. 2009. "Simulations of the 2004 North American Monsoon: NAMAP2." *Journal of Climate* 22: 6716–40.

Hagos, S. M., L. R. Leung, J. H. Yoon, J. Lu, and Y. Gao. 2016. "A Projection of Changes in Landfalling Atmospheric River Frequency and Extreme Precipitation over Western North America from the Large Ensemble CESM Simulations." *Geophysical Research Letters* 43:1357–63.

Hallegatte, S., J. Hourcade, and P. Ambrosi. 2007. "Using Climate Analogues for Assessing Climate Change Economic Impacts in Urban Areas." *Climatic Change* 82:47–60.

Harpold, A. A., and P. D. Brooks. 2018. "Humidity Determines Future Snowpack Ablation." *Proceedings of the National Academy of Sciences of the USA* 115:1215–20.

Harris, R., M. Grose, G. Lee, N. Bindoff, L. Porfirio, and P. Fox-Hughes. 2014. "Climate Projections for Ecologists." *WIREs Climate Change* 5:621–37. http://doi.org/10.1002/wcc.291.

Hawkins, E., and R. Sutton. 2011. "The Potential to Narrow Uncertainty in Projections of Regional Precipitation Change." *Climate Dynamics* 37:407–18.

Hayhoe, K., J. Edmonds, R. E. Kopp, A. N. LeGrande, B. M. Sanderson, M. F. Wehner, and D. J. Wuebbles. 2017. "Climate Models, Scenarios, and Projections." In *Climate Science Special Report: Fourth National Climate Assessment*. Vol. Edited by D. J. Wuebbles et al., 133–160. Washington, D.C.: U.S. Global Change Research Program. http://doi.org/10.7930/J0WH2N54.

Hayhoe, K., D. J. Wuebbles, D. R. Easterling, D. W. Fahey, S. Doherty, J. Kossin, W. Sweet, R. Vose, and M. Wehner. 2018. "Our Changing Climate." In *Impacts, Risks, and Adaptation in the United States: Fourth National Climate Assessment*. Vol. 2. Edited by D. R. Reidmiller et al., 72–144. Washington, D.C.: U.S. Global Change Research Program. http://doi.org/10.7930/NCA4.2018.CH2.

Held, I. M., and B. J. Soden. 2006. "Robust Responses of the Hydrological Cycle to Global Warming." *Journal of Climate* 19:5686–99.

Heller, N. E., and E. S. Zavaleta. 2009. "Biodiversity Management in the Face of Climate Change: A Review of 22 Years of Recommendations." *Biological Conservation* 142:14–32.

Higgins, R. W., Y. Yao, and X. L. Wang. 1997. "Influence of the North American Monsoon System on the U.S. Summer Precipitation Regime." *Journal of Climate* 10:2600–22.

Hinojosa-Huerta, O., P. L. Nagler, Y. K. Carrillo-Guererro, and E. P. Glenn. 2013. "Effects of Drought on Birds and Riparian Vegetation in the Colorado River Delta, Mexico." *Ecological Engineering* 51:275–81.

Holling, C. S. 1973. "Resilience and Stability of Ecological Systems." *Annual Review of Ecology and Systematics* 4:1–23.

Huning, L. S., B. Guan, D. E. Waliser, and D. P. Lettenmaier. 2019. "Sensitivity of Seasonal Snowfall Attribution to Atmospheric Rivers and Their Reanalysis-Based Detection." *Geophysical Research Letters* 46:794–803. https://doi.org/10.1029/2018GL080783.

Hurd, B. H., and J. Coonrod. 2007. *Climate Change and Its Implications for New Mexico's Water Resources and Economic Opportunities*. Las Cruces: New Mexico State University.

INEGI (Instituto Nacional de Estadística y Geografía). 2013. Conjunto de datos vectoriales de uso del suelo y vegetación escala 1:250 000, serie V (Capa unión).

Ingram, M. 2009. Practicing Ecological Restoration: Climate Change in Context. *Ecological Restoration* 27 (3): 235–37.

IPCC. 2012. *Managing the Risks of Extreme Events and Disasters to Advance Climate Change Adaptation*. A Special Report of Working Groups I and II of the Intergovernmental Panel on Climate Change. Cambridge: Cambridge University Press.

IPCC. 2013. *Climate Change 2013: The Physical Science Basis*. Contribution of Working Group I to the Fifth Assessment Report of the Intergovernmental Panel on Climate Change. Cambridge: Cambridge University Press.

IPCC. 2014. "Summary for Policymakers." In *Climate Change 2014: Impacts, Adaptation, and Vulnerability. Part A: Global and Sectoral Aspects*. Contribution of Working Group II to the Fifth Assessment Report of the Intergovernmental Panel on Climate Change, edited by C. B. Field et al., 1–32. Cambridge: Cambridge University Press.

Janetos, A., L. Hansen, D. Inouye, B. P. Kelly, L. Meyerson, B. Peterson, and R. Shaw. 2008. "Biodiversity." In *The Effects of Climate Change on Agriculture, Land Resources, Water Resources, and Biodiversity in the United States*, edited by the U.S. Climate Change Science Program and the Subcommittee on Global Change Research. Washington, D.C.

Jha, M., J. G. Arnold, P. W. Gassman, F. Giorgi, and R. R. Gu. 2006. "Climate Change Sensitivity Assessment on Upper Mississippi River Basin Streamflows Using SWAT." *Journal of the American Water Resources Association* 42 (4): 997–1015.

Jiang, T., Y. D. Chen, C. Xu, X. Chen, and V. P. Singh. 2007. "Comparison of Hydrological Impacts of Climate Change Simulated by Six Hydrological Models in the Dongjiang Basin, South China." *Journal of Hydrology* 336 (3–4): 316–33.

Jiménez Cisneros, B. E., T. Oki, N. W. Arnell, G. Benito, J. G. Cogley, P. Döll, T. Jiang, and S. S. Mwakalila. 2014. "Freshwater Resources." In *Climate Change 2014: Impacts, Adaptation, and Vulnerability. Part A: Global and Sectoral Aspects*. Contribution of Working Group II to the Fifth Assessment Report of the Intergovernmental Panel on Climate Change, edited by C. B. Field et al., 229–69. Cambridge: Cambridge University Press.

Jones, P. D., and T. M. L. Wigley. 2010. "Estimation of Global Temperature Trends: What's Important and What Isn't." *Climatic Change* 100:59–69.

Kalnay, E., M. Kanamitsu, R. Kistler, W. Collins, D. Deaven, L. Gandin, M. Iredell, et al. 1996. "The NCEP/ NCAR 40-Year Reanalysis Project." *Bulletin of the American Meteorological Society* 77:437–71.

Kareiva P., C. Enquist, and A. Johnson. 2008. "Synthesis and Conclusions." In *Preliminary Review of Adaptation Options for Climate-Sensitive Ecosystems and Resources.* A Report by the U.S. Climate Change Science Program and the Subcommittee on Global Change Research. Washington, D.C.: U.S. Environmental Protection Agency.

Karl, T. R., J. M. Melillo, and T. C. Peterson, eds. 2009. *Global Climate Change Impacts in the United States.* New York: Cambridge University Press.

Kennard, M. J., B. J. Pusey, J. D. Olden, S. J. Mackay, J. Stein, and N. Marsh. 2010. "Classification of Natural Flow Regimes in Australia to Support Environmental Flow Management." *Freshwater Biology* 55:171–93. http://doi.org/10.1111/J.1365-2427.2009.02307.X.

King, J., and C. Brown. 2006. "Environmental Flows: Striking the Balance Between Development and Resource Protection." *Ecology and Society* 11 (2): 26–47.

Klein, R., R. Nicholls, S. Ragoonaden, M. Capobianco, J. Aston, E. Buckley. 2001. "Technological Options for Adaptation to Climate Change in Coastal Zones." *Journal of Coastal Research* 17 (3): 531–43.

Knowles, N., M. D. Dettinger, and D. R. Cayan. 2006. "Trends in Snowfall Versus Rainfall in the Western United States." *Journal of Climate* 19:4545–59.

Knutson, T. R., J. J. Sirutis, M. Zhao, R. E. Tuleya, M. Bender, G. A. Vecchi, G. Villarini, and D. Chavas. 2015. "Global Projections of Intense Tropical Cyclone Activity for the Late Twenty-First Century from Dynamical Downscaling of CMIP5/RCP4: 5 Scenarios." *Journal of Climate* 28:7203–24.

Koirala, S., Y. Hirabayashi, R. Mahendran, and S. Kanae. 2014. "Global Assessment of Agreement Among Streamflow Projections Using CMIP5 Model Outputs." *Environmental Research Letters* 9. http://doi .org/10.1088/1748-9326/9/6/064017.

Kopf, S., M. Ha-Duong, and S. Hallegatte. 2008. "Using Maps of City Analogues to Display and Interpret Climate Change Scenarios and Their Uncertainty." *Natural Hazards and Earth System Sciences* 8:905–18.

Kossin, J. P., T. Hall, T. Knutson, K. E. Kunkel, R. J. Trapp, D. E. Waliser, and M. F. Wehner. 2017. "Extreme Storms." In *Climate Science Special Report: Fourth National Climate Assessment.* Vol. 1. Edited by D. J. Wuebbles et al., 257–76. Washington, D.C.: U.S. Global Change Research Program. http://doi .org/10.7930/J07S7KXX.

Kotamarthi, R., K. Hayhoe, L. Mearns, D. Wuebbles, J. Jacobs, and J. Jurado. 2020. *Downscaling Techniques for High-Resolution Climate Projections: From Global Change to Local Impacts.* Cambridge: Cambridge University Press.

Kotamarthi, R., L. Mearns, K. Hayhoe, C. L. Castro, and D. Wuebbles. 2016. *Use of Climate Information for Decision-Making and Impacts Research: State of Our Understanding.* Prepared for the Department of Defense, Strategic Environmental Research and Development Program. https://www.serdp-estcp .org/.

Lawler, J. J., C. A. Schloss, and A. K. Ettinger. 2013. "Climate Change: Anticipating and Adapting to the Impacts on Terrestrial Species." In *Encyclopedia of Biodiversity*, Vol. 2, 100–14. https://doi.org/10 .1016/B978-0-12-384719-5.00327-0.

Leigh, C., A. Bush, E. T. Harrison, S. S. Ho, L. Luke, R. J. Rolls, and M. E. Ledger. 2015. "Ecological Effects of Extreme Climatic Events on Riverine Ecosystems: Insights from Australia." *Freshwater Biology* 60:2620–38.

Lettenmaier, D., D. Major, L. Poff, and S. Running. 2008. "Water Resources." In *The Effects of Climate Change on Agriculture, Land Resources, Water Resources, and Biodiversity in the United States.* A Report by the U.S. Climate Change Science Program and the Subcommittee on Global Change Research. Washington, D.C.: U.S. Climate Change Science Program.

Liang, X.-Z., J. Zhu, K. E. Kunkel, M. Ting, and J. X. L. Wang. 2008. "Do GCMs Simulate the North American Monsoon Precipitation Seasonal-Interannual Variability?" *Journal of Climate* 21:4424–48.

Lin, J. L., B. E. Mapes, K. M. Weickmann, G. N. Kiladis, S. D. Schubert, M. J. Suarez, J. T. Bacmeister, and M. I. Lee. 2008. "North American Monsoon and Convectively Coupled Equatorial Waves Simulated by IPCC AR4 Coupled GCMs." *Journal of Climate* 21:2919–37.

Lindenmayer, D. B., W. Steffen, A. A. Burbidge, L. Hughes, R. L. Kitching, W. Musgrave, M. S. Smith, and P. A. Werner. 2010. "Conservation Strategies in Response to Rapid Climate Change: Australia as a Case Study." *Biological Conservation* 143 (7): 1587–93.

Livneh, B., E. A. Rosenberg, C. Lin, B. Nijssen, V. Mishra, K. M. Andreadis, E. P. Maurer, and D. P. Lettenmaier. 2013. "A Long-Term Hydrologically Based Dataset of Land Surface Fluxes and States for the Conterminous United States: Update and Extensions." *Journal of Climate* 26: 9384–92. https://doi.org/10.1175/JCLI-D-12-00508.1.

Llewellyn, D., and S. Vaddey. 2013. *West-Wide Climate Risk Assessment: Upper Rio Grande Impact Assessment.* U.S. Department of the Interior Bureau of Reclamation. https://www.usbr.gov/library/reclamationpubs.html.

Lukasiewicz, A., J. Pittock, and M. Finlayson. 2015. "Institutional Challenges of Adopting Ecosystem-Based Adaptation to Climate Change." *Regional Environmental Change.* https://doi.org/10.1007/s10113-015-0765-6.

Luong, T. M., C. L. Castro, H. I. Chang, T. Lahmers, D. K. Adams, and C. A. Ochoa-Moya. 2017. "The Extreme Nature of North American Monsoon Precipitation in the Southwestern United States as Revealed by a Historical Climatology of Simulated Severe Weather Events." *Journal of Applied Meteorology and Climatology* 56:2509–29.

Mantua, N., I. Tohver, and A. Hamlet. 2010. "Climate Change Impacts on Streamflow Extremes and Summertime Stream Temperature and Their Possible Consequences for Freshwater Salmon Habitat in Washington State." *Climatic Change* 102 (1–2): 187–223.

Margolis, E., D. Meko, and R. Touchan. 2011. "A Tree-Ring Reconstruction of Streamflow in the Santa Fe River, New Mexico." *Journal of Hydrology* 397 (1): 118–27.

Matthews, J. H., and A. J. Wickel. 2009. "Embracing Uncertainty in Freshwater Climate Change Adaptation: A Natural History Approach." *Climate and Development* 1:269–79. https://doi.org/10.3763/cdev.2009.0018.

Matthews, J. H., A. J. Wickel, and S. Freeman. 2011. "Converging Streams in Climate-Relevant Conservation: Water, Infrastructure, and Institutions." *PLOS Biology* 9 (9). https://doi.org/10.1371/journal.pbio.1001159.

McCarthy, L., and L. Head. 2001. "Holocene Variability in Semi-arid Vegetation: New Evidence from Leporillus Middens from the Flinders Ranges, South Australia." *Holocene* 11 (6): 681–89. https://doi.org/10.1191/09596830195708.

McRoberts, B., and J. Nielsen-Gammon. 2010. *Historic and Future Droughts in the Big Bend Region of the Chihuahuan Desert.* WWF Project Final Report, Washington D.C.

Meehl, G. A., and C. Tebaldi. 2004. "More Intense, More Frequent, and Longer Lasting Heat Waves in the 21st Century." *Science* 305 (5686): 994–97. https://doi.org/10.1126/science.1098704.

Meixner, T., A. H. Manning, D. A. Stonestrom, D. M. Allen, H. Ajami, K. W. Blasch, A. E. Brookfield, et al. 2016. "Implications of Projected Climate Change for Groundwater Recharge in the Western United States." *Journal of Hydrology* 534:124–38.

Merritt, D. M., and N. L. R. Poff. 2010. "Shifting Dominance of Riparian *Populus* and *Tamarix* Along Gradients of Flow Alteration in Western North American Rivers." *Ecological Applications* 20:135–52. https://doi.org/10.1890/08-2251.1.

Mesinger, F., G. DiMego, E. Kalnay, K. Mitchell, P. C. Shafran, W. Ebisuzaki, D. Jović, et al. 2006. "North American Regional Reanalysis." *Bulletin of the American Meteorological Society* 87:343–60. https://doi.org/10.1175/BAMS-87-3-343.

Metcalfe, S., M. Jones, S. Davies, A. Noren, and A. MacKenzie. 2010. "Climate Variability over the Last Two Millennia in the North American Monsoon Region, Recorded in Laminated Lake Sediments from Laguna de Juanacatlán, Mexico." *Holocene* 20 (8): 1195–206. https://doi.org/10.1177/09596 83610371994.

Milly, P. C. D., J. Betancourt, M. Falkenmark, R. M. Hirsch, Z. W. Kundzewicz, D. P. Lettenmaier, and R. J. Stouffer. 2008. "Stationarity Is Dead: Whither Water Management?" *Science* 319:573–74.

Mitchell, T. D., and P. D. Jones. 2005. "An Improved Method of Constructing a Database of Monthly Climate Observations and Associated High-Resolution Grids." *International Journal of Climatology* 25:693–712. https://doi.org/10.1002/joc.1181.

Mote, P., L. Brekke, P. B. Duffy, and E. Maurer. 2011. "Guidelines for Constructing Climate Scenarios." *Eos* 92 (31): 257–64.

Muñoz-Arriola, F., R. Avissar, C. Zhu, and D. P. Lettenmaier. 2009. "Sensitivity of the Water Resources of Rio Yaqui Basin, Mexico, to Agriculture Extensification Under Multiscale Climate Conditions." *Water Resources Research* 45 (11). https://doi.org/10.1029/2007WR006783.

Naiman, R. J., and H. Décamps. 1997. "The Ecology of Interfaces: Riparian Zones." *Annual Review Ecological System* 28:621–58. https://doi.org/10.1146/annurev.ecolsys.28.1.621.

National Park Service. 2013. *Using Scenarios to Explore Climate Change: A Handbook for Practitioners.* Fort Collins, Colo.: National Park Service Climate Change Response Program.

National Research Council. 2001. *A Climate Services Vision: First Steps Toward the Future.* Washington, D.C.: National Academy Press.

The Nature Conservancy. 2006. *Indicators of Hydrologic Alteration.* Version 7.1. User's Manual. USA (https://www.conservationgateway.org/Files/Pages/indicators-hydrologic-altaspx47.aspx).

Nerem, R. S., B. D. Beckley, J. T. Fasullo, B. D. Hamlington, D. Masters, and G. T. Mitchum. 2018. "Climate-Change-Driven Accelerated Sea-Level Rise Detected in the Altimeter Era." *Proceedings of the National Academy of Sciences of the United States of America* 115 (9): 2022–25.

Niraula, R., T. Meixner, F. Dominguez, N. Bhattarai, M. Rodell, H. Ajami, D. Gochis, and C. Castro. 2017. "How Might Recharge Change Under Projected Climate Change in the Western U.S.?" *Geophysical Research Letters* 44(10): 407–10, 418.

Nover, D., J. Witt, J. Butcher, T. Johnson, and C. Weaver. 2016. "The Effects of Downscaling Method on the Variability of Simulated Watershed Response to Climate Change in Five U.S. Basins." *Earth Interactions* 20 (11): 1–27.

O'Donnell, A., L. Cullen, W. McCaw, M. Boer, and P. Grierson. 2010. "Dendroecological Potential of *Callitris preissii* for Dating Historical Fires in Semi-arid Shrublands of Southern Western Australia." *Dendrochronologia* 28 (1): 37–48. https://doi.org/10.1016/j.dendro.2009.01.002.

Ohlemüller, R., E. S. Gritti, M. T. Sykes, and C. D. Thomas. 2006. "Towards European Climate Risk Surfaces: The Extent and Distribution of Analogous and Non-analogous Climates, 1931–2100." *Global Ecology and Biogeography* 15:395–405.

Palutikof, J. P., D. Rissik, S. Webb, F. N. Tonmoy, S. L. Boulter, A. M. Leitch, A. C. Perez Vidaurre, and M. J. Campbell. 2018. "CoastAdapt: An Adaptation Decision Support Framework for Australia's Coastal Managers." *Climatic Change* https://doi.org/10.1007/s10584-018-2200-8.

Park Williams, A., E. R. Cook, J. E. Smerdon, B. I. Cook, J. T. Abatzoglou, K. Bolles, S. H. Baek, A. M. Badger, and B. Livneh. 2020. "Large Contribution from Anthropogenic Warming to an Emerging North American Megadrought." *Science* 368:314–18.

Perry, L. G., D. C. Andersen, L. V. Reynolds, S. M. Nelson, and P. B. Shafroth. 2012. "Vulnerability of Riparian Ecosystems to Elevated $CO_2$ and Climate Change in Arid and Semiarid Western North America." *Global Change Biology* 18: 821–42. https://doi.org/10.1111/j.1365-2486.2011.02588.x.

Perry, L. G., L. V. Reynolds, T. J. Beechie, M.J. Collins, and P. B. Shafroth. 2015. "Incorporating Climate Change Projections into Riparian Restoration Planning and Design." *Ecohydrology* 8:863–79.

Pervez, M. S., and G. M. Henebry. 2015. "Assessing the Impacts of Climate and Land Use and Land Cover Change on the Freshwater Availability in the Brahmaputra River Basin." *Journal of Hydrology* 3:285–311.

Peterson, T. C., and R. S. Vose. 1997. "An Overview of the Global Historical Climatology Network Temperature Database." *Bulletin of the American Meteorological Society* 78:2837–49.

Pittock, J. 2009. "Lessons for Climate Change Adaptation from Better Management of Rivers." *Climate and Development* 1:194–211.

Poff, N. L., D. Allan, M. B. Bain, J. R. Karr, K. L. Prestegaard, B. D. Richter, R. E. Sparks, and J. C. Stromberg. 1997. "The Natural Flow Regime: A Paradigm for River Conservation and Restoration." *BioScience* 47 (11): 769–84.

Poff, N. L., M. I. Pyne, B. P. Bledsoe, C. C. Cuhaciyan, and D. M. Carlisle. 2010. "Developing Linkages Between Species Traits and Multiscaled Environmental Variation to Explore Vulnerability of Stream Benthic Communities to Climate Change." *Journal of North American Benthological Society* 29 (4): 1441–58.

Pratchett, M. S., L. K. Bay, P. C. Gehrke, J. D. Koehn, K. Osborne, R. L. Pressey, H. P. A. Sweatman, and D. Wachenfeld. 2011. "Contribution of Climate Change to Degradation and Loss of Critical Fish Habitats in Australian Marine and Freshwater Environments." *Marine and Freshwater Research* 62:1062–81.

Prieto-González, R., V. E. Cortés-Hernández, and M. J. Montero-Martínez. 2011. "Variability of the Standardized Precipitation Index over México Under the A2 Climate Change Scenario." *Atmósfera* 24 (3). http://www.scielo.org.mx/scielo.php?script=sci_arttext&pid=S0187-62362011000300001.

Rahel, F. J., and J. D. Olden. 2008. "Assessing the Effects of Climate Change on Aquatic Invasive Species." *Conservation Biology* 22:521–33.

Redmond, K. T., and J. T. Abatzoglou. 2014. "Current Climate and Recent Trends." In *Climate Change in North America*, edited by G. Ohring, 53–94. Cham: Springer International Publishing. https://doi .org/10.1007/978-3-319-03768-4.

Rissik, D., S. Boulter, V. Doerr, N. Marshall, A. Hobday, and L. Lim-Camacho. 2014. *The NRM Adaptation Checklist: Supporting Climate Adaptation Planning and Decision-Making for Regional NRM.* CSIRO and NCCARF. https://www.nccarf.edu.au/.

Rivas Acosta, I., and M. J. Montero Martínez. 2013. "Downscaling Technique to Estimate Hydrologic Vulnerability to Climate Change: An Application to the Conchos River Basin, Mexico." *Journal of Water and Climate Change* 4 (4): 440–57. http://jwcc.iwaponline.com/content/4/4/440.abstract.

Rocha, F. 2005. *Programa de Manejo Integral de la Cuenca del Río Conchos.* Grupo Interinstitucional de Trabajo (GIT). México. https://www.agua.org.mx/wp-content/uploads/2009/05/07_fernando_rocha.pdf.

Rogers, K. 2016. *A Case Study of Good Coastal Adaptation on the Hunter River, NSW.* Case Study for CoastAdapt, National Climate Change Adaptation Research Facility, Griffith University, Gold Coast, Australia. https://www.nccarf.edu.au/.

Rogers, K., N. Saintilan, and C. Copeland. 2013. "Managed Retreat of Saline Coastal Wetlands: Challenges and Opportunities Identified from the Hunter River Estuary, Australia." *Estuaries and Coasts* 37:67–78.

Routson, C., C. Woodhouse, and J. Overpeck. 2011. "Second Century Megadrought in the Rio Grande Headwaters, Colorado: How Unusual Was Medieval Drought?" *Geophysical Research Letters* 38. https://doi.org/10.1029/2011GL050015.

Rowland, E. R., M. S. Cross, and H. Hartmann. 2014. *Considering Multiple Futures: Scenario Planning to Address Uncertainty in Natural Resource Conservation.* Washington, D.C.: U.S. Fish and Wildlife Service.

Ryu, J. H., and K. Hayhoe. 2017. "Observed and CMIP5 Modeled Influence of Large-Scale Circulation on Summer Precipitation and Drought in the South-Central United States." *Climate Dynamics* 49:4293–310. https://doi.org/10.1007/s00382-017-3534-z.

Sabater, S., and K. Tockner. 2010. "Effects of Hydrologic Alterations on the Ecological Quality of River Ecosystems." In *Water Scarcity in the Mediterranean: Perspectives Under Global Change,* edited by S. Sabaterand and D. Barcelo, 15–39. Berlin: Springer Verlag.

Sadoti, G., M. E. Gray, M. L. Farnsworth, and B. G. Dickson. 2017. "Discriminating Patterns and Drivers of Multiscale Movement in Herpetofauna: The Dynamic and Changing Environment of the Mojave Desert Tortoise." *Ecology and Evolution* 7 (17): 7010–22.

Sanderson, B. M., M. Wehner, and R. Knutti. 2017. "Skill and Independence Weighting for Multi-model Assessments." *Geoscientific Model Development* 10:2379–95.

Seager, R., M. Hoerling, S. Schubert, H. Wang, B. Lyon, A. Kumar, J. Nakamura, and N. Henderson. 2015. "Causes of the 2011–14 California Drought." *Journal of Climate* 28:6997–7024. https://doi.org/10.1175/JCLI-D-14-00860.1.

Seavy, N. E., T. Gardali, G. H. Golet, F. T. Griggs, C. A. Howell, R. Kelsey, S. L. Small, J. H. Viers, and J. F. Weigand. 2009. "Why Climate Change Makes Riparian Restoration More Important than Ever: Recommendations for Practice and Research." *Ecological Restoration* 27 (3): 330–38.

Serrat-Capdevila, A., J. B. Valdes, J. G. Perez, K. Baird, L. J. Mata, and T. Maddock. 2007. "Modeling Climate Change Impacts—and Uncertainty—on the Hydrology of a Riparian System: The San Pedro Basin (Arizona/Sonora)." *Journal of Hydrology* 347:48–66.

Shepard, D. S. 1984. "Computer Mapping: The SYMAP Interpolation Algorithm." In *Spatial Statistics and Models,* edited by G. L. Gaile, C. J. Willmott, and D. Reidel, 133–45.

Skidmore, P. B., C. R. Thorne, B. L. Cluer, G. R. Pess, J. M. Castro, T. J. Beechie, and C. C. Shea. 2011. "Science Base and Tools for Evaluating Stream Engineering, Management, and Restoration Proposals." U.S. Department of Commerce, NOAA Technical Memorandum. NMFS-NWFSC-112.

Slangen, A. B. A., M. Carson, C. A. Katsman, R. S. W. van de Wal, A. Köhl, L. L. A. Vermeersen, and D. Stammer. 2014. "Projecting Twenty-First Century Regional Sea-Level Changes." *Climatic Change* 124:317–32.

Solander, K. C., K. E. Bennett, and R. S. Middleton. 2017. "Shifts in Historical Streamflow Extremes in the Colorado River Basin." *Journal of Hydrology: Regional Studies* 12:363–77.

State of Victoria Department of Environment, Land, Water and Planning. 2016. *Guidelines for Assessing the Impact of Climate Change on Water Supplies in Victoria.* Victoria State Government. https://water.vic.gov.au/__data/assets/pdf_file/0014/52331/Guidelines-for-Assessing-the-Impact-of-Climate-Change-on-Water-Availability-in-Victoria.pdf.

Stein, B. A., P. Glick, N. Edelson, and A. Staudt, eds. 2014. *Climate-Smart Conservation: Putting Adaptation Principles into Practice.* Washington, D.C.: National Wildlife Federation.

Stewart, I. T., D. R. Cayan, and M. D. Dettinger. 2005. "Changes Toward Earlier Streamflow Timing Across Western North America." *Journal of Climate* 18: https://doi.org/10.1175/JCLI3321.1.

Stoffels, R. J., K. E. Weatherman, and S. Allen-Ankins. 2017. "Heat and Hypoxia Give a Global Invader, *Gambusia holbrooki*, the Edge over a Threatened Endemic Fish on Australian Floodplains." *Biological Invasions* 19:2477–89.

Stromberg, J. C., V. B. Beuchamp, and M. D. Dixon. 2007. "Importance of Low-Flow and High-Flow Characteristics to Restoration of Riparian Vegetation Along Rivers in Arid South-Western United States." *Freshwater Biology* 52:651–79.

Stromberg, J. C., R. Tiller, and B. Richter. 1996. "Effects of Groundwater Decline on Riparian Vegetation of Semiarid Regions: The San Pedro, Arizona." *Ecological Applications* 6:113–31.

Sun, C., and L. Ren. 2013. "Assessment of Surface Water Resources and Evapotranspiration in the Haihe River Basin of China Using SWAT Model." *Hydrological Processes* 27 (8): 1200–22.

Svoboda, P. 2017. *Kooragang Wetlands: Retrospective of an Integrated Ecological Restoration Project in the Hunter River Estuary.* Proceedings of NSW Coastal Conference, Port Stephens. http://www.coastalconference.com/2017/papers2017/Peggy%20Svoboda%20Updated.pdf.

Sweet, W. V., R. Horton, R. E. Kopp, A. N. LeGrande, and A. Romanou. 2017. "Sea Level Rise." In *Climate Science Special Report: Fourth National Climate Assessment.* Vol. 1. Edited by D. J. Wuebbles et al., 333–63. Washington, D.C.: U.S. Global Change Research Program. https://doi.org/10.7930/J0VM49F2.

Taylor, K., R. Stouffer, and G. Meehl. 2012. "An Overview of CMIP5 and the Experiment Design." *Bulletin of the American Meteorological Society* 93 (4): 485–98.

Thomson, J. R., N. R. Bond, S. C. Cunningham, L. Metzeling, P. Reich, R. M. Thompson, and R. Mac Nally. 2012. "The Influences of Climatic Variation and Vegetation on Stream Biota: Lessons from the Big Dry in Southeastern Australia." *Global Change Biology* 18 (5): 1582–96.

Thompson, R. S., and K. H. Anderson. 2000. "Biomes of Western North America at 18,000, 6000 and 0 $^{14}$C yr $_{BP}$ Reconstructed from Pollen and Packrat Midden Data." *Journal of Biogeography* 27:555–84. https://doi.org/10.1046/j.1365-2699.2000.00427.x.

Udall, B., and J. Overpeck. 2017. "The Twenty-First Century Colorado River Hot Drought and Implications for the Future." *Water Resources Research* 53:2404–18. https://doi.org/10.1002/2016WR019638.

Ullrich, A., and M. Volk. 2009. "Application of the Soil and Water Assessment Tool (SWAT) to Predict the Impact of Alternative Management Practices on Water Quality and Quantity." *Agricultural Water Management* 96 (8): 1207–17.

USAID. 2014. A Review of Downscaling Methods for Climate Change Projections.

U.S. Army Corps of Engineers, Albuquerque District. 2011. *Environmental Assessment for the Middle Rio Grande Bosque Restoration Project.* http://www.spa.usace.army.mil/Portals/16/docs/environmental/fonsi/MRG%20Bosque%20Final%20Environmental%20Assessment.pdf.

USGCRP. 2017. *Climate Science Special Report: Fourth National Climate Assessment*. Vol. 1. Washington, D.C.: U.S. Global Change Research Program. https:/doi.org/10.7930/J0J964J6.

Van Mantgem, P. J., N. L. Stephenson, J. C. Bryne, L. D. Daniels, J. F. Franklin, P. Z. Fulé, M. E. Harmon, A. J. Larson, J. M. Smith, A. H. Taylor, and T. T. Veblen. 2009. "Widespread Increase of Tree Mortality Rates in the Western United States." *Science* 323:521–24.

Vose, R. S., D. R. Easterling, K. E. Kunkel, A. N. LeGrande, and M. F. Wehner. 2017. "Temperature Changes in the United States." In *Climate Science Special Report: Fourth National Climate Assessment*. Vol. 1. Edited by D. J. Wuebbles et al., 185–206. Washington, D.C.: U.S. Global Change Research Program. https:/doi.org/10.7930/J0N29V45.

Walsh, J., D. Wuebbles, K. Hayhoe, J. Kossin, K. Kunkel, G. Stephens, P. Thorne, et al. 2014. "Appendix 3: Climate Science Supplement." In *Climate Change Impacts in the United States: The Third National Climate Assessment*, edited by J. M. Melillo, T. C. Richmond, and G. W. Yohe, 735–89. Washington D.C.: U.S. Global Change Research Program. https://doi.org/10.7930/J0KS6PHH.

Wasko, C., and R. Nathan. 2019. "Influence of Changes in Rainfall and Soil Moisture on Trends in Flooding." *Journal of Hydrology* 575:432–41.

Wehner, M. F., J. R. Arnold, T. Knutson, K. E. Kunkel, and A. N. LeGrande. 2017. "Droughts, Floods, and Wildfires." In *Climate Science Special Report: Fourth National Climate Assessment*. Vol. 1. Edited by D. J. Wuebbles et al., 231–56. Washington, D.C.: U.S. Global Change Research Program. https:/doi.org/10.7930/J0CJ8BNN.

Wilby, R., S. Charles, E. Zorita, P. Whetton, and L. Mearns. 2004. *Guidelines for Use of Climate Scenarios Developed from Statistical Downscaling Methods*. N.p.: IPCC Data Distribution Center. http://www.ipcc-data.org/guidelines/dgm_no2_v1_09_2004.pdf.

Wiles, G. J., and K. S. Kalasz. 2017. *Status Report for the Yellow-Billed Cuckoo in Washington*. Olympia, Wash.: Washington Department of Fish and Wildlife.

Williams, A. P., R. Seager, J. T. Abatzoglou, B. I. Cook, J. E. Smerdon, and E. R. Cook. 2015. "Contribution of Anthropogenic Warming to California Drought During 2012–2014." *Geophysical Research Letters* 42: 6819–28. https://doi.org/10.1002/2015GL064924.

Woodhouse, C., S. T. Gray, and D. M. Meko. 2006. "Updated Streamflow Reconstructions for the Upper Colorado River Basin." *Water Resources Research* 42. https://doi.org/10.1029/2005WR004455.

Woodhouse, C., D. Meko, G. MacDonald, D. Stahle, and E. Cook. 2009. "A 1,200-year Perspective of 21st Century Drought in Southwestern North America." *Proceedings of the National Academy of Sciences of the United States of America* 107 (50): 21283–88. https://doi.org/10.1073/pnas.0911197107.

Woodhouse, C., and J. Overpeck. 1998. 2000 "Years of Drought Variability in the Central United States." *Bulletin of the American Meteorological Society* 79 (12): 2693–714.

Wuebbles, D. J., D. R. Easterling, K. Hayhoe, T. Knutson, R. E. Kopp, J. P. Kossin, K. E. Kunkel, A. N. LeGrande, C. Mears, W. V. Sweet, P. C. Taylor, R. S. Vose, and M. F. Wehner. 2017a. "Our Globally Changing Climate." In *Climate Science Special Report: Fourth National Climate Assessment*. Vol. 1. Edited by D. J. Wuebbles et al., 35–72. Washington, D.C.: U.S. Global Change Research Program. https://doi.org/10.7930/J08S4N35.

Wuebbles, D. J., D. W. Fahey, K. A. Hibbard, B. DeAngelo, S. Doherty, K. Hayhoe, R. Horton, J. P. Kossin, P. C. Taylor, A. M. Waple, and C. P. Weaver. 2017b. "Executive Summary." In *Climate Science Special Report: Fourth National Climate Assessment*. Vol. 1. Edited by D. J. Wuebbles et al., 12–34. Washington, D.C.: U.S. Global Change Research Program. https://doi.org/10.7930/J0DJ5CTG.

Zavala-Hidalgo, J., and J. L. Ochoa-de la Torre. 2015. "Observaciones oceánicas." In *Reporte Mexicano de cambio climático, grupo I: Bases científicas; Modelos y modelación*, edited by C. Gay, C. Rueda, and B. Martínez, 35–54. Universidad Nacional Autónoma de México, México City, México. www .unam.mx.

Zhu, C., and D. P. Lettenmaier. 2007. "Long-Term Climate and Derived Surface Hydrology and Energy Flux Data for Mexico: 1925–2004." *Journal of Climate* 20:1936–46.

# 5

# Quantifying and Securing Environmental Flow

*Amy McCoy, Patrick B. Shafroth, Mark K. Briggs, Karen J. Schlatter,*
*Lindsay White, Francisco Zamora, Mauricio de la Maza Benignos, Jennifer Pitt,*
*Paul Tashjian, and Yamilette Carrillo*

## INTRODUCTION

No matter the objectives of your stream restoration project, your stream's flow regime will strongly influence how well they are achieved. During the past 150 years, many streams of the arid regions of Australia, Mexico, and western United States have been tapped to grow alfalfa and other water-intensive crops, quench the thirst of large and rapidly growing cities, generate electricity, and sustain recreational fisheries—and do it all without flooding. The natural boom-and-bust hydrological cycles of these streams have been buffered and managed to meet human needs, not those of native freshwater species. Indeed, many arid river basins in Australia, northern Mexico, and the western United States are overappropriated and overallocated, which is to say that the demand for water exceeds the amount of water that is typically available in the system. This is a significant reason why rivers run dry and why protecting and enhancing flow to better meet the needs of native stream ecosystems can be critical to the success of your stream corridor restoration program.

In developed arid watersheds, flow can be thought of as a plumbing system that begins with a large basin that collects natural contributions of snow and rain at the top of the watershed (see figure 5.1). As flow emanates from mountain regions to the desert floor, and ultimately to the river's terminus, water supplies are typically consumed in three main ways: through irrigated agriculture, evaporation (from reservoir and other open water surfaces), and municipal/industrial uses. In many semiarid and arid basins, not only is the system overallocated, but the strain on the remaining drops is likely to become more severe as populations grow and climate change intensifies (Dudgeon et al. 2006; Avril et al. 2017).

In response to these dramatic changes in streamflow, river advocates around the world are pursuing creative efforts to protect streamflow for native species (Glenn et al. 2017; Horne et al. 2017). These efforts include improving management of dams and irrigation canals, forging

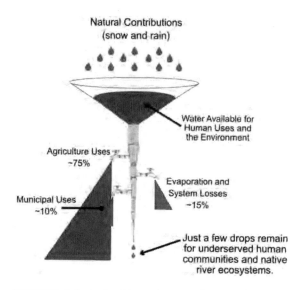

Natural Contributions
(snow and rain)

Water Available for
Human Uses and
the Environment

Agriculture Uses
~75%

Evaporation and
System Losses
~15%

Municipal Uses
~10%

Just a few drops remain
for underserved human
communities and native
river ecosystems.

**FIGURE 5.1** A plumbing schematic depicting how surface water is consumed in an arid basin as it descends from the mountains to the desert floor. Percentages are representative of several basins in the western United States, northern Mexico, and Australia. The large pool of natural water contributions at the top is progressively consumed by irrigated agriculture, evaporation, and municipal users until only residual drops remain for native stream ecosystems and underrepresented human communities. Diagram by Amy McCoy.

stronger links between groundwater and surface water use, initiating enhanced monitoring of changes in conditions, protecting free-flowing stream segments, and restoring streamside habitats and upland landscapes (Poff 2014). As defined by scientists, economists, engineers, resource managers, and policy makers from fifty nations at the 2007 International River Symposium in Brisbane, Australia, environmental flow is "*the quantity, timing, and quality of water flows required to sustain freshwater and estuarine ecosystems and the human livelihoods and well-being that depend upon these ecosystems*" (Brisbane Declaration 2007). In this book, *environmental flow* refers to efforts that aspire both to protect a stream's current flow regime from further human interferences as well as to secure additional water or change the management of flow to support the realization of stream restoration objectives.

Given that stream restoration practitioners are often looking at bare-bones project budgets, the idea of designing and incorporating an environmental flow program may seem far-fetched. However, if you are tempted to skip to the next chapter, keep two things in mind. First, an environmental flow program does not automatically mean a big expense. By identifying near-term opportunities to secure or protect small quantities of streamflow, you may be able to support key hydroecological thresholds that are important to realizing your stream restoration objectives with minimal cost and effort (the "low-hanging fruit" concept). Second, stream restoration is inherently long term. The hope is that whatever you are modestly able to accomplish today will lead to more impressive and consequential accomplishments tomorrow. You plant some native riparian trees, today, and then augment and/or protect the flow needed for their long-term survival, tomorrow.

Strategies on how to protect or enhance the streamflow of your target stream are presented in the following sections:

- Assessing the need for an environmental flow program
- Conducting a water inventory
- Environmental flow strategies and lessons learned from the field

# ASSESSING THE NEED FOR AN ENVIRONMENTAL FLOW PROGRAM

The central questions that we address in this section are: Will your stream's current flow regime likely support the species and high-quality habitat that are essential to realizing your stream restoration objective? And, if not, what might be needed as part of an environmental flow program to improve the likelihood of achieving your stream restoration objectives?

What we propose is a first-cut assessment approach that leans heavily on expert opinion complemented by a basic analysis of available streamflow data. The six steps that constitute this assessment are cost-effective and will provide insights into the potential need for an environmental flow program as well as the important gaps in knowledge that might need to be addressed later by more intensive investigatory approaches. The aim of the first-cut approach is to place stream restoration projects into one of three environmental flow categories or camps:

- *Camp One (not needed):* An environmental flow program is not needed to achieve stream restoration objectives, at least at present or in the near term (i.e., up to a decade in the future). In other words, your stream's flow regime appears conducive to the realization of the stream restoration objectives that you and your team have put forward.
- *Camp Two (needed):* An environmental flow program is needed and should be an integral component of the restoration response in order to achieve stream restoration objectives as they are currently stated. The camp will likely also include situations where important gaps exist in knowledge and data regarding current streamflow conditions compared to what is needed to realize restoration objectives.
- *Camp Three (Holy smokes):*\* The "Holy smokes!" camp covers situations where potentially glaring gaps exist between current streamflow conditions and what is needed to realize restoration objectives. These gaps may be so eye-opening as to force practitioners to reassess the entire design of their stream corridor restoration project.

Before we embark on the first initial assessment, we encourage attention to three pieces of background information that are needed to make the first-cut assessment effective. First, make sure you have developed a sound and realistic stream corridor restoration objective or objectives (see chapter 2). Whether your restoration objective is focused on physical, chemical, biological, or socioeconomic parameters or a mix thereof, a sound and realistic restoration objective is essential to the first-cut approach, providing the lens that will allow you to narrow a multitude of questions related to environmental flow needs to those focused specifically on what is most critical to realizing your stream restoration objectives.

Second, have a sound understanding of the current hydrological condition of your stream and its watershed. Background information and strategies for evaluating the hydrological condition of your stream and its watershed are the focus of chapter 3 and provide the foundation for any assessment of environmental flow needs. In addition, identifying key ecological

---

\* For readers not familiar with this term, "holy smokes" is an informal, relatively polite way of expressing astonishment at a surprising situation. Like, "holy smokes, you cut your own hair!"—often exclaimed with a hand on forehead.

thresholds that are related to your restoration objective and how they could be affected by changing streamflow conditions is essential. Will average streamflow conditions foster an environment where key ecological thresholds related to your restoration objectives are routinely compromised? This is one of the key questions that needs to be answered as part of understanding environmental flow needs.

If you are not familiar with the concept of ecological thresholds, a couple paragraphs on this topic with a few references may be helpful as background. An ecological threshold is the point where a small change or disturbance in external conditions causes a dramatic change in an ecosystem. One of the important aspects of ecological thresholds is that once they have been breached, the ecosystem may no longer be able to return to its previous state by means of its inherent resilience (Huggett 2005; Groffman et al. 2006). Much research has focused on defining and identifying ecological thresholds and describing processes of transition to alternative stable states once these breakpoints are breached (Bestelmeyer et al. 2003; Huggett 2005). In the context of considering your stream's flow regime, the key question we need to address is whether changes in the flow regime (as a result of human interferences, including climate change) have breached ecological thresholds associated with your restoration objectives, leading to alternative species and habitat in the stream community. For example, if reestablishing a native fish species is central to your stream restoration objective, have human manipulations altered your stream's flow regime to a point where your target fish species can no longer survive (e.g., due to reduced flow, lack of timely flooding, deterioration in water quality)?

Thresholds are, however, difficult to predict, as they depend on a number of factors, including landscape characteristics, species traits, and nonlinear relationships between species and the environment (Lindenmayer and Luck 2005; Roni et al. 2008; Suding and Hobbs 2008). In addition, the interactions among environmental drivers may affect threshold values and produce complex responses in species distribution and complicate the outcomes of restoration (Pittman and Brown 2011; Trigal and Degerman 2015). For example, as part of reintroducing a native fish species, your restoration program may aim to increase habitat heterogeneity by adding large woody debris. Whether or not such restoration tactics are successful will depend on how well those tactics will actually improve habitat heterogeneity as well as numerous regional variables, such as the flow regime of your stream.

This is where species distribution models can come into play as a potentially helpful tool. Such models evaluate habitat suitability and the existence of thresholds in species occupancy over large spatial and temporal scales and often include nonlinear relationships between species occurrence or abundance and habitat variability. However, developing such models can be time-consuming and costly. This is one of the reasons we advocate conducting the first-cut approach, which relies heavily on expert opinion. That noted, habitat distribution models may be something to look into in the future depending on the results of the first-stage approach and other considerations. If you want to learn more, the following references provide additional background and examples of how distribution modeling can aid conservation and restoration decision-making: Elith and Leathwick 2009, Guisan et al. 2006, Guisan et al. 2013. For now, the salient point is simply to understand in general terms what ecological thresholds are and what your technical experts may be considering as part of the first-cut assessment.

Finally, third, before undertaking any assessment of environmental flow needs (first cut or otherwise), it may be helpful to familiarize yourself with some of the literature regarding quantification of environmental flow. This does not have to be a deep dive at this point; just gain enough familiarity to provide context for the first-cut approach that we are supporting.

The quantification of environmental flow needs is supported by a trove of research and literature. Over the past fifteen years, well over 250 methodologies have been developed that offer a wide range of options for determining environmental flow needs in a diversity of ecological and social landscapes (Tharme 2003; Horne et al. 2017; Poff 2018). Below, we provide additional paragraphs as background. As an entrée, begin with the three aforementioned references, which provide an excellent and comprehensive summary of quantification methodologies, many of which are hydrological characterizations of the natural flow regime (or simply hydrological characterization methodologies).

Most hydrological characterization methodologies are built upon advances in river science regarding (1) the *natural flow* regime of a river, which can be defined as the magnitude, frequency, duration, timing, and rate of change of a stream (Poff et al. 1997); (2) the *connection* between flow variability and ecosystem processes (Naiman et al. 2008); and (3) the *relationships* between river structure, physical processes, and biological communities (Fremier and Strickler 2010; Wohl 2012; Yarnell et al. 2015). Early approaches were aimed at quantifying target flow recommendations or minimum deviations from average flows over time. These methodologies include the Instream Flow Incremental Methodology (IFIM) and the Physical Habitat Simulation (PHABSIM) system (Petts 2009; Bouwes et al. 2011). IFIM points to relationships between habitat suitability variables for target fish populations (most often velocity, depth, and substrate or cover) and streamflow variability (Bovee et al. 1998). PHABSIM provides inputs for IFIM by producing "weighted usable areas" that predict the carrying capacity of the stream for certain aquatic species based solely on physical conditions (Bovee 1978). These methodologies are rather narrowly focused and do not incorporate the full range of relevant stream dynamics. However, some federal agencies continue to use these models to establish minimum instream flow volumes to maintain life cycle stages of target species. They are mentioned here for restoration practitioners who are working closely with these agencies on recovery efforts for endangered fish species (e.g., minimum flow requirements for coastal and tributary streams in support of salmon migrations).

Over the past decade, hydrological characterization methodologies have shifted toward more integrated and holistic assessments that incorporate a wide range of influential factors on the flow regime, such as human interference, climate patterns, species recovery, surface water and groundwater interactions, socioeconomic considerations, and relevant legal frameworks. For example, data-driven efforts such as the Ecological Limits of Hydrologic Alteration can help prescribe a detailed and integrated view of flow needs (Poff et al. 2010), while other strategies aim to downscale the flow regime to sustain seasonal patterns at lower magnitudes (Hall et al. 2011). Methods such as these assume that the connection between flows and ecological processes can be recovered. Such data-driven efforts can be expensive and time consuming, potentially putting them beyond the reach of stream restoration projects where data, funding, or other resources may be limited. An additional challenge for using these methodologies in highly altered river systems is the potential of spending significant amounts of money and time to quantify flow needs or prescriptions that may never be achieved given the political, cultural, and socioeconomic landscape that your stream passes through. This may especially be the case for streams whose hydrology has been significantly altered by dams, diversions, groundwater pumping, and development. The bottom line: when evaluating flow needs, recognize the amount and quality of water that is realistically available.

More recently, another type of approach, Designer and Functional Flow Assessment (DFFA) methodologies, is being used to quantify environmental flow needs. DFFA methodologies acknowledge the often-insurmountable challenges associated with restoring a stream's flow regime to its pre-disturbance condition. They also recognize the great value of (1) taking advantage of observations made by natural resource experts (biologists, hydrologists, etc.) versus extensive data collection and modeling (at least as a first step to understanding environmental flow needs), and (2) focusing on key elements of a stream's flow regime that are critical to realizing restoration objectives (Acreman et al. 2014). The first-cut strategy that we put forward below is essentially a modified DFFA approach.

## A Proposed Strategy

With the above homework in hand, below are the five steps that constitute the first-cut environmental flow assessment. There is heavy reliance on technical expertise as well as streamflow data. If your stream is not gauged, there are methods for characterizing your stream's flow regime when flow data are not available (see chapter 3). It is likely going to be an iterative process where you find yourself going through the five steps a couple times over a period of a year or so. For example, you could decide to do a first pass through the steps completely on your own during the course of a couple days (e.g., a day collecting background info and data, maybe another doing some basic analysis, a third day in the field). Based on the insights you gained during the first pass through the five steps, you will probably need to do a second pass (maybe a couple months later) with several members of your technical team that involves a deeper dive into collecting, analyzing, and discussing background information and data. The first pass addresses some questions, raises others. The second pass provides clarity while likely identifying grey areas where further work is needed. Regardless, the intention of our proposed strategy is to provide a relatively straightforward, inexpensive, and rapid process that allows you to gauge potential environmental flow needs for your stream.

### Step One: Understand Your Watershed's Land-Use History

In the context of assessing environmental flow needs, of particular interest to us when it comes to the land-use history of your watershed is pinpointing to the extent possible the time period when human development in your watershed likely began to dramatically influence the flow regime of your stream. If you are able to do this (and if you have streamflow data), it offers the opportunity to divide and analyze your streamflow record into pre- and postdevelopment periods. Of course, the grand majority of watersheds in the world have experienced human development for hundreds if not thousands of years. What we are looking for here are the dramatic human impacts (often contemporary) that likely have had a dramatic impact on your stream's flow regime. In some cases, this may be relatively straightforward as in the construction completion date of a large dam or water diversion project (see "Basic Analysis of Streamflow Data from Multiple Gauges" in chapter 3, which provides an example of analyzing and comparing long-term streamflow data between unimpounded and impounded periods along the Rio Grande/Bravo). In other cases, development may be much more gradual and more difficult to pinpoint (e.g., gradual urbanization or deforestation over a long time period). In such cases, a comparative hydrological analysis of pre- and postdevelopment periods may not be possible, and the analysis may have to focus on the entire flow record.

One additional point on land use. Do not restrict your investigation to your watershed's land-use history; also investigate potential or planned changes in water use and management. If dramatic groundwater pumping is being planned that could impact the flow regime of your target stream, such information could be vital for not only determining the potential need for an environmental flow program but assessing the vitality of your entire stream restoration response. For example, even if the first-cut evaluation places your stream restoration project in Camp One (no need), the prospect of significant groundwater pumping taking place in the near future may move you to Camp Two and underscore the need to place greater attention on developing an environmental flow program in the future (to reduce or eliminate the impacts of groundwater pumping and/or to secure additional drops to compensate for water lost to pumping). Chapter 3 provides greater detail on the importance of and methods for understanding your watershed's land-use history.

### Step Two: Conduct a Basic Analysis of Streamflow Data

As discussed in chapter 3, there are several basic analyses that can be done relatively easily and cheaply (given data availability) that will provide information pertinent to this first-cut assessment. Developing an annual hydrograph is a productive way to start, as it will allow you to see how your stream's flow regime changes over the course of the year (e.g., by season: summer versus winter, wet season versus dry season, etc.). If you are able to divide your flow record into pre- and postdevelopment periods (per Step One), comparing annual hydrographs depicting pre- and postdevelopment periods can provide informative insights into how dramatically streamflow has changed between two periods of time (i.e., between a period relatively unaffected by human interferences and one that has been dramatically affected). An excellent example of such an analysis is Schmidt et al. 2003, who used such an assessment to quantify hydrological changes along the RGB between prelarge dam and postlarge dam periods. Depending on your restoration objective, developing a flood frequency curve can also be valuable. Both of the above tools are reviewed in chapter 3. In addition, as noted already, if your stream is not gauged and does not have a flow record, chapter 3 reviews methods for assessing flow for ungauged streams.

### Step Three: Identify Parts of Your Stream's Flow Regime That Are Most Critical to Realizing Restoration Objectives

Certainly, all streamflow that occurs during the course of the year is important for stream biota and in supporting various aspects of your stream restoration objective. But some aspects of your stream's flow regime will stand out not only for their importance in realizing your stream restoration objectives but also for the extent they have already been altered by human impacts or are likely to be impacted in the future. Using your stream restoration objectives as a filter or lens, the technical experts on your team will likely be able to identify the parts of your stream's flow regime that are most critical to realizing your restoration objectives. Narrowing to the critical parts of your stream's flow regime is the objective here. Use your stream restoration objectives along with the technical expertise to help you narrow assessment, and then get to the field* with your technical experts to assess conditions.

For example, if your restoration objectives include reestablishing native fish, the fish biologists on your team can identify times of the year when target fish species may be most vulnerable to changing flow conditions as well as when specific flow conditions are essential to realizing long-term objectives (e.g., high flows during spring to spark reproduction or low flow thresholds to get the species through the hot, dry season). If the low-flow period during the hot, dry months is identified by your technical team as particularly critical, heading to the field with your team (maybe consisting of a fish biologist, hydrologist, and local manager or land owner—that is, someone who knows the area very well) can potentially provide great insight as to the streamflow needed to maintain high quality habitat in key pools and backwater areas during the period when precipitation is scarce. While in the field, the fish biologist will have the expertise to identify key pools and other types of habitat. The hydrologist may be able to estimate the number of days of no-flow if your stream is intermittent (which you will know based on the preliminary hydrologic assessment you have done) that will make the pools go dry or dip below important water-quality thresholds. The rancher whose land the stream passes through may be able to describe specific hydrological conditions with confidence (for example, how many days without streamflow will it take for pools identified as critical by your fish biologist to go dry during the dry season). Such insights, along with an analysis of the streamflow record on frequency and length of no-flow periods, can allow you to begin to describe and even estimate minimum-flow requirements—the flow needed to maintain pools at desirable levels and quality during the hot, dry season. If average streamflow characteristics during the dry season appear in the ballpark of maintaining threshold pool conditions (as illuminated by analysis conducted as part of Step Two, above), you may be in Camp One. If there's a wide gap, you may be in Camp Three.

On the other hand, if your stream restoration objective focuses on reestablishing native riparian trees, you probably want to get a riparian ecologist involved, and he or she will likely be interested in the high-flow period during the spring when trees are producing seed and the low-flow periods in the summer when low moisture availability may be an issue. If reestablishing desirable channel morphological conditions is a restoration objective, the geomorphologist on your team may focus on the occurrence of flood events of estimated discharge and duration deemed capable of moving and distributing channel alluvium. Stream restoration objectives that focus on water quality may highlight low-flow periods as well when key water-quality parameters of concern become most concentrated or periods when contaminated runoff is most worrisome (such as when agricultural return-flow contaminants are most concentrated). And so forth. Use your stream restoration objectives to narrow the focus of analysis.

## Step Four: Narrow Further Based on What Is Realistic

Next, even at this early stage in assessing the potential need for environmental flow, give some thought as to what potentially can *realistically* be addressed by an environmental flow program given the socioeconomic and political landscape that your stream passes through. Of the

---

* Please make sure to consult chapter 3 (specifically, the section on field reconnaissance), which summarizes a variety of strategies for getting the most out of field visits to your stream.

parts of your stream's flow regime that may need to be supported by an environmental flow program (in order to meet your stream restoration objectives), there may be parts that could be addressed by an environmental flow program, and others not. Focusing on the realistic parts will allow you to spend your time and money on evaluating the flow needs associated with these critical sections of the hydrograph.

For example, if your experts have identified high-magnitude flows (such as those needed for reproduction of native fish or riparian plants) as a critical part of the annual hydrograph, discuss frankly with your team whether securing and protecting such high-flow events as part of an environmental flow program is realistic. An environmental flow program focused on high-flow events will likely require securing large volumes of water for environmental purposes, and there may be significant challenges in getting a high-magnitude discharge from its release point to where it is needed (if your stream is impounded, for example). These challenges could include infrastructure limitations that cannot accommodate the desired discharge, potential threats posed by a large flow event to downstream users, or the timing of release with respect to conflicting needs of other water users.

This is not to say that high-flow events can't be part of an environmental flow program (see, for example, the Bill Williams case study, Case Study 5.2; and the Colorado River Delta Pulse Flow study, Case Study 5.5), but we promote a more strategic process where practitioners focus their evaluation energies first on low-hanging fruit before trying to reach the highest orange. In this vein, maybe instead of trying to secure large volumes of water needed for a pulse or flood flow, begin by considering what might be needed to augment flow during the hottest, driest months of the year in order to maintain key physical, biological, and chemical thresholds that are central to your stream restoration objective. In comparison to orchestrating a large flood event, securing a small bump in discharge during the hot, dry months may be more politically and socioeconomically realistic.

With that in mind, it is understood that you may simply not have sufficient information at this point to identify the wanting parts of your steam's flow regime that could be realistically addressed by an environmental flow program. But it is prudent to begin thinking about the *realities* of securing or protecting flow as you assess the significance of the potential gap between current streamflow conditions and what is needed to realize your stream restoration objectives.

### Step Five: Focus Further Hydrological Analysis on the Portions of Your Stream's Flow Regime Deemed Most Critical and Realistic

To answer the question, Is an environmental flow program necessary to realize stream restoration objectives? your technical team may identify the need for additional investigations. Avoid laundry lists of investigation needs. Be targeted and keep it simple. Don't go immediately to modeling and other time-consuming and expensive endeavors. Additional investigations that you and your team feel are needed should be focused on the portions of your stream's flow regime deemed most critical to realizing your stream restoration objectives and most realistic. Additional investigations could include further analysis of the average flow during the critical time period (e.g., standard deviation around mean discharges), the length and frequency of no-flow periods (or frequency of occurrence of a minimum discharge during a critical time period), the recurrence interval of specified discharges, and maxima or minima discharges. The computation of these and other selected parameters are reviewed in chapter 3 and can often be conducted with minor cost by a trained hydrologist.

In addition to the parameters mentioned above, consider computing a simple index that describes the consistency of flow during the critical and realistic parts of your stream's flow regime that you have identified. This can be particularly important during low-flow periods, when ecological thresholds for key species may be compromised if flows drop below minimum thresholds for key species for extended periods of time. The example below is for a time period of low flow, but can be computed for any season or time of year that practitioners feel is critical.

$$\text{Baseflow index} = \text{lowest mean daily flow during low-flow}$$
$$\text{period mean flow for low-flow period} \times 100$$

Flows are expressed as average discharges in $m^3$ per second. An index value near 1 indicates a fairly constant flow during the period of interest, whereas a value near zero is indicative of flow with high variability (after Hamilton and Bergersen 1984).

To offer an example using figure 5.2, for the years 2007–2017, the lowest minimum daily flow from mid-April to July 1 along the RGB (per flow data collected at Rio Grande Village streamflow gauge) is 1.3 $m^3s^{-1}$, while the mean flow for that time period is 7.9 $m^3s^{-1}$, giving a very high variability index of 0.2. Given how low the average flow is along this reach of the basin during the hot, dry season, such high variability of flow means that there will be years when flow during this critical season frequently drops below average, underscoring the potential need for supplementing environmental flow in support of key species to achieve stream restoration objectives during this critical time of the year. In the Rio Grande Basin, an international plan is focusing on how to store a conservation pool of water in an upstream reservoir (water legally held by an NGO) and call for the water during the low-flow season.

Finally, do not forget about water quality. Although water quality data are often not as readily available as streamflow data, there are many regions where water-quality measurement has been conducted over an extended time. The focus again is to characterize your stream's water quality (for whatever parameters are available) during the time periods that you and your team have identified as critical.

Remember, this is a first-stage assessment. If you wind up in Camp Two with key gaps in knowledge identified, the information gathered here can help to prioritize the investigations needed to address those gaps (such as hydraulic modeling to accurately quantify the discharge required to inundate key backwater areas). Regardless, conducting such a DFFA-oriented analysis is an effective way of providing the information needed for a thoughtful assessment of environmental flow needs.

If results of the above investigations put you in Camp One, you can at least breathe easy about the immediate need to pursue an environmental flow program. However, per chapter 4, do not forget to consider the impacts of climate change. Depending on climate forecasts for your region and your stream's current flow regime, the impacts of a rapidly changing climate (particularly in areas where climate forecasts predict a hotter, drier climate, such as the southwestern U.S., northern Mexico, and Australia) could easily—and rapidly—push you from Camp One to Camp Two or from Camp Two to Camp Three. Chapter 4 provides a review of climate change, a summary of methodologies to better understand what climate change means

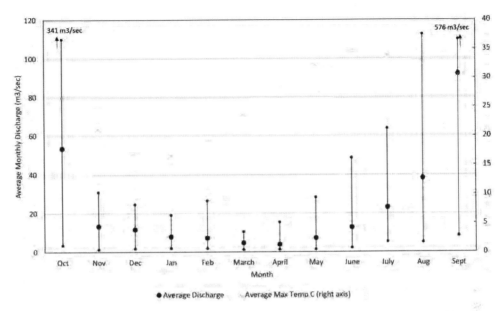

**FIGURE 5.2** Graph of average monthly discharge (with maximum and minimum monthly averages) for 2007 to 2017, combined with average monthly maximum temperatures for Rio Grande Village, RGB, along the U.S.-Mexico border. Basic hydrological information organized in such a manner can be useful to hone environmental flow information needs. In this case, a priority is to assess environmental flow needs for key native fish during the hot, dry season of April to June, when low flows can drop critical biophysical and chemical variables below the thresholds needed for the survival of key native fish. Graph by Mark Briggs.

for your region, and the types of adaptation strategies to improve the likelihood for stream restoration success in a rapidly changing climate.

If results of above strategy put you in the "Holy smokes!" Camp Three, several options are available, including stopping the project altogether, changing restoration objectives, and changing restoration tactics. Not coincidentally, these same approaches are reviewed in chapter 4 as potential ways to strengthen the climate-adaptive capacity of your stream restoration project.

If your results place you in Camp Two, you will probably feel comfortable moving forward with your stream restoration project while pursuing investigations to fill priority gaps in knowledge to better quantify environmental flow needs as well as to identify strategies and opportunities for protecting or securing environmental flow. This is the focus of the rest of this chapter.

## CONDUCTING A WATER INVENTORY

Conducting a water inventory provides the information needed to describe the entities who have the greatest impact on water use and management (for both surface water and groundwater) in your watershed (Poff 2018). A water inventory provides insights into how water resources in your basin might be managed differently or where additional drops might be found in support of your stream restoration objectives. The greater your understanding of the water users and water managers impacting your reach of water, the easier it will be to develop the relationships necessary to approach, develop, and execute your vision. We'll look at three elements of a water inventory: (1) the water users, (2) the water managers, and (3) ways to change management of, or obtain legal permission to use, water that is already allocated.

## Water Users

Water users typically fall into one of three sectors: agricultural, industrial, and municipal. An inventory of users of surface water and groundwater describes how water is legally allocated and accounted for in a basin or watershed (including the types and priorities of water rights—or concessions, in the case of Mexico), who uses the most water, and which water rights might be best to pursue as part of an environmental flow program. Specifically, an inventory of water users will identify (1) the water rights owners (from oldest right to the most recent claims), (2) the location in the watershed where water is being withdrawn to satisfy their water rights, and (3) the total volume of water they are using.

For example, in the western United States, surface water is managed according to the doctrine of prior appropriation, which gives those who first made "beneficial" use of the water the highest priority right, allowing them to receive their full measure of allotted water each year before lower priority users. In this context, identifying water users who have the oldest rights will give you a sense of the rights in your basin that will be fulfilled first during times of drought or shortage. Sources of information on surface water users vary with country, but in general, information on both water users and managers can be found with the entities in charge of managing water.

In Australia, surface water and groundwater management falls to the six states and two territories of the Commonwealth, which means that there are eight different approaches to granting permission to use water resources. While there have been recent efforts to coordinate and align rights-to-use across state boundaries, the governments of each state own the rights to all water (surface water and groundwater) and have authority to grant access to use it. The states and territory water managers house much of the information on water users. This is more or less the situation in Spain and other European countries. For Australia, additional information on water users can be found at the federal level via Commonwealth agencies such as the Murray-Darling Authority.

In Mexico, CONAGUA is a key source of information on water users in all the basins this federal agency manages. At state and local scales, river basin councils have been created to allow the voice of local water users (agriculturists, municipalities, and industry) to be heard in water-management decisions. In 1997, thirteen hydrological administrative regions were formed to integrate socioeconomic issues across hydrological basins and state and municipal boundaries (Arias-Rojo and Salmón-Castelo 2019). Surface water (which provides roughly 80 percent of the water consumed in Mexico) and groundwater are managed separately (Arias-Rojo and Salmón-Castelo 2019). This effectively means that any water inventory will be situated within the unique dimensions of each administrative region and will most likely focus predominantly on surface water use.

In the western United States, Reclamation and the USACE are the primary federal agencies that oversee the management of interstate rivers and account for water allocations and deliveries. In addition to managing water deliveries, Reclamation can hold water allocations on interstate rivers and is thus a water user as well. States are also responsible for allocating and certifying surface water rights and can hold rights as a water user.

An inventory should identify users of groundwater as well as surface water; this is important, given how the overpumping of groundwater can and does impact surface flow. Gathering information on groundwater use is often more complicated because in many regions of the world such use is not allocated by rights or claims. Without an official tracking or allocation system, anyone can install a well and pump as much groundwater as they would like. In addition, in places without rigorous oversight and monitoring, people can and do install wells and pump water illegally or without reporting their use to the management authorities. In all of these instances, unregulated, unreported, or unrestricted groundwater pumping can increase the rate of aquifer drawdown, impact rivers and riparian habitat, and jeopardize the reliability and durability of supplies for everyone.

In the western United States, legal constraints on the use of groundwater vary by state. Some states have developed basin-specific management plans where surface water and groundwater sources are managed conjunctively and where groundwater use permits are required. Such permitting allows relatively easy access to information on the location of pumping and who is pumping, along with how much they are pumping and for what purpose. Such basin-specific management plans are often associated with urban parts of the state where groundwater decline has become more apparent and thus an issue, leaving the rural parts of the states (and the agricultural sector) relatively unregulated. From a stream restoration perspective, this effectively means that anyone can sink a new well and start pumping, even if it negatively impacts stream conditions. Consequently, it is difficult to account for groundwater pumping in a water-use inventory. In regions where such use is unregulated, you may need to develop relationships with the local community to learn who is pumping, where, and for what purpose. In Australia, groundwater is a public resource and is managed within each state. Groundwater entitlements are granted in the form of a license, which is like a water property right and has historically been linked to the overlaying land property right. However, more recently under the National Water Initiative, land and water rights have been separated to allow for transfers and trading among groundwater users. Typically, a groundwater license specifies the condition of use, length of tenure, and volume that is permitted for extraction. In Mexico, groundwater is managed at the national level by CONAGUA. The National Water Law established hydrological basins, within which groundwater entitlements are granted and managed. The groundwater entitlement system is based on reasonable use and is managed conjunctively with surface water. Sources of information on groundwater use are summarized in Table 5.1.

## Water Managers

Water management entities develop, measure, and deliver water supplies almost solely in response to human needs, which is reflected in the fact that streams are essentially used as conveyance channels to get water from point A to point B. Water management typically

**TABLE 5.1** Selected sources of information on surface water and groundwater users and managers in Australia, Mexico, and the western United States as part of developing a water inventory.

| Topic | Source of Information |
|---|---|
| **Australia** | |
| Surface water use | Australian Government, National Surface Water Information site: http://www.ga.gov.au/scientific-topics/national-location-information/national-surface-water-information |
| Groundwater use | Australian Government, Groundwater Use site: http://www.ga.gov.au/scientific-topics/water/groundwater/basics/groundwater-use |
| Management | Murray-Darling Basin Authority: https://www.mdba.gov.au; Australian Water Markets Dashboard: http://www.bom.gov.au/water/dashboards/#/water-markets/national/state/at; Commonwealth Environmental Water Office: http://www.environment.gov.au/water/cewo |
| **Mexico** | |
| Surface water use | CONAGUA: https://www.gob.mx/conagua |
| Groundwater use | |
| Management | SEMARNAT: https://www.gob.mx/semarnat |
| **Western United States** | |
| Surface water use | Arizona Department of Water Resources: https://new.azwater.gov; New Mexico State Engineers Office, Water Rights Reporting System: http://www.ose.state.nm.us/WRAB/index.php; Texas Water Development Board, Surface Water Resources Division: http://www.twdb.texas.gov/surfacewater/index.asp |
| Groundwater use | USGS Groundwater Information site: https://water.usgs.gov/ogw/ |
| Management | USGS, daily streamflow data: https://waterdata.usgs.gov/nwis/rt; Reclamation: https://www.usbr.gov/main/water/; USACE: https://www.usace.army.mil |

*Note*: The western states listed above are provided as selected examples. Other western states have similar and comparable departments.

combines top-down policies and institutions with more localized and basin-specific water-governance procedures. This human-centric approach to water management is one of the most important reasons why environmental flow programs are needed and why securing environmental flows are such a challenge.

The Australian Constitution states that individual states and Commonwealth territories are primarily responsible for water management and the creation of water rights, which are defined as permanent water-access entitlements (the right to receive water each year) and temporary water allocations (the physical water available for use) (Garrick et al. 2018). There is some variation across states and territories as to how each manages, distributes, accounts for, and reports on water use. From late 1996 to mid-2010, the Murray-Darling Basin in southeastern Australia endured the most severe drought in the country's historical record (i.e., since the late nineteenth century [Leblanc et al. 2012]). In response to this drought and the stress it exerted on farmers, cities, and the environment, Australia invested in environmental flow programs that utilize market-based sales and leases of water rights. These programs could be developed in part because the Murray-Darling Basin is shared across five states and has a long history of joint water and financial management to support water distribution and the water-management infrastructure (Garrick et al. 2018).

In Mexico, CONAGUA is the federal agency responsible for managing almost all large dams and water diversions, while irrigation districts and farmers work together to distribute and maximize releases. Regionally, most states in Mexico have state water commissions (CEAs), autonomous entities under the authority of the State Ministry of Public Works; the CEAs provide technical assistance on water supply, distribution, and sanitation to municipalities and irrigation districts.

As noted above, in the western United States, federal agencies like Reclamation and the USACE manage most large dams (typically those considered to be over 50 ft. [15 m] in height), though there are also a significant number of dams that are managed by private entities. In some U.S. states, state agencies such as the Central Arizona Project or the New Mexico Office of the State Engineer have a strong say in how water is managed for delivery to key stakeholders, such as municipalities and irrigation districts.

For transboundary rivers, water management and allocation are often set through international treaties. For example, along the border between the United States and Mexico, waters of the Colorado River and RGB are distributed between the two countries under the 1944 U.S.-Mexico Water Treaty, with refinement of obligations taking place via amendments ("Minutes") to the treaty. The treaty is administered binationally, in the United States by the State Department through the U.S. sector of the International Boundary and Water Commission (IBWC-US) and in Mexico by the Secretaria de Relaciones Exteriores through the Mexican sector of IBWC (CILA).

The bottom line for practitioners is the importance of identifying (and getting to know) the agencies and organizations—from federal governments to local water utilities and irrigation districts—that will need to be consulted in some fashion when a change to the volume, timing, location, or purpose of water deliveries is made.

## Strategies to Change Flow Management and/or to Obtain Permission to Use Water

Typically, the path toward obtaining legal permission to use water starts with discussions and negotiations involving your organization, targeted water users in the basin, local water managers, regional (and sometimes national) regulating agencies, and funders who might help pay for the water. Among other benefits, such discussions provide opportunities for you to share your stream restoration goals and objectives with managers and users, including the flows that are needed to support them, and set the stage for the collaboration required to develop legally binding water agreements built on common interests for river flows and conditions. Such discussions take time! For example, it took over a decade of such discussions to orchestrate the pulse flow in support of human and environmental objectives in the Colorado River Delta (see Colorado River Delta Pulse Flow, Case Study 5.5).

National and state water policies will determine how a public or private entity (such as an environmentally focused NGO) can apply to relevant water management entities to obtain legal permission to use water that is already allocated or work to change its management for environmental benefit. Several types of formal agreements may be required to legally recognize a change in water use or management; examples are provided below. Although it is useful to work with an attorney to ensure that you are not missing important details in developing legal agreements or contracts, the best agreements are written in clear and descriptive language that outlines what you hope to accomplish in identifying, obtaining, and using water (Aylward 2013). While typically the details of these agreements are confidential and maintained between the water rights holder, the state approving agency, and the organization that helped to broker the deal, some useful resources can help guide you through the legal and policy pathways to conceptualize an actual agreement. In the United States and in the Colorado River Delta in Mexico, the River Network (https://www.rivernetwork.org/our-impact /how-we-help/ample-water/best-practices/exploring-voluntary-water-transactions) provides a very useful summary of resources, including the *Environmental Water Transactions Practitioner's Handbook* (Aylward 2013). The Murray-Darling Basin Authority in Australia maintains information on water markets and trades (https://www.mdba.gov.au/managing-water /water-markets-and-trade).

## Agreements Between the Current Water Rights Holder and Your Organization

An agreement (or multiple agreements) between the current water rights holder who is working with you and your organization serves to define and confirm the actions that the water rights holder will take to make water available in support of your restoration or environmental flow goals (see Case studies 5.2 and 5.5). Actions summarized in the agreement that the current water rights holder will need to make could include (but are not limited to):

- filing a change application to permanently dedicate water to an environmental need;
- articulating a leasing agreement whereby a specific amount of water for a predetermined amount of time is routed to a restoration need;
- shutting down or reducing the rights holder's irrigation operation on an agreed-upon date to temporarily reroute water to river flows or a restoration site;
- monitoring and accounting to verify that the water moving from the water rights holder to the environmental need is the agreed-upon amount.

## Agreements with a Funding Entity

In situations where an outside funding entity is paying costs associated with altering a water right (e.g., purchasing the water right or water itself, or costs associated with developing the agreement), an agreement outlining the relationship between the entity representing the environment and the source of funding may be needed. The funder could be a private foundation, public agency, or individual, any of whom could be interested in investing in environmental flows and river restoration outcomes. For example, in the Verde River in Arizona, several corporations have worked with the Bonneville Environmental Foundation to improve the efficiency of irrigation infrastructure and agricultural water use to restore and sustain flows in the river. Their interest in supporting flow restoration stems from corporate sustainability goals to invest in surface water and groundwater replenishment projects.

The key elements of such an agreement include an assessment of the costs associated with the agreement (including the water or infrastructure costs associated with changed management procedures), process requirements for disbursing funds, periodic financial reporting, potential matching funds, taxes, and so on. It should be noted that determining these costs, including the amount of money that should be paid to the water user, may require an official appraisal, which is a process that exceeds the focus of this chapter. A thorough summary of the appraisal process can be found in Aylward 2013.

### Agreements with the Water Management Agency or Agencies

An agreement may be required with the water management authority that focuses on the administrative process for reviewing and approving the legal transfer from one use of water to another (e.g., from irrigation to minimum flow for restoration). Transfers can be either temporary (lease) or permanent. This type of agreement allowed water rights to be moved from irrigated agriculture in support of riparian restoration efforts in the Colorado River Delta, for example (Carrillo-Guerrero et al. 2013; see also Case Study 5.2).

In addition, make sure to reach out to the agencies in your watershed that are in charge of water management to see if there are established protocols for securing environmental flow. For example, in Mexico, La Norma Mexicana de Caudal Ecológico (NMX-AA-159-SCFI-2012), or the Ecological Flows Policy of Mexico, provides a process to calculate flow requirements to maintain ecosystem services for watersheds in Mexico. Ecological flow needs calculated through this process can be used to justify the protection of water resources through establishment of a Water Reserve, a legal mechanism in Mexico that requires federal protection of water for environmental uses. Such water reserves are being established in key river basins throughout Mexico to sustainably manage water resources for people and the environment (see Case Study 5.1).

## ENVIRONMENTAL FLOW STRATEGIES AND LESSONS LEARNED FROM THE FIELD

A variety of strategies can be pursued as part of an environmental flow program. Which strategies you select for your restoration project will depend on the water management and water use landscape that your stream passes through and the stream restoration objectives of your project. Keep in mind that regardless of the suite of strategies you ultimately identify as best for your restoration project, progress in developing an environmental flow program will typically be incremental and will potentially involve multiple strategies. In this final section, we present case studies that underscore critical points learned in the field on differing aspects of protecting and augmenting streamflow in the name of realizing stream restoration objectives. Themes highlight the need to (1) protect streamflow from further impact; (2) alter reservoir management; (3) implement other restoration tactics (in addition to developing an environmental flow program); (4) change infrastructure to create more flows; (5) engage in

## CASE STUDY 5.1
Applying La Norma Mexicana de Caudal Ecológico (Ecological Flows Policy of Mexico) to Establish a Water Reserve in the Río Hardy, Mexico

Francisco Zamora

The Ecological Flows Policy of Mexico was established in 2012 by CONAGUA as the method to determine the amount of river flow needed to maintain ecosystem services in diverse river basins in Mexico. The process is forward-looking in that it recognizes the ecological significance of various aspects of the hydrological regime and aims to provide recommendations for the sustainable administration, conservation, and/or recovery of water resources for environmental use. The policy promotes the strategic conservation of biodiversity through watershed protection as well as socioecological resiliency in the face of climate change impacts.

The Hardy River is a tributary of the Colorado River in Mexico and is a critical source of freshwater for the Delta region and estuary—where many marine species utilize brackish waters for reproduction (see place map in Case Study 5.5). The Hardy has a long history of cultural importance in the region and remains a popular area for fishing, hunting, water sports, and tourism today. Hardy River water comes primarily from agricultural return flows as well as effluent flows from the Las Arenitas Wastewater Treatment Plant. As water continues to get scarcer in the Delta region, farmers and other water users are increasingly using water that was once dedicated to instream flows in the Hardy. The Sonoran Institute is working with partners to establish a federally protected water reserve on the Hardy River so that a portion of the flows continue to be used for instream flows—which benefits not only wildlife species but also the local communities that rely on the river for their livelihoods. Additionally, the Hardy and Colorado Rivers in Mexico are RAMSAR wetland-management areas, which helps to facilitate the establishment of the water reserve, as they are high ecological priorities.

In order to establish a water reserve, the Norma Mexicana must be used to develop an Ecological Flows Evaluation, followed by Technical Justification Studies for establishing the water reserve. The Sonoran Institute and partners have developed the evaluation and studies and are in the process of review and approval by CONAGUA.

environmental water transactions; (6) develop a long-term perspective; (7) persevere and don't be daunted; and (8) establish diverse and novel partnerships.

## Protect Streamflow from Further Impact

Depending on what you learn from evaluating land-use history and potential land-use changes in the future, there may be opportunities to protect your stream's flow regime. Addressing impacts to your stream's flow regime that are caused by human interferences should get the highest priority. When possible, nipping an impact in the bud before it takes effect or reducing or eliminating an impact that is already affecting flow is likely to be far more effective in achieving substantial ecological improvement than any other measure. It is true that addressing many human interferences on flow can be difficult if not impossible to address; for example, dams can be essential to water supply and flood control, and water diversions for irrigated agriculture serve a basic human need. However, opportunities—sometimes surprising ones—often present themselves as you conduct your water inventory and reach out to water managers and users. The examples below illustrate directly addressing the root causes of streamflow change.

## Reduce Groundwater Pumping

Shallow groundwater sustains streamflow in many streams of arid and semiarid regions, particularly during dry months (Stromberg et al. 1996). Groundwater pumping near streams can create a localized cone of depression (i.e., a localized lowering of the water table) that draws down river flows and affects native stream biota. Reducing the pumping of shallow aquifers near streams can reverse negative trends and improve long-term streamflow conditions required for many native riverine species. In the San Pedro River in southeastern Arizona, the San Pedro Partnership and the Cochise Conservation and Recharge Network—a public/private network consisting of county and city government, NGOs, and agencies—rallied local citizens around streamflow protection, including the importance of reducing the pumping of shallow groundwater aquifers to protect streamflow along this important river (see figure 5.3). See also Case Study 7.4, which focuses on monitoring the hydrological response of reduced pumping.

## Prevent Construction of Undesirable New Dams or Remove Unneeded or Unsafe Dams

Large dams are commonplace in much of the southwestern United States, northern Mexico, and Australia; many of them were constructed during the dam-building era from the 1930s to roughly 1980. In the western United States, Reclamation has overseen dam construction on almost every major river basin (Benson 2018). Given how dramatically dams alter natural streamflow, there may be nothing more important to protecting or improving streamflow conditions than preventing new dam construction or removing an old, unsafe dam.

An example of an ongoing controversial dam construction project is in western New Mexico and is associated with the 2004 Arizona Water Settlements Act (U.S. Public Law 108–451), which authorized $66 million to build a diversion dam along the Upper Gila River that would retain flows in New Mexico. Given that the Upper Gila is free-flowing with much of the watershed designated as wilderness, the proposed dam has become a political lightning rod. Such controversial new dam proposals are worthy of debate, providing potential opportunities for practitioners and conservationists to weigh various perspectives on river management.

## Get Riverside Communities Involved

Whether you are addressing impacts of dams, groundwater pumping, pollution, or other impacts on your river system, having the support of riverside citizens will be key. They are often the ultimate stakeholders in river conservation efforts and can become advocates for protecting streamflow. There are a variety of ways to involve communities, from involving

**FIGURE 5.3** As part of efforts by both the Upper San Pedro Partnership and the Cochise Conservation and Recharge Network, voluntary retirement of irrigation pumping, municipal water conservation measures, and managed effluent and stormwater recharge in the upper subwatershed has resulted in a 5,100 acre-feet (629 hectare-meter) per year reduction in net human water use from 2002 to 2012 (Gungle et al. 2016). Such reductions in human water use are seen as key to protecting threshold levels of baseflow during the hot, dry months of the year. Photo by Brooke Bushman, The Nature Conservancy.

citizens directly in restoration efforts, inviting the participation of citizen-scientists to help with ecological or water-quality monitoring, or opening your restoration sites to guided tours and bird watching. Please also see chapter 6, which includes an entire section on community-based restoration. In the context of engaging riverside communities in environmental flow efforts, a couple of efforts are worth highlighting. Along the Darling River, Australia, local residents of nearby communities played cricket in the dry bed of the Darling River in April 2018 to call attention to lack of flow (see figure 5.4 [top]). Such community events were critical to garnering the support needed to successfully implement flow releases as part of the Northern Connectivity Event, which returned flow to the river channel on June 13, 2018 (See figure 5.4 [bottom]). In another example in Mexico, a few days prior to the release of the 2014 pulse flow in the Colorado River Delta, citizens of San Luis organized an effort to clean up the dry riverbed. So, when the pulse flow arrived at the community it would not be laden with trash as would have been the case without the cleanup effort (see Case Study 5.5).

**FIGURE 5.4** Local citizens play cricket in the dry Darling River streambed to draw attention to the need for environmental flow releases (left) (Tim Lee, "Landline," ABC TV). Compare this with the same view of the river channel in June 2018, when flow was released as part of the Northern Connectivity Event (right). Photos from Commonwealth Environmental Water Office.

*Collaborate with Entities Whose Mission Is to Keep Streams Free-Flowing*
Partnerships with other river conservation and restoration organizations can help leverage broader public support for your efforts. Several organizations are worth contacting for partnership opportunities, including American Rivers (U.S.), the Audubon Society (U.S., and also related to international migrating birds), Pronatura (Mexico), The Nature Conservancy (international), and the World Wildlife Fund (WWF) (international). In addition to large organizations, smaller, locally based organizations often have networks that can provide volunteers, fundraising, or general awareness and support. The aim is to enhance public awareness of the tradeoffs between the benefits and drawbacks of human actions that negatively affect your stream's flow regime.

## Alter Reservoir Management to Better Meet Restoration Objectives

Altering the timing, volume, and duration of releases of water from dams upstream or downstream of your restoration project may help establish a flow regime better suited to native ecosystems and so realize your stream corridor restoration objectives. Note that altering dam releases can potentially be achieved without having to secure additional water. For example, changing flow management from a trickle throughout the year to releasing the same volume of water as a short-duration, high-magnitude flow can move your stream's flow regime from one that supports invasive species to one that better supports native species. For example, if conducted during the time of year when desired native riparian plants are disseminating seed, a short-duration, high-magnitude release has the potential of reworking channel alluvium and establishing surfaces needed to promote natural regeneration.

Changing dam management often involves specific agreements between dam operators and river restoration groups. For example, in Australia, the National Water Institute has produced a series of guidance documents intended to help jurisdictions consider and incorporate possible impacts from climate change and extreme events into water management (Australian Government 2017). In Mexico, CONAGUA has become a global leader in adjusting water management to address a range of future challenges associated with climate change (Herron 2013). The bottom line: if your restoration site sits upstream or downstream of a large dam, some initiatives may already be underway to revise management; these could be amplified by the participation of voices representing the stream restoration community. In the southwestern United States on the Bill Williams River in Arizona, collaborative work between the USACE, the Nature Conservancy, the USFWS, and other organizations created opportunities to shift the timing, volume, and duration of releases from the Alamo Dam to benefit both a downstream wildlife refuge and offstream riparian habitat (see Case Study 5.2). More generally, Reclamation launched its Reservoir Operations Pilot Initiative in 2014 to evaluate opportunities to shift reservoir management from a sole focus on providing and conveying water to principal users to a broader mandate that also considers climate adaptation, native river ecosystems, aspirations of riverside citizens (including native peoples), and recreation (Benson 2018).

## Implement Stream Restoration Tactics in Addition to Environmental Flow

While protecting and securing additional flow for your stream may be critically important to realizing your stream restoration objectives, other complementary restoration tactics may also be needed (see figure 5.5). These may include

- reintroducing or augmenting populations of native flora and fauna;
- reducing the extent and distribution of nonnative, invasive species; and
- working with local riverside communities on projects that bring both environmental and socio-economic benefit.

For example, prior to the 2014 pulse flow release in the Colorado River Delta, land managers removed nonnative salt cedar and strategically graded and contoured land in the Laguna Grande restoration site to expose bare ground. This active management technique created areas where wind- and water-dispersed native cottonwood and willow seed could germinate and thrive after bare soils were wetted by the pulse flow, which proved to be a successful strategy (Schlatter et al. 2017). See chapter 6 for more on implementing restoration plans.

## Change Water Infrastructure to Secure Additional Flow

Three strategies for dealing with water infrastructure are highlighted here that can put more water in a stream by physically moving it from where it is currently used to where it is needed for environmental benefit: changing the point of diversion, switching the water source, and increasing the efficiency of the diversion.

### *Point-of-Diversion Change*

On rivers that are diverted directly upstream of a restoration site or reach, a straightforward management change is to physically move the point of diversion (POD) downstream of the restoration site, enhancing flow past the restoration site before it is removed for human uses. Such strategies are typically put into effect on a permanent basis through either the voluntary action of the water user or by compensating a water user to make infrastructure improvements.

In the state of Colorado, near the town of Crested Butte, a point of diversion change was successfully implemented to increase flow of the Slate River. A downstream water supplier, the Colorado Water Trust (a local NGO), and the Colorado Water Conservation Board together purchased a senior water right and moved the POD for that right downstream of a critical four-mile reach of the river (see figure 5.6).

### *Source Switch*

A source switch refers to changing the source of water used for a POD. The alternate source can be deep groundwater (e.g., water stored in the ground through aquifer recharge or injection) or water from (typically) a larger stream or river. One commonly used source switch involves irrigators turning on groundwater pumps (using groundwater instead of surface water) to meet their irrigation requirements during critical low-flow periods. This strategy was very effectively implemented in Sisters, Oregon, on Whychus Creek, a tributary to the Deschutes River which flows into the Columbia River. To meet growing water demands while also sustaining river flows for migrating salmon and other fish species, the city of Sisters tapped into groundwater and used its surface water rights to offset impacts from groundwater pumping (Kendy et al 2018). Another option is to shift water users from using a small tributary to a larger river, if their lands are in proximity to both water sources. This can maintain surface flow at desirable levels in the more vulnerable water course.

# CASE STUDY 5.2
Coordinating Reservoir Management and Environmental Flows on the
Bill Williams River, Arizona

Amy McCoy and Patrick B. Shafroth

The Bill Williams River, a tributary to the Lower Colorado River in western Arizona (see map in this case study), has had an environmental flow program for decades, has provided an excellent setting for testing and applying a variety of quantitative hydrologic-characterization methodologies, and serves as a reference point for highly coordinated, comprehensive, and thorough scientific approaches. Various field and modeling studies have been employed to understand changes in streamflow and ecological conditions caused by the construction and management of Alamo Dam, as well as to forecast various dam-management scenarios that could be leveraged to meet environmental flow requirements to ensure the health of riparian and aquatic species (Shafroth and Beauchamp 2006; Shafroth et al. 2010). The articles referenced in this case study can help restoration practitioners learn more about field-tested models and multiyear monitoring approaches.

In 1968, Alamo Dam was completed on the Bill Williams River and flows were regulated primarily to control floods on the Lower Colorado River. Historically, the Bill Williams watershed was well adapted to large floods and prolonged droughts, but the dam significantly altered the ecological processes and riverine habitat of the thirty-six-mile stretch of the river that extends from the dam to its confluence with the Lower Colorado River. The most significant impacts were evident at the Bill Williams River National Wildlife Refuge, which was created in 1941 and currently serves to protect the last and most extensive native riparian forest and plant community along the Lower Colorado River. Ecohydrological studies showed that damming and river regulation had a significant impact on the extent, structure, species composition, and dynamics of stream and riparian habitat (Shafroth et al. 2002; Kui et al. 2017).

Recognizing an opportunity to align dam management efforts with restoring the flow and habitat along the Bill Williams River, a technical committee was formed in the early 1990s to merge state-of-the-science information with policies to provide for multiple benefits downstream of the Alamo Dam. In the mid-2000s, a group of more than fifty participants representing twenty organizations collaboratively estimated environmental flow requirements for the Bill Williams River (Shafroth and Beauchamp 2006). Changes in dam operations were suggested to mimic more natural flooding conditions, in part to create suitable moist sites for native tree seed germination and seedling establishment, and to provide baseflows throughout the year to sustain riparian plants. Studies of several experimental flow releases revealed that variations in the size and duration of floods and baseflows had a significant effect on riparian systems (Shafroth et al. 1998; Andersen and Shafroth 2010; Shafroth et al. 2010; Wilcox and Shafroth 2013; Kui et al. 2017). These insights confirmed that an effective environmental flows prescription must be built upon an integrated understanding of linkages among flow events, geomorphic processes, and biotic responses (Shafroth et al. 2010; Wilcox and Shafroth 2013). While modeling can assist with this understanding, more budget-friendly approaches can be effective as well, such as articulating relationships between flow and ecological responses through environmental flow implementation and monitoring (Shafroth et al. 2010).

## Lessons Learned
- Strong collaboration among land and water managers, other stakeholders, and river and riparian scientists was essential to moving the environmental flow program on the Bill Williams River forward.

Idealized map showing the Bill Williams River below Alamo Dam (from Shafroth et al. 2010).

Among other outputs, the collaborative process identified "areas of incompatibility" as constraints that needed to be addressed.

- A deep understanding of the river system's streamflow hydrology was key to developing an environmental flow regime. The science team took advantage of five environmental flow events between 2005 and 2010 to study responses to various high and low flows.
- The ecological goals of flow restoration incorporated multiple ecological needs, including those of riparian vegetation, aquatic species, birds, and other animals. An understanding of a broad range of ecological water requirements provided a foundation for developing both flood and baseflow prescriptions (Konrad 2010).
- Hydrologic-simulation models helped the partnership assess the relationship between release volumes and predicted ecological responses. Unlike the Colorado River Delta, where informed expert opinions were key to negotiating the terms of the treaty minutes, the Bill Williams collaboration among agencies, academics, and nongovernmental organizations found common ground in technical modeling, research, and monitoring.

**FIGURE 5.5** As long-term, politically challenging binational efforts were rolled out to secure streamflow for environmental purposes in the Colorado River Delta (CRD), small-scale and less politically challenging restoration projects, such as this revegetation effort at Laguna Grande, were conducted. Results of this and similar revegetation projects in the CRD demonstrated restoration potential, garnered goodwill, strengthened collaboration, and ultimately, provided the foundation for the Colorado River pulse flow that occurred in 2014. Photo by Rabí Hernández, Sonoran Institute.

## Improved Diversion Efficiency

Diversion efficiency refers to the efficiency with which water is diverted from the stream through canals or pipes from the point of delivery to point of use (as opposed to improving irrigation efficiency, which may not be as useful in providing additional water to streams). In many parts of Australia, Mexico, and the western United States, aging water delivery systems lose significant amounts of water via seepage into the soil during transport and evaporation into the air. In situations where diversions are not well metered, measured, or managed, the formal legal diversion rates are often exceeded to provide the necessary irrigation water for crops. Improving diversion efficiency can reduce losses, and when combined with a suite of other actions, such as improved water accounting and imposing a limit on new uses, can help to keep saved water in the stream downstream of the diversion (Grafton et al. 2018).

This approach can be a win-win for farmers and restoration practitioners. For example, the Verde River is one of the few perennial and free-flowing rivers in Arizona and is a critical focus of local water-management efforts, river conservation initiatives, and economic development in the region (see figure 5.7). Antiquated and inefficient irrigation ditches and gates have fostered increasing water losses and greatly diminished flow on the Verde River over the last century. Over the past decade, The Nature Conservancy has been successfully leading an

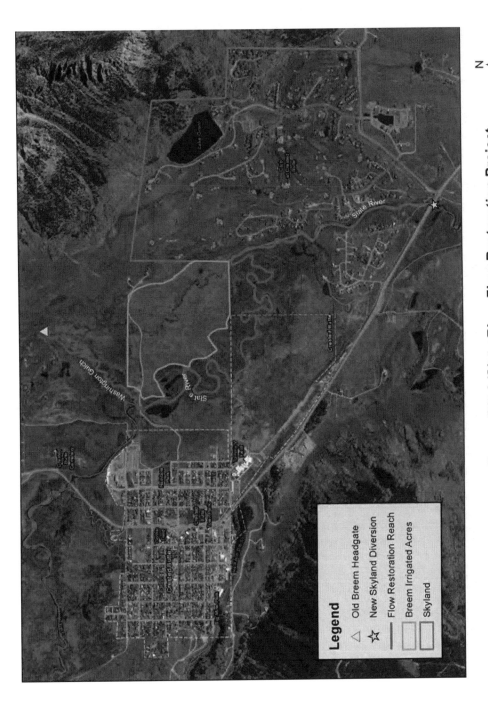

**Breem Ditch/Slate River Flow Restoration Project**

0.5    0.25    0    0.5 Miles

N

**FIGURE 5.6** Map showing POD change along Slate River that allows water to stay in the river through a reach deemed critical for native species and recreation. The critical reach is the 6.4 km (4 mile) reach from the old Breem headgate (indicated with a triangle) to the new Skyland diversion (star).

**FIGURE 5.7** Improving the function and efficiency of irrigation infrastructure, such as this headgate on Clear Creek in Arizona, benefits irrigators and improves downstream flows to the Verde River to meet environmental objectives. Photo from the Bonneville Environmental Foundation.

effort to improve irrigation-diversion infrastructure to increase the volume, consistency, and duration of flows to the Verde River. In addition to improving the effectiveness of irrigation deliveries, this work has infused energy into stakeholder participation within the region by providing opportunities for partnerships among the local communities, including farmers, residents of expanding area municipalities, and the emerging wine industry.

## Engage in Environmental Water Transactions

Environmental water transactions are contractual agreements that reallocate water rights from an existing water use to an environmental or instream water use (Garrick et al. 2011). Such transactions can be spurred by conditions of overallocation where water has become a scarce and valuable commodity traded between willing buyers (or leasers) and sellers. Examples of water-transaction frameworks already in existence include the Murray-Darling River Basin in Australia (Garrick et al. 2009; McCoy et al. 2018), the Columbia River Basin in the northwest United States (Garrick et al. 2009; Aylward 2013), and the Colorado River Basin (Carrillo-Guerrero et al. 2013; Szeptycki et al. 2018).

Initially, environmental water transactions were used in targeted tributary streams, where small-scale and often temporary agreements helped to restore flows for the benefit of fish and aquatic species. They have expanded into more coordinated and integrated programs for restoring environmental flows (Garrick et al. 2009). Environmental water transactions can span a range of options, including fair-market purchase of water rights by a conservation organization from a current holder, such as an irrigation district, rancher, farmer, or other water-management agency.

Regardless of the water-transaction strategy that is implemented, the intent is to secure and augment flows at your site during specific periods of time to further your restoration objectives. Environmental water transactions can be in perpetuity or temporary, as for a full irrigation season, part of the season, or only when certain conditions exist, such as during drought.

How an environmental water transaction is conducted depends on several key elements that are identified in your water inventory, including the water rights holder (i.e., the seller), how the water right is currently being applied, the water right or water use, the location of use, and the set of agreements that make it permissible to move the water to an environmental use.

A temporary transaction may take the form of a lease, while a permanent transaction is often done by physically changing the location of the POD, the place of use, or the manner of use of a water right without altering the source of water. Transfers can be associated with a change in the type of water right or amount or season of use. Typically, a water user must make a capital expenditure to change the POD, place of use, or manner of use, and the transaction contracts also include one or more funding agreements between the water user and environmental buyer. In the Colorado River Delta, for example, water transfers allow the Colorado Delta Water Trust (Case Study 5.3) to lease water from local farmers, changing water use from agriculture to a restoration or instream flow use that benefits the Delta's riparian and aquatic ecosystems.

Environmental water transactions have also been effectively used along the Middle Rio Grande in New Mexico to support environmental restoration efforts to offset or mitigate new municipal and industrial groundwater pumping permits. (In this context, mitigation refers to reducing or fully offsetting the impacts of new or existing groundwater pumping on connected aquifers or surface water sources or both [Cronin et al. 2017]). To effectively process a transfer in New Mexico, a change request must be filed with the New Mexico Office of the State Engineer and an official legal process initiated to transfer the use associated with the water right. While there is a clear legal process for transfers in New Mexico, a variety of political, social, and infrastructure challenges have tempered how many transfers have been successfully completed. This is both an invitation for entrepreneurial restoration practitioners to look for innovative ways to use New Mexico's transfer policies and a disadvantage, as there are not many precedent-setting case studies to lead the way.

For more detailed information on the specifics of water-transaction agreements, Aylward (2013) offers an excellent resource for the framework and implementation of environmental water transactions.

## Develop a Long-Term Perspective

### *Be Strategic*

Streamflow restoration needs abound and there are only so many resources that can be leveraged. For environmental flow, it is the rare restoration project that will secure in a single step all the flow needed to realize its restoration objectives. The environmental flow strategy that is ultimately developed will likely be a combination of identifying the critical parts of the flow regime as well as the opportunities that can be realistically supported by the community of water users and managers in your watershed. As emphasized in the first section of this chapter, the "low-hanging fruit" options should be explored before pursuing options that involve greater amounts of water or necessitate addressing entrenched and challenging political issues. Even if these easier options are temporally or spatially limited, they can provide important near-term victories that may pave the way for more complicated and long-lasting environmental flow opportunities. An example of this is an effort in Chihuahua, Mexico, to protect flow for the Julimes pupfish, one of the most threatened fish species in the world (Case Study 5.4). Although not 100 percent successful, this precedent-setting effort is serving as a springboard for conducting additional environmental flow efforts throughout Mexico.

# CASE STUDY 5.4
Protecting Flow for a Critically Endangered Species: Establishing Environmental Precedent for the Hottest Fish in the World

Mauricio de la Maza Benignos

Coined "the hottest fish in the world," the Julimes pupfish (*Cyprinodon julimes*), a member of the killfish family, is endemic to El Pandeño de los Pandos hot spring in the municipality of Julimes, Chihuahua, Mexico (see map below), and thrives in water temperatures that often reach 46°C (114°F). These diminutive fish are fascinating to observe. Female and young pupfish often travel in a single line while feeding on detritus, algae, and invertebrates such as the microendemic Julimes snail (*Tryonia julimensis*). Dominant males establish territories of about 0.5 m (1.6 ft.) in diameter around sunken branches, rocks, or pieces of travertine from where they court passing females.

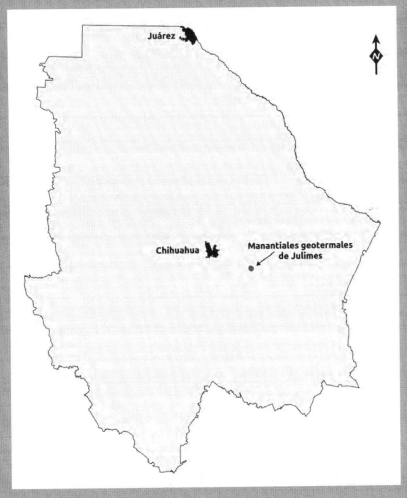

Idealized map showing location of the Julimes hotsprings in the state of Chihuahua, Mexico. Map by G. Rendon, Pronatura.

(*continued*)

Once you arrive at El Pandeño hot springs, it is often easy to spot the pupfish, as the thermal springs where the fish resides occupy just 600 m². The combination of the pupfish's narrow distribution and increasing demands on water resources of the region place this species at high risk of extinction. It may have been one of the world's most threatened fish species prior to the onset of the conservation actions described here. The pupfish is currently listed under federal Mexican law as "critically endangered."

To protect the Julimes pupfish population in the long term, two central actions were required: (1) spring flow had to be secured to guarantee water of sufficient quantity and adequate quality for both the native spring species (including the pupfish) and the local citizens who also depend on the water; and (2) the spring habitat itself needed to be protected from contamination, destruction, invasive species, and other impacts. Although this case study focuses on the legal protection of spring flow, several complementary milestones were also achieved that have brightened the long-term prognosis of the Julimes pupfish and the spring waters on which it depends:

- In January 2013, 300 m² of dried lower marsh habitat was restored with support from the NGO Pronatura Noreste, which more than doubled the area of suitable habitat for the pupfish.
- In November 2013, El Pandeño hotspring was nominated as a global Ramsar Site ("Manantiales Geotermales de Julimes" [Geothermal Julimes Springs]) by the Ramsar Convention on Wetlands.

Regarding the legal protection of spring flow, the rights to the waters that emanate from El Pandeño are owned by the San José de Pandos Irrigation Society (SJPIS), which uses the water for agricultural and recreational activities. Deeply concerned about the long-term viability of their water source, in 2008 the farmers of the SJPIS, together with Julimes residents, chartered the grassroots non-profit organization Amigos del Pandeño, A.C., ("Friends of El Pandeño"). With support from partnering NGOs, including Pronatura Noreste, Biodesert, Profauna, and the WWF-Chihuahuan Desert Program, Amigos del Pandeño implemented an integrated spring-management program that, among other things, allocated spring flow both for sustainable agricultural and recreational activities and helped maintain the natural environment of the spring and the Julimes pupfish.

Environmental flow, such as what is described under the Julimes spring-management plan, is recognized under the Mexican National Water Law (LAN) of 1992 as "the flow or minimum volume necessary for water bodies, including diverse forms of currents or reservoirs, or the minimum natural discharge flow of an aquifer that must be conserved to protect environmental conditions and ecological equilibrium of the system."

The "hottest fish" in the world, the Julimes pupfish. Photo by Maurico de la Maza, Pronatura.

In May 2009, the newly formed Amigos del Pandeño asked CONAGUA to grant it custody of the spring area for conservation purposes and give the spring nonconsumptive environmental water rights for the preservation of its endemic fauna. CONAGUA subsequently granted Amigos del Pandeño the custody, use, and management of the federal lands of the thermal spring but denied its request to recognize the natural environment as a user of the surface waters of the spring (CONAGUA 2010). In essence, CONAGUA granted rights to use of the land but denied the water title.

Although a partial win, this precedent-setting decision provided the legal basis for establishing and enforcing an environmental flow requirement for the Río Bravo, independent of whether it is established as a prior right (before other allocations) or as an allocation in competition with other users. It also set the basis for a later decision by CONAGUA in August 2013 that, for the first time in the history of Mexico, legally recognized water for environmental use. In this decision, a portion of the surface waters that emanate from Pozas Azules Ecological Reserve, Coahuila, was awarded to Pronatura Noreste for "environmental use" to conserve biodiversity and valuable and unique ecosystems.

## Key Lessons Learned

Although not entirely successful, the precedent-setting application by the Amigos del Pandeño to CONAGUA established the basis for implementing the principles of environmental protection set forth in relevant national and international laws, fostering the creation of a legal instrument that can protect valuable ecosystems against acts by third parties, including the government, that could endanger their environmental conditions and ecological balance.

This also fostered significant progress toward recognizing the environment as a legally binding user of water, including the establishment of a legal mechanism that allows the granting of a national water concession for environmental use based on the human right to a healthy environment (as recognized in Mexico's Constitution). Based on the legal gains associated with the Julimes pupfish, other legal cases are underway. For example, in Cuatro Ciénegas, Coahuila, an environmental flow case was brought by Pronatura Noreste and admitted this year into an amparo court (an extraordinary constitutional appeal).

The other key benefit of this precedent-setting experience is that CONAGUA has expressly recognized the social and environmental importance of water as an essential component for social welfare and economic development. Such heightened recognition has increased national attention on environmental governance, adequate planning, and management of water resources to reduce water waste, addressing competition for its use as well as water pollution challenges.

Central lessons learned during this process, include the following:

1. Participation and recognition of the needs of local citizens in the development of the management plan was essential to the plan's adoption in 2009 and to the environmental flow application made by Amigos del Pandeño.
2. The formation and chartering of the NGO Amigos del Pandeño, A.C., facilitated the involvement of Pronatura Noreste, WWF, and other NGOs in this important conservation initiative, allowing these outside groups to assist the newly formed A.C. to conduct the needed scientific and legal investigations, as well as the community-based environmental education and outreach programs, and develop the sustainable water-management program. Since then, the model has been replicated and enhanced in other sections of the Río Bravo, where overallocation of water and water stress is the norm.

Rewatering the Colorado River Delta: The Pulse Flow of 2014

Francisco Zamora, Jennifer Pitt, Karen J. Schlatter, and Amy McCoy

As one of the most managed, monitored, and carefully controlled rivers in the world, the Colorado River system includes a vast network of pipes and canals that convey water to support nearly forty million people. In a highly developed and elaborately managed system, it can be particularly challenging to envision ways to restore environmental flows.

Since the 1960s, as the river flowed south through the border region between the United States and Mexico, impacts from drought, overallocation, and water shortages were obvious in the diminished and disappearing flows of the river as it neared its terminus in the Sea of Cortez. In the 1990s, several conservation groups began to focus on bringing the Colorado River Delta back to life (Briggs and Cornelius 1998). Environmental flow needs for the Colorado River Delta were determined using several different approaches that relied heavily on expert opinion, including tapping into lessons learned and insights gained from other experiences, such as that involving the Bill Williams River (see Case Study 5.2) (Lueke et al 1999; Zamora-Arroyo et al. 2005). Based on restoration priorities established in the early 2000s, as well as careful and consistent observations, a team of experts from conservation organizations and universities worked on estimating optimal base and pulse flow volumes. First, the team evaluated the biophysical characteristics of the Delta and developed high and low estimates of the amount of water required to flow through the system to maintain existing, remnant native freshwater and estuary ecosystems. Second, the team correlated recent flood events to current stands of riparian forest to estimate volumes of water needed to support riparian habitat (Lueke et al 1999; Zamora-Arroyo et al. 2005). Third, the team mapped how water was managed and distributed throughout the

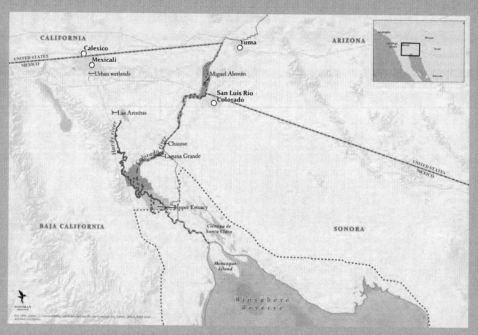

Map of the Colorado River Delta showing the river south of the international border and key restoration sites. Map from the Sonoran Institute.

Local citizens celebrate the arrival of the Colorado River pulse flow. Photo by Jennifer Pitt.

region to understand what volumes of water might be available to direct to the Colorado River through allocation, purchase, or management. Throughout, the team engaged Colorado River water managers in both countries to ensure transparency and highlight the scientific underpinnings of their work. Results of these evaluations indicated that 61.7 m³ (50,000 acre-feet) per year would meet base flow needs and 308.4 m³ (250,000 acre-feet) every four to five years would meet pulse flow requirements. Notably, this amounts to less than one percent of the Colorado River's annual average flow.

Conservation NGOs achieved initial recognition of the Colorado River Delta's importance by the United States and Mexico in 2000 with the adoption of Minute 306 (the two countries negotiate consensus-based Minutes to add greater definition to their implementation of the 1944 Treaty by which they share the Colorado River). Small victories accrued through riparian revegetation projects, scientific journal articles, media attention, and community work. With each success, the profile of the Colorado River Delta increased. This momentum, and the modest volume identified as the environmental flow need, made it possible to enter higher-level diplomatic efforts to include water for the Delta as part of broader binational reforms to water management. By 2017, two additional Minutes (Minute 319 and Minute 323) had been adopted in support of these broader binational efforts.

Minute 319 was formalized by the United States and Mexico in March 2012. That agreement set forth new specifications for how both countries would work together to share and manage the waters of the Colorado River (Pitt et al. 2017). Among several new management approaches, the two countries and a coalition of NGOs committed themselves to deliver environmental flows to the Colorado River Delta as part of a pilot project to provide both regular base flows and a one-time pulse flow to the Delta, which occurred in the spring of 2014. The volumes of flows were informed by research and on-the-ground experience established in the previous fifteen years, but ultimately were defined in a political process that also defined how the two countries share water shortages and other management reforms. In September 2017, formalization of Minute 323 extended the commitment of both countries as well as NGOs to continue providing environmental flows to the Colorado River Delta (Kendy et al. 2017).

*(continued)*

## Lessons Learned

- A critical mass of players armed with a firm restoration goal and a notable capacity for patience allowed the diplomatic hurdles associated with the 1944 Water Treaty to be successfully overcome to the betterment of both the environment and wildlife. The precedent-setting pulse flow demonstrates that environmental flow successes are possible even in complicated and challenging political landscapes.
- Early expert opinion on quantifying flow needs (more than a focus on developing expensive hydrological models) played a key role in moving negotiations on Minute 319 forward. Although hydrological models were eventually developed, the Delta experience demonstrates the value of using experts, particularly when budgets are thin.
- By engaging water managers in the United States and Mexico in a conversation about the benefits of binational collaboration, NGOs gained the attention of federal leaders and embedded the topic of environmental restoration in a negotiation that focused on water supply for people.
- Existing infrastructure for irrigated agriculture was successfully used to convey a portion of the pulse flow release to targeted restoration sites in the central Delta, bypassing a dry reach and allowing for greater ecological benefit to be achieved with the limited water resources available.
- While the volume of water delivered via the pulse flow was a significant diplomatic achievement, it was not sufficient to spark significant natural regeneration of native plants (Mueller et al. 2017; Shafroth et al. 2017; Schlatter et al. 2017). Sustained restoration requires consistent and sufficient baseflows, as well as pulse flows of sufficient magnitude and frequency, particularly along reaches that have experienced significant dewatering because of upstream manipulations and groundwater pumping.
- Social benefits must be considered along with biophysical needs and can be even more important. During the 2014 pulse flow, the flowing Colorado River captured the attention and good will of local communities in Mexico, many of whose residents remembered their tangible connection to the river in the 1960s, before it dried up. Ultimately, the diplomatic and social aspects of the pulse flow were the most important reminder that river management, at its core, is fundamentally rooted in human relationships and experiences.

## Don't Be Daunted: Environmental Flow Can Be Successful Even in Politically Challenging Regions

It is important to emphasize that environmental flow can be successfully achieved even in the most politically challenging landscapes, such as areas characterized by water scarcity with a growing gap between water supply and demand. However, in such circumstances, success will often require fortitude, a track record of smaller-scale project successes that benefit wildlife and people, and dedication by a critical group of stakeholders that can keep momentum going during the inevitable ebbs and flows of water politics. The environmental pulse flow delivered to the Colorado River Delta, Mexico, which took place during March 2014 (Case Study 5.5), was successfully orchestrated in one of the most water-challenged landscapes in the world. After over a decade of work by dedicated NGOs and scientists from both Mexico and the United States, water-management agencies and water users from both sides of the border ultimately provided the essential foundation for success.

# CASE STUDY 5.6
Unique Conservation Partnerships to Support Native Freshwater Ecosystems Along the Middle Rio Grande

Paul Tashjian

The RGB in New Mexico has been altered by a variety of human activities, including the construction of dams, diversion of water, habitat fragmentation, channelization, and groundwater pumping, among other things. As a result, over 80 percent of the historic wetland and riparian habitats are estimated to have been lost and more than one-third of the river's 748 km (465 mile) length through New Mexico frequently dries up during the summer. Despite these changes, the river remains an important migratory wintering and nesting corridor, supporting around two hundred thousand waterfowl, eighteen thousand greater sandhill cranes, and tens of thousands of other water birds and shorebirds. In addition, the Rio Grande Delta above Elephant Butte Reservoir is home to the largest number of contiguous breeding territories for both the endangered southwestern willow flycatcher and the threatened yellow-billed cuckoo in their entire ranges (see map of Middle Rio Grande in New Mexico in Case Study 4.3).

## The Partnerships
In September 2016, Audubon New Mexico and four Pueblos of the Middle Rio Grande region—Sandia, Isleta, Santa Ana, and Cochiti—joined in a unique partnership, wherein each Pueblo supplied 123,348 m³ (100 acre-feet) of stored water to Audubon New Mexico for supporting flows in a reach of the Middle Rio Grande that had cultural and ecological importance but was susceptible to summer drying. The partnership fostered an additional water transaction supplied by the Club at Las Campanas, a private country club in Santa Fe, New Mexico, which donated an additional 0.5 m³ (about 399 acre-feet) of water, bringing the total to almost 1 m³ (799 acre-feet). Release of the environmental water supplemented flows for almost twenty-four days during the driest time of the year along a critical thirty-five-mile reach that is designated Wild and Scenic by the U.S. federal government in recognition of the river's outstanding natural, cultural, and recreational values. Successful delivery of the water included agreement and coordination with the Middle Rio Grande Conservancy District (MRGCD), Reclamation, USACE, and the Albuquerque-Bernalillo County Water Utility Authority.*

This marked the first time in the state's history that a conservation organization stored environmental water in a reservoir for delivery when the river and the river's native species need it most. The partnership between Audubon New Mexico and the Middle Rio Grande Pueblos is a confluence of aspirations and vision for the Rio Grande. The mission of Audubon New Mexico is to both maintain and bring back the state's great diversity of resident and migratory birds. Given that many birds depend on wetland and riparian habitat for their survival, securing environmental flows to restore damaged habitat has become a key priority. For the Middle Rio Grande Pueblos, the Rio Grande is considered sacred, with river flows carrying great cultural importance.

The 2016 partnership laid the groundwork for future nongovernmental water transactions in the Middle Rio Grande. In 2018, Audubon New Mexico acquired 1.2 m³ (998 acre-feet) of water from the municipalities of Belen, Los Lunas, and Bernalillo, and the Club at Las Campanas. The Club at Las Campanas's interest in supplying water stemmed from its own emphasis on conservation; it maintains habitat that supports over two hundred species of migrating and breeding birds. It is home to the only golf course in New Mexico that is certified by Audubon International (an independent organization not

*(continued)*

205

affiliated with National Audubon Society) as a cooperative sanctuary. The additional water supported environmentally critical flow for fifty-five days in July, August, and early September 2018.

## Current and Next Steps

These water flows are a key component of a larger effort by Audubon New Mexico to bring back flow in the Middle Rio Grande that also includes (1) securing funds for additional leases; (2) assisting the New Mexico Department of Game and Fish with wetland management and securing water rights at La Joya and Bernardo Wildlife Management Areas; and (3) conducting a range of public education and outreach activities. Future work will focus on bringing environmental water to key habitat-restoration locations for maximum conservation benefit.

## Lessons Learned

- The strong partnerships that Audubon New Mexico formed with water managers, such as Reclamation, MRGCD, and water rights holders, was essential to the successful delivery and use of the acquired water.
- Having a third party (like an NGO such as Audubon New Mexico) facilitate transactions streamlined the process by dividing some of the tasks that would normally fall solely on agencies.
- Economic incentives are not always needed to catalyze water transactions. The municipalities, Pueblos, and private stakeholders involved in this effort cite the inherent value of water in the river as motivation for participating in these transactions. This work demonstrates the willingness of local communities to invest in the environment.
- Over the long term, water transactions will probably not be effective if they are isolated and not implemented holistically with other complementary conservation actions. However, temporary water transfers can help set the stage in the region for a more comprehensive water market in which such transactions become commonplace. For example, the combined water provided by the municipalities is expected to support fifty-five days of flow.

## Establish Diverse and Novel Partnerships

Successful implementation of environmental flows often hinges on diverse partnerships (as is highlighted in each of the case studies). In New Mexico, a precedent-setting and creative environmental flow program (Case Study 5.6) is made possible by a unique partnership involving Audubon New Mexico (an NGO), four Middle Rio Grande Pueblos, agencies, and businesses; it can serve as a reference point for future projects. Establishing diverse and novel partnerships is also more generally covered in chapter 8 as a critical element to expanding the temporal and spatial footprint of your restoration effort for greater long-term impact.

The conclusion of this chapter is an important benchmark in the arc of developing your stream restoration response. We now move from an emphasis on gathering the information about your stream and its watershed needed to effectively plan your stream restoration response to an emphasis on identifying and implementing the restoration tactics required to

Water secured as part of this conservation partnership can help to augment flow along the Middle Rio Grande during severe drought periods, such as the drought that was occurring when this photograph was taken on July 24, 2018. Photo by Paul Tashjian.

- Public outreach through regular press releases strengthens support to initiate and expand the participation of municipalities and private entities in water transactions that are beneficial to the river and native species.
- Environmental water transactions can make stream restoration a more promising investment by mitigating some of the uncertainty associated with relying on unpredictable and diminishing flows for long-term restoration success.

*Reclamation has operational responsibilities at various reservoirs upstream of the MRGCD. ABCWUA is the owner of Abiquiu Reservoir, which is operated by the USACE.

realize the objectives of your stream restoration effort. In the following chapter, "Implementation: Putting Your Stream Corridor Restoration Plan into Action," we summarize the nuances of implementing restoration tactics, including a checklist of tasks that need to be considered prior to hitting the implementation "start button," lessons learned from implementing various types of stream restoration tactics, and some of the pitfalls that need to be avoided to improve likelihood of success.

## REFERENCES

Acreman, M. C., I. C. Overton, J. King, P. J. Wood, I. G. Cowx, M. J. Dunbar, E. Kendy, and W. J. Young. 2014. "The Changing Role of Ecohydrological Science in Guiding Environmental Flows" *Hydrological Sciences Journal* 59 (3–4): 433–50, https:/doi.org/10.1080/02626667.2014.886019.

Allen, D. C., D. A. Kopp, K. H. Costigan, T. Datry, B. Hugueny, D. S. Turner, G. S. Bodner, and T. J. Flood. 2019. "Citizen Scientists Document Long-Term Streamflow Declines in Intermittent Rivers of the Desert Southwest, USA." *Freshwater Science* 38 (2). https://doi.org/10.1086/701483.

Andersen, D. C., and P. B. Shafroth. 2010. "Beaver Dams, Ecological Thresholds, and Controlled Floods as a Management Tool in a Desert Riverine Ecosystem, Bill Williams River, Arizona." *Ecohydrology* 3:325–38.

Arias-Rojo, H. M., and R. F. Salmón-Castelo. 2019. "Mexican Water Sector: A Brief Review of Its History." In *Water Policy in Mexico, Global Issues in Water Policy 20*, edited by H. R. Guerrero García Rojas, 19–51. Cham: Springer International Publishing. https://doi.org/10.1007/978-3-319-76115-2_2.

Aylward, B., ed. 2013. *Environmental Water Transactions: A Practitioner's Handbook*. Bend, Ore.: Ecosystem Economics.

Benson, R. 2018. "Reviewing Reservoir Operations in the North American West: An Opportunity for Adaptation." *Regional Environmental Change* 18:1633–43. https://doi.org/10.1007/s10113-018-1330-x.

Besttelmeyer, B., K. M. Havstad, R. Alexander, and J. R. Brown. 2003. "Development and Use of State-and-Transition Models for Rangelands." *Journal of Range Management* 56 (2): 114–26.

Bouwes, N., J. Moberg, N. Weber, B. Bouwes, S. Bennett, C. Beasley, C. Jordan, et al. 2011. *Scientific Protocol for Salmonid Habitat Surveys Within the Columbia Habitat Monitoring Program*. Technical Report. https://doi.org/10.13140/RG.2.1.4609.6886.

Bovee, K. D. 1978. "The Incremental Method of Assessing Habitat Potential for Coolwater Species, with Management Implications." *American Fisheries Society Special Publication* 11:340–46.

Bovee, K. D., B. L. Lamb, J. M. Bartholow, C. B. Stalnaker, J. Taylor, and J. Henriksen. 1998. *Stream Habitat Analysis Using the Instream Flow Incremental Methodology*. U.S. Geological Survey, Biological Resources Division Information and Technology Report USGS/BRD-1998–0004. National Technical Information Service, Springfield, Virgina.

Briggs, M. K., and S. Cornelius. 1998. "Opportunities for Ecological Improvement Along the Lower Colorado River and Delta." *Wetlands* 18 (4): 513–29.

Brisbane Declaration. 2007. "*Environmental Flows Are Essential for Freshwater Ecosystem Health and Human Well-Being*." http://www.indiawaterportal.org/sites/indiawaterportal.org/files/Brisbane_Declaration.pdf.

Carrillo-Guerrero, Y., E. P. Glenn, and O. Hinojosa-Huerta. 2013. "Water Budget for Agricultural and Aquatic Ecosystems in the Delta of the Colorado River, Mexico: Implications for Obtaining Water for the Environment." *Ecological Engineering* 59:41–51.

CONAGUA. 2010. *Financing Water Resources Management in Mexico*. Comisión Nacional del Agua, Organization for Economic Cooperation and Development, Instituto Mexicano de Tecnología del Agua (IMTA). Distrito Federal de la Ciudad de México. http://www.conagua.gob.mx/CONAGUA07/Contenido/Documentos/OECD.pdf.

Cronin, A. E., J. Gibbon, and D. Pilz. 2017. "Groundwater Mitigation: Piloting Groundwater Mitigation in Arizona's Verde Valley." *Water Report* 162:12–20.

Dudgeon, D., A. H. Arthington, M. O. Gessner, Z. I. Kawabata, D. J. Knowler, C. Lévêque, R. J. Naiman, A. H. Prieur-Richard, D. Soto, M. L. J. Stiassny, and C. A. Sullivan. 2006. "Freshwater Biodiversity: Importance, Threats, Status and Conservation Challenges." *Biological Reviews* 81:163–82. https://doi.org/10.1017/S1464793105006950.

Elith, J., and J. R. Leathwick. 2009. "Species Distribution Models: Ecological Explanation and Prediction Across Space and Time." *Annual Review of Ecology, Evolution, and Systematics* 40:677–97. https://doi.org/10.1146/annurev.ecolsys.110308.120159.

Fremier, A. K., and K. M. Strickler, eds. 2010. *Topics in BioScience: River Structure and Function*. American Institute of Biological Sciences. https://www.aibs.org/home/index.html.

Garrick, D., N. Hernández-Mora, and E. O'Donnell. 2018. "Water Markets in Federal Countries: Comparing Coordination Institutions in Australia, Spain and the Western USA." *Regional Environmental Change* 18:1593–606. https://doi.org/10.1007/s10113-018-1320-z.

Garrick, D., C. Lane-Miller, and A. L. McCoy. 2011. "Institutional Innovations to Govern Environmental Water in the Western United States: Lessons for Australia's Murray-Darling Basin." *Economic Papers* 30:167–84.

Garrick, D., M. A. Siebentritt, B. Aylward, C. J. Bauer, and A. Purkey. 2009. "Water Markets and Freshwater Ecosystem Service: Policy Reform and Implementation in the Columbia and Murray-Darling Basins." *Ecological Economics* 69:366–79.

Glenn, E. P., P. L. Nagler, P. B. Shafroth, and C. J. Jarchow. 2017. "Effectiveness of Environmental Flows for Riparian Restoration in Arid Regions: A Tale of Four Rivers." *Ecological Engineering* 106:695–703.

Grafton, R. Q., J. Williams, C. J. Perry, F. Molle, C. Ringler, P. Steduto, B. Udall, S. A. Wheeler, Y. Wang, D. Garrick, and R. G. Allen. 2018. "The Paradox of Irrigation Efficiency." *Science* 361 (6404): 748–50.

Groffman, P. M., J. S. Baron, T. Blett, A. J. Gold, I. Goodman, L. H. Gunderson, B. M. Levinson, et al. 2006. "Ecological Thresholds: The Key to Successful Environmental Management or an Important Concept with No Practical Application?" *Ecosystems* 9 (1): 1–13. https://doi.org/10.1007/s10021-003-0142-z.

Guisan, A., A. Lehmann, S. Ferrier, M. Austin, J. M. C. Overton, R. Aspinall, and T. Hastie. 2006. "Making Better Biogeographical Predictions of Species' Distributions." *Journal of Applied Ecology* 43 (3): 386–92. https://doi.org/10.1111/j.1365-2664.2006.01164.x.

Guisan, A., R. Tingley, J. B. Baumgartner, I. Naujokaitis-Lewis, P. R. Sutcliffe, A. I. T. Tulloch, T. J. Regan, et al. 2013. "Predicting Species Distributions for Conservation Decisions." *Ecology Letters* 16 (12): 1424–35.

Gungle, B., J. B. Callegary, N. V. Paretti, J. R. Kennedy, C. J. Eastoe, D. S. Turner, J. E. Dickinson, L. R. Levick, and Z. P. Sugg. 2016. *Hydrological Conditions and Evaluation of Sustainable Groundwater Use in the Sierra Vista Subwatershed, Upper San Pedro Basin, Southeastern Arizona*. USGS Scientific Investigations Report 2016–5114. https://doi.org/10.3133/sir20165114.

Hall, M. R., J. West, B. Sherman, J. Lane, and D. de Haas. 2011. *Long Term Trends and Opportunities for Managing Regional Water Supply and Wastewater Greenhouse Gas Emissions*. New South Wales: CSIRO.

Hamilton, K., and E. P. Bergersen. 1984. *Methods to Estimate Aquatic Habitat Variables*. U.S. Govt. Printing Office: 1984-779-756. Colorado Cooperative Fishery Research Unit and Bureau of Reclamation. https://www.usbr.gov/library/bresource.html.

Herron, C. A. 2013. *Agua y cambio climático en México 2007–2012: Análisis y recomendaciones a futuro*. Con el apoyo de Comisión Nacional del Agua (CONAGUA) y Proyecto de Fortalecimiento del Manejo Integrado del Agua (PREMIA). https://www.gwp.org/globalassets/global/toolbox/references/water-and-climate-change-in-mexico-2007-2012.-analysis-and-future-recommendations-conaguapremia-2013-spanish.pdf.

Hinojosa-Huerta, O., H. Iturribarría-Rojas, E. Zamora, A. Calvo-Fonseca. 2008. "Densities, Species Richness and Habitat Relationships of the Avian Community in the Colorado River, Mexico." *Studies in Avian Biology* 37:74–82.

Horne, A., A. Webb, M. Stewardson, B. Richter, and M. Acreman. 2017. *Water for the Environment: From Policy and Science to Implementation and Management*. Elsevier Inc. https:/doi.org/10.1016/C2015-0-00163-0.

Huggett, A. 2005. "The Concept and Utility of 'Ecological Thresholds' in Biodiversity Conservation." *Biological Conservation* 124:301–10.

Kendy, E., B. Aylward, L. S. Ziemer, B. D. Richter, B. G. Colby, T. E. Grantham, L. Sanchez, et al. 2018. "Water Transactions for Streamflow Restoration, Water Supply Reliability, and Rural Economic Vitality in the Western United States." *Journal of the American Water Resources Association* 54 (2): 487–504.

Kendy, E., K. W. Flessa, K. J. Schlatter, C. A. de la Parra, O. Hinojosa-Huerta, Y. K. Carrillo-Guerrero, and E. Guillen. 2017. "Leveraging Environmental Flows to Reform Water Management Policy: Lessons Learned from the 2014 Colorado River Delta Pulse Flow." *Ecological Engineering* 106:683–69.

Konrad, C. P. 2010. *Monitoring and Evaluation of Environmental Flow Prescriptions for Five Demonstration Sites of the Sustainable Rivers Project*. U.S. Geological Survey Open-File Report 2010–1065. https://pubs.er.usgs.gov/.

Kui, L., J. C. Stella, P. B. Shafroth, P. K. House, and A. C. Wilcox. 2017. "The Long-Term Legacy of Geomorphic and Riparian Vegetation Feedbacks on the Dammed Bill Williams River, Arizona, USA." *Ecohydrology* 10. https:/doi.org/10.1002/eco.1839.

Leblanc, M., S. Tweed, A. Van Dijk, and B. Timbal. 2012. "A Review of Historic and Future Hydrological Changes in the Murray-Darling Basin." *Global and Planetary Change* 80:226–46. https://doi.org/10.1016/j.gloplacha.2011.10.012.

Leucke, D. F., J. Pitt, C. Congdon, E. Glenn, C. Valdés-Casillas, and M. Briggs. 1999. *A Delta Once More: Restoring Riparian and Wetland Habitat in the Colorado River Delta*. Environmental Defense Fund. https://www.edf.org/sites/default/files/425_delta.pdf.

Lindenmayer, D. B., and G. Luck. 2005. "Synthesis: Thresholds in Conservation and Management." *Biological Conservation* 124 (3): 351–54.

McCoy, A. L., S. R. Holmes, B. A. Boisjolie. 2018. "Flow Restoration in the Columbia River Basin: An Evaluation of a Flow Restoration Accounting Framework." *Environmental Management* 61 (3): 506–19.

Mueller, E. R., J. C. Schmidt, D. J. Topping, P. B. Shafroth, J. E. Rodriquez-Burqueno, P. E. Grams. 2017. "Geomorphic Change and Sediment Transport During a Small Artificial Flood in a Transformed Postdam Delta: The Colorado River Delta, United States and Mexico." *Ecological Engineering* 106:757–75.

Naiman R. J., J. J. Latterell, N. E. Pettit, and J. D. Olden. 2008. "Flow Variability and the Biophysical Vitality of River Systems." *Comptes Rendus Geoscience* 340:629–43.

Petts, G. E. 2009. "Instream Flow Science for Sustainable River Management." *Journal of the American Water Resources Association* 45:1–16.

Pitt, J., E. Kendy, K. Schlatter, O. Hinojosa-Huerta, K. Flessa, P. B. Shafroth, J. Ramírez-Hernández, P. Nagler, and E. P. Glenn. 2017. "It Takes More Than Water: Restoring the Colorado River Delta." *Ecological Engineering* 106:683–94.

Pittman, S. J., and K. A. Brown. 2011. "Multi-scale Approach for Predicting Fish Species Distributions Across Coral Reef Seascapes." *PLoS One* 6. https://doi.org/10.1371/journal.pone.0020583.

Poff, N. L. 2014. "Rivers of the Anthropocene?" *Frontiers in Ecology and the Environment* 12 (8): 427. https://doi.org/10.1890/1540-9295-12.8.427.

Poff, N. L. 2018. "Beyond the Natural Flow Regime? Broadening the Hydro-ecological Foundation to Meet Environmental Flows Challenges in a Non-stationary World." *Freshwater Biology* 63:1011–21. https://doi.org/10.1111/fwb.13038.

Poff, N. L., J. Allan, M. Bain, J. Karr, K. Prestegaard, B. Richter, R. Sparks, and J. Stromberg. 1997. "The Natural Flow Regime." *Bioscience* 47:769–84.

Poff, N. L., B. D. Richter, A. H. Arthington, S. E. Bunn, R. J. Naiman, E. Kendy, M. Acreman, et al. 2010. "The Ecological Limits of Hydrologic Alteration (ELOHA): A New Framework for Developing Regional Environmental Flow Standards." *Freshwater Biology* 55:147–70. https:/doi.org/10.1111 /j.1365-2427.2009.02204.x.

Postel, S. 2013. "A Water Bank Helps Revive the Colorado Delta Willows and Wetlands." *Changing Planet* (blog). https://blog.nationalgeographic.org/2013/04/23/a-water-bank-helps-revive-colorado-delta -willows-and-wetlands/.

Roni, P., K. Hanson, and T. Beechie. 2008. "Global Review of the Physical and Biological Effectiveness of Stream Habitat Rehabilitation Techniques." *North American Journal of Fish Management* 28:856–90.

Schlatter, K. J., M. R. Grabau, P. B. Shafroth, and F. Zamora-Arroyo. 2017. "Integrating Active Restoration with Environmental Flows to Improve Native Riparian Tree Establishment in the Colorado River Delta." *Ecological Engineering* 106:661–74.

Schmidt, J. C., B. L. Everitt, and G. A. Richard. 2003. "Hydrology and Geomorphology of the Rio Grande and Implications for River Rehabilitation." *Museum of Texas Tech University Special Publications* 46:25–45.

Shafroth, P. B., G. T. Auble, J. C. Stromberg, and D. T. Patten. 1998. "Establishment of Woody Riparian Vegetation in Relation to Annual Patterns of Streamflow, Bill Williams River, Arizona." *Wetlands* 18:577–90. http://www.fort.usgs.gov/products/3732.

Shafroth, P. B., and V. B. Beauchamp. 2006. *Defining Ecosystem Flow Requirements for the Bill Williams River, Arizona.* U.S. Geological Survey Open File Report 2006–1314. http://www.fort.usgs.gov /products/publications/21745/21745.pdf.

Shafroth, P. B., K. J. Schlatter, M. Gomez-Sapiens, E. Lundgren, M. R. Grabau, J. Ramirez-Hernandez, and E. Rodriquez-Burgeueno. 2017. "A Large-Scale Environmental Flow Experiment for Riparian Restoration in the Colorado River Delta." *Ecological Engineering* 106:645–60. https://doi.org/10.1016 /ecoleng.2017.02.016.

Shafroth, P. B., J. C. Stromberg, and D. T. Patten. 2002. "Riparian Vegetation Response to Altered Disturbance and Stress Regimes." *Ecological Applications* 12 (1): 107–23. https://doi.org/10.1890/1051 -0761(2002)012[0107:RVRTAD]2.0.CO;2.

Shafroth, P. B., A. C. Wilcox, D. A. Lytle, J. T. Hicky, D. C. Andersen, V. B. Beauchamp, A. Hautzinger, L. E. McMullen, and A. Warner. 2010. "Ecosystem Effects of Environmental Flows: Modelling and Experimental Floods in a Dryland River." *Freshwater Biology* 55 (1): 68–85. https://doi.org/10.1111/j .1365-2427.2009.02271.x.

Stromberg, J. C., P. B. Shafroth, and A. F. Hazelton. 2012. "Legacies of Flood Reduction on a Dryland River." *River Research and Applications* 28: 143–59. https://www.fort.usgs.gov/products/22817.

Stromberg, J. C., S. J. Lite, R. Marler, C. Paradzick, P. B. Shafroth, D. Shorrock, J. M. White, and M. S. White. 2007. "Altered Stream-Flow Regimes and Invasive Plant Species: The *Tamarix* Case." *Global Ecology and Biogeography* 16:381–93. https://doi.org/10.1111/j.1466-8238.2007.00297.x.

Stromberg, J. C., R. Tiller, B. Richter. 1996. "Effects of Groundwater Decline on Riparian Vegetation of Semiarid Regions: The San Pedro, Arizona." *Ecological Applications* 6 (1): 113–31.

Suding, K. N., and R. J. Hobbs. 2008. "Threshold Models in Restoration and Conservation: A Developing Framework." *Trends in Ecology and Evolution* 24 (5): 271–79.

Szeptycki, L., D. Pilz, R. O'Connor, and B. Gordon. 2018. *Environmental Water Transactions in the Colorado River Basin: A Closer Look.* Stanford Woods Institute for the Environment. https://waterinthe west.stanford.edu/publications/environmental-water-transactions-colorado-river-basin.

Tharme, R. 2003. "A Global Perspective on Environmental Flow Assessment: Emerging Trends in the Development and Application of Environmental Flow Methodologies for Rivers." In "Selected papers from the Joint Meeting on Environmental Flows for River Systems and 4th International Ecohydraulics Symposium, Cape Town, March 2002" (special issue). *River Research and Applications* 19 (5–6): 397–441.

Trigal, C., and E. Degerman. 2015. "Multiple Factors and Thresholds Explaining Fish Species Distributions in Lowland Streams." *Global Ecology and Conservation* 4:589–601.

Wilcox, A. C., and P. B. Shafroth. 2013. "Coupled Hydrogeomorphic and Woody-Seedling Responses to Controlled Flood Releases in a Dryland River." *Water Resources Research* 49:2843–60. https://doi.org /10.1002/wrcr.20256.

Wohl, E. 2012. "Identifying and Mitigating Dam-Induced Declines in River Health: Three Case Studies from the Western United States." *International Journal of Sediment Research* 27:271–87.

Yarnell, S. M., G. E. Petts, J. C. Schmidt, A. A. Whipple, E. E. Beller, C. N. Dahm, P. Goodwin, and J. H. Viers. 2015. "Functional Flows in Modified Riverscapes: Hydrographs, Habitats, and Opportunities." *BioScience* 65 (10): 963–72.

Zamora-Arroyo, F., J. Pitt, S. Cornelius, E. Glenn, O. Hinojosa-Huerta, M. Moreno, J. García, P. Nagler, M. de la Garza, and I. Parra. 2005. *Conservation Priorities in the Colorado River Delta, Mexico and the United States.* Prepared by the Sonoran Institute, Environmental Defense, University of Arizona, Pronatura Noroeste Dirección de Conservación Sonora, Centro de Investigación en Alimentación y Desarrollo, and World Wildlife Fund—Gulf of California Program, Sonoran Institute, Tucson, Arizona.

# Implementation

PUTTING YOUR STREAM CORRIDOR RESTORATION
PLAN INTO ACTION

*Daniel Bunting, Jeffery Bennett, Todd Caplan, Gary Garrett, Randy Gimblett,
Michael Hammer, Chris Hoagstrom, Robert Itami, Kirk C. McDaniel, Joaquin Murrieta,
Javier Ochoa-Espinoza, Jeff Renfrow, Aimee Roberson, Alfredo Rodriguez, Joe Sirotnak,
Nick Whiterod, and Mark K. Briggs*

## INTRODUCTION

The transition from planning and designing a stream restoration project to implementing restoration tactics is an important milestone for you and your restoration team. The restoration tactics that you and your team employ will be based in large measure on your restoration objectives and can take on many forms. In this chapter, our focus is on stream restoration projects that used restoration tactics that directly manipulated stream conditions. As defined here, such "on-the-ground" restoration tactics include reintroducing native species, eradicating invasive species, constructing gabions, and so on, and are in contrast with stream restoration projects that focus more on policy, governance, and management. Our focus on the implementation of "on-the-ground" restoration tactics is based on assumption that, even if small in scale (e.g., a pilot project), the successful implementation of such on-the-ground restoration actions can be critical to providing the foundation for future restoration efforts that will have even greater impact. No guarantees, of course. Your stream may experience a one hundred–year flood the day after your riparian revegetation is completed. But the more that we are able to address practical implementation considerations and avoid common pitfalls, the greater likelihood for success and that our initial efforts will lead to greater things (e.g., a small scale restoration project becomes a true stream corridor restoration project that may involve not only the implementation of on-the-ground tactics at a larger scale but also policy efforts that affect the entire watershed).

In this chapter, we will review main considerations and examine lessons learned from the implementation of a diversity of stream restoration projects that involved a variety of objectives and on-the-ground restoration tactics. As we implement restoration tactics, keep in mind the iterative nature of stream restoration. Whatever restoration action is pursued, implementation should not be thought of as a straight line to an envisioned result. Remember that your stream corridor restoration project is essentially an experiment. Much insight will be gained throughout the process of implementing restoration tactics as well as through evaluating results via monitoring. Paying attention to these insights will help ensure long-term success. Trial and error is a reality of all restoration programs as you adapt restoration tactics to the sociopolitical and biophysical characteristics of your watershed; remaining flexible will allow you to adapt your restoration plan to the realities of the stream environment that you are working in and discovering. Evaluation and adjustment are key to restoration success, particularly if initial progress toward restoration objectives is unsatisfactory.

Although the stream environment, restoration objectives, and tactics employed to realize objectives may differ significantly from one restoration project to the next, some commonalities exist that allow general lessons learned in one type of restoration effort to apply in other settings. Other lessons are more specific to the type of restoration that is being implemented (such as revegetation projects, native fish reintroduction, or battling nonnative species). Of the multitude of types of stream corridor restoration objectives and tactics that can be selected from, we selected five themes that reflect both the commonalities and inherent differences among different types of stream corridor restoration efforts. These five themes are reflected in the following chapter sections of this chapter:

Overarching Lessons Learned Implementing Stream Restoration Tactics
Enhancing Community Involvement in Stream Corridor Restoration
Managing Non-native Invasive Riparian Plants
Conducting Native Riparian Revegetation
Implementing Native Fish Conservation and Reintroduction

## OVERARCHING LESSONS LEARNED IMPLEMENTING STREAM RESTORATION TACTICS

Before the first shovel hits the ground, we strongly encourage practitioners to develop a stream restoration implementation plan that summarizes objectives, who will be doing what, when, and where, as well as how they will do it. The plan should be a succinct, back-pocket document (no more than ten pages) whose development can be a team-building exercise between all those directly involved in implementing restoration tactics, providing everyone the opportunity for input and to come to agreement on project logistics, roles and responsibilities, finalization of implementation schedules, and general operating procedures and safety concerns.

Suggested main components of a stream restoration implementation plan are summarized below, with general points to consider and lessons learned from past restoration projects discussed as well.

## Main Components of an Implementation Plan

### The Introduction Section
The introduction section of the implementation plan should succinctly review (in a few paragraphs) the type of restoration project being conducted; qualitative project goals and quantitative objectives; the need or importance of the project; project location, size, or scale as

well as landownership; who the main implementers are; stakeholders who will benefit; and a general time frame. These aspects of your stream corridor restoration project are critical to success and were covered in chapter 2. Regarding developing the implementation plan, the introduction basically should provide enough context for those not directly involved in the restoration planning to understand why the project is being conducted, desired results, who is conducting it, where it is taking place, and when.

## Roles and Responsibilities

Stream corridor restoration projects can be complicated and involve a variety of different players. The roles and responsibilities of those implementing the project should be clearly defined at the start to avoid a sloppy rollout, confusion, or delays. Who is doing what? Key personnel should be identified with contact information to outline the main players on the restoration team, including land and water managers, scientists, consultants, landowners, funders, and contractors.

Identify an implementation leader or program manager who will be the go-to person for decision-making. If the nominal project manager is office-bound and only indirectly involved in day-to-day events, another qualified manager or field technical lead should be appointed to act as implementation leader who can deal with daily implementation concerns and has the authority and ability to make decisions as needed. As a trusted associate to the program manager, the implementation leader should also be able to quickly receive feedback from the program manager when multiple or contrasting needs arise among restoration partners.

What happens if you do not have a good leader or a liaison among your restoration team? A restoration project and design may look great on paper, but to implement or actualize the plan is a different story. In efforts that span multiple years with multiple partners, it is likely that project success will require strict adherence to the project/contract timeline as well as biological windows. A lot can go wrong among a contractor preparing a site, a field technician evaluating site conditions, a biologist analyzing precipitation and flow gages and tracking weather forecasts, a project manager facilitating the schedule and budget, and the program manager coordinating with partners. Further, when phased projects have prerequisite tasks that hold up other activities, the entire project timeline could be shifted, complicating both logistics and availability. Did you build in resiliency in the case of team-member or key personnel turnover? What if key personnel are sick or on leave when needing approval to proceed? What if the gate is locked when you are escorting heavy machinery to the site at 5:00 a.m.? Therein lies the importance behind having an implementation leader, someone who sees the big picture and can facilitate communication and coordination among all players. A go-to person for such critical logistic questions as, Where is the gate key (or who can authorize the use of the universal spare key—the lock cutter)?

## Permitting

Though mundane, the importance of acquiring necessary permits before implementing restoration actions cannot be overstated. While restoration inherently seeks to improve the condition of the stream environment, the implementation of restoration tactics can impact natural, cultural, and physical (water and soil) resources. For this reason, regulatory programs and requirements (such as pertain to equipment use, distance from water's edge, and types of materials used) are common at federal and state levels in many countries and can affect implementation (FISRWG 1998). Although highlighted in this chapter, accessing the potential permit needs of your stream restoration project should be considered early on in the restoration process. Do not overlook this step, as significant delays can be expected if necessary permits are not secured. State and local laws may be similar to national laws but provide additional protection for issues of more local concern. If the project involves federal or state funding or is occurring on federal or state lands, compliance with all applicable federal and state environmental regulations is mandatory. In these types of situations, the state or federal agency often handles the permitting processes internally. In other cases, a qualified consultant may be tapped to conduct the appropriate clearances.

In the United States, projects involving ground disturbance that take place on state or federal lands may require both biological and archaeological clearances. Generally, biological clearances are facilitated by National Environmental Policy Act consultation, whereby Endangered Species Act compliance and state and federal agency concurrence is carried out through a scoping process. Cultural resource clearances are mandated by the National Environmental Policy Act and the National Historic Preservation Act and are typically coordinated through the local State Historic Preservation Office (SHPO).

When stream restoration has the potential to impact Waters of the United States—a term defined by the EPA as waters protected by the Clean Water Act and administered by the USACE—a Section 404 permit and, in other situations, 401 certification will be required. Examples of relevant disturbances include the discharge of dredge or fill material, installation of bank protection, use of materials to create riffles or pools, or any efforts that impact over 0.1 acres of these waters or significantly alter the channel alignment. Types of Section 404 permits are nationwide permits, general regional permits, and general state programmatic permits. While nationwide permits assist in streamlining the permitting process (allowing a project to carry on without further agency consultation or review), not knowing how to weave through the process could delay your project initiation date. The Nationwide Permit 27, which pertains to aquatic-habitat restoration, establishment, and enhancement activities, is the most common authorization for stream restoration projects (USACE 2017).

The EPA also administers the National Pollutant Discharge Elimination System (NPDES) Storm Water Program, which restoration practitioners may need to pay attention to if ground disturbance during implementation is likely to be a major pollutant. If that is the case, the regulated entity will be required to develop a storm water pollution prevention plan to establish various control measures (e.g., best management practices) to keep sediment in place, prevent harmful pollutants from entering water bodies, and reduce wind erosion to limit fugitive dust. Other common state and local level restrictions may apply or permits could be required when working near developed areas or roads; these include sound and traffic ordinances that determine noise-level restrictions, traffic-control requirements, and time restrictions on when heavy machinery can be operated. Selected permits, certifications, and reports that may affect stream restoration projects in the western United States are reviewed in Table 6.1.

For Australia, consultation on potential environmental assessment and approval requirements is done through the Department of the Environment and Energy, which has developed a streamlined "One-Stop-Shop" process that also provides information on best practice approaches for restoration and monitoring. The actual permits and approvals required for

**TABLE 6.1** Common permits, certifications, and reports required by stream restoration projects in the western U.S.

| | Federal Act | Permit, Certification, Report | Administering Agency |
|---|---|---|---|
| | National Environmental Policy Act (1969) Section 2 | | EPA |
| | [Title 40, Chapter V, Part 1508] | Biological Evaluation | USFWS/EPA/others |
| | [Title 40, Chapter V, Part 1508] | Categorical Exclusion | USFWS/EPA/others |
| | [Title 40, Chapter V, Part 1501] | Environmental Assessment | USFWS/EPA/others |
| | [Title 40, Chapter V, Part 1502] | Environmental Impact Statement | USFWS/EPA/others |
| Biological Resources | Endangered Species Act (1973) Section 7 | | USFWS |
| | | Biological Assessment | USFWS |
| | | Biological Opinion | USFWS |
| | | Incidental Take | USFWS |
| | Wilderness Act (1964) | | |
| | | general restrictions (mechanical) | USDA/USFS/others |
| | National Historic Preservation Act (1966) Section 106 | | SHPO |
| | | Class I (archival research) | SHPO |
| Cultural Resources | | Class II (reconnaissance survey) | SHPO |
| | | Class III (intensive survey) | SHPO |
| | Antiquities Act (1906) | | |
| | | State Permit (e.g., AZ Antiquities Act) | SHPO/(e.g., ASM) |
| | Clean Water Act (1972) Section 404 | | USACE |
| | | Nationwide (27, 26, 13, 3) | USACE |
| | | Regional General Permits | USACE |
| | | Individual | USACE |
| | Section 402 | | |
| Physical Resources | | NPDES permit | EPA |
| | | Self-certification | USACE |
| | Section 401 | | |
| | | Water Quality Certification | USACE |
| | Wild and Scenic Rivers Act (1968) | Section 7 | BLM, NPS, USFWS, USFS |
| | Rivers and Harbors Act (1849) | Sections 9, 10, 13 | USACE |
| | Clean Air Act (1970) | Section 309 | EPA |

the implementation of stream restoration in Australia will depend on the location and type of restoration that is being planned. For instance, in southern Australia, all activities that affect surface water (including wetlands and watercourses) as well as groundwater will require approval in accordance with the Natural Resources Management Act 2004. For projects that potentially impact species of national conservation significance under the Environment Protection and Biodiversity Conservation Act 1999, an EPBC referral may be required to ensure the restoration project does not pose a significant impact. For restoration projects that involve environmental flow, it may be necessary to consult with the Commonwealth Environmental Water Holder (and state-based equivalents such as the Victorian Environmental Water Holder) to secure the supply of water.

Links to the above referenced sources of information are:

- One Stop Shop: https://www.environment.gov.au/epbc/one-stop-shop
- EPBC referral: https://www.environment.gov.au/protection/environment-assessments/assessment-and-approval-process
- NRM Act: https://www.legislation.sa.gov.au/LZ/C/A/NATURAL%20RESOURCES%20MANAGEMENT%20ACT%202004/CURRENT/2004.34.AUTH.PDF

In Mexico, consultation prior to stream restoration should be completed through CONAGUA. (See appendix A for a review of pertinent U.S., Australian, and Mexican agencies).

Regardless of country, we strongly encourage doing the consultation early in the development of the restoration project to ensure that restoration objectives are feasible, align with broader management programs, and adhere to all relevant legislation.

## Implementation Schedule

Creating a schedule that is agreed to by all involved in implementing the restoration project is a critical part of the implementation plan. An implementation schedule describes when the main restoration tactics as well as pre- and postimplementation monitoring and reporting of results will take place, and the estimated time needed to complete each task. Responsible parties should be assigned to each task, which will provide transparency and accountability. Stream restoration requires a coordinated effort, often within a multidisciplinary framework. While delays often occur, an implementation schedule will allow managers to track progress, establish benchmarks or milestones, and initiate conflict resolution early in the process, should anything happen that could potentially impact the project's timeline or results.

The implementation schedule should build in sufficient flexibility to allow for delays from uncontrollable circumstances (such as extreme weather events or government shutdowns), but milestones or benchmark dates should also be specified. Opportune biological windows (such as eradicating invasive species before they go to seed or harvesting poles* before bud break) are examples of benchmark dates that, if missed, could result in a significant schedule delay (e.g., until the following year, when the biological window presents itself again). Just

---

* In the parlance of riparian revegetation, "pole" refers to branches or stems of cottonwood (*Populus* spp.) or willow (*Salix* spp.) species, and other woody riparian species that form adventitious roots. Pole planting is a technique that places unrooted tree branches or stems in direct contact with the water table or saturated soils.

as farmers know which crops to grow each season and when it is too late to sow their fields, restoration practitioners should know how to optimize their chances for success by working with natural processes. A start date that specifies a time threshold for success should be strictly enforced to prevent delays of a year or more. However, if a benchmark date is missed, budgetary needs or pride should not interfere with the difficult decision to put off a project until the appropriate conditions again present themselves (e.g., putting off harvesting of cottonwood poles until the next dormant season). A delay of a year is better than project failure. Requesting a contract modification or no-cost extension could make the difference between success and failure.

While routine meetings with your restoration team should be conducted throughout the implementation process, it is important to include in the implementation schedule formal meetings where everyone involved gathers to discuss progress. Such meetings can be extremely beneficial to accomplishing restoration objectives, allowing teams to jointly address challenges and solutions and adjust the implementation plan if necessary.

## Site Access

All parties involved in implementing the stream corridor restoration effort need to have access to the project site. However obvious this point appears, securing access (ideally in a manner that can accommodate changing implementation timetables) can be quite complicated. Depending on jurisdiction and landownership, securing access to a site might require permission through a right of entry. In other situations, property easements or temporary construction easements are required to define the location, time frame, and purpose for which the property is used. Ensuring that all parties have directions and a key to the infamous locked gate can prevent unnecessary project delays. In the most extreme cases, conflicts that arise between current and planned land uses may require property to be purchased outright to implement the restoration action. All access issues should be worked out and succinctly described in the implementation plan. This also includes the compilation of relevant phone numbers and permission to contact staff or landowners outside business hours in case access issues arise, or in the case of emergencies.

## Constraints

A variety of constraints (social as well as biophysical) inherent to the project site could hamper the rollout of implementation and limit progress toward reaching stated objectives. Constraints should be summarized within the implementation plan. While developing realistic restoration objectives, you and your team will have already considered constraints related to the biophysical condition of your stream, current trends, main drivers of deterioration, and the socioeconomic and political environment. As you begin implementation, other constraints associated with on-site conditions, project coordination, budget, or other unforeseen circumstances may also need to be addressed. When working on private lands, for example, landowners may have restrictions on materials or treatments used, while work on state or federal lands may require standardized methods or protocols, and may also require certified materials, such as local, native, weed-free plant materials, or use of fill dirt from a certified borrow pit (an aggregate operation that has biological and cultural clearance for their source material). Buying or renting needed equipment may be a challenge for remote sites. Channel conditions may limit access by large equipment. Noise during implementation may restrict implementation actions to sites located a designated distance from homes. Visiting the restoration site as often as possible prior to implementation is key to identifying constraints before boots hit the ground, providing opportunities to address them before implementation actions begin.

## Pilot Projects

No matter how much planning you do or the number of experienced professionals with whom you consult, you and your team will almost certainly learn significantly more about both your stream and the people you work with when you transition from planning to implementation. Surprises—some good, some not so good—will always present themselves. In this light, including a pilot project in your implementation plan as a small-scale experiment can be very useful to test the overall feasibility of the restoration objectives that you have put forward, as well as the rollout of tactics and their effectiveness. A pilot project, by definition, is an activity planned as a test or a trial, allowing new challenges not considered prior to implementation to be identified and addressed before larger-scale efforts are undertaken. Just as it is useful to try a new paint color on one wall before painting the entire house, a pilot project will allow your team to roll out and adapt the implementation plan to the unique circumstances of the project site before tackling the entire project area. For example, in the Colorado River Delta, numerous small-scale, community-based restoration projects provided a wealth of information on the effectiveness of such efforts to improve environmental conditions for people and wildlife at larger scales (Briggs and Flores 2003).

## Project Budget

The implementation plan needs to summarize the budget that, in a concrete manner, impacts your project's objectives, tactics, time frame, footprint, and other factors. For the implementation plan, general budget categories outside labor and oversight should include fees, contractors, supplies, and equipment. While restoration project leaders should have a detailed budget or cost breakdown for the entire project, the implementation plan should focus on the main expenditures or other direct costs most important for carrying out the implementation phase. If heavy machinery and large mechanical equipment are needed, the budget should put together a quote that details costs for machine rentals, including costs for mobilization of equipment, the overall time frame, and subcontractors who may be required to complete tasks. Heavy machinery is expensive, and when rentals are required, managers should take into account the rental time frame, subcontractor availability, as well as fees associated with contractor licensing, insurance, transporting equipment to the site, and taxes, to name a few. Diving deep into budget details at the start may also lead to the identification of related yet nonfiduciary issues that are also important to address prior to implementation. For example, projects funded by federal and state agencies may come with inherent definitions (e.g., construction versus services contracts) that dictate specific requirements related to the cost of the project and whether a general contractor's license, performance bond, or line of credit are required. While this may be unclear at first, it is much better to research these topics before you find yourself having to let go of your preferred contractor and scramble to find a licensed and bonded general contractor with the funder's required level of liability insurance.

In addition to making certain the overall budget adequately covers project costs, cash flow is another important fiduciary consideration. If subcontractors are implementing a component of the restoration plan, keep in mind the challenge—particularly when subcontractors are small companies—of supporting up-front costs or securing a line of credit for equipment and materials. For example, some grants and contracts provide only minimal up-front funding, covering costs only after they have been incurred. In such situations, the prime contractor may not be able to pay a subcontract until being paid by the client. Identifying and addressing such cash-flow challenges can be key to avoiding delays or hurting relationships that could cascade through the entire implementation timetable. Discussing, agreeing to, and signing off on a payment plan prior to implementation is an effective approach to avoid cash-flow issues. This attention to detail requires going a step beyond simply signing a boilerplate subcontract or subagreement that may overlook payment clauses or schedules.

Increasing the flexibility of how your budget can be used (as for labor, supplies and equipment, travel, etc.) is another way to ensure an unhindered implementation process. Certain tasks may require more or less funding than originally anticipated; having the flexibility to move funding from one budget category to another to cover shifting priorities can be beneficial. Putting aside contingency funds to cover cost overruns and unplanned needs is always a good idea. If you get to the end of the implementation phase with contingency funding intact, use it to cover additional implementation and monitoring.

### Communicating Results

Also consider including as part of your implementation plan someone on your restoration team who can serve (or be) a designated communication spokesperson. The planning process typically requires outreach to a variety of entities who may not be directly involved in the restoration effort yet could benefit from project results or whose participation could benefit the restoration effort itself. These entities may include funders, local citizens and citizen groups, natural resource managers, scientists, politicians, students and teachers, among others. A designated spokesperson can take charge of such tasks as reporting results to funders, taking photographs, and giving presentations and interviews to these entities and other interested parties. If funding allows, hiring a public communications firm to educate and disseminate results to a broader audience can be an invaluable investment. As noted previously as one of the overarching lessons from past stream restoration experiences, you want to be in a position to highlight success (see also chapter 8).

## ENHANCING COMMUNITY INVOLVEMENT IN STREAM CORRIDOR RESTORATION

At the global scale, tens of thousands of stream miles pass through privately owned lands. If stream restoration is to be a viable global response to the deterioration of our freshwater ecosystems, private citizens must be strongly engaged. Indeed, the success of stream restoration often hinges on sincere and effective community engagement, with "community" defined as the citizens (or public) who use, own, or manage the landscape that your target stream passes through, upstream as well as downstream of your restoration site. The community can be one landowner or a group of citizens representing a variety of local sectors, societies, associations, or interests; these may include farmers, ranchers, local businesses, recreationists, interest groups, even federal and state managers. Ultimately, it is the local citizens who are most affected by stream conditions and restoration results; they are the ultimate stakeholders.

Unfortunately, too often the public or local community is left out of the stream restoration discussion. The challenge for restoration practitioners is how best to inspire and involve streamside residents so they are directly engaged in the stream restoration effort from the beginning. In this section, we select a few lessons learned from past stream restoration efforts whose results were significantly improved (or made possible) by the participation of the local community in stream restoration.

## Introduce Restoration Projects at Planned Public Events

Most communities hold public forums with a variety of themes throughout the year; many of these events can provide ideal opportunities for your team to introduce a local restoration project. Participating in a community event that is already planned (or, even better, a regularly occurring event, such as a biannual community planning meeting) will require little organization and advertisement on your part—given that the event would likely be advertised and organized by another entity. This can introduce your restoration project to the local community and lay the groundwork for an expanded and focused demonstration later that invites questions and elicits support.

In 2016, for example, Stream Dynamics Inc.—a restoration consulting firm in Silver City, New Mexico, that specializes in turning runoff and erosion problems into water harvesting opportunities using earthworks—used the annual Gila River Festival as a platform to introduce several restoration projects that they had completed. A community bike tour hosted by the firm allowed local citizens to visit twenty restoration and water-harvesting sites (see figure 6.1). As part of this event, the public was able to see firsthand how earthwork structures can be used to dissipate stormwater runoff and how curb cuts (which augment local street infrastructure to convey stormwater from roads into designed basins) are constructed to retain runoff to benefit native vegetation growth and town aesthetics. It also was an opportunity to share the history of Silver City's Arroyo San Vicente, a severely incised drainage (i.e., a stream whose channel has cut deeply into valley alluvium) running through downtown Silver City. In the late 1800s, overgrazing in the upstream watershed and poor town design resulted in runoff tearing through Main Street, eventually carving an arroyo through downtown. The public heard and could see how humans had impacted the environment and how seasonal runoff, which had caused large-scale flooding and damage to houses and buildings, could be not only controlled, but used to promote green spaces. Today, Silver City boasts the most rainwater-harvesting sites per capita in the United States, and restoration projects are welcomed and supported by the local community.

## Solicit Public Involvement at the Onset

Although this chapter focuses on restoration tactics during the implementation phase, remember that effective community involvement should begin at the very beginning of your project when the goals and objectives of your stream restoration project are being developed. Com-

FIGURE 6.1 Participants in a bike tour during the 11th Annual Gila River Festival visited local restoration sites and learned about the benefits of rainwater harvesting. Several treatments constructed throughout Silver City attenuate flows, harvest rainwater, and create green spaces. Photo from Stream Dynamics, Inc.

municating goals and objectives that have been developed without community participation will almost always backfire and lead to a lack of "participatory culture" (Nardini 2008). Particularly in the early stages of planning, local citizens can help identify and address diverging objectives and conflicting interests (Nardini 2008).

Providing opportunities for local citizens to work with the restoration team to develop restoration objectives and overall design early in the restoration process will garner trust and strong community involvement throughout the entire duration of the project. Inviting citizens to the restoration site to discuss objectives, tactics, and approach can be a sound first step. All public comments need to be welcomed and the voice of the community empowered through meaningful and thoughtful dialogue that addresses disagreements and provides opportunities for suggestions and commentary. Restoration project leaders should be open-minded, amenable to compromise, speak in layman's terms (i.e., without jargon), and foster compromise when conflicts or disagreements arise.

Below are examples of two stream restoration efforts conducted by the U.S. Fish and Wildlife Service USFWS and U.S. Bureau of Land Management (BLM) near the Arizona–New Mexico border that benefited from public involvement during the project planning phase. Working with landowners from the onset not only improved relationships and built trust but provided opportunities for input on project design.

The first project (see figure 6.2, top) involved the construction of over thirty rock structures to provide grade stabilization, water retention, and vegetation enhancement in the Guadalupe watershed. The second (see figure 6.2, bottom) involved the installation of a "plug and pond"

**FIGURE 6.2** A successful collaboration with a local landowner stabilized head cuts (top) and road crossings along a private ranch access road in the Guadalupe Canyon watershed of Arizona and New Mexico. The work was expanded to include thirty other projects in the same watershed. Similar projects in the Silver Creek watershed in Arizona (bottom) involved implementing upstream treatments to slow runoff velocity and redistribute flows into multiple pathways at varying runoff stages. The result was improved overall water availability for many native species and reduced scour threats to ranch infrastructure downstream. Photos by Daniel Bunting, Harris Environmental Group, Inc.

design to redirect water from a severely incised drainage back onto the historic floodplain to support native vegetation growth in the Silver Creek watershed. In the first case, local residents benefited from improved road conditions; in the second, stream treatments reduced runoff threats to ranchers and ranching infrastructure downstream. These two efforts underscore the importance of (a) beginning restoration with small-scale pilot projects that demonstrate results and strengthen confidence between all players, and (b) emphasizing benefits of private-federal collaboration to produce win-wins for both the environment and landowners.

## Consensus Building When Thorny Issues Arise

For stream restoration projects that take place in areas affected by a variety of human uses (for example, a combination of grazing and recreation activities), successful restoration may hinge on reducing or even eliminating the uses that cause the most harm to the stream eco-system. Such situations can be complicated and contentious from a sociopolitical standpoint, and may require consensus before implementation can proceed, with "consensus" defined as a positive outcome where the position of all stakeholders is strengthened (Susskind 1999). Many consensus-building approaches have been developed and used over the years. To be suc-cessful, consensus-building needs to produce sustained benefits by building social capital and promoting collective action through mutual learning (Lewins 2001). Forming a community stakeholder group that fosters broader consensus and collaboration on a broad range of issues can be effective (Keough and Blahna 2006). The next case study incorporates a management-planning approach called Level of Sustainable Activity, which has been used in water-planning contexts where consensus among a diverse range of competing interests was required for long-term sustainable management (Itami et al. 2017a). This type of approach can be adapted to the stream restoration context when consensus on restoration objectives and tactics is required in order for the project to move forward.

## "Your" Stream Restoration Project Needs to Be the Community's Restoration Project

Community involvement is predicated on a sense of ownership by the community in the res-toration project. The project should not be seen as "your" restoration project (i.e., a project completed by an outside entity), but the community's. Inherent to restoration projects that local citizens point to as their own is a sense of pride that carries with it multiple benefits that potentially include long-term care and maintenance of the project years after completion. Practitioners should always strive to emphasize the benefits of the restoration effort to the public and avoid the one-and-out approach that produces a project that is not understood or supported by the community.

Making a strong connection with a local community takes time. But, if done effectively, the results can benefit both the community and the restoration project itself, as was the case with stream restoration work conducted with the Ejido of Miguel Hidalgo, along the Santa Cruz River in northern Mexico. In this example, a conservation NGO worked with local ranchers to install cattle exclusion fences and rock dams that increased both water availability and forage—a golden combination for ranchers in semiarid climates. Once community members

**Two Rivers Traffic Management Plan: A Strategy for Integrating Riverside Citizens in the Development of Sustainable Management of Melbourne's Rivers and Bays**

Randy Gimblett and Robert Itami

The Yarra River and Maribyrnong Rivers join together in Melbourne, Victoria, Australia, and their waters flow into Port Phillip Bay, a major port for shipping, luxury cruise vessels, and passenger ferry services, and as a home for rowing, sailing, canoeing and motor yacht clubs, private marinas, and associated marine industries. A traffic-management plan for the "Two Rivers" was required to address increasing conflicts between commercial and recreational traffic, and local riverside citizens. To be effective, the traffic plan needed to involve all interests in a rigorous, transparent, and participatory process that considered natural resource conditions, quality of service and experience, and the well-being and aspirations of riverside citizens (Parks Victoria 2006).

As part of developing the traffic plan, a Level of Sustainability Approach (LSA) was used to quantify and integrate input from all river users on existing and future river use, river capacity, and issues related to impacts that various uses have on river conditions (GeoDimensions Pty Ltd. 2006; Itami 2008; Itami et al. 2017a). LSA is inherently bottom-up, with communities and river users driving the data collection process as well as modeling future scenarios, interpreting impacts on users and on the environment, then providing advice and options for river management (Poe and Gimblett 2017). This approach has also been used in Prince William Sound Alaska (Itami et al. 2017b) and the Verde River in Arizona to quantify desirable river flows for high-quality recreation opportunities.

In the Two Rivers project, four main metrics were incorporated: (1) classification and inventory of waterways; (2) selection of stakeholders; (3) identification of core use issues and patterns of use; and (4) forecasting use trends. A wide range of options for managing commercial and recreational use along waterways was introduced to the stakeholders to allow them to visualize how different users and levels of use would impact quality of service, riverine natural resources, and the well-being of riverside citizens. Quantitative traffic data, used to simulate future conditions, supplemented qualitative data on user needs, safety, and quality of experience. Users were asked to judge preferred densities for exclusive use of the waterway and also the impact of traffic densities of competing vessels on their quality of experience. Face-to-face interviews identified clear conflicts between rowers, passenger ferries, and motorized recreation vessels.

Main conflicts associated with different uses and levels of use were identified by asking each stakeholder group such questions as the following: (1) What type of experience is desirable for water users in the study area? (2) What are the key factors that you look for to attain this experience? (3) What

understood the potential economic value of the project, the NGO's project became the community's project (see Case Study 6.2).

## Pilot Projects Can Help Elicit Public Participation

A large-scale, complicated project does not have to be tackled right out of the starting gate. Pilot projects provide opportunities to enhance community interest, support, and participation in is the "ideal" LSA level? (4) What is the "maximum tolerable" LSA? (5) What is the LSA during peak-summer "non-event" days? And (6) What is the LSA during peak event days? As each stakeholder group answered independently of one another, the process allowed a comparison for consistency and conflicts between groups. Finally, stakeholders were asked to provide ideas to mitigate or manage river traffic to maintain or improve quality of service. Comparing responses of different groups allowed different tolerances to uses and levels of use to be determined.

LSA results revealed that boating clubs were most sensitive to traffic, commercial operators were least sensitive to traffic, and marina managers felt that the traffic was already nearing capacity. The process made it possible to determine and simulate the capacity limits for each group and to better understand the factors that create conflicts between them. Management options identified by stakeholders to address different uses and levels of use included designing and configuring new marinas, limiting and enforcing traffic flow and speed limits, enhancing public education, altering activity schedules, rezoning water during events, and locating commercial berths in less sensitive areas. Results were directly relevant to the planning and management of the river and are being incorporated into the traffic-management plan (Parks Victoria 2006). While not specific to stream restoration, LSA is a participatory process that could be utilized in restoration planning when several stakeholders are involved in a complex project.

### Lessons Learned

The key take-home lessons from developing the Two Rivers traffic-management plan were the following:

- A community-based, bottom-up approach is essential to (1) better understand community values, (2) identify levels of use most suitable for meeting long-term community needs, and (3) make informed management decisions on capacity and use.
- True community engagement leads to responsible self-management with a vested interest in protecting inherent values. An unexpected outcome of the Two Rivers study was that deep engagement of stakeholder groups in the LSA process led to responsible self-regulation and management. For example, local boating groups approached other river users that were violating the new river-management policies and educated them as to proper behavior on the river. Other river users engaged in volunteer shoreline-management activities such as picking up trash, removing dead or dying brush, or other tasks to improve the surrounding river environment.
- Working closely with communities provided managers information essential to exploring new recreational uses for the area as well as detailed guidance to the development of new facilities and infrastructure.
- Using a structured approach—like LSA—improved communication between managers and stakeholders and in turn led to long-term sustainable conservation of river natural resources.

your stream restoration effort. There is no better way to enhance understanding and support for your restoration effort than to have local citizens directly involved. This also allows citizens who participate early in the process to become lead voices for both the immediate restoration effort and for continuing stream restoration efforts (see also chapter 8).

A pilot project can be seen as a dress rehearsal of sorts for the full-scale effort. If the aim of your project is to plant five thousand native riparian trees, getting out to the site with local citizens a few weeks beforehand to plant fifty trees can be fun and rewarding for all.

# CASE STUDY 6.2
Making Stream Restoration Alluring for Ranchers: Ejido Miguel Hidalgo,
Sonora, Mexico

Joaquin Murrieta

*In memory of our good friend, Ventura Rivera,*
*leader of Group 5 of Ejido Miguel Hidalgo*

## The River

Stream restoration actions were conducted with the ranchers of the Ejido of Miguel Hidalgo along the Santa Cruz River in Sonora, Mexico. The river is a tributary of the greater Colorado Watershed with an extension of 402 km (250 mi.), occupying an area of around 31,000 km$^2$ (12,000 sq. mi.). Average annual precipitation in the region is between 200 mm (8 in.) at lower elevations to 553 mm (22 in.) in higher parts of the watershed, such as in the Santa Rita and Santa Catalina ranges of Arizona. The Santa Cruz River is rather unique in that it crosses the U.S.-Mexico border twice. With its headwaters in the grasslands of the San Rafael Valley in Arizona (east of the Santa Rita Mountains), the Santa Cruz River meanders south into the town of Santa Cruz, Sonora, where it meets the Sierra de Los Picos, whose geological structures make the river bend north toward the twin cities of Nogales ("Ambos Nogales," in Sonora and Arizona). Just east of Ambos Nogales, the river crosses the international border and continues north along Interstate-19 through Tubac, Tucson, and Marana, Arizona. The binational population of the watershed is around 1.3 million.

Along the horseshoe shape of the river in the Sonoran side are several ranching and agricultural towns (ejidos) such as Santa Cruz, Mascareñas, and Miguel Hidalgo. Within the communal land tenure known as the ejido system in México, the Ejido of Miguel Hidalgo is one of the largest in northern Sonora, supporting a population of around two thousand with an economy that is largely based on ranching, farming, and, more recently, mining.

## Community and Science

When the Watershed Management Group and the Sonoran Institute—two Tucson-based NGOs—began this project in 1998, organizers knew little about the condition of the reach of the Santa Cruz River in Mexico, nor much about the towns and people who lived there. The first two years were primarily spent on getting to know the watershed and its people. We began by organizing community meetings and introducing ourselves. One of the first lessons learned was that "the community" does not necessarily come to community meetings. To more directly involve community members, we initiated actions that were of mutual interest. We launched a community and science campaign with the University of Sonora in Hermosillo that began with monitoring the quality and quantity of water along the Santa Cruz River. Students from the local middle school helped monitor the river waters with portable kits measuring temperature, pH, dissolved solids, and fecal coliforms. One of the most important findings was the presence of *E. coli* in some of the drinking wells of the town. In response to this finding, we launched a health campaign in the communities of San Lazaro and Santa Cruz that focused on removing cattle from reaches of the river that passed through the towns. The success of this campaign was due to local students, not outsiders, presenting their monitoring results directly to the community.

Following the water-monitoring campaign, one of the most rewarding community experiences of this project was the creation of Los Halcones, a club comprised mostly of students who monitored birds along the binational Santa Cruz River. Club members monitored the water for over five years

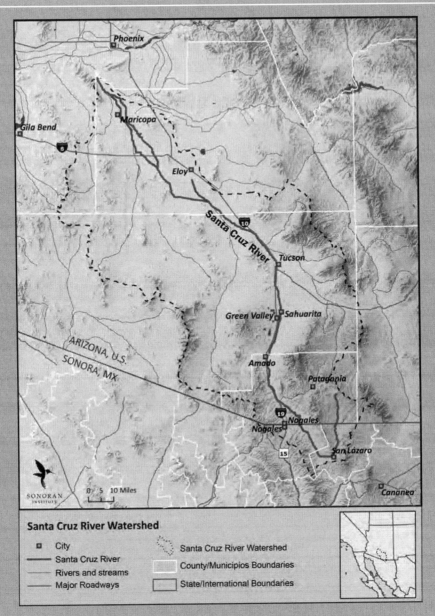

Base map of the Santa Cruz watershed, showing the outline of the watershed in the context of the international border, selected towns and cities and tributaries (the town of San Lazaro is located just across the Arizona-Sonora border). Map by Claire Zugmeyer, Sonoran Institute.

with expert guidance from the Sonoran Joint Venture (a partnership of diverse organizations and individuals from the southwestern United States and northwestern Mexico that share a common commitment to the conservation of birds). Together, over two hundred species of riparian and upland birds were identified. The effort provided important information on habitat quality and use and built community acceptance and trust. The initial phase of Los Halcones culminated with two river festivals in the community of San Lazaro that celebrated community and science.

(*continued*)

Another important step was the establishment of the Community Center for the Conservation of the Santa Cruz River (CCC). Staffed with a field coordinator, the CCC proved to be a very effective outreach mechanism to the region. It developed a portfolio of community and science activities to present results to the community as an educational tool for understanding the watershed and its importance to the well-being of the people who lived there, and helped develop a collaborative long-term vision for the well-being of the river and its communities.

## Ranching and Conservation

After two years of meetings and monitoring, and having a few beers with local ranchers well into the evenings, sufficient trust was established to start working with them on the most significant land use in the area that was impacting the river: cattle grazing. The discussion prompted some powerful observations from the ranching community of San Lazaro, such as: "The trainings and presentations on possible restoration actions are good, but who is going to do the work? Do you want us to do the work? We make a living by ranching the land; do you like cows?" These pointed questions led to an open dialogue with a group of Miguel Hidalgo's ranchers known as Grupo Cinco (G5). We spent considerable time discussing with G5 the common ground between conservation and ranching for example, the importance of both healthy, fat cows and yellow-billed cuckoos. Ranchers were interested in the productivity of their cattle and we (the visiting conservation community) were interested in the health of the ecosystem. By finding the overlap between these two objectives, we hoped to accomplish great things together. A major threat to river conditions was extensive vegetation and soil loss from surrounding uplands as a result of years of overgrazing. Many of the local tributaries to the Santa Cruz River were incised and had little capacity to retain water for the benefit of wildlife or cattle. Here was the conservation/ranching overlap we were looking for: restoring local tributaries for the benefit of both the ranching community and riparian flora and fauna.

Having the support and participation of G5 leadership was essential for initiating a ranching conservation program. With the assistance of Ventura Rivera—one of the key leaders of G5—we were able to move forward with the ranching community to implement best-management practices for enhancing productivity and conservation of the land. This goal was furthered by conducting tours and informational exchanges between ranchers in Arizona and Sonora on a variety of topics, including the following:

- Introducing best cattle-management practices (such as dividing up the land into smaller parcels and rotating the cattle use between different parcels);
- Planting "living fences" (tight rows of willow [*Salix* spp.] and cottonwood [*Populus fremontii*] cuttings) close to and parallel to the river channel to guide flow and retain sediment; and
- Constructing gabions (rock structures) in small incised channels to retain soil and water.

The first project with G5 that emerged from these exchanges was to install new barbwire and posts that allowed them to divide their parcels and improve cattle management. Implementing this initial project significantly strengthened our relationship with the ranchers and provided an opportunity for all to work together for a common goal, understand the importance of project management, and explore additional grants to do more. Dividing G5 parcels and rotating cattle improved grassland

Alfonso González with Grupo 5 of the Ejido Miguel Hidalgo after building numerous rock gabions to slow water runoff, increase infiltration, and enhance water availability for cattle and wildlife.

conditions, giving G5 the impetus to move forward to create a twenty-five-acre cattle exclusion parcel in an effort to establish a "native-grass-seed bank" on their land. Once the cattle exclusion was complete, we celebrated with a meal and talked about the possibility of registering the property as a sustainability ranch or Environmental Management Unit—a federal recognition in Mexico that opens doors for federal-private support.

Not long after the celebration of this initial work, G5 was ready to do more and funding was secured to initiate a three-year gabion construction program to control erosion and increase water retention and ground cover. During this period, over two thousand gabions were installed by G5 ranchers, who took turns building them with some level of compensation for each gabion built.

In a short time, the gabions produced positive results that included more water and grasses in the tributaries, an important win-win for ranchers and the environment. With increased confidence, G5 initiated a gabion construction training workshop that extended to ranchers from other ejidos in the region. Hundreds of additional gabions have since been constructed on lands managed by G5 and in other locations in the Mexican portion of the Santa Cruz watershed.

## Lessons Learned

- Ranchers and local communities are more apt to work collaboratively on conservation efforts if they understand the tangible and direct benefits these efforts will have on their lives.
- It takes time to develop the necessary trust, understanding, and local leadership that is needed for community-based projects to be effective and produce viable, long-term results.
- Providing a local consistent presence via the CCC office and staff was key to moving our work forward with the Ejido. This process cannot be rushed and can only be accomplished via invitation from the local community.

(continued)

Much like a developer constructing a model home prior to the whole subdivision, a restoration pilot project provides a public demonstration of the large-scale plans, how goals will be accomplished, and what results might look like. Citizen involvement at this point—well ahead of the main implementation phase—provides opportunities for residents to make comments and suggestions and become invested in and committed to the larger effort.

The Yuma East Wetland restoration project along the lower Colorado River between the states of California and Arizona illustrates the benefits of pilot restoration. A 10 ha (25 acre) pilot project in 2004 demonstrated success in converting agricultural lands and dense tamarisk monocultures (stands of the single species) to native riparian habitat and was a springboard for fostering continued funding for restoration. Building off the success of the pilot demonstration, the city of Yuma partnered with the native Quechan tribe and the Yuma Crossing National Heritage Area to convert a 574 ha (1,418 acre) landscape of largely nonnative plants that had several illegal dumpsites and homeless encampments to a thriving riparian and wetland community. The homeless were not uprooted from the location and were given the opportunity to help return the degraded reach to a productive riparian forest. The Yuma East Wetlands, once viewed by locals as a dump and an area even considered dangerous for public recreation, is today easily accessed by the public for bird-watching, fishing, recreation, and beach access (see figure 6.8 in section on native riparian revegetation).

## Sharing Local History of Your Target Stream Can Broaden Support for Restoration

Most towns or cities have a river, stream, creek, or main drainage running through the heart of the community; its existence may have been a historical factor sparking settlement by native and European cultures. Many streams, unfortunately, have run dry from water extraction or are in a deteriorated condition, possibly to the point that many local residents do not even know the stream exists (it may have become a cement-lined culvert). As such, the historical image of the once-pristine riparian corridor—as well as the resources and ecosystem services that it once provided—is often lost to younger generations. Learning about the local heritage of your stream and relaying this information back to the community can provide a springboard for enhancing involvement of local citizens in stream restoration. Even when restoration of predevelopment conditions is impossible, creating a vision of past conditions can be the impetus for developing community-based restoration objectives that increase local stewardship and promote stream improvements with long-lasting benefits to the local community and environment.

An example is the management of the Lower Santa Cruz River, which runs through the city of Tucson in Pima County, Arizona. The once-thriving river has become ephemeral since the early 1900s due to water extractions that lowered water tables. It has been dry for so long that most citizens of eastern Pima County did not know that the once lush river used to flow for much of the year and, indeed, was a central reason for the city's establishment. In 2013, upgrades to two wastewater treatment facilities provided opportunities for significant amounts of treated effluent to be discharged into the ephemeral Santa Cruz River. While the effluent-dependent system is far from natural, aquatic and riparian flora have dramatically rebounded. Pima County and its partners are now using the river's improved condition as a stage for community outreach that educates the public about the river's natural and cultural history as well as its current importance to the community (see figure 6.3). These efforts fostered public concern and thirst for more knowledge about the river. An annual Living River report (a collaborative effort of the Sonoran Institute, Pima County, and other partners) is now disseminated to the local community to summarize monitoring efforts as well as current conditions and improvements to the river. Such an undertaking is likely beyond the means of most stream restoration efforts. However, restoration practitioners should be alert to ongoing public outreach and education programs that may facilitate public awareness, support, and participation in their restoration projects.

## Paying Local Citizens to Work on Stream Restoration

Many co-authors of this guidebook work in regions that are not prospering economically, which can pose additional challenges in securing local citizen involvement. While we are not advocating for restoration or conservation groups to expand their missions to include community development, it nonetheless is critical for restoration practitioners to recognize the socioeconomic environment they are working in. In areas where work is hard to find, paying local citizens to help implement various aspects of restoration can benefit both the community and the project.

Hiring locally is often less expensive than bringing in labor from outside the community (because of lower travel, lodging, and food costs). In addition, paying local citizens provides them hands-on stream restoration experience and invests them in the restoration project, potentially fostering a community of stream leaders who can enhance understanding, support, and long-term involvement throughout the community (see figure 2.1 as a binational example). As mentioned in the Ejido de Miguel Hidalgo case study, community members were paid to help construct the first series of rock dams, providing opportunities for their immediate involvement and rapid completion of a pilot project. This laid the foundation for longer-term restoration actions that have continued since 2010.

**FIGURE 6.3** The Santa Cruz River through Tucson, Arizona, has been revitalized thanks to treated effluent discharged into the ephemeral riverbed. Pima County and its partners regularly conduct public tours that highlight the river's resiliency as well as the river's historic and cultural importance. Much of the river has high bird diversity and has become a hot spot for birders. Photos from Pima County.

## Emphasize Ecosystem Services

Stream restoration is not just about critters or even streamflow. Although projects that conservation or restoration groups conduct often focus on improving habitat for native species, practitioners should emphasize other benefits to the local community. Why should the public care and want to be involved? Many will be motivated by the plight of native species, but augmenting the environmental message with information on how stream restoration will benefit them socio-economically can greatly incentivize local support and involvement in the restoration effort.

Even if the goals and objectives of your project were not initially built around meeting community needs, restoration results will likely directly or indirectly strengthen a variety of ecosystem services that the stream provides to the public. Ecosystem services, in this context, are the many and varied benefits that the public could gain from a local restoration effort. These may include flood control, better water quality and quantity in the local stream, and recreation and aesthetic values.

Several studies have quantified increased housing sales prices due to proximity to the riparian corridor. For example, a study of 786 home sales along the Santa Cruz River in southern Arizona found that proximity to the riparian corridor increased sale prices by an average of 3.1 percent (Bourne 2007). This and similar studies reveal the economic value for protecting ecosystem services of a waterway despite apparent lack of traditional contact recreation

(Weber et al. 2016), offering the potential that stream restoration efforts could be self-financing (Bark-Hodgins and Colby 2006; Bark et al. 2009).

Underscoring the ecosystem service benefits of restoration may also bring long-term benefits. For example, a diverse coalition of groups, organizations, agencies, and private citizens joined forces to support restoration actions in the upper watershed around the city of Santa Fe, New Mexico, to protect that city's water supply (see Case Study 8.3). The bottom line is that local citizens are more likely to become engaged if they understand and, even better, see firsthand how restoration will benefit their communities.

If the link between your restoration project and the local community is weak, consider altering the restoration game plan to increase local community benefit. When the Sonoran Institute first began working with the members of the Ejido de Miguel Hidalgo (see Case Study 6.2), the focus was on the mainstem of the Santa Cruz River. The geographic focus expanded to local tributaries of the river when community members expressed greater support for efforts along those tributaries that would improve both native habitat conditions and conditions for their livestock.

## MANAGING NONNATIVE INVASIVE RIPARIAN PLANTS

Nonnative, invasive species have proliferated along rivers in many parts of the world for a hundred years or more. *Nonnative species* (sometimes referred to as "exotic," "alien," or "nonindigenous" species) are species that have been introduced (intentionally or unintentionally) into regions that were not historically part of their native range (Smithsonian Marine Station at Fort Pierce 2017). An *invasive species* is any nonnative species whose introduction causes or is likely to cause economic or environmental harm or harm to human health (Presidential Executive Order 2016). Note that the definition of "invasive" carries with it two central elements: (1) the species has established and is reproducing outside its natural distribution range, and (2) the species is causing harm to native species or humans. With this definition in mind, we use (for the most part) the term 'invasive' alone with the implication that the descriptor refers to an exotic or nonnative species. The mode of introduction and establishment varies with the invasive species being considered. Not all species introduced outside their natural environment become invasive (i.e., become a problem). In river ecosystems, a principal factor that has increased the severity of the invasive species problem are changes in the river ecosystem brought on by numerous human-related factors (such as river impoundment), changes in management, and flow diversion, which make the river environment much more conducive to the establishment and proliferation of nonnative species.

After habitat destruction, the spread of invasive species is the most significant threat to biodiversity, causing enormous harm by displacing native species and adding to the deterioration of native ecosystems and the ecosystem services they provide (McGinley and Duffy 2012). Invasive species are responsible for the extinction or decline of many native species and continue to pose a huge threat to many more, often in combination with other human-related impacts. While it is difficult to estimate how many invasive plants have become established in any particular region, Pimentel and colleagues (2001, 2005) and Burnham (2004) estimated that over five thousand invasive species have become established across the United States, with the invaded footprint affecting approximately 700,000 hectares (around 1.73 million acres) per year of wildlife habitat (Babbit 1998). The annual cost of invasive species within the United States is estimated at $120 billion to $137 billion (Pimentel et al. 2001; Pimentel et al. 2005; EPA 2017) and $13.6 billion in Australia (Australia Academy of Science [https://www.science .org.au/curious/earth-environment/invasive-species]).

With invasive species threatening not only the environment but the economy and human health, it is not surprising that millions of dollars are spent every year to control them. Costs related to controlling invasive species vary with species, methods, and region. A study in Australia comparing the 2001 and 2011 financial years estimated that costs due to combined economic losses and control of invasive species increased from $9.8 to $13.6 billion per year, with control expenditures increasing from $2.3 to $3.8 billion per year (Hoffman and Broadhurst 2016). The cost associated with controlling invasive aquatic plants in the United States is estimated at over $100 million per year, and some species alone cost several millions of dollars per year (Office of Technology Assessment 1993). To address costs more specifically associated with managing invasive plants in stream ecosystems, the 2006 Saltcedar and Russian Olive Control and Demonstration Act (U.S. Public Law 109–320) authorized $80 million for five years of planning and management of these two species alone. With so much money being thrown at invasive species control, many have questioned whether the efforts are effective. Unfortunately, many projects lack the quantified monitoring needed to understand the level of success. Moreover, other skeptics wonder if unintended negative consequences could arise when eradication efforts are not planned in conjunction with restoration (Zavaleta et al. 2001; Sogge et al. 2008; Stromberg et al. 2009a).

The central debate for stream restoration practitioners is not so much whether invasive species have spread, but the level of harm they do or can impose on native habitat if left unchecked, as well as what realistically can be done about it. Invasive species threaten the entire stream environment (benthic, aquatic, and riparian ecosystems). For purposes of narrowing the discussion, we will focus our attention in this section to the management of invasive riparian plants. Similar to what occurs thoughout the stream environment, invasive riparian plants can outcompete native species if they are better adapted to the altered hydrological characteristics of the stream. In such situations, simply removing invasive riparian plants will not necessarily promote reestablishment of native species that may no longer be adapted to the current physical and chemical conditions of the river and its adjoining floodplain (Briggs 1996; Brown and Amacher 1999; Nagler et al. 2005; Nagler and Glenn 2013a). For this reason, restoration projects that aim to reduce the extent and distribution of invasive riparian plants need to consider restoration tactics that go beyond eradicating these species. A multi-faceted approach is often needed for long-term control. For example, combining targeted eradication with an environmental flow program that reestablishes critical parts of the natural hydrograph (e.g., through a spring pulse flow to promote natural recruitment of native species) may produce more effective results (González et al. 2018) (see also chapter 5).

The debate on controlling invasive species must avoid the simplistic argument that nonnative equates to bad and native equates to good; it is more complicated and nuanced than that. The removal of tamarisk (*T. ramosissima, T. chinensis, T. aphylla*, and others), an invasive species native to Eurasia that has proliferated across over 1.5 million hectares of riparian habitat in the United States, is a case in point (Sher and Quigley 2013). When diversity and structure of native riparian vegetation are lacking, the presence of tamarisk provides important habitat for key species of wildlife, such as for the endangered southwestern willow flycatcher, *Empidonax traillii extimus* (Sogge et al. 2008). The flycatcher is one of the few riparian midstory species adapted to current floodplain hydrological and soil chemistry conditions that dominate many rivers in the southwestern United States and northern Mexico.

Before embarking on a concerted eradication program (whether focused on tamarisk species or other riparian invasive plants), stream practitioners need to have the "Is it appropriate and worth the effort?" debate. Summarized below are important questions that practitioners should consider as part of this deliberation. If the majority of answers to the questions is yes, then a concerted effort to control an invasive riparian plant may be sensible. If most answers are no, it may be wise to alter the restoration program to include other objectives.

- *Are objectives concerning invasive species management part of a broader restoration response that includes other complementary objectives?* We encourage practitioners to go beyond objectives that are solely focused on reducing the extent and distribution of an invasive species and consider the broader purpose of why managing invasive species might be a priority in the first place. Is management of a particular invasive species important not only for improving wildlife habitat but for other objectives as well, including recreation experience, channel morphological conditions, biodiversity, the well-being of riverside human communities, or other? If so, invasive species management may be a viable response as it becomes a means to multiple end goals (one tool in the restoration tool box) and not the end goal itself. What we want to avoid is a myopic focus on managing invasive species simply because the species is present. Practitioners need to be able to elucidate what the greater purpose of invasive species control is (per Clark et al. 2019).

- *Are eradication objectives realistic, given funding levels, personnel, resources, the time frame, and other potential constraints?* Please see chapter 2 to review considerations for developing realistic stream restoration objectives. Avoid the scattered approach of undertaking unfocused eradication until the budget is exhausted. Know the detailed labor, material, supply, and transport costs for managing invasive species and quantify exactly how much your exotic management program can realistically accomplish with available resources. Exactly what invasive management "bang" will you get for the money spent? The answer to that question needs to be put on the table as part of the invasive species management debate.

- *Is the objective for managing a particular nonnative species sustainable in the long term?* A thoughtful answer to this important question requires understanding of the following:
  - The long-term invasive plant-management objective: The amount of resources needed to maintain the extent and distribution of an invasive plant at desirable levels following initial interventions can vary dramatically with target species, how rapidly the target species reinvades following initial management, site conditions, the cost of eradication tactics, among other factors. Ultimately, accurate quantification of long-term costs will be determined with experience, which can be gained by initially implementing several small-scale pilot projects. (See Case Study 6.5.)
  - The likely response or recolonization potential of the targeted plant following initial treatment: Does the plant reestablish quickly and strongly following initial treatment, or is regrowth following initial treatment spotty and slow? This is a critical ques-

tion for managing invasive riparian plants when practitioners are not addressing the underlying reasons why the stream environment has deteriorated and become conducive for the proliferation of the target invasive species in the first place. For example, the biophysical and chemical condition of the river may have allowed an exotic species to outcompete the native plant community and become invasive. If removal is not combined with other management goals (such as changes in flow management or revegetation of natives adapted to the site), the stream will retain the same physical template that nurtured the proliferation of the invasive plant, and the target species may simply reestablish rapidly after treatment. If such is the case, long-term control may be unsustainable. As with the above point, beginning management at a small scale and monitoring results can provide the data and information practitioners need to answer this question.

- The long-term maintenance budget: It is likely that the target invasive species will reestablish at some level following initial treatment, potentially requiring follow-up management actions to maintain reestablishment at a desirable level. The cost of follow-up management will vary with the specific species being controlled, stream conditions following initial treatment, the region, level of access, and a variety of other factors. Practitioners will gain insight into long-term maintenance costs with experience, but the more resources are available for follow-up treatment, the more likely practitioners will be able to meet long-term objectives. If the resources you have for managing invasive species only cover costs associated with initial treatment, give pause to embarking at all.

- *Are methods for managing and treating the invasive plants effective and appropriate?* The more effective a given method is at controlling a target invasive species, the more likely that practitioners will be able to maintain the species at desired levels in the long term. However, on the heels of the effectiveness question is whether the control methods will do more harm than good. This is particularly an issue when herbicide is used as the main method of control. Despite EPA-approved products and labeling by the manufacturing company, it is valid to question the value of introducing potentially significant quantities of herbicide into a riverine environment. Can management objectives for the target invasive species be accomplished by staying within recommended volumes of active ingredient per area or by using other, nonchemical approaches? These are key questions that practitioners will need to address, as they touch on environmental impacts, human safety, and the sustainability of an invasive species management program.

- *Does the increase in the extent and distribution of the target invasive plant contribute to processes that are further exacerbating biophysical conditions?* In some cases, nonnative species that are invading our riparian ecosystems form monotypic (or nearly monotypic) stands. When this occurs, loss of biodiversity and reduced habitat quality are often inevitable (dense stands of a single species lack the vertical and longitudinal complexity that is so important to biological diversity). In addition, many invasive species that are now prevalent along rivers in Australia, Mexico, and the United States are directly impacting channel morphological processes to the detriment of many native aquatic and riparian species. For example, dense stands of *Arundo donax* (giant cane) along channel margins tend to hold sediment in place and increase sediment accumulation. When this occurs along rivers that already have a sediment surplus (too much sediment without sufficient flow for consistent evacuation), a wide, shallow, and complex channel morphology can transition to one that is narrow, deep, and simple, with the accumulating sediment burying high-quality aquatic and riparian habitat. Along the Rio Grande / Río Bravo, dense stands of giant cane appear to be exacerbating such sediment aggradation, making the management of this invasive species more of a priority than if the species was having little impact on channel morphological processes (see Case Study 6.5).

- *Will the management methods that are being proposed complement biocontrol agents that may have been purposefully introduced to battle the target invasive species?* Biocontrol agents have been introduced in many countries to control invasive riparian species. The tamarisk leaf beetle (*Diorhabda* spp.) to control tamarisk and the gall wasp (*Tetramesa romana*) to control giant cane are two examples from northern Mexico and the southwestern United States. If a biocontrol agent has made it to your target stream, its impacts on the target invasive species need to be assessed. If effective, biocontrol may reduce the need for other control methods (such as treatment with herbicide or mechanical means).

As noted previously, consider including restoration tactics in addition to actions associated directly with invasive species management in your overall response if, after reviewing the above criteria and having "the debate," you decide that managing a particular invasive riparian species should be undertaken. Developing an environmental flow program that addresses streamflow regulation—which likely led to the proliferation of the invasive species in the first place—should be high on the list. For example, altering flow management to foster the release of high magnitude spring flows can promote native plant recruitment and sustainability. Such was the goal behind the binational pulse flow into the Colorado River Delta (see Case Study 5.5). And/or conduct targeted revegetation with native species. Eradicating an invasive plant can reduce competition and improve the potential for establishing native species via targeted artificial plantings. How appropriate the inclusion of additional restoration tactics is will depend on a variety of factors, including your restoration objectives, historical and current conditions of your stream, the reasons behind the spread of target invasive species, and others. But a multipronged approach toward a desired end is often more effective than a single-pronged one.

In the context of dryland riparian ecosystems across the globe, it is important to note that established native and nonnative plant assemblages vary from continent to continent. For example, cottonwood and willow species, which are focal to riparian restoration in the United States, are by and large nonnative and problematic in Australia. The ability of nonnative willows in Australia to hybridize and disperse rapidly via seed, adventitious roots, or vegetatively has displaced natives, disrupted natural flows, and reduced biodiversity (Cremer 2003; Pope et al. 2006). In fact, the Australian government now restricts willow importations and, with the exception of a few willow species, it is illegal to sell or plant them (Rivers of Carbon 2012). Whereas *Eucalyptis* spp. and *Acacia* spp. are native to Australia, they are highly invasive in South Africa, leading to reductions in water quantity and quality while altering the fire regime (Chamier et al. 2012). Working for Water, a national program in South Africa, was established to control invasive alien plants, many of which are closely related to native plants within Australia and the United States (Van Wilgen et al. 2011). With the understanding that restoration priorities and targeted species differ among regions continentally, the remainder of this section focuses on lessons gained from managing three nonnative, invasive riparian plants of importance in northern Mexico and the southwestern United States: tamarisk, Russian olive, and giant cane. These species have been highly successful at either replacing or outcompeting native species across this geography; tamarisk is also a problematic invasive species in Australia.

## Tamarisk (*Tamarix* spp.)

Of the eight or so tamarisk species (*Tamarix* spp.) introduced to North America, some of the most common and problematic species naturalized across the southwestern U.S. and northern Mexico are *T. ramosissima*, *T. chinensis* (some authorities consider *T. ramosissima* synonymous with *T. chinensis*, *T. pentandra*, and *T. gallica*), and *T. aphylla*. *T. ramosissima* and *T. chinensis* are deciduous species native to China, although the former has a distribution extending into the Middle East. Both species have similar structure, can form dense thickets, and are known to hybridize. For the purposes of this review, we combine and use the common name "tamarisk" to refer to these two species. *T. aphylla* is a larger evergreen native to Eurasia (and can grow over eighteen meters tall [sixty feet]), and can easily be distinguished by its smoother, pine-like needles. We refer to this species as the Athel tree and simply use the term "tamarisk" when discussing information relevant to all three species.

Tamarisk was introduced to the western United States as an ornamental plant and also was widely used as a shade tree or windbreak and to stabilize eroded banks. It quickly escaped cultivation and currently is naturalized along most river systems in the southwestern United States and northern Mexico. Tamarisk has become the third-most common tree along rivers in the western United States, occupying an estimated 1.5 million hectares (around 3.7 million acres) (Friedman et al. 2005; Tamarisk Coalition 2017). It also was introduced to Australia in the 1930s and, as in North America, has become a serious environmental issue there. The distribution of tamarisk ranges from below sea level to 2,400 meters (roughly 8,000 feet), but it is particularly problematic at lower elevations where it sometimes forms monocultural thickets (Zavaleta 2000).

Native riparian trees such as cottonwoods can outcompete tamarisk when water is sufficient and soil chemistry is within tolerable levels (Sher et al. 2002; Sher and Marshall 2003; Bunting et al. 2011). Shading or crowding by other plants such as cottonwood can also hinder tamarisk growth (Bunting et al. 2011). However, research over the past couple of decades indicates that tamarisk inhabits a broader environmental niche leading to the displacement of cottonwoods and willows in areas where these native plants are no longer adapted to the conditions imposed by river management (Anderson 1995; Di Tomaso 1998; Taylor and McDaniel 2004; Nagler and Glenn 2013a). Once tamarisk is established in significant numbers it is difficult to control and almost impossible to eradicate completely (Pearce and Smith 2003). Physiological adaptations that have allowed tamarisk to proliferate include (1) high, year-round seed production with distribution by wind and water, (2) rapid germination and seedling establishment across a broad range of environmental conditions (Brotherson and Winkel 1986), (3) extreme tolerance to high salinity and low soil moisture (Glenn and Nagler 2005), and (4) the ability to resprout after stems are cut to ground surface, following burning, or from severed stems (DiTomaso 1998). Tamarisk tends to proliferate when riparian conditions are open with little overstory, and where its root systems are able to access saturated soils. As a facultative species, tamarisk can establish off-channel on disconnected floodplains, but it also thrives in-channel where it can stabilize active channel bars, promote sediment deposition, and prevent bank erosion and widening of channels during flood events (Keller et al. 2014, Dean and Schmidt 2011). These changes in geomorphology have contributed to the channelization of once dynamic systems (Birken and Cooper 2006). Native species that historically required flooding and associated fluvial processes for natural recruitment are not adapted to these changing environments.

## Management

Decades of research, field trials, and management have provided numerous resources to help understand when, where, why, and how to manage tamarisk (see, for example, Anderson et al. 2004; Sher and Quigley 2013). Commonly cited objectives associated with tamarisk management include (1) augmenting streamflow, based on the assumption that tamarisk consumes large amounts of water (Di Tomaso 1998; Bay 2013;); (2) protecting and enhancing cottonwood/willow communities where tamarisk has established (Barrows 1998); (3) improving wildlife habitat (Engel-Wilson and Ohmart 1978; Ellis 1995; Bailey et al. 2001); and (4) decreasing riparian forest fire frequency and severity (Shafroth et al. 2005). Researchers continue to evaluate the amount of water (particularly streamflow) that may be saved by controlling tamarisk. Water consumption in tamarisk, like native cottonwoods and willows, has been correlated to leaf area and is highly variable as it is influenced by a host of environmental factors (Nagler and Glenn 2013a; Cleverly 2013). Some control efforts have quantified immediate water savings (Hatler and Hart 2009; Doody et al. 2011), but long-term impacts to streamflow are dependent on several factors, including whether water used by tamarisk was directly connected to the river (or residual waters associated with floodplain surfaces) as well as how vegetation establishment following treatment impacts the water balance (Yu et al. 2017). While salvaged water may contribute to aquifer recharge or may increase water availability in floodplain soils, one should not count on a dramatic and long-term increase in streamflow (Sheng et al. 2014).

Tamarisk-control efforts should be evaluated on a site-by-site basis to weigh the benefits and disadvantages that may result from the specific actions proposed. If establishing native species is improbable due to site constraints (such as low water availability, lack of flooding, high salinity, or an absence of source trees), the value of controlling tamarisk should be questioned, as it may result in an overall decrease in available habitat for native wildlife species, underscoring the importance of accompanying tamarisk-management actions with native plant revegetation. The bottom line for practitioners, however, is that unless management of tamarisk is accompanied by efforts to reestablish physical and chemical conditions more favorable to native riparian plants, it is likely that tamarisk will recolonize areas where it is removed (Briggs 1996; Glenn and Nagler 2005; González et al. 2017a; González et al. 2017b).

Another factor that complicates the question of whether or not to control tamarisk is that tamarisk species provide habitat for several native species. In fact, riparian habitat can still be considered high quality even if tamarisk becomes a dominant part of a mixed-plant community that includes native species (Sogge 2008; van Riper et al. 2008). Researchers have found that habitat value for many native species does not significantly decrease until tamarisk achieves 90 percent cover, and some estimates indicate a native tree composition of 10 to 40 percent provides optimal habitat (Cohn 2005; Sogge et al. 2008; van Riper et al. 2008).

The decision to embark on a concerted tamarisk-management program can be even further complicated in situations where tamarisk is providing habitat for protected wildlife species. For example, in the western United States and northern Mexico, the endangered southwestern willow flycatcher now breeds in tamarisk throughout much of the bird's range, as the invasive tree has similar characteristics to the bird's native habitats, including a dense structure, high canopy cover, tall stature, and mesic or wetland habitat (Sogge et al. 2008). It is generally accepted that the bird will select tamarisk when native species are not present or do not meet structural requirements for successful nesting. If your restoration project is taking place on lands managed by the federal government and is within the known distribution range or federally designated critical habitat of an endangered species, you will be required by ESA and NEPA to analyze the potential impacts that your management activities may have on the species and its habitat.

## Methods

A variety of methods have been used over the years to control tamarisk. Equipment availability, financial constraints, available personnel, time needed to complete the task, environmental restrictions, biological limitations, and numerous other factors enter into the decision-making process when selecting a control procedure (McDaniel and Duncan 2000). Methods summarized below for tamarisk control can be used alone or in combination depending on the restoration objective and characteristics of the site.

### MANUAL

Manual control consists of the use of hand tools (shovels, hoes, pick axes, etc.) to remove plants. Typically, manual control is only effective during the seedling stage when the entire plant with root can be pulled out of the ground. While this method is not used for large-scale control of tamarisk, it can be effective during postproject maintenance when seedlings begin to germinate (USDA 2010).

### MECHANICAL

Mechanical control includes the use of mechanized tools to cut or remove stems or entire trees. Chainsaws are often used to cut stems or trees just above ground level, and heavy machinery equipped with implements can mow, rip, disk, or root-plow portions of or the entire plant. Bulldozers with a front-mounted dirt blade are commonly used to remove above-ground biomass (using a front-mounted dirt blade), to clear large fields (with a front-mounted rake), and to pile brush (using a hydraulic thumb or articulating loader). Root-plowing, however, is the most effective mechanical technique for controlling tamarisk, because root plows target roots 30 cm (about 1 ft.) or more below the ground surface (Di Tomaso et al. 2013). In fact, grubbing (i.e., digging up the root or plowing to uproot plant materials) is the only effective mechanical method, due to the ability of tamarisk to resprout. Disadvantages of mechanical methods include disturbance of surface soils leading to wind and soil erosion and the production of tamarisk fragments that can resprout adventitiously (Di Tomaso et al. 2013). A synopsis of several restoration projects also showed that high-disturbance techniques, such as mechanical removal or fire, result in the highest abundance of noxious weeds compared to other techniques (Gonzáles et al. 2017a). Therefore, it is cautioned that restoration planning should discuss potential implications that mechanical treatments alone may have on weedy plant succession. Follow-up maintenance with selective grubbing or a spot herbicide will often improve effectiveness, and supplemental seeding or planting with native species can further increase diversity.

PRESCRIBED BURNING

Conducting a prescribed fire—a planned, controlled burn conducted to meet specific management objectives—can effectively reduce the above-ground live biomass, potentially facilitating and improving the effectiveness of follow-up treatments. It is important to emphasize that although burning tamarisk can kill it outright (when the burn is sufficiently hot and the trees are already stressed by environmental conditions), burning alone without follow-up management is often not effective, as it often promotes resprouting and flowering (Di Tomaso et al. 2013). Strategic cutting and placement of tamarisk branches and other woody debris can also increase fuel loads, promote fire propagation, and increase fire severity (Hohlt et al. 2002). Regardless, conducting a prescribed fire followed by selective herbicide treatments (Di Tomaso et al. 2013) or follow-up mechanical control strategies such as grubbing or root-plowing (Barranco 2001) appears to be most effective. As noted above, high-disturbance control techniques, such as fire, can induce secondary invasions of nontarget invaders (Gonzáles et al. 2017b). Typically, use of prescribed burning would be followed by active revegetation and maintenance of undesired species. Planting or seeding a cover or nurse crop (e.g., an early successional herbaceous plant or subshrub) alongside targeted riparian shrubs and trees is another technique used to suppress undesired, weedy nonnatives (Eubanks 2004). In particular when intense fires result in burning to mineral soil, cover crops will limit erosion and rebuild soils by increasing soil organic matter and essential nutrients (Jones 2016). More information on revegetation strategy, techniques, and lessons learned can be found in the next section ("Conducting Native Riparian Revegetation").

CHEMICAL

Chemical control methods include the application of herbicides. No matter what herbicide or application method is used for tamarisk control, resprouting is to be expected, which often necessitates follow-up treatment. An evolution of herbicide trials and the banning of some chemicals by federal agencies has narrowed the list of effective herbicides to a common few: triclopyr as a bark penetrant (Neill 1988), imazapyr for foliar application (Taylor 1987), and glyphosate, which is typically used for broadleaf plants and grasses as it is absorbed through foliage and minimally through roots (see table 6.2). Aerial application of herbicide (imazapyr alone or in combination with glyphosate) with a fixed-wing aircraft or helicopter is a viable choice when the objective includes large-scale eradication of monotypic tamarisk stands. Using such an approach, aerial treatments in New Mexico have achieved 90 to 99 percent control over two years (Barranco 2001). However, aerial application (as opposed to ground application) must be conducted with great care, given the potential for herbicide drift to nontarget areas.

Basal-bark application and cut-stump methods are often used when aerial spraying is not feasible, when tamarisk is intermixed with native plants, or when the control program imple-

## CASE STUDY 6.3
Controlling Tamarisk Monocultures at the Bosque Del Apache National Wildlife Refuge: Lessons Along the Middle Rio Grande, New Mexico

Kirk C. McDaniel and Daniel P. Bunting

Established in 1939 and located along the Middle Rio Grande in the U.S. state of New Mexico (see place map in Case Study 4.3), the Bosque del Apache National Wildlife Refuge (Bosque NWR) was established, at least in part, to provide critical stopover habitat for migrating waterfowl. Although tamarisk was part of the Bosque NWR's landscape when it was established in 1939, efforts to manage its spread became more of a priority as dense monotypic stands of this invasive species began to take over marsh and other valuable wetland waterfowl habitat (Taylor and McDaniel 1998).

Over the years, several tamarisk-control methods have been implemented at the Bosque NWR, including herbicide spraying and prescribed fire. In 1960, a root plow pulled by a bulldozer was developed to clear tamarisk stands and was standardized and adopted by river-management agencies, including the U.S. Bureau of Reclamation (Reclamation) (Taylor and McDaniel 1998).

Beginning in the late 1980s, methods and costs were quantified to document levels of success as well as cost per hectare. In most cases, tamarisk-control actions spanned multiple years and involved a combination of methods that included repeat treatments (e.g., follow-up herbicide, root plowing, and burning) to improve results. Numerous technical reports and articles have documented the strategies used at the site; the table in this case study provides a synthesis of costs by control method for each (Taylor 1987; Taylor and McDaniel 1998; McDaniel and Taylor 2003). While most of the estimates are associated with 1990s dollars, research and quantification by McDaniel and Taylor provide the most detail and consistency of the literature reviewed. A typical inflation calculator from 1990 to 2020 roughly doubles the costs (i.e., $1.00 = $2.02).

Chemical methods included aerial application of imazapyr using a fixed-wing aircraft and spot herbicide applications using a portable backpack sprayer to apply triclopyr for cut-stump and basal-bark treatments. Prescribed fire methods included broadcast controlled burning of live monotypic stands and brush-pile burning of bulldozed debris. Mechanical methods used a combination of bulldozing, cutting, root plowing, raking, chaining, and piling live and dead biomass.

### Lessons Learned
Of all the methods used to control tamarisk over a fifty-year period at the Bosque NWR, the most success involved the following steps:

- Pushing down the aboveground growth of tamarisk with a front-mounted dirt blade attached to a D-7/D-8 bulldozer (see photos in this case study);
- Raking debris into piles—using a front-mounted rake implement attached to a bulldozer—and then burning them;
- Plowing roots (i.e., grubbing) and then raking root debris using implements attached to heavy machinery;
- Burning accumulated debris;
- Smoothing the entire area using a rail iron dragged over the surface, and
- Revegetating with native plants.

## Tamarisk-control methods and cost per acre implemented at Bosque NWR

| Control method | Cost per hectare in dollars | When carried out and reference source |
|---|---|---|
| Herbicide | | 1987–91 |
| Broadcast application | 141–91 | (Taylor and McDaniel |
| Aerial application | 13–20 | 1998) (1993 dollars) |
| Resprout spot treatments | 8–94 | |
| Prescribed Burning | | |
| Chaining and firelines | 72 | |
| Broadcast burn (no piling) | 15–25 | |
| Aerial debris burning (with piling) | 99–334 | |
| Mechanical | | |
| Root plowing | 75–364 | |
| Root raking and piling | 235–317 | |
| Dragging and smoothing | 38–112 | |
| Revegetation | | |
| Plant materials (local collection by practitioners versus purchase from nursery) | 3.75–7.75/plant | |
| Planting (cottonwood/willow poles, shrub seedlings) | 556–890 | |
| Bulldozing, brush piling, and burning; root plowing, raking, and burning, few resprouts and no herbicide application; 6 months (1991) | 4,600 | 1987–91 (Taylor and McDaniel tech report; Taylor and McDaniel 1998) (1993 dollars) |
| Aerial herbicide application; chaining, brush piling, and burning; root plowing, raking, and burning; and follow-up herbicide application (cut-stump, foliar); 4 yrs. (1987–91) | 6,300 | |
| Aerial herbicide application; broadcast burn; chaining, brush piling, and burning; root plowing, raking, and burning; and follow-up herbicide application; 3 yrs. (1988–91) | 7,900 | |
| Herbicide | | 1995–2000 |
| Manual broadcast and spot treatments | 1,250 | (McDaniel and Taylor |
| Aerial applications (fixed-wing aircraft) | 115 | 2003) (2000 dollars) |
| Prescribed burning | | |
| Broadcast burns | 385 | |
| Mechanical | | |
| Aerial growth cutting, raking, piling, and burning | 2,965 | |
| Root plowing, raking, stacking, and pile burning | 2,400–3,700 | |
| Combinations | | |
| Herbicide and burning | 3,460 | |
| Mechanical aerial biomass and root treatments | 3,630 | |

Costs associated with this method amount to roughly $1,850 per hectare ($750 per acre) (Taylor and McDaniel).

Other findings and lessons gained from applied experiments and research on tamarisk control at Bosque NWR include the following:

*(continued)*

- Clearing of aboveground biomass should be completed in cooler temperatures (winter-spring), and root plowing should be completed in the hottest months (e.g., June, when plants are most stressed).
- Root grubbing appears to produce less resprouting than chemical applications and prescribed burning.
- Large-scale aerial herbicide applications followed by prescribed burning was not effective in controlling tamarisk and was abandoned at the Bosque NWR in favor of the root-grubbing technique highlighted above (Dello Russo 2013).
- Chemical and mechanical treatments were less effective within areas of higher soil moisture and finer soil texture (silt/clay), which favored tamarisk resprouts (Bosque NWR unpublished data).
- Several techniques were effective in areas where there was a mix of tamarisk and desirable plants, including the cut-stump method for larger isolated trees, foliar and basal-bark herbicide application on resprouts, and mechanical removal using an excavator with a hydraulic thumb (i.e., a front-mounted implement for grabbing large, heavy debris; Dello Russo 2013).
- Regardless of the method used to control tamarisk, 100 percent mortality was never achieved, and follow-up treatment was always needed (Duncan and McDaniel 1998; McDaniel and Taylor 2003).

Restoration at Bosque NWR included the use of D7 or D8 bulldozers (left) to push down aboveground vegetation, root plow the belowground biomass, and pile debris for burning (right). Photos by Kirk McDaniel, New Mexico State University.

ments selective control. The basal-bark treatment involves applying an herbicide mixture (such as triclopyr) to the lower 45 cm (18 in.) of younger plants with a basal diameter of less than 10 cm (4 in.). Basal-bark treatments are not as effective on older trees with thick, furrowed bark (USDA 2010). The cut-stump method involves a combination of removing as much of the above-ground biomass as possible (usually with a chainsaw) followed immediately by the application of an effective herbicide directly on the cut stump. Stumps with less than a 10 cm diameter should be thoroughly wetted to kill roots, while stumps greater than 10 cm should ensure herbicide application of the cambial layer (the growing part of the trunk) (USDA 2010). Foliar treatments should be made in late summer or early fall when plants are preparing to shut down for the winter season and translocating carbohydrates to the below-ground tissues (Di

Tomaso et al. 2013). Cut-stump treatments can be made year-round, but will not be effective during droughts, since the method relies on the herbicide to be drawn into the plant tissue. In addition, glyphosate applications are more lethal to tamarisk after recent rains have removed the buildup of salt from leaf excretions, which can otherwise reduce herbicide absorption (Di Tomaso et al. 2013).

BIOLOGICAL

Biological control is defined as the control of a pest by introduction of a natural enemy or predator. In 2001, after years of research, the USDA released the northern tamarisk leaf beetle (*Diorhabda carinulata*) to control *T. ramosissima*, *T. chinensis*, and other related tamarisk species (but not *T. aphylla*, whose leaf morphology reportedly is not attractive to the released insects). Four species of tamarisk leaf beetle have been released in North America (Di Tomaso et al. 2013; Tamarisk Coalition 2017; Bean and Dudley 2018) (see table 6.3): (1) *D. carinulata* from China and Kazakhstan (where it has successfully suppressed tamarisk over large areas of the Great Basin desert), (2) *D. elongata* from Greece (which is establishing well in California and parts of west Texas and is best suited for Mediterranean habitats), (3) *D. carinata* from the Uzbekistan region (which appears best suited for the Great Plains grasslands and the Mojave and northern Chihuahuan deserts), and (4) *D. sublineata* from northern Africa and Spain (and is showing potential for the Sonoran and southeastern Chihuahuan deserts). A fifth species, *D. meridionalis* from Syria/Iran/Pakistan, has not yet been safely tested or released but may be suited to subtropical maritime deserts.

Tamarisk leaf beetles feed on the leaves of tamarisk, slowly reducing plant vigor. The greatest defoliation occurs the first few years after beetle introduction. Although tamarisk does not typically die and often resprouts after being initially defoliated by the beetle, repeated defoliation of individual tamarisk trees has led to severe dieback and death of some tamarisk within several years (Di Tomaso et al. 2013). A synthesis of fifty-five sites where tamarisk defoliation was occurring found an approximately 50 percent reduction in tamarisk abundance (Gonzales et al. 2017a). Other biocontrol monitoring studies have shown tamarisk suppression up to 85 percent (Di Tomaso et al. 2013), with mortality ranging from 20 to 60 percent after three to four years of defoliation (Di Tomaso et al. 2013; Bean and Dudley 2018). Still, biological control will not completely eradicate tamarisk, and other uncertainties complicate the quantification of long-term biocontrol benefits. The mechanisms by which regrowth occurs after repeated defoliation are not completely understood, which opens the door to the question of whether tamarisk populations may be adapting or producing novel genotypes to tolerate beetle infestations or other environmental disturbances (Hultine et al. 2015).

One controversial aspect of the release of the tamarisk beetle is that defoliation can locally reduce nesting habitat for riparian woodland birds. The spread of the tamarisk leaf beetle into habitat used by the federally endangered southwestern willow flycatcher underscores this controversy. Although research from USDA's Animal and Plant Health Inspection Service suggested that the ranges of the different species of tamarisk leaf beetle would be confined latitudinally, several of the released species have expanded outside their predicted range to unintentionally affect riparian habitat along several watercourses across the southwestern United States (see figure 6.4). Nagler et al. (2017) estimated the leaf beetle defoliation progressing from the Upper Basin to Lower Basin states of the United States at a rate of 40 km/yr. For

**TABLE 6.2** Summary of chemical name, application rate, treatment formula, and time of application of three common herbicides used to eradicate tamarisk (from USDA 2010, and data by McDaniel and Taylor 2003, and Taylor 1987).

| Common chemical name | Product name (examples) | Rate (per acre) | Treatment formula | Time of application |
|---|---|---|---|---|
| Triclopyr ester | Garlon 4, Remedy, Turflon, Weed-B-Gone | N/A | 50:50 mixture of triclopyr and crop oil / diesel oil and blue indicator dye | Anytime |
| Imazapyr | Arsenal, Chopper, Stalker, GroundClear | 1–2 quarts | 1:100, with 0.25 percent surfactant and blue indicator dye (1 percent mixture for foliage spray) | Late summer to early fall |
| Imazapyr + Glyphosate | Arsenal + Roundup, Rodeo | 1–1.5 quarts each | 50:100 + 50:100 (1–2 lbs + 2–4 lbs per 100 gal of water) with 0.25 percent surfactant and blue indicator dye | Late summer to early fall |

**TABLE 6.3** Tamarisk leaf beetles actively used or considered for research and testing

| Scientific name | Common name | Region of origin | Release date | Current distribution (in the U.S.) |
|---|---|---|---|---|
| *D. carinulata* | Northern | China, Kazakhstan | 2001 | CA, NV, UT, WY, CO, AZ, NM |
| *D. elongata* | Mediterranean | Greece | 2004 | CA, TX |
| *D. carinata* | Larger | Uzbekistan | 2007 | TX |
| *D. sublineata* | Subtropical | N. Africa, Spain, Tunisia | 2009 | TX |
| *D. meridionalis* | Southern | Iran, Pakistan | not released | - |

example, beetles have defoliated flycatcher nesting sites in tamarisk nest trees along the Virgin River in southern Utah. Defoliation exposes the birds and eggs not just to predators, but to the sun (Bean and Dudley 2018). The subsequent increase in temperature and decrease in humidity decreases reproductive success. In some areas, tamarisk may be replaced by grasslands or shrublands, resulting in a loss of riparian forest habitat for birds (Di Tomaso et al. 2013). That noted, recent research on the response of saltcedar to beetle infestation over a longer time period (ten to eleven years) noted initial dramatic decline (around 50 percent) but then strong recovery afterward (González et al. 2020).

In addition to the tamarisk leaf beetle, other insect species that are being considered in the United States as biocontrol agents for tamarisk include the mealybug (*Trabutina mannipara*) and a weevil (*Coniatus tamarisci*) (RIVRLab 2012). Interestingly, another species, the splendid tamarisk weevil (*Coniatus splendidulus*), has recently been observed in Arizona, California, New Mexico, and northern Mexico (RIVRLab 2012).

## Management Costs

Costs of tamarisk management and control depend on numerous variables, including the method of removal (mechanical, chemical, biological), targeted tamarisk species, extent of coverage and level of control desired, and current and past conditions (e.g., fire history and past management). Unfortunately, cost comparisons among watersheds have little management value, because cost-per-acre analyses are complicated by cost variables such as the density and distribution of the invasive species, the control method used, the tools and machinery available, access to and remoteness of the site, and land ownership and management. Skidmore (2017) estimated costs ranging from $1,200 to $12,000 per hectare ($500 to $5,000 per acre) for treatments within the Colorado River Basin, depending on methods used and whether revegetation was included as a component. The Gila Watershed Partnership estimated costs greater than $17,000 per hectare (more than $7,000 per acre) for combined methods (biomass mulching, chainsaw cutting, pre- and posttreatment herbicide applications) and more than $25,000 per hectare (more than $10,000 per acre) when including the planning and permitting phases

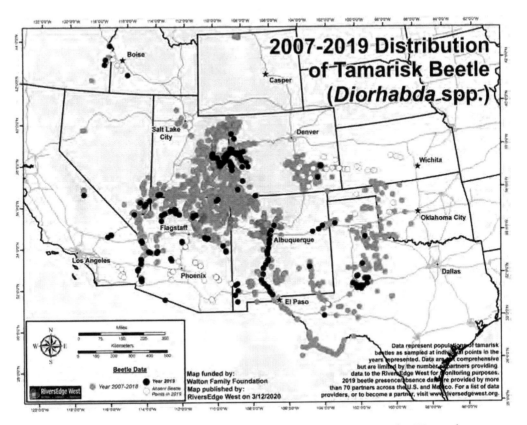

**FIGURE 6.4** Tamarisk leaf beetle distribution in 2019, as tracked by RiversEdge West and partners. Diagram by Ben Bloodworth, RiversEdge West.

(Shawn Stone, pers. comm., 2017). RiversEdge West (a nonprofit conservation organization based in Grand Junction, Colorado) provides a "Restoration Calculator" to assist practitioners in estimating costs when site characteristics such as percentage of tamarisk canopy cover are known. It is available online at https://riversedgewest.org/resource-center/documents /riparian-restoration-cost-calculator.

## Russian Olive (*Elaeagnus angustifolia*)

Russian olive was introduced to North America in the early 1900s as an ornamental plant and windbreak. It is a Eurasian tree that has become widely naturalized within riparian areas throughout the United States. Russian olive is found in all but thirteen U.S. states, is naturalized in seventeen, and due to its relatively wide ecological amplitude, has been reported as the fifth-most dominant woody species in western riparian areas (Friedman et al 2005). Russian olive appears to have greater impacts on upper watersheds at higher latitudes and elevations compared to tamarisk, which is more problematic along lower portions of watersheds. Within the Colorado River Basin, for example, it has been estimated that Russian olive occupies over 16,000 hectares (40,000 acres), primarily in the Upper Basin states, while tamarisk covers over 250,000 acres across the Upper and Lower Basin states (Tamarisk Coalition 2009). Russian olive is present in Mexico and Australia, but with far fewer occurrences and is not a target for invasive control. Several publications describe the biological traits and invasion ecology of Russian olive, most of which focus on impacts in the western United States (e.g., Christensen 1963; Carman and Brotherson 1982; Bertrand and LaLonde 1985; Shafroth et al. 1994; Lesica and Miles 2001; Katz and Shafroth 2003; DeCant 2008); others discuss management options and treatment success (e.g., Brock 1998; Caplan 2002; McDaniel et al. 2002; Stannard et al. 2002).

A primary concern expressed by managers regarding Russian olive is that it forms dense thickets beneath the canopy of native cottonwood gallery forests (Shafroth et al. 1995), where it becomes a hazardous fuel source and fire risk to fire-intolerant native riparian trees and shrubs. Russian olive is considered to be an invasive species in many places in the United States because it thrives on poor soil, has low seedling mortality rates, matures in a few years, propagates vegetatively* (by root crown buds and root suckers), and outcompetes wild native vegetation (Tamarisk Coalition 2009). That stated, managers are also reluctant to completely remove Russian olive, owing to its abundant, edible fleshy fruits (the plant is also known as "silver berry"), which are considered a valuable food source for wildlife. It is particularly important in areas where this invasive species is one of the few fruit-bearing riparian tree-shrubs present (Edwards et al. 2014; Zouhar 2005).

Despite its fruiting capabilities, research indicates that bird species richness and density is lower in Russian olive stands compared to native-dominated riparian communities (Knopf and Olson 1984; Brown 1990). For example, with few exceptions, when it is in the understory of native-dominated cottonwood stands along the Gila River, birds prefer to nest in the native

---

* Vegetative reproduction is one form of asexual reproduction (other examples of vegetative reproduction include spores and budding). Its complement is sexual reproduction, which requires transfer of male and female gametes (pollination) that leads to fertilization with ovules growing into seeds within a fruit.

riparian trees (Stoleson and Finch 2001). Furthermore, the populations of roughly a third of the native bird species that depend on cottonwoods for cavity nesting and insect prey would be negatively impacted if mixed cottonwood–Russian olive stands were to convert to monotypic Russian olive (F.L. Knopf, personal communication [cited in Shafroth et al. 1994]). Most floodplain rehabilitation projects in the Middle Rio Grande therefore focus on at least partial removal of Russian olive to reduce fire hazard, followed by aggressive revegetation using native fruit-bearing riparian shrubs.

The most common control method for Russian olive is chemical. Many practitioners have tested several herbicide applications with varying levels of success (see Case Study 6.4). Research is currently being conducted to identify biological control agents for Russian olive, but none have been approved at this time (USDA 2014). Grazing by mature, trained goats on seedlings and younger trees is a promising component for successful control (USDA 2014). Russian olive is also susceptible to *Verticillium* wilt and *Phomopsis* canker; both are disease-causing fungi that cause gradual dieback (Worf and Stewart 1999).

## Giant Cane (*Arundo donax*)

Giant cane (*Arundo donax*) is a native of eastern Asia (Polunin and Huxley 1987), but it has been cultivated and has spread widely in many parts of southern Europe, northern Africa, and the Middle East for thousands of years (Perdue 1958; Zohary 1962; Zohary and Willis 1992). An excellent review of giant cane physiology, genetic variation, physical structure, biomass and density, growth rate, and reproduction and spread is provided by Giessow et al. (2011). More recently, giant cane was introduced in areas of the western United States and northern Mexico to control erosion along agriculture drainage canals and was also used as thatching for roofs. It was intentionally introduced to the Los Angeles area in the 1820s for roofing material and as erosion control (Bell 1997) and is now a nuisance within coastal streams and agricultural canals from central to southern California (Goolsby and Moran 2009). In the United States and northern Mexico, giant cane has become most problematic along the lower Rio Grande, where monocultures of giant cane line the river bank for several hundreds of river miles. Giant cane occupies an estimated 28,000 to 40,000 ha (70,000 to 100,000 acres) in the Rio Grande Basin (Goolsby 2017), much of which straddles the international border along the Rio Grande / Río Bravo. Encroachment into the Río Nadadores in Cuatro Cienegas, Mexico, was so rampant that it caused the extinction of endemic fish species after the river stopped flowing (Goolsby 2017). Giant cane is also invasive in Australia where it is found in every state, including the Northern Territory.

The rapid increase in the extent and distribution of giant cane is believed to be due to a variety of factors, including altered river hydrology that favors the spread of this species over native riparian species. The plant's high tolerance for and adaptation to fire allows it to quickly resprout after wildfire, often crowding out native species and resulting in a giant cane monoculture (Bell 1997). Giant cane is not known to provide a food source or nesting habitat for native animals (Bell 1997; Herrera et al. 2003; Lambert et al. 2010); however, it competes with native species such as willows (*Salix* spp.) and cottonwoods (*Populus* spp.), which provide high quality habitat to many species of wildlife, including nesting habitat for such federally endangered birds as the least Bell's vireo (*Vireo bellii pusillus*) and southwest

# CASE STUDY 6.4

Controlling Russian Olive (*Elaeagnus angustifolia*) in Riparian Floodplains: Lessons and Field Observations along the Middle Rio Grande, New Mexico

Todd Caplan

Over the past fifteen years, we have experimented with a variety of mechanical and chemical treatments for controlling Russian olive in the Middle Rio Grande floodplain (see map of Middle Rio Grande region in Case Study 4.3). Although mechanical treatments can quickly reduce standing biomass, Russian olives root-sprout vigorously from buried lateral roots following physical disturbance (including fire), making herbicide treatment more effective in preventing rapid reestablishment. This case study summarizes results from different herbicide treatments conducted to control Russian olive trees in the Middle Rio Grande floodplain near Albuquerque, New Mexico.

## Methods

Masticating tractors (150 hp front-end loaders with flail mowing heads) and chainsaw crews were utilized in the winter of 1999 to clear dense thickets of Russian olive trees growing below a mature cottonwood gallery forest (*Populus deltoides*, ssp. *wislizeni*) near Bernalillo, New Mexico. Russian olive trees felled by chainsaw crews received a cut-stump application of triclopyr (ester formulation, 50% solution) within five minutes of cutting. Treated stumps ranged in size between approximately 10 to 90 cm (around 4 to 36 inches). All cutting and masticating treatments were done in the winter.

A block-plot design was subsequently established to evaluate the effectiveness of early and late summer herbicide treatments on the Russian olive root-sprouts, using four different herbicide formulations: glyphosate (5%), imazapyr (1%), metsulfuron (1 gm product / gal water), and the amine formulation of triclopyr (25%). All four herbicide formulations were mixed with water and a 0.25 percent nonionic surfactant.

For each treatment block, herbicide was applied with backpack spray units fitted with nozzles that delivered fine- to moderate-sized spray droplets that completely wet the foliage. Russian olive root-sprouts in each treatment block were counted during spraying and averaged about 150 per plot. Counts of live versus dead root-sprouts were made prior to treatment, and at the beginning (May) and end (September) of the following summer to determine plant control.

## Results

By early summer (around four months after clearing), root-sprouts of Russian olive were found throughout the treatment area. Study results indicate that glyphosate, triclopyr, and imazapyr were highly effective in controlling Russian olive root-sprouts, although treatment timing appeared to influence effectiveness (see table in this case study). By the end of the experiment, imazapyr had effectively controlled Russian olive sprouts when applied in August (88%), but control was poor when applied in June (40%). Conversely, triclopyr gave better control of Russian olive root-sprouts when applied in June (91%) than August (78%). Glyphosate provided high root-sprout control following both June (91%) and August (93%) treatments. Metsulfuron showed equivalent control on both spray dates (75%), but the rate we applied was probably too low for effective plant control.

Based on these plot results, we implemented an early summer (June 15–July 15) foliar application of 25 percent triclopyr (amine) formulation to Russian olive root-sprouts throughout the larger two hundred–acre treatment for three seasons following mechanical clearing. Although not quantified, root-sprouts by the end of the third summer were less than three per acre. However, larger stumps

Russian olive root-sprout control with different early and late summer herbicide applications.

| Herbicide | Concentration | June 2000 treatment mortality (%) | | August 2000 treatment mortality (%) | |
|---|---|---|---|---|---|
| | | 5/23/2001 | 9/26/2001 | 5/23/2001 | 9/26/2001 |
| Glyphosate | 5.0 | 90 | 91 | 84 | 93 |
| Imazapyr | 1.0 | 34 | 40 | 63 | 88 |
| Triclopyr | 25.0 | 89 | 91 | 79 | 78 |
| Metsulfuron | 2.5 | 50 | 56 | 76 | 75 |

Herbicide application to Russian olive in the understory of Rio Grande cottonwoods (left), and an example of Russian olive root-sprout (right). Photos by Todd Caplan, GeoSystems Analysis, Inc.

(greater than 30 cm [12 in.] diameter) continued to produce root-sprouts following three seasons of herbicide application. To treat these larger stumps, we uprooted them with a tractor and treated root-sprouts that emerged from adventitious roots remaining below ground. These sprouts were sufficiently controlled after one additional early summer herbicide application.

## Lessons Learned

The results from this project demonstrated several important lessons that have been carried forward to other projects in the Middle Rio Grande:

- Treating cut stumps that are less than 20 cm (around 8 in.) in diameter with herbicide resulted in far fewer root-sprouts compared to stumps that were not treated.
- Early summer foliar herbicide applications to Russian olive root-sprouts using triclopyr (25%) produced high mortality (>90%) after one year. Treating with glyphosate herbicide (5%) can provide similar levels of control following early or late summer foliar treatments.
- Root-sprouts from Russian olive trees with basal stem diameters greater than 20 cm may be more effectively controlled (with fewer herbicide applications) if the tree is first uprooted to remove the stump and the larger attached roots.
- Although high rates of mortality can be achieved with one herbicide treatment following mechanical clearing, we recommend treating resprouts with herbicide for three summer seasons to achieve nearly 100 percent control.

A Collaborative, Binational Effort to Manage Giant Cane (*Arundo donax*) Along the Rio Grande / Río Bravo in Big Bend, U.S.-Mexico Border

Joe Sirotnak, Javier Ochoa-Espinoza, and Jeff Renfrow
*In memory of Joe*

The Rio Grande / Río Bravo Basin covers over 870,000 square kilometers (336,000 square miles) of the southwestern United States and northern Mexico. Winding its way through this arid landscape, the Rio Grande—called the Río Bravo in Mexico—forms a natural boundary between the two countries that provides fresh water resources for millions of people. A centerpiece of the river along the border is the Big Bend binational area, which is flanked by U.S. national and state parks and Mexican protected areas (see map in this case study). Most of the Big Bend reach of the Rio Grande is designated as wild and scenic by both the U.S. and Mexican governments. With very few road access points, it is considered one of the great wilderness rivers in North America.

By the early 2000s, dense, almost impenetrable stands of giant cane had become established along many parts of the Rio Grande / Río Bravo in Big Bend. In response, a binational partnership consisting of federal and state land and water managers, NGOs, private foundations, businesses, and riverside citizens designed and implemented a program whose aim was to significantly reduce the extent and dominance of giant cane along the Big Bend reach of the river. A variety of factors made the management of giant cane along this reach a priority. These factors included the following:

- The potential that dense stands of cane narrow channels by anchoring deposited sediment and promoting sediment deposition during high flow (Dean and Schmidt 2011). As the channel narrows, riparian and aquatic backwater habitats are lost, affecting many native species, including the endangered Rio Grande silvery minnow (*Hybognathus amarus*) and other native fish (Garrett and Edwards 2004);
- The potential that reduced channel capacity was leading to increased flooding (and flood damage) of riverside towns and infrastructure;
- Reduction of both structural and biotic diversity of the riparian plant community and degradation of wildlife habitat;
- Reduced recreation experience (e.g., loss of camp sites);
- Reduced river access for managers, recreationists, and communities; and
- Increased overall evapotranspiration and resulting reduced flow during low-flow periods.

Our binational partnership in Big Bend has now carried out giant cane management activities along the river every year since 2008. Although the methodology of managing giant cane has changed in subtle ways since 2008, seven main steps have consistently been included as the footprint of treatment increased: (1) finalizing a binational giant cane management plan the beginning of each year that describes the when, where, how, and who of all actions planned during the course of the year; (2) completing required environmental and cultural resource compliance, organizing work crews, logistics, materials, and supplies; (3) conducting pretreatment monitoring of channel morphology and riparian vegetation conditions; (4) carrying out prescribed burns (when and where possible) to reduce giant cane biomass prior to herbicide treatment; (5) treating giant cane resprouts with EPA-approved herbicide (imazapyr) four to five weeks after the prescribed burn is completed; (6) retreating giant cane resprouts with herbicide one year after initial herbicide treatment; and

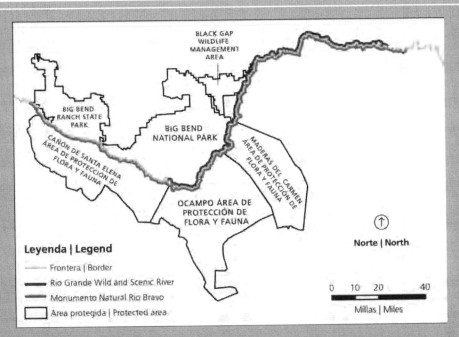

Map of the Rio Grande / Bravo through Big Bend, with delineation of bordering federal and state protected areas. Map by Marie Landis, Big Bend National Park.

The effort to control giant cane along the Rio Grande/Bravo is a true collaborative binational effort consisting of Mexican and U.S. state and federal agencies, local citizens, NGOs, and consultants. Left to right: Miguel Jurado (NPS), Jeffery Bennett (NPS), Francisco Torralba González (CONANP), Joe Sirotnak (then with NPS), Javier Ochoa-Espinoza (CONANP), Jeff Renfrow (Río Bravo Restoration), Mark Briggs (REW), Carlos A. Sifuentes (CONANP), Billy Brauch (NPS), and Javier Lombard (Profauna). Photo by Jeff Renfrow, Río Bravo Restoration.

*(continued)*

(7) posttreatment monitoring at one and two years after initial treatment, and subsequently at three-year intervals.

### Results

To date, over 100 km (62 mi.) of both sides of the Rio Grande / Río Bravo in Big Bend have been treated. Our work has shown that giant cane can be effectively reduced to a negligible component of the riparian community with just two treatments. Although the cost of managing cane varies with location and conditions, costs to date have not been prohibitive, with total costs (logistical support, equipment, materials, and labor for conducting prescribed burns and treatment) averaging between $2,500 and $5,000 per acre for the first treatment. Follow-up treatments are much less expensive. Although a definitive correlation between cane eradication and channel morphological response remains elusive, we do see desirable channel-bank sloughing along treated reaches in some areas, and studies are underway to quantify the extent of channel change since treatment was initiated. In general, we are excited with the results: the river through Big Bend looks like a river again.

### Lessons Learned

Take-home lessons from our experiences to manage giant cane along the river in Big Bend include the following:

* *Maintenance Costs.* We initially feared that giant cane would bounce back almost immediately after treatment, making the long-term cost of limiting the extent and distribution of giant cane prohibitively expensive. But although posttreatment riparian and transitional plant communities tend to be a diverse mix of native and exotic plant species, long-term maintenance costs have been relatively inexpensive. In general, we are finding that a two- to three-week retreatment trip is required every three years to spray resprouts. One caveat to this is that the effectiveness of treatment declines in areas where giant cane has continual access to saturated soils (e.g., in areas receiving spring/groundwater). In such situations, additional follow-up treatments are required, which increase annual maintenance costs.

willow flycatcher (Franzreb 1989). In addition, giant cane uses phenomenal amounts of both shallow and deep water (Moore at al. 2016)—as much as 2.4 m (8 ft.) per year—to support its incredible growth rate, which can be as much as 5 cm per day (about 2 in.). It can also produce more than 20 tons per ha (about 50 tons per acre) of aboveground dry mass (Perdue 1958; Iverson 1998).

In the southwestern United States and northern Mexico, the plant's large plume-like flower cluster (panicle) seldom, if ever, produces fertile seeds. Instead it spreads vegetatively from fragmented stem nodes and rhizomes, allowing it to proliferate along disturbed waterways (Bell 1997). In addition to negatively impacting biodiversity, other impacts from giant cane invasion include altered channel morphology (it is a perfect sediment filter that can promote significant sediment deposition during high flow events), reduced visibility, and loss of access

- *Work Crews.* Using paid, well-trained, binational work crews composed of local riverside citizens who are overseen and managed by agency, consultant, and NGO personnel is an effective model that is producing positive results at both the environmental and community levels.
- *Prescribed Burning.* Conducting prescribed burns of dense stands of giant cane prior to treating regrowth four to five weeks later greatly reduces the quantity (and therefore cost) of herbicide needed as well as increases the overall effectiveness of treatment.
- *Clearly Defined Objectives.* In making decisions about priority areas to treat, retreatment intervals, and other key management decisions, it is important to define a desired or acceptable future outcome. It may not be necessary or possible to eradicate giant cane completely from even a short segment of the river. Compared to the high extent and distribution of giant cane that existed when we began this binational project and zero giant cane, what is desirable? Years of debate among our binational team members coupled with careful field observations helped us better define the desired future condition of the Big Bend reach's riparian zone, providing a stronger foundation for a more focused and cost-effective effort.
- *Public-Private Partnership.* Private funding that matches federal and state in-kind support was essential for this effort. On the United States' side, such private sources of funding as the Coca-Cola Foundation, Wildlife Conservation Society, and Apple Earth were instrumental to success. On the Mexican side, the Carlos Slim Foundation (Alianza para la Naturaleza) is a key private source of funding that is matching agency funding on the Mexican side of the river.
- *Local Consultants.* The hiring of local consultants on both sides of the border to carry out implementation (planning, organizing work crews, purchasing supplies and materials, finalizing schedules, collecting monitoring data) has been key to success. Managing an exotic species as invasive as giant cane along an extensive river reach is a significant undertaking that can be overwhelming to under-funded and understaffed federal and state agencies. Hiring locally provided the assistance required to carry out required actions as well as opportunities to strengthen relations between agencies and local communities.
- *Binational Collaboration.* The strengthened binational collaboration that has occurred as a result of this work is providing a positive international story that underscores the virtues of working closely on managing shared resources of mutual importance.

and recreation sites. These collective impacts have provided the impetus for several giant cane eradication and control programs.

While management of established giant cane is challenging, several methods (mechanical, prescribed fire, and chemical) are promising. The overall aim of mechanical cutting and prescribed burning of giant cane is simply to reduce aboveground biomass in a manner that allows more effective subsequent application of herbicide. Mechanical cutting can be carried out with chainsaws or steel-blade brush cutters. If access by heavy equipment is possible, mowing or bulldozing may be used. Prescribed burning is conducted during the times of year with appropriate weather conditions (dry and warm, with not too much prevailing wind). Prescribed burns can be initiated in a variety of ways, including using fuses or torches. Depending on extent, distribution, and density of giant cane, as well as weather conditions, once fire has been initiated, numerous hectares of giant cane can be burned in minutes (see figure 6.5). One of the great advantages of burning—over mechanical cutting—is that practitioners will not have to deal with potentially vast volumes of cut foliage. Both mechanical cutting and prescribed burns typically do not kill giant cane and, actually, can promote significant and rapid regrowth. A month to six weeks after cutting or burning, giant cane will be waist high, but in reduced densities that provide improved access for targeted treatment with herbicide. Not only is herbicide treatment following mechanical cutting or burning more effective, the reduced biomass means less herbicide will be required overall. Foliar applications (i.e., direct herbicide application to leaves) have shown greater success (almost 100 percent) than cut-stem treatments (5 to 50 percent) (Bell 1997). Successful management along the Rio Grande/ Río Bravo has used a combination of prescribed burning with imazapyr application about a month after the prescribed burn is completed, followed by repeat treatment of resprouts one year later (see Case Study 6.5).

In addition to mechanical, prescribed burning and herbicide treatment, biocontrol is also being tested. Three biocontrol species have been approved and released in the United States: (1) the arundo wasp (*Tetramesa romana*), released in April 2009; (2) the arundo scale (*Rhizaspidiotus donacis*), released in December 2010; and (3) the arundo leafminer (*Lasioptera donacis*), released in December 2016. The arundo fly (*Cryptonevra* spp.) is currently being

**FIGURE 6.5** A prescribed burn of giant reed (*Arundo donax*) along the Rio Grande/Bravo, 2018. Photo by Kelon Crawford, RGSSS Consultants.

tested in Europe. Preliminary impacts of the release of the arundo wasp appear to show a reduction of giant cane biomass by 22 percent along 558 miles of the Rio Grande over a five-year period (Goolsby et al. 2015). Some preliminary research also appears to show water savings, socioeconomic benefits ($4 to $8 benefit per $1 biocontrol cost), and recovery of native plants as a result of the biocontrol program (Goolsby 2017). It needs to be stressed that the results of biocontrol are preliminary, with many outstanding questions remaining to be addressed. In general, the three biocontrol species may offer a good complement to ongoing mechanical, prescribed burning, and chemical management of giant cane, particularly with respect to maintaining the extent and distribution of giant cane at desirable levels following an initial treatment phase.

## NATIVE RIPARIAN REVEGETATION

Revegetation, in the simplest sense, is the process of regenerating plants on disturbed lands. Revegetation strategies, however, come in many forms, including passive and active revegetation and numerous methods in between. Passive revegetation—often interchanged with passive restoration—is a technique that involves site preparation without actively planting focal species. Typically, this approach relies on recruitment from seed banks or natural seed dispersal. If legacy site conditions are intact (e.g., healthy soils, good natural seedbed, hydrological regime), then in some cases simply identifying and treating the problem (e.g., removing grazing livestock, preparing a retired agricultural field, removing a levee) will result in natural regeneration and recovery of ecological processes. Many methods also are available to encourage or enhance natural regeneration, including the reduction of weed competition or improving the seedbed (Greening Australia Victoria 2003). Passive restoration techniques are most effective in climates and soil conditions that have an inherent resiliency (e.g., when the soil is relatively healthy and in wetter climates); in arid climates, recovery will take longer (Eubanks 2004).

Unfortunately, most targeted restoration locations will be impacted by multiple biological or physical constraints that will require intensive site preparation followed by active planting. Active revegetation—often interchanged with active restoration—is an artificial technique to accelerate the process of plant establishment on disturbed soils. Active revegetation on highly disturbed soils, or at locations where historical and natural conditions are drastically altered, will likely require long-term management and maintenance, sometimes in perpetuity. For example, efforts to revegetate areas that lack stream permanence or do not undergo natural fluvial processes (e.g., disconnected floodplains) will need follow-up irrigation and maintenance. In many cases, these sites will have limited natural recruitment and will require a phased planting strategy to establish multiple cohorts to promote structural diversity. Regardless of strategy, the motivations for revegetation should be carefully thought out not only to allow for appropriate selection of methods and planting materials but to increase the likelihood for success in the short and long term.

Thousands of riparian revegetation projects have been conducted globally, involving the use of many different types of species, planting methods, and propagules (a vegetative structure that can become detached from a plant and give rise to a new plant), including cuttings, poles, seedlings, and seeds. In this section, the focus will be on revegetating with *obligate* riparian plants such as *Populus fremontii, Salix* spp., *Fraxinus velutina,* and *Typha* spp. Often referred to as phreatophytes (which means "water-loving"), obligate riparian plants are those whose roots must extend into the water table (or saturated zone of the soil profile, where all easily drained pores between soil particles are filled with water) for survival. As most obligate riparian plants have shallow root systems, they can only survive in areas where the elevation of the saturated zone of the soil profile is shallow (e.g., less than 3 m below the surface of the soil), where a supplemental source of water is available (such as in a perched aquifer or confining layer), or where water is provided to them artificially (as through irrigation). Obligate riparian species are in contrast to *facultative* riparian species, which can withstand periods of dry soils along intermittent or ephemeral streams and can survive within uplands.

Riparian revegetation efforts can produce dramatic results, helping to replace lost riparian vegetation and stabilize deteriorating conditions, thereby initiating recovery of the riparian ecosystem (Briggs 1996). Reestablishing native obligate riparian plants through revegetation efforts continues to be commonplace globally. Restoration project goals dictate which species are selected for the planting effort, while specific project objectives describe the types of plants, plant densities, and scale of the planting effort. Greening Australia Victoria (2003) provides an extensive overview on methods and techniques to reestablish native vegetation.

Riparian revegetation projects fall into two broad categories: *in-channel* and *off-channel.* For in-channel riparian revegetation projects, planting takes place on surfaces within the active bottomland of the channel (i.e., on surfaces frequently influenced by the current stream-flow regime). The results of revegetation efforts that are conducted in-channel can be greatly affected by the stream's flow regime: occurrences of high-magnitude flooding can scour out newly planted vegetation, and periods of no-flow can leave establishing plants high and dry. In contrast, off-channel riparian revegetation efforts refer to plantings that take place on surfaces outside the active bottomland environment (e.g., on abandoned terrace surfaces) that are hydrologically disconnected from the river. Particularly in arid and semiarid climates, off-channel surfaces often have much lower moisture availability than surfaces that remain hydrologically connected to the stream. In these situations, when the aim of revegetation is to establish phreatophytes, irrigation is often required both to establish and maintain plantings in the long term.

Lessons learned from riparian revegetation experiences can apply to both in-channel and off-channel situations or be category-specific. For the remainder of this section, we will specify "in-channel" or "off-channel" when riparian revegetation lessons and considerations are category-specific. If lessons or considerations can be applied to both categories, we will simply use *riparian revegetation* without specifying in-channel or off-channel.

Riparian revegetation is conducted to meet a variety of restoration objectives, including restoring or enhancing habitat, stabilizing banks and controlling erosion, fulfilling mitigation obligations, and providing recreation and community amenities.

Along the lower Colorado River in the United States, the objective of several off-channel revegetation projects spearheaded by federal agencies was to restore wildlife habitat to meet a mitigation program requirement. Reclamation has mass-transplanted thousands of rooted plant cuttings onto retired agricultural fields that can be supplementally watered using existing agricultural infrastructure and water rights associated with the planted areas. This technique has been used at several conservation areas (such as Cibola Valley Conservation Area and the Palo Verde Ecological Preserve) to create cottonwood willow habitat with over three million cottonwood and willow trees planted on over 400 ha (1,000 ac) (see figure 6.6). The approach uses dormant plant cuttings (10 cm [4 in.] in length by 1.3 cm [0.5 in.] in diameter) and requires a preparation phase in a controlled environment to allow roots to establish within soil plugs that will be transplant-ready. This approach, using rooted cuttings, was designed to take advantage of farming equipment such as tomato planters or modified vegetable planters for mass transplanting thousands of plants at high densities (up to 5,500 plants per acre) in just a few days. Note that this method is different than traditional live cuttings (also called stakes, whips, and poles), which often target pole lengths of 2.5 to 5 m (8 to 16 ft.) in length and 2.5 to 7.5 cm (1 to 3 in.) in diameter that are placed directly in the saturated zone or capillary fringe (the soil layer above the saturated zone that is filled with water due to tension saturation).

Seeding cottonwoods and willows within off-channel basins with subsequent irrigation also has successfully created habitat. Seed feasibility studies initiated in 2007 showed high success for cottonwood and limited success for willow seed germination and establishment (Bunting et al. 2011). It also was acknowledged in this study that willows (i.e., *Salix goodingii, S. exigua*) did germinate successfully, but rodent herbivory was responsible for a high number of willow seedlings compared to Fremont cottonwood (*Populus fremontii*). Several 6 × 12 m (20 × 40 ft.) basins (furrowed and nonfurrowed) were constructed to test different irrigation regimes and seeding rates. Ten years later, these seeded plots have healthy, dense cottonwoods, some well over 6 m (20 ft.) in height. A rooting survey (undertaken by excavating a trench to investigate root growth and depth of penetration from the surface) a couple years after seeding also showed evidence of the larger cottonwood trees reaching the capillary fringe (around five or six feet below ground surface). This suggests cottonwoods could survive without supplemental irrigation once tap roots reach the groundwater table but also assumes healthy groundwater exists without the need for leaching salts. It should also be mentioned that high-density seeding of locally sourced species will improve genetic diversity and can be used as a nursery stock to supply plant poles for other restoration activities.

**FIGURE 6.6** An example of mass planting of cottonwoods and willows using rooted soil plugs at Cibola Valley Conservation Area (left) (photo by Reclamation). Also, a 10-year-old cottonwoods and high-density seedling establishment (hydroseeded) at Cibola National Wildlife Refuge Unit #1 (right) (photo by Daniel Bunting, Harris Environmental Group, Inc.)

Riparian revegetation projects also are conducted for stream-bank stabilization. However, keep in mind that alluvial stream channels adjust their channel form in a variety of ways as sediment is evacuated and deposited. In the context of stream restoration, "healthy" means dynamic and is the opposite of "stable." Therefore, any attempt to artificially stabilize alluvial channels in wildland settings should be discouraged. However, situations do arise where channel-bank migration threatens riverside towns and infrastructure. Planting vegetation along stream banks may provide the same function as hardened stabilization structures (such as lining stream banks with concrete or soil cement or installing stabilization jacks). Further, numerous bioengineering techniques have been used that take advantage of earthworks or organic materials to stabilize channels. Revegetating can increase surface roughness, slow flow velocity, and decrease erosional processes while offering the advantages of providing habitat and maintaining resiliency by allowing the system to withstand a variety of environmental conditions (Elmore and Beschta 1987). Ancillary benefits from the use of plants include their ability to neutralize pollutants; provide shade, habitat, and food sources; and improve aesthetics. A recent project on the Animas River included stream-bank revegetation to fix highly eroded banks, improve trout habitat, and provide access to the shoreline (see figure 6.7). In addition, revegetation is usually a less expensive option and, if establishment is successful, revegetation inherently increases biodiversity while requiring less maintenance than installing in-stream structures.

Numerous riparian revegetation efforts also provide amenities for streamside towns and citizens, such as serving as focal points of public parks or as neighborhood green spaces (see figure 6.8). Water and riparian vegetation attract wildlife and humans alike: wildlife seek them for refuge, shelter, and forage, and humans appreciate them for recreational opportunities and their experiential value. Even when the stream restoration goal is purely focused on improving native habitat, restoration practitioners should not be dissuaded from becoming involved in riparian revegetation efforts that take place in more public settings (like parks). If designed correctly, such efforts can improve wildlife habitat and connectivity. Furthermore, by working with community leaders, managers, and planners to address such public issues as improving public parks (or flood control or public safety), you can help to educate the public on

**FIGURE 6.7** Before (left) and after (right) photographs of the Animas River Bank Stabilization Project, which used bioengineering techniques to recontour bank slope, install erosion cloth, and plant willow and cottonwood trees. Photos from Aqua-Hab Inc.

**FIGURE 6.8** The Yuma East Wetlands in Arizona was designed for public access, including Sunrise Point Park, which includes a medicinal herb garden requested by tribal elders and a circular plaza in the shape of a medicine wheel, symbolizing the sacred nature of the river. Photo by M. K. Briggs.

the importance of their stream (González et al. 2017c). This can be undertaken while garnering long-term support and participation by community citizens for larger-scale stream restoration efforts in both urban and wildland settings. Conducting riparian revegetation in a public setting is an attractive opportunity that should be embraced.

## Considerations and Lessons from Past Riparian Revegetation Efforts

In general, riparian revegetation is most effectively used when (1) the potential for natural regeneration of desirable species along your target reach is low, and (2) the vegetation that is artificially planted is likely to survive. *Are revegetation actions truly needed and, if so, will plantings have a strong likelihood of establishing and maturing?* That is the central question that restoration practitioners must ask themselves before conducting revegetation. If there is little likelihood that the natural regeneration of desired species will be robust, artificial planting efforts may be needed. Whether the revegetated plants survive depends on the physiological requirements of the species and the ability of the stream environment to meet those requirements, particularly regarding water availability, soil salinity, vulnerability to flood scour, direct impacts from land-use activities, and other factors (Briggs 1996; Hupp and Osterkamp 1996; Kauffman et al. 1997; Anderson et al. 2004; Stromberg et al. 2007; Bay and Sher 2008). The lessons from past riparian revegetation projects, summarized below, were selected to help answer both parts of this key question.

### Understand Your Stream's Flow Regime

Your stream's flow regime refers to the amount, distribution, and movement of water along a channel reach and is the principal variable of bottomland ecosystems. Streamflow affects basic habitat function because flow velocity, flow depth, floodplain inundation, sediment fluxes, and shallow groundwater movement and fluctuation determine the channel morphology and aquatic and riparian ecosystem health. For riparian revegetation efforts, understanding streamflow is key to determining whether the channel surface you are revegetating will provide the conditions suitable to the establishment and long-term survival of the planted vegetation. Two questions are of particular importance: (1) Are average flow conditions during the driest, hottest time of the year adequate to provide the water and soil moisture needed to prevent newly planted vegetation from desiccating? (2) Are the surfaces you are planting susceptible to high-intensity flooding? In other words, will plants that need time for root development be scoured away? Chapter 3 goes into great detail on the importance of and methods for evaluating streamflow.

## *Revegetation Does Not Address Underlying Causes of Ecological Decline*

A principal reason that riparian revegetation often is only marginally successful is that factors originally responsible for ecological deterioration often also hamper or prevent establishment of the artificially planted native vegetation. Of the numerous riparian revegetation projects evaluated by Briggs (1996), the majority of successful projects addressed the causes of site degradation (indirectly or directly) by including secondary recovery techniques such as bank-stabilization structures, check dams, irrigation, and improved land-management strategies. The ability of these secondary strategies to overcome the causes of site degradation appeared to have a more significant impact on the overall results of the projects than did revegetation (Briggs 1996). The bottom line: do not conduct riparian revegetation unless the main causes of ecological deterioration have been addressed. On the other hand, if the main reasons behind the ecological decline of your target stream have been addressed, strong regeneration of desired species may occur (see "Section C," below). This is the paradox of riparian revegetation. If the site conditions for desired plants are not good, revegetation may not be a good idea as planting survival will likely be low. On the other hand, if conditions are good for desired plants, they may reestablish naturally, thus reducing the need for artificial revegetation. Gonzalez et al. 2018 offers a decision tree that defines conditions when revegetation was necessary.

In the context of addressing the main causes of ecological decline before revegetation activities get underway, two activities that directly affect stream (and riparian) conditions—livestock grazing and recreation—need to be highlighted, owing to the impacts these activities have on streams in general and more specifically on riparian revegetation projects.

### Livestock Grazing

Livestock-management practices should be changed and improved if livestock are contributing to the deterioration in stream conditions. This strategy is preferable to riparian revegetation actions because it directly addresses the causes of site degradation (Briggs 1996). The primary objective should be to create a situation where livestock use the land in a sustainable manner. The type of livestock (e.g., cattle or sheep), class (e.g., calves, steers, or cows), number, as well as season and intensity of use can be changed. Also, herding can be introduced, fences installed, or upland water sources developed. Numerous studies have examined the development of appropriate livestock-management options in response to particular riparian conditions on rangelands (Agouridis et al. 2005; Armour et al. 1991; Belsky et al. 1999; Claire and Storch 1983; Elmore and Beschta 1987; Fleischner 1994; Kauffman and Krueger 1984; Lucas et al. 2004; Meehan et al. 2011; Mendez-Estrella et al. 2016; Platts 1979; Winegar 1977; and many others). The website Global Rangelands (https://globalrangelands.org/) offers access to a variety of resources on the management of rangelands, including a section focused on riparian ecosystems.

Alarm bells should go off if you are contemplating riparian revegetation efforts along a stream reach where livestock have unfettered access. Depending on the species you are planting, a riparian revegetation project can be like opening up a restaurant for livestock. Even a single cow (much less a herd) can cause considerable damage to a newly planted site. For stream reaches that continue to experience livestock pressure, a common practice is to install exclusion fencing temporarily (for several growing seasons following planting) or permanently, to keep livestock out of areas where riparian revegetation efforts are taking place (Kauffman and

Krueger 1984). Planting cottonwood and willow poles that are over 3.5 m in length has also been attempted with varying degrees of success in reducing vulnerability to trampling and grazing (Briggs 1996). This strategy requires auguring deep holes, which may limit its use to sites that allow access by mechanized equipment and to projects that have the necessary resources (i.e., those that already have the mechanized equipment or funds to rent the needed equipment).

**Recreation**

As with livestock grazing, pressure from recreational activities (particularly of the mechanized variety, such as all-terrain vehicles) may not cause significant regional channel instability yet can dramatically impact local riparian ecosystems as well as riparian revegetation efforts. Other recreational enthusiasts such as hikers, cyclists, picnickers, or campers, are also attracted to streamsides. While these activities are certainly less impactful than ATVs, popular trails or use areas attract many visitors, resulting in cumulative impacts to vegetation (as by trampling and firewood gathering), soils (through compaction), and water (from pollution) (Cole and Landres 1995). Understand which recreational activities are engaged in along your target stream and work with managers to reduce pressure for the sites that will be planted. Posting signage or providing educational materials that describe the restoration project, its objectives, importance, and schedule may help reduce pressure on the revegetation site, prevent the need for more severe measures, and augment involvement by the recreation community in the restoration project itself. If there is an ATV outfit, hunting club, or local community that regularly uses your target stream, make a presentation to its members and ask for their support and participation.

## *Riparian Revegetation Will Be Wasted Along Streams Where Natural Regeneration Is Strong*

Riparian revegetation efforts may be squandered in areas that experience strong natural regeneration of desirable species. Natural regeneration is the best possible result and the restoration practitioner's true ally; fostering this process should be the aim of most riparian restoration projects. Natural regeneration can be robust along streams where

- physical and chemical conditions remain conducive to the establishment of the desirable species (particularly regarding water availability, flooding, and the presence of native species);
- a significant impact has been addressed (e.g., livestock pressure reduced); or
- a significant flood event has recently occurred.

An example of a riparian revegetation effort whose results were overwhelmed by natural regeneration is Aravaipa Creek in southeastern Arizona. Aravaipa Creek flows through Aravaipa Canyon, draining portions of the Galiuro and Pinaleño Mountains. In October 1983, a high-magnitude, statistically rare flood (roughly a five hundred–year event) passed through the canyon, removing significant amounts of streamside vegetation. The damage was considered so severe that many thought it would take generations for the streamside vegetation communities to recover. To encourage recovery, a large revegetation project was completed a year after this destructive flood. Over two thousand cottonwood (*Populus fremontii*), Goodding's willow (*Salix gooddingii*), and Arizona walnut (*Juglans major*) propagules were planted.

Seven years later, a site evaluation showed that the natural regeneration of native vegetation (i.e., those not planted) was so extensive that the results of the artificial revegetation effort could not be found (see figure 6.9).

The Aravaipa Canyon example underscores the resiliency of riparian ecosystems and their ability to revive naturally and dramatically following perturbation. In this case, the perturbation was natural flooding, but practitioners should also recognize the potential for strong regeneration after a significant driver of deterioration has been addressed (e.g., improved livestock management per the examples provided above). In either category, the best option is to wait a few growing seasons to better gauge how much natural regeneration is occurring and identify other sites where riparian revegetation resources are less likely to be wasted. In Aravaipa, waiting this amount of time before initiating revegetation would likely have revealed significant natural recruitment of desirable riparian plants, possibly making human interaction unnecessary. Similar results occurred along the Blanco River, Texas, where revegetation was conducted to complement natural reestablishment following a five hundred–year flood event (Meitzen et al. 2017).

## Plant on Sites That Offer Both Sufficient Water Availability and Protection from Flood Scour

The two factors that often lead to high mortality of planted obligate riparian species (especially for in-channel revegetation efforts) are desiccation and flood scour. Reichenbacher (1984) described riparian communities in terms of a continuum of stability. The most unstable communities are on surfaces that experience frequent flooding; typically, these are active floodplain surfaces that are closest to and only slightly higher in elevation than the stream channel. In contrast and on the stable end of the continuum are surfaces that are higher in elevation and further removed from the stream channel (such as terraces), where flood disturbances are relatively infrequent. The challenge for revegetating with obligate riparian plants in this more stable environment is that less water is generally available for them the farther or higher from the stream channel one plants. (Water availability is a function of flood regime, low-flow levels, groundwater dynamics, and precipitation.) Unstable surfaces that are frequently inundated will tend to have higher water availability and low soil salinity, which can make them suitable for obligate riparian plants. Stable surfaces that may no longer be inundated by surface flow are generally characterized by lower water availability and higher salinity, making them more suitable for nonobligate riparian plants (Shafroth et al. 2005; Merritt and Shafroth 2012).

For in-channel riparian revegetation projects, a challenge is to find the sweet spot between planting on active floodplain surfaces where seedlings or poles are vulnerable to scour by frequently occurring flow events and planting on higher elevated surfaces that are seldom inundated and do not offer sufficient water availability. By conducting an in-channel riparian revegetation project you have become a riparian farmer whose success is greatly dependent on the weather, particularly during the first few years after planting, when newly planted vegetation is most vulnerable to both scour and desiccation. Considerations that are presented below are intended to help you find that optimum location that will increase the likelihood that the vegetation you plant will get through those first few vulnerable years and survive to maturity.

**FIGURE 6.9** Aravaipa Creek as it appeared less than one year after it experienced flooding that exceeded the magnitude of a five hundred–year event and during artificial revegetation efforts (top). And the same view, eight years after flooding (bottom). All regrowth in the photo is a result of natural regeneration, not artificial plantings. Photos by Albert R. Bammann, BLM.

## Flood Scour

If you think of your in-channel riparian revegetation site as your occupational real estate, "Location, location, location!" should be your refrain for reducing the vulnerability of your plants to flood scour. Based on your understanding of your stream's flow regime, look for sites that offer some degree of protection from flood scour and consider the following recommendations.

AVOID PLANTING IN INCISED CHANNELS

Channel incision in alluvial channels occurs as the channel bed is lowered by erosion when sediment transport capacity has exceeded the sediment supply delivered to the channel (see figure 6.10 as well as discussion on this topic in chapter 3). Erosion of either the bed or banks or both may occur, depending on the relative erosion resistance of the two (Simon and Darby 1997; Bledsoe 1999). As channel incision continues, bank heights increase, and flow events of increasingly greater magnitudes are contained within the channel banks. Such channel deepening can create a positive feedback process wherein shear stress on the bed and toe of the banks is increased as larger flow events are contained within the banks (Watson et al. 2002). A channel becomes incised when its former floodplain becomes a terrace (Pickup and Warner 1976). That is, the active depositional surfaces adjoining a river channel that are overflowed at times of high discharge (i.e., floodplain surfaces) become "abandoned" and are only affected by relatively rare, high-magnitude flooding. Bed degradation usually precedes channel widening, which occurs when stream banks fail (Schumm 1984; Harvey and Watson 1986) after a critical bank height has been exceeded (Elliott et al. 1999).

Channel incision affects stream biota in a variety of ways. For example, it can significantly affect the amount of water that is available to bottomland vegetation. As channel incision proceeds, runoff concentrates in incised channels rather than being dispersed onto adjacent floodplains (Elliott et al. 1999). Since the water spends less time in the channel, aquifer recharge rates may also decline, lowering the local water table to approximately the depth of incision in the main channel (Van Haveren and Jackson 1986). Chapter 3 reviews causes of channel incision processes and methods for measuring and assessing channel morphological change.

To develop an appropriate and effective stream restoration response, therefore, it is critical to identify the evolutionary trends in the channel system and select rehabilitation measures that complement the morphological phase (Watson et al. 2002). For practitioners who are identifying suitable revegetation sites, incised channels provide very little diversity of channel surfaces along the continuum between the higher surfaces that are not frequently scoured but have low water availability and the lower surfaces with good water availability that are frequently scoured. In essence, you are limited to two options that are far from ideal: the abandoned floodplain where vegetation will likely desiccate or in the incised channel where the vegetation will likely be scoured out during the next high-flow event.

For these and other reasons, off-channel riparian revegetation in irrigated agricultural fields has become common along the lower Colorado River in the southwestern United States, resulting in hundreds of hectares of native riparian tree "farms" that are disconnected from the river's hydrological regime and forever dependent on agricultural irrigation (Grabau et al. 2013). We strongly encourage practitioners to not plant in incised channels. Consider other

**FIGURE 6.10** Examples of incised channels along the lower Santa Cruz River in Tucson, Arizona (top), and the Rio del Oso, a tributary of the Chama River, New Mexico (bottom), which was scoured and incised after rainfall events followed a 2013 fire in the Jemez Mountains. Photos by Daniel Bunting, Harris Environmental Group, Inc.

parts of the same stream where channel incision is not occurring or wait until channel incision and subsequent channel widening processes have slowed to a new dynamic equilibrium more conducive to planting success, which could take years.

### PLANT ON THE INSIDE OF MEANDER BENDS

Surfaces on the outside of a meander bend can be particularly prone to flood scour and should be avoided. Water velocity is greatest when approaching the outside meander and the force inevitably increases sediment evacuation that can, in turn, lead to cut banks and vegetation scour. Conversely, sediment deposition occurs on the inside of meanders, where calmer water allows heavier sediments to settle out. (The kinetic energy of the streamflow carrying the sediment load is reduced and is overcome by the forces of gravity and friction.) Figure 6.11 shows how land is lost on the outer banks and gained on the inner meander. In general, newly deposited sediment following a high flow event can provide ideal seeding or planting sites because both water availability and sediment needs are met for successful germination and plant establishment, and calmer waters reduce the likelihood of scour.

### TAKE ADVANTAGE OF OTHER NATURAL FEATURES THAT OFFER PROTECTION FROM FLOOD SCOUR

Many types of channel surfaces and channel characteristics can protect planted vegetation from flood scour. These include wide and topographically diverse surfaces as well as surfaces just downstream of large boulders or other impediments that offer lee-side flood protection. As mentioned above, sediment is deposited at slower water velocities, as heavier materi-

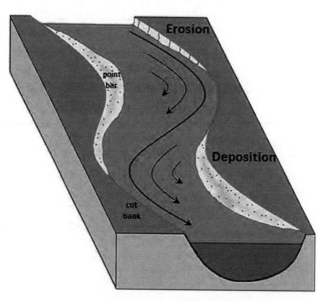

**FIGURE 6.11** Idealized graphic depicting erosion and deposition zones as they relate to meander bends. Planting near outside bends where flow energies are concentrated increase likelihood that vegetation will be removed by flood scour. Diagram by Daniel Bunting, FWS.

als such as sand and cobble can settle out. Fallen trees, logs, displaced boulders, and other obstructions provide the roughness to reduce the speed of streamflow and promote interspersed areas of aggradation. No site within an active bottomland environment is immune to scour. If your stream experiences a one hundred–year flow event the week after you plant, it is likely that most, if not all, of your plantings will be washed downstream. If you plant across a diversity of sites that offer increased protection from flood scour, you will improve the likelihood that at least some will survive. Secondary channels and oxbows offer refuges for native vegetation as well as aquatic species when main channels undergo scouring events. These areas offer off-channel depressions that can retain water and provide refugia during periods of high disturbance as well as drought (Stella et al. 2011).

## Water Availability

Several other factors beyond streamflow and flow regime determine the amount of water that will be needed or available for plant use. These include the water-absorptive capabilities of the plant species, age of the plants, climate, soil type, soil salinity, and depth to groundwater. Of these factors, depth to groundwater is probably the most important for many native obligate riparian trees, due to the need for their typically shallow root systems to have access to saturated soils or available water in the capillary fringe above the water table. Practitioners can use the following strategies to identify sites where depth to saturated soils are suitable for the establishment of native obligate riparian trees.

LOOK FOR SITES WHERE NATIVE OBLIGATE RIPARIAN PLANTS CURRENTLY EXIST

Obviously, native riparian obligate plants that are surviving in their natural settings are alive because they have an available water source. One strategy is to look for native trees that are surviving and focus revegetation actions just upstream or downstream of where phreatophytes currently exist. However, there are a couple things to watch out for with this approach. First, if the native desired phreatophytes along your stream are in stands that have a diversity of age classes, it means they are regenerating and may not require help from artificial revegetation. Understanding how the extent and distribution of desired species has changed over the years can help practitioners to better assess this need (via interviews with long-time landowners or managers, or examination of ground and aerial photographs from different years). Second, the presence of mature obligate riparian species does not necessarily mean that conditions are conducive for the establishment of newly planted vegetation. Mature trees have deep root systems capable of accessing water at much lower elevations than seedlings. And mature trees, with their mature root systems, can withstand flood scour much more readily than newly planted propagules.

DIG A HOLE

If you have heavy equipment or if digging by hand is not too arduous, excavating holes at potential planting sites to a depth below the upper portion of the zone of soil saturation can provide valuable information (see Case Study 4.4). The surface elevation of pooled water in an excavated hole is essentially the elevation of the soil saturated zone. As with other groundwater evaluations (see "Install Piezometers," below and in chapter 7), observing how the surface elevations change during the time of the year when water availability is likely to be the lowest can provide insights into the most severe conditions (at least with regard to water availability) that newly planted vegetation will have to endure. In much of northern Mexico and the southwestern United States, that time is early summer, prior to the onset of the monsoon. During this time of the year, lack of precipitation and cloudless, hot days create a situation where water availability is lowest and water demands of many plant species are highest. A safe rule of thumb is that phreatophytes will need to have their roots in the saturated zone of the soil profile in order to survive this dry period. Research has been conducted on species-specific groundwater depth thresholds along groundwater gradients extending away from the main channel (Stromberg et al. 2009b). The bottom line is that the lower the saturation zone of the soil retreats, and the more rapidly it retreats, the more likely that the root systems of newly planted vegetation will be left high and dry.

TAKE ADVANTAGE OF EXISTING WELLS

A well that is located on or adjacent to the proposed revegetation site is an obvious prize that practitioners should take full advantage of. Wells can be common in even relatively remote locations (as for watering cattle), and ecosystem managers should explore the revegetation site and bordering areas to see if any exist. Measuring the water depth is usually a straightforward process that can be accomplished in a variety of ways. An automated water-level meter (also called a water-level sounder) is a measuring tape with a weighted sensor that can be lowered into a well and will alert the user with a beep, alarm sound, or LED light when it contacts water. Some advanced sensors also can detect temperature and the bottom of a well. A less expensive, simple, and accurate option is to lower a steel tape or a weighted tag line into a well. Adding chalk to the measuring device produces an obvious submersion line, providing the distance from the top of the well to the water surface. A reasonably accurate picture of groundwater conditions can be developed by adjusting surface elevation differences between points of known groundwater depth (i.e., wells) and planting sites. Obviously, this is much easier to do for sites with little topographical variation than those with rugged terrain. The more wells included in the evaluation, the more accurately groundwater conditions throughout the site can be determined (see Case Study 7.4).

INSTALL MONITORING WELLS OR PIEZOMETERS

If no wells exist on site, installing piezometers to monitor groundwater conditions may be a viable option. A piezometer is a tube that is placed in the soil to depths below the water table and extends to the soil surface and is open to the atmosphere to allow easy access for measuring depth to the water table. In some cases, the piezometer is made of polyvinyl chloride (PVC) or chlorinated polyvinyl chloride (CPVC) and is housed inside a metal casing that protects the piezometer and allows the entire unit to be driven into the ground. Installing perforated PVC or CPVC pipes can be done inexpensively and, along with monitoring existing wells, is reviewed in chapter 7.

## Avoid Planting Obligate Riparian Species at Sites with Elevated Soil Salinity

Most obligate riparian plants have low tolerances to soil salinity. Typically, elevated soil-salinity levels are not a problem along healthy streams where frequent flooding prevents the buildup of salts. However, soil salinity can become elevated where human activities (such as agriculture) and altered hydrological processes have diminished water quality and reduced or eliminated flooding and the associated leaching or flushing of salts (Vandersande et al. 2001; Anderson 2017). Elevated soil-salinity levels can also occur where groundwater is consistently near the surface and where stream waters are high in total dissolved salts (Anderson and Ohmart 1982). This has become a problem in Victoria, Australia, for example, where abnormally high levels of soil salinity have contributed to the decline of native riparian communities along the Murray-Darling River (Hart et al. 1990; Jolly et al. 1993). Agricultural practices in the Colorado River Basin in the southwestern United States have greatly affected water quality, with the river picking up an estimated 7.7 million tons of salt per year (USBR 2013).

In a review of revegetation experiences at twenty-nine sites along the lower Colorado River over a two-decade period, Anderson (2015) concluded that assessing soil type and measuring electroconductivity* (EC) levels—a proxy for soil salinity—in the first and fourth quarters of the soil profile (between the surface and just above the zone of saturation) were key variables in determining the potential success of establishing cottonwoods and willows. He concluded that a twofold increase in salinity (i.e., EC measurements increasing from 1 to 2 mmho/cc) resulted in a 20 percent drop in potential cottonwood and willow growth. It should be cautioned that in situ measurements of EC can vary by method and may not be directly comparable between sites (e.g., portable meters and field sensors are analyzed differently than an augured core sample sent to laboratory for saturated paste extract). Briggs (1996) also summarized tolerances of selected riparian plants to soil salinity, reviewed soil sampling methods to map soil-salinity levels at potential riparian revegetation sites, and summarized several planting techniques that may help establish plant species with low salinity tolerances in areas characterized by elevated soil salinity.

## Use Plant Species and Planting Techniques That Are Adapted to Site Conditions

Due to their smaller tissue mass and less-developed root systems, seedlings are especially vulnerable to low water availability, scour, and a multitude of other impacts associated with in-channel riparian revegetation projects (Briggs 1996; Walters et al. 1980). Some planting techniques are described below that may be helpful in promoting establishment of newly planted vegetation in challenging sites. Stream practitioners need to recognize, however, that these techniques have inherent limitations and should be used either on a small scale (to determine their effectiveness before large amounts of money and time are wasted) or in combination with efforts that address the principal causes of site decline (e.g., land- or water-management changes).

---

* Conductivity is a measure of water's capability to pass electrical flow, which is directly related to the concentration of ions (dissolved salts and inorganic materials such as alkalis, chlorides, sulfides, and carbonate compounds) in the water. The more ions present, the higher the conductivity of water. Conductivity is usually measured in micro- or millisiemens per centimeter (uS/cm or mS/cm) or micromhos or millimhos/centimeter (umhos/cm or mmhos/cm). One siemen is equal to one mho. Salinity is derived from the conductivity measurement and is often given as parts per million or parts per thousand or grams/kilogram (1 ppt = 1 g/kg). In some freshwater sources, salinity values are reported using the unitless Practical Salinity Scale (sometimes denoted in practical salinity units as "psu").

## Using Large Poles

Long (over 7 m) cottonwood (*P. fremontii*) and willow (*S. gooddingii*) poles have been successfully planted in areas where groundwater elevations are 3 m or less from the soil surface as well as in sites where livestock continue to have access (Swenson and Mullins 1985) (see figure 6.12, left photo). This technique, however, has at least two limitations. First, mechanized equipment is often necessary to either auger holes to reach deep-lying water tables or to punch poles through the soil. Consequently, the site must be accessible to mechanized equipment. Second, it can be difficult, if not impossible, to drill or dig deep holes in riparian alluvial areas with high compositions of cobble and rock. In such situations, revegetation may have to be abandoned or adapted to include plants that can become established in areas with low-lying groundwater. Other techniques such as seeding, which is generally a good supplemental approach to any revegetation effort, may be more appropriate when site conditions preclude pole planting.

Several practitioners have produced resources explaining in detail how to implement successful pole-planting methods (Dreesen et al. 2002; Los Lunas Plant Materials Center 2005; Tamarisk Coalition 2014). Below is a brief summary of methods to deal with large poles, including steps for harvesting, soaking, and planting.

HARVESTING

Poles can be obtained from plantations or nurseries or naturally collected, but should be harvested while they are dormant, typically from January until bud break (March–April). Only healthy trees (nonstressed trees, without infections or fungus growth) that have at least reached their second growing season should be selected. In situations where there are high

FIGURE 6.12 Long cottonwood poles being planted along Terlingua Creek in western Texas to reach increasing depth to saturated soils at planting sites (left). Cottonwood poles are soaked after dormant pole collection and prior to planting (right). Photos by Jeff Renfrow, Río Bravo Restoration.

densities of seedlings, the entire trunk of a seedling can be harvested, harvesting one out of three or every other seedling in a source area that has high densities of seedlings. Trunks of two-year old cottonwoods (*Populus fremontii*) and some species of willow (*S. gooddingii*) can yield poles between 4 to 8 m long (12 and 25 ft.) with diameters exceeding 7 cm (3 in.); they can be cut to smaller lengths to increase the total number of poles harvested. Branches can also be harvested, providing numerous poles from one tree. Whether the pole is a trunk or a branch, all smaller diameter branches that emanate from it need to be removed to reduce ET losses. Poles collected closer to the revegetation site will be better acclimated to the local environment and have genetics most suitable for the site. That noted, collecting suitable plant materials from areas warmer than the region you are working in can be an important climate-adapted situation if your region is rapidly warming due to climate change (Grady et al. 2011). When collecting from natural sources, no more than 40 percent of a given stand should be collected.

SOAKING

The base of the poles (or ideally the entire pole) should be stored in water immediately after cutting until just prior to planting in order to saturate plant tissues and reduce the likelihood of desiccation (see figure 6.12, right photo). If required due to logistics of transport, pole cuttings can tolerate being out of water for up to one day, but it is best to minimize exposure to drying or stress. Poles can be stored in field settings in a stock pond or tank, agricultural canal or ditch, or a stream, provided they are weighted down to keep the base of all poles moist. When they need to be stored for longer periods of time (e.g., weeks to months), poles should be kept in cold storage between 0°C to 4°C (32°F to 39°F). Successful pole planting has been documented with poles soaked from one day up to three months (Tamarisk Coalition 2014).

PLANTING

Poles should be planted while still dormant, in late winter to early spring. They can be planted directly after cold storage. Poles can tolerate some drying when transport makes it necessary, but this should be minimized to the extent possible (another benefit of harvesting poles as close as possible to the planting location). The base of poles can be re-cut to encourage water uptake and should be re-cut if there is evidence of either pronounced drying or, conversely, rotting.

Depending on depth to saturated soils, holes can be augered manually or mechanically using a gas-powered auger, stinger bar, backhoe, or other machinery that allows deep holes to be excavated rapidly (see figure 6.12, left photo). When cobble is present that precludes the use of an auger, a stinger bar (a heavy steel implement attached to a tractor) can be used. Heavy machinery is useful to dig small trenches if sandy soils cave in when using augers. Poles should maintain contact with the soil and can be tamped to compact soil and displace air pockets. Table 6.4 highlights site characteristics and other considerations for suitable planting conditions.

Optimal planting density and location are dependent on the soils and water availability and the objectives of the revegetation project. Often it is best to plant large obligate riparian tree species like *Populus fremontii* and *Salix gooddingii* at over a one meter spacing (9–10 ft.), while other obligate phreatophytes of less stature (e.g., *Salix exigua*) can be planted in bundles or planted 1/3 to 1 meter apart (1–3 ft.). If beavers, feral pigs, cattle, or other species are a concern, planted poles may need to be caged to protect from herbivory (see figure 6.13).

**TABLE 6.4** Site characteristics and considerations for pole planting

| | Soils | | Groundwater |
|---|---|---|---|
| Texture | • *coarse, loamy sands to sandy loams* are ideal texture (use auger)<br>• in *sand and gravel*, augured holes may collapse (use heavy machinery to prepare trenches)<br>• *cobbles* may preclude use of augers (use stinger bar)<br>• *finer soils* with high percentage of clay and silt may not have high aeration and will minimize root development | Depth from surface to zone of saturation | • dependent on pole length<br>• poles should be in direct contact in at least 0.6–1.0 m (2–3 ft.) of saturated zone, but ideally still have 6 to 8 ft. of stem above ground surface<br>• base of pole should be in groundwater for the entire growing season, even when the water table drops |
| Depth from surface to zone of saturation | • 1 to 1.2 m (3 to 4 ft.) for cottonwood species<br>• 0.5 m (1.5 to 2 ft.) for willow species | Salinity | • EC < 3–4 dS/m (mmho/cm)<br>• Total dissolved solid ~2000 mg/L |

**FIGURE 6.13** Cages are installed around individual pole plantings to protect them from predation by beavers. Photo by Todd Caplan, GeoSystems Analysis, Inc.

## Plant Greater Quantities of Propagules for a Range of Site Conditions

If sufficient planting materials are available, increasing the planting footprint to encompass different channel surfaces along the wet-dry or stable-unstable continuum can improve success. Planting on a variety of channel surfaces that vary in elevation above the channel bed can increase the likelihood that at least some of the plantings will establish and survive. If weather conditions for the first several years following revegetation are wet, plantings closest to the active channel may be swept away by flooding, while those planted along higher elevations may survive. Even if survival rates are low, the ability for even a few propagules to survive can be considered a success, particularly if surviving plants become a seed source that was not present prior to restorative actions. On the other hand, if weather conditions are extraordinarily dry, plantings on the higher elevations may desiccate, but those in the active channel and lower elevations may be able to acquire sufficient water. Along Terlingua Creek in western Texas, for example, cottonwood and willow poles were planted along a continuum that extended away from the channel to floodplain surfaces that were over two meters in elevation above the thalweg (centerline of the deepest part of the channel) (see figure 6.14). Note that once propagules, other planting supplies, equipment, and crews have been secured and organized, increasing the number of plantings at a given site can often be done relatively cheaply. Planting

**FIGURE 6.14** Cottonwood and willow poles planted on a streambank perpendicular to the channel on surfaces with ever-increasing elevation above the channel bed can increase the likelihood that at least some of the planted vegetation will survive high-magnitude scouring flow. Photo by Jeff Renfrow, Río Bravo Restoration

more the first time is often less expensive than returning to replant a site if establishment rates are not adequate.

## Augering and Excavating

One technique that has been used successfully to plant obligate riparian species at sites with low water tables is to auger holes (18 cm in diameter) to the saturated soil zone, refill holes with displaced alluvium or mulch, and then plant seedlings on top of the refilled holes (Anderson et al. 2004). Tilling breaks up compacted soil and clay lenses, allowing irrigation water to flow to the water table with less resistance. The developing roots will follow the moisture gradient to the water table, often producing taproot-like development in species like cottonwood and willow that typically have shallow lateral root systems. Holes should be backfilled so that air spaces, which seem to be detrimental to developing vegetation, are kept to a minimum. As implied above, this technique was tested in concert with drip irrigation. Anderson and Ohmart (1982) noted that the cost of augering to the water table (about two dollars per hole in 1982 dollars) is justified by the increase in growth and survival of planted species, decreased irrigation requirements, and reduced labor costs. Trees planted after augering to the water table were significantly taller, had greater total growth and foliage volume, and experienced lower mortality rates than trees planted without using this technique (Anderson and Laymon 1988).

Physically removing soils to lower the elevation of planting surfaces in order to reduce the distance between the root systems of planted vegetation and saturated soils may also be a viable technique for planting obligate riparian plants at sites where depth to saturated soils is significant. By lowering the elevation of the surface that is being planted, this technique has the potential added value of increasing the frequency that the planted sites are inundated by flooding (thus increasing overall water availability). However, that potential added benefit may be compromised completely if increased frequency of inundation carries with it an increased likelihood that planted vegetation will be scoured by flooding. This potentially limits this technique to planting along regulated rivers where impoundment has greatly reduced large flood events. Excavation can be accomplished with a front loader, backhoe, or a variety of manual instruments. Both augering and excavation are restricted by some of the same factors that limit where large poles can be planted: the consistency of overlying alluvium can prevent drilling and excavation and site location may prevent access by heavy mechanized equipment. In addition, excavation of soils via heavy mechanized equipment can be prohibitively expensive.

# THE INTERSECTION OF STREAM CORRIDOR RESTORATION AND THE CONSERVATION OF NATIVE FRESHWATER DESERT FISH

In this section, we turn our attention to the conservation of native freshwater fish of the dryland regions of Australia, northern Mexico, and the southwestern United States. Although the emphasis is on freshwater fish of dryland regions, a few examples and references do stray into freshwater fish conservation taking place in montane and coastal areas. Our goals here are to (1) highlight the critical need for freshwater fish conservation, particularly freshwater fish of dryland regions, (2) provide selected references on the decline and conservation of freshwater fish (with emphasis on dryland regions), and (3) provide information on selected efforts and groups involved in saving these important species. With the inherent challenges in accomplishing the above objectives within a single section of a chapter, we can only dive superficially into these objectives, with hope that interested readers will go deeper by pursuing the sources of information that are cited.

By highlighting native fish conservation in a guidebook on stream restoration, we want to underscore the need for closer coordination and collaboration among those involved in stream corridor restoration and the conservation of freshwater fish. It seems that initiatives that come out of the stream corridor restoration community are often isolated from—or at least not well communicated to—the freshwater fish conservation community, and vice versa. This is not to say that collaboration does not occur; examples of such collaboration are highlighted in the case studies that are included in this section. However, more collaboration has great potential for mutual benefit. We hope the background information that is provided here will help to stimulate and encourage collaborative conservation and restoration efforts of Earth's stream ecosystems.

Many native freshwater fish species that are found in the dryland regions of the world are unique to the river basins where they occur. Mexico supports over 500 freshwater fish species, many of them characterized by high local and regional endemism (uniqueness to a specific geographic location) due to the country's pronounced physiographic and climatic diversity, long and complex geological history, large latitudinal extent, and long isolation of key geographies (e.g., the Mesa Central) (Miller et al. 2005). Australia's fifty million–year geographic isolation from the rest of the terrestrial world has also produced a great degree of endemism, with 38 percent of this continent's estimated 345 species of freshwater fish categorized as endemic (www.fishbase.org).

Desert fish can possess interesting and unusual adaptations to the challenges of living in a desert stream that allow them to thrive in silt-laden waters as well as withstand extreme fluctuations in water temperatures and streamflow. In Australia, the desert goby (*Chlamydogobius eremius*) occupies bedroom-sized pools that can be three times as salty as the ocean and where the temperature can top 40°C (104°F). This small fish forsakes the safety of permanent spring-fed waterholes to puddle-hop across the desert, taking advantage of rainstorms that create fleeting pools (Mossop et al. 2015). In the southwestern United States and northern Mexico, the native freshwater fish community includes minnows (Cyprinidae), suckers (Catostomidae), catfish (Ictaluridae), pupfish (Cyprinodontidae), and a few native trout species (Salmonidae), with many of these species found only in the drainages, springs, and cienegas in this part of the world (Minckley and Marsh 2009). Prior to significant human disturbance, some, like the longfin dace (*Agosia chrysogaster*), were found in a variety of habitats typically associated with low-to-mid-elevation streams, while others, including the Colorado squaw fish (*Ptychocheilus lucius*) and razorback sucker (*Xyrauchen texanus*) were abundant in such strongly flowing, silt-laden rivers as the mainstem of the Colorado River in the southwestern United States (Minckley and Marsh 2009).

In spite of being well-adapted, many native freshwater fish species are threatened or endangered, often due to human actions that have altered their natural habitat. Several factors impact fish habitat quality including, but not limited to, water quality (temperature, turbidity), water chemistry (pH, dissolved solids, dissolved oxygen), water quantity (e.g., streamflow, permanence), physical attributes (riffle-run-pool complexity, substrate), and food availability (factors influencing nutrient cycling and healthy vertebrate and invertebrate communities). Particularly for lotic* systems, a critical factor affecting habitat for freshwater fish is streamflow. Changes in flow dynamics as a result of such human impacts as dam-building, surface water diversion, groundwater pumping, as well as the impacts of climate change, are having dramatic effects on aquatic habitat and the fishes that depend on them (Walker et al. 1995; Meyer et al. 1999). Reduced streamflow as a result of these impacts can lead to increases in water temperature (Buisson et al. 2008) as well as the drying of wetlands and formerly perennial water holes in intermittent and ephemeral streams (Kennard et al. 2010).

Changes in streamflow often produce changes in sediment transport that in turn modify the channel geomorphology of alluvial streams and can greatly affect aquatic habitat. Scour often occurs along reaches that have a deficit of sediment (such as immediately downstream of dams), because flow that is relatively sediment-free picks up and conveys sediment downstream, producing reaches of the stream that are channelized and disconnected from the floodplain. River segments with a surplus of sediment occur along reaches where the contribution of sediment from tributaries remains high, but the high-magnitude pulse flows needed to evacuate accumulated sediment have often been reduced in magnitude and frequency by upstream impoundments and diversions. The end result of both situations is a loss of geomorphic complexity as critical aquatic habitat becomes abandoned due to channel incision (in cases of sediment deficit) or buried under accumulating sediment (in cases of sediment surplus).

The impact of groundwater extraction has also had a dramatic impact on fish and is particularly severe in arid zones. For example, in Mexico, near Torreón, Coahuila, and on the coastal plain of Sonora, groundwater pumping has led to a deterioration in water quality and, in some areas, has dried springs altogether (Flores 1990 [cited in Miller 2005]; Contreras-Balderas and Lozano-Vilano 1994).

Another important contributor to the decline of native freshwater fish is the introduction of exotic species. Minckley and March (2009) noted that nonnative fish are the single-most important factor in the decline and ultimate extinction of native fish of the southwestern United States. Because of their antiquity and isolation, many native western fish species do not possess the competitive abilities and predator defenses developed by fishes in more species-rich areas and, as a result, are highly susceptible to introductions of exotics (Minckley and Douglas 1991). It can be argued that in many situations—even in streams that have been dramatically impacted by human activities—native fish would continue to persist if not for the presence of nonnative fish (Minckley and Marsh 2009).

---

* Lotic ecosystems are those affected by unidirectional free-flowing water (like streams and rivers). In contrast, lentic ecosystems are terrestrial water bodies with flow of less energy that often moves in multiple directions (e.g., ponds and lakes).

**TABLE 6.5**  Selected literature summarizing the decline and conservation challenges associated with inland, freshwater fishes in Australia, Mexico, and western U.S.

| North American continent and/or global | |
|---|---|
| Conservation and conflict between endangered pupfish | Gumm et al. 2008 |
| Conservation status of freshwater and diadromous species | Jelks et al. 2008 |
| **Australia** | |
| General overview | Pollard et al. 1990 |
| Conservation action planning | Wager and Jackson 1993 |
| Status of galaxiid fishes in Tasmania* | Hardie et al. 2006 |
| Impacts of climate change on fish habitat | Prachett et al. 2011 |
| Conservation | Ebner et al. 2016 |
| **Mexico** | |
| Fish at risk or extinct | Contreras-Balderas et al. 2002 |
| Overview | Miller et al. 2005 |
| Freshwater fish in biosphere reserves | Pino del Carpio et al. 2011 |
| **United States** | |
| Decline of native fish in California | Moyle and Williams 1990 |
| Overview in American West | Minckley and Deacon 1991 |
| Overview in western U.S. | Moyle et al. 1996 |
| Broad-scale assessment of aquatic habitat | Lee et al. 1997 |
| Homogenization of fish faunas | Rahel 2000 |
| Impacts of roads | Trombulka and Frissell 2000 |
| Lower Colorado River Basin | Mueller and Marsh 2002 |
| Conservation plan for Lower Colorado River Basin | Minckley et al. 2003 |
| Status and issues of fire and fuels management | Rieman et al. 2003 |
| Southwestern U.S. | Minckley and Marsh 2009 |
| North American plains | Hoagstrom et al. 2011 |
| Conservation priorities in the southwestern U.S. | Propst et al. 2020 |

*From mostly montane streams.

As a result of these impacts, native fish have become some of the most imperiled biological taxa on the planet, with numerous examples of local and regional extinctions of native fish occurring over the past century (Minckley and Deacon 1991; Frissell 1993; Lee et al. 1997; Ricciardi and Rasmussen 1999; Propst et al. 2020). The American Fisheries Society has published three editions on the conservation status of imperiled or extinct freshwater and diadromous (those that live part of the year in saltwater) North American fish. Its latest list (Jelks et al. 2008) noted a 92 percent increase in the species listed as imperiled since the preceding list in 1989. Of those listed in 1989, 89 percent were in the same or worse conservation status.

Of the remaining populations of freshwater fish, many are restricted to small and often isolated remnants of a much larger and more continuous historical range (Moyle and Williams 1990; Minckley and Deacon 1991; Lee et al. 1997). In Australia, of the forty-six species listed as extinct, critically endangered, endangered, or vulnerable, 56 percent are from freshwater habitats (Environmental Protection and Biodiversity Conservation Act 1999). Selected literature on the decline of native freshwater fish is summarized in table 6.5.

## Conservation Response: Selected Examples

> Only if and when the worth of such organisms and the places they need to survive are recognized will they persist, and only through persisting can they and other aquatic organisms continue to act as sentinels to changes in the quality and quantity of resources equally critical to humans.
>
> —W. L. MINCKLEY, IN THE PREFACE TO
> *INLAND FISHES OF THE GREATER SOUTHWEST*

In this section, we present a few selected lessons learned from past efforts to protect and bring back the freshwater fishes of the arid regions of Australia, northern Mexico, and the southwestern United States. The conservation responses of course vary by country, river basin, focus species, and general approach. Obviously, many more fish conservation efforts are taking place in these countries. Nonetheless, we hope the brief overview provides insight into the commonalties and differences in the challenges, strategies, and lessons of these efforts. In Australia, we look at four small freshwater specialists that are restricted to the Lower Murray; in Mexico, we review a community-based effort to protect the Julimes pupfish; and in the southwestern United States, we look at conservation efforts associated with two species of minnow.

### *Conservation Efforts in Australia*

Australia is the driest inhabited continent and not surprisingly has a paucity of freshwater fish fauna relative to other similar-sized regions of the world. The commonwealth has approximately three hundred native species and, due to its geographic isolation, almost two-thirds of its freshwater fish fauna are not found elsewhere. As in other regions of the world, freshwater fishes of Australia are threatened by river regulation, flow alteration, alien species, and changing climates. Selected publications that are cited below are an introduction to the diverse and broad conservation response to the decline of inland freshwater fish in Australia.

- Environmental flow release impacts on native fish (King et al. 2009);
- Native cod (*Maccullochella* spp.)—status and management (Forbs et al. 2015; Growns et al. 2004; Humphries 2005; Ingram et al. 2011; Lintermans et al. 2016; Simpson and Mapleston 2002; Trout Cod Recovery Team 2004);
- Overview of status of Australian freshwater fish and their conservation (Australian Society of Fish Biology 2004; Lyon et al. 2012; Rowland 2013; Ebner et al. 2016);
- Murray-Darling Basin freshwater fish—status, management, and conservation (Gilligan and Schiller 2003; MDBC 2004; Crook et al. 2010; Crook et al. 2016);
- Specific restoration actions (Brooks et al. 2004; Howell et al. 2012; Paice et al. 2016).

As with the freshwater fish conservation efforts we highlight in Mexico and the United States, there is a wealth of examples in Australia to choose from. Case Study 6.6 highlights con-

**FIGURE 6.15** The Hunter River is a regulated gravel bed stream on the east coast of Australia. As one of the earliest developed areas of Australia, the Hunter River has been subject to significant human impacts (including the targeted removal of structural woody habitat [SWH]), which led to widespread streambank erosion and channel homogenization. The experimental reintroduction of SWH was aimed at increasing stream channel diversity and habitat availability for fishes. Photo by Timothy Howell, Freshwater Ecology Consulting.

servation actions to protect native small-bodied fish in the Lower Lakes of the Murray-Darling Basin during extreme drought periods. Another example from Australia that we selected to highlight focuses on reintroducing structural woody habitat (SWH) in streams where such important habitat has been removed by human interferences. Although the example highlighted comes from the Hunter River, which is located on the east coast of Australia (and, therefore, not a dryland stream), the reintroduction of SWH has potential for broad application to other stream systems (see figure 6.15).

## Conservation Efforts in Mexico
Although Mexico is only one-fifth the size of the continental United States, it is home to nearly two-thirds as many freshwater fish species as those that swim the waters of the United States and Canada combined. Mexico's diverse freshwater fauna can be attributed to the country's highly varied physical geography, a wide latitudinal range, the largest river system in middle America, and, ironically, its oceans: many marine groups left the brine for the inland springs

# CASE STUDY 6.6

Conserving Native Small-Bodied Freshwater Fishes During Extreme Drought in the Lower Murray-Darling Basin, Australia

Nick Whiterod and Michael Hammer

## Overview

Recent experience from the expansive Murray-Darling Basin illustrates the challenges facing freshwater fishes in Australia. The Murray-Darling Basin—the twentieth-largest river basin in the world—drains over one million square kilometers of southeastern Australia, or 14 percent of Australia's total surface area (see map in this case study). The lower temperate reaches of the system (the "Lower Murray") include a lowland section of the River Murray that flows into the large terminal freshwater waterbodies known as Lake Alexandrina and Lake Albert (the "Lower Lakes") before entering the Murray Estuary and adjacent coastal lagoons of the Coorong and then flowing out to sea via the Murray Mouth. The Lower Murray is a long-term freshwater refuge and biodiversity hotspot supporting a diversity of plants and animals, including more than half of the forty-nine freshwater fishes present in the Murray-Darling Basin and 80 percent of the small fishes (Ye and Hammer 2009). The Lower Lakes / Murray Mouth / Coorong area is recognized as a Wetland of International Importance under the Ramsar convention.

The Murray-Darling Basin is a land of droughts and floods, characterized by erratic to seasonal rainfall and highly variable flows that once shaped diverse and heterogeneous waterways and wetlands (see paired photos on facing page). Now it is one of the most heavily modified river systems in the world, due to widespread withdrawals of water and its regulation through upland impoundments, lowland locks, weirs, and barrages. In the Lower Murray, river flows are now much diminished (mean annual discharge to the sea has been reduced by two-thirds as compared to predeveloped times) and natural variability has been replaced by stability in water level. Changes in hydrology have acted to reduce the biologically complexity of the system. Freshwater generalists (fish that can live in a variety of freshwater habitats) are now most common, whereas freshwater specialists are increasingly fragmented and scarce, and alien species are prolific (Wedderburn et al. 2017). Four small freshwater specialists that are restricted to the Lower Murray include an atherinid Murray hardyhead (*Craterocephalus fluviatilis*, internationally critically endangered), two percichthyids—the southern pygmy perch (*Nannoperca australis*, locally endangered) and Yarra pygmy perch (*Nannoperca obscura*, internationally endangered)—and an eleotrid southern purple-spotted gudgeon (*Mogurnda adspersa*, locally critically endangered) (see photos at the end of this case study).

The enduring impacts of river regulation and water withdrawals along the Lower Murray were exacerbated by a prolonged and severe drought from 1997 to 2010 (the "Millennium Drought"). Acute water shortages between 2007 and 2010 resulted in broad-scale loss and desiccation of aquatic habitat. Rapid recession of water levels of the lower-most River Murray and Lower Lakes (to one meter below sea level) virtually eliminated all habitat suitable for the small fishes and deteriorated and fragmentated the few habitats that remained. The Millennium Drought added pressure on the small fishes already under threat. In fact, two (the southern purple-spotted gudgeon and Yarra pygmy perch) were considered to have become regionally extinct in the wild at this time (Wedderburn et al. 2019).

The response to the challenges facing small fishes of the Lower Murray can be characterized in four phases: (1) recognition of the decline in 2007, (2) a period of prolonged stress (2008–10), (3) the return to favorable conditions (2011–14; described in Hammer et al. 2013), and (4) efforts to secure long-term viability (2014 to present). During the initial decline there was limited capacity and

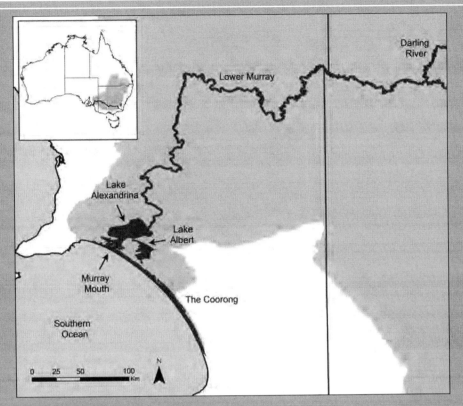

The Murray-Darling Basin is a one-million sq km watershed in southeastern Australia.
Map by Nick Whiterod.

Dried (left) and typical fringing habitat (right) of Lake Alexandrina. Photos by Michael Hammer.

inertia for managers to respond, so urgent responses—fish rescues and captive maintenance and breeding—fell largely to dedicated individuals and singular managers with appropriate expertise. As the drought intensified, greater appreciation of the problem prompted a mutlistakeholder response by a consortium of government agencies, NGOs, and schools. Central to this was a Drought Action Plan (DAP), which provided a coordinated decision-making framework for continuing and enhancing

(*continued*)

## CASE STUDY 6.6
*(continued)*

Rescued southern purple-spotted gudgeon (left), and Yarra pygmy perch (right). Photos by Michael Hammer.

the initial urgent responses and identifying and addressing ongoing issues and logistics (Hammer et al. 2013). The DAP framework put forward fifty-two wide-ranging conservation actions, including expanded captive breeding, in situ habitat works, environmental watering, and targeted alien-species control. A key outcome was the creation of surrogate refuges that included seminatural holding ponds and dams capable of producing higher numbers and fitter individuals compared to captive breeding facilities. A parallel collaborative research project to guide captive breeding efforts using genetics was also enacted (Attard et al. 2016).

### Results

With the return of water and favorable habitats in 2011, fish that had been bred in surrogate refuges were reintroduced into their former habitats. Initially 15,840 fish of the four small fish species were released into ten former habitats. Persistence at eighteen months following reintroduction as well as wild recruitment was demonstrated for all four species (Bice et al. 2014). However, with the "passing of the drought," the management team responsible for the conservation of these small fishes was disbanded in 2014. Consequently, sustained recovery conservation actions for these threatened fish, including maintenance of surrogate refuges, continued reintroductions, and monitoring, are now

and never returned to the sea (Miller et al. 2005). Selected references on native inland freshwater fishes of Mexico and their conservation status are provided below.

- Overview of diversity, status and conservation in Mexico (general) (Williams et al. 1989; Soto-Galera et al. 1999; Contreras-Balderas et al. 2003; Miller et al. 2005; Contreras-Balderas et al. 2008; Valdez-Moreno et al. 2009; Pino del Carpio et al. 2011; Contreras-MacBeath 2014).
- Invasive fish management (Contreras-MacBeath et al. 1998; Zambrano et al. 1999).
- Wild trout populations—conservation and status (Ruiz-Campos et al. 2003; Mayden 2004).

being carried out in a less coordinated manner and without a long-term vision to secure the future of these small fishes. An additional 20,000 fish were released to eight former locations between 2014 and 2018 with mixed results.

A decade after the first urgent interventions, threatened small fishes continue to face an uncertain future in the Lower Murray. There is little doubt that without the conservation actions undertaken as part of the four phases of response to the Millennium Drought, some, if not all, of these ecological assets would have already been lost. Most notably, the fish rescues that enabled captive maintenance and breeding as well as establishment of surrogate refuges provided a second chance for the fish when favorable conditions returned. These actions helped the Murray hardyhead and southern pygmy perch to show gradual, although partial, signs of recovery. For the two species thought to have become regionally extinct during the severe drought, less progress has been made. Recently, wild southern purple-spotted gudgeon have been detected, but the species remains in low number and restricted to a single wetland (Hammer et al. 2015). The status of Yarra pygmy perch remains extremely precarious, with no wild individuals detected since 2015 despite recent extensive surveys, prompting fears of the first extinction of a freshwater fish in the Murray-Darling Basin (Wedderburn et al. 2019).

### Lessons Learned
We have reached a critical moment in time: without decisive actions, these small fishes may be lost to the Lower Murray. The experience of the past conservation actions has given us greater clarity of what is needed as well as several key lessons learned. These include the following:

- Urgent interventions, such as rescuing fish from drying habitats and establishing refuge pools, can avert catastrophe.
- As important as these interventions are, proactive, long-term programs that focus on vulnerable taxa and ecosystems are needed to match the scale and severity of the problem facing these small fishes. Sustained reintroductions (more fish, more often) will be required to reestablish wild self-sustaining populations. In addition, complementary actions, such as habitat restoration, improved flow regimes, and alien species control, will also be required.
- Conservation of small fishes in drought-prone regions requires preparedness, persistence, and commitment to thoughtful, long-term, coordinated, and collaborative responses that are guided by sound planning, actions, and monitoring. Thought has been given to including the conservation needs of these small fishes in broader management plans, such as the visionary but contentious environmental watering plan for the Murray-Darling Basin ("The Basin Plan"), but much more deliberation is required.

The impact of groundwater pumping has dried up countless springs and small streams with particularly severe repercussions on the fish fauna of the arid regions of Mexico (Contreras-Balderas and Lozano-Vilano 1994). In Case Study 6.7 we highlight conservation efforts to protect what arguably may be the most threatened fish in North America, the Julimes pupfish.

### Conservation Efforts in the Western United States
In the western United States, remnant populations and remaining strongholds for freshwater native fish are often found on public lands (Lee et al. 1997). Over the last couple of decades,

## CASE STUDY 6.7
Protecting the Hottest Fish in the World: The Julimes Pupfish

Alfredo Rodríguez

Rarely more than two inches long, the Julimes pupfish (*Cyprinodon julimes*) lives in hot-spring waters that reach 45.5°C (114°F), earning it the title of the "hottest fish in the world." The Julimes pupfish is endemic to the hot springs of the small municipality of Julimes, where a population of a several thousand fish—along with a tiny snail that is its main food source—survives in an area of about 745 sq m (8,000 sq ft) (see place map in Case Study 5.4). This area includes the original spring and a man-made canal. Due to high water temperatures, the pupfish has no natural predator other than humans. Now, thanks to the efforts of the farmers who use the springs to irrigate their farms and a number of conservation groups, including the World Wildlife Fund-Mexico (WWF-Mexico), the pupfish's habitat is protected.

Before efforts to save the Julimes pupfish were initiated, local farmers used to clear the bottom of the spring to increase flow, which greatly disturbed the fish's habitat. Thanks to the collaborative conservation efforts, this practice has stopped, and Julimes hot spring is now a sanctuary. Locals feel very proud of this little fish, because it is now internationally known as the fish that lives in the hottest water in the world. With the farmers, WWF-Mexico has set up three water-monitoring stations in the springs to measure seasonal changes in temperature and flow. The site is protected by a chain-link fence and kept locked, but tours are welcome.

Eduardo Pando, who owns the land where the pupfish lives, is also the president of the irrigation unit of fifteen farmers who receive water from the springs. He helped create a conservation group, called Amigos del Pandeño, to seek resources from the government.

"We have a different attitude toward the fish," Pando said. Not disturbing the sediment (which is about knee-deep at the bottom of the shallow spring) preserves the fish but reduces the water flow for irrigation. To meet this challenge, Pando, the irrigation unit, and partners have been working for several years to cement the irrigation channels that are not key to the fish's habitat, which improves flow for irrigated agriculture and compensates for what is lost by not touching the springs.

the federal agencies that manage these lands have undertaken major assessments of aquatic ecosystems, habitats, species, and the processes that influence them (e.g. Forest Ecosystem Management Team 1993; Quigley and Arbelbide 1997), and have proposed initiatives to recognize, restore, and conserve sensitive populations and critical habitat (e.g., USDA 1995; USDA/USDOI 1995; NWPPC 2000; USFWS 2015).

As with Australia and Mexico, the conservation response to the decline of native fish in the western United States has been diverse. A few selected examples, include the following:

- Large-scale stream restoration (Romanov et al. 2012; Albertson et al. 2013);
- Dam decommissioning and removal of exotic fish (Marks et al. 2010);

Known as the hottest fish in the world, the Julimes pupfish (*Cyprinodon julimes*) (photo at left by Day's Edge Productions) is found only in hot springs near the town of Julimes, Chihuahua, Mexico (photo on right by Alfredo Rodriguez, WWF).

"It is not easy for other farmers to understand the importance of the little fish," Pando said. He added that he must work with the other farmers to come to a consensus.

The springs serve three purposes for the area. Besides serving as a now-protected and expanded habitat for the pupfish, the water feeds a popular park, Los Manantiales, that, especially in the autumn months, attracts many visitors who believe the hot springs have healing qualities. Water passes through the park to irrigate seventy hectares of alfalfa, chili peppers, oat, cantaloupes, watermelon, and pecans.

## Lessons Learned

- A concerted community-based conservation effort can be developed around a single species, benefiting the target species, providing a greater environmental good, and contributing to the well-being of local citizens.
- Achieving these multiple benefits took time and required long-term investment by the NGO community.
- Local citizens' concerns about the plight of the Julimes pupfish grew as they were educated about the fish, but linking the pupfish's conservation to the socioeconomic benefits of protecting the fish's hot springs was key to fomenting action and long-term success.

- Removal of tamarisk spp. (Kennedy et al. 2005; Keller et al. 2014); and
- Implementation of environmental flow on a regulated stream (Kiernan et al. 2012).

Below, we highlight conservation efforts associated with the bluntnose shiner (*Notropis simus*) and the Rio Grande silvery minnow (*Hybognathus amarus*).

## The Bluntnose Shiner

The bluntnose shiner is a small, freshwater minnow belonging to the Cyprinidae family, which is a diverse family that comprises over two thousand species (Campbell and Dawes 2004). Its common name refers to its distinctive blunt snout and a shimmering silver streak that runs from the head to the tail. The fish was once widespread in the Rio Grande and Pecos River of

# CASE STUDY 6.8
Pecos River Restoration and the Threatened Pecos Bluntnose Shiner, New Mexico

Chris Hoagstrom

Findings from studies instigated by the listing of the Pecos bluntnose shiner as threatened indicated that downstream displacement of its eggs and larvae may limit recruitment (Dudley and Platania 2007) and adults are limited to unchannelized reaches where streamflow is perennial and adequate to provide fluvial habitats (Hoagstrom et al. 2008a). Studies also indicated a lack of alternative suitable river reaches to establish new populations (Hoagstrom et al. 2008a) and confirmed the vulnerability of the remaining shiner population to streamflow intermittence (Hoagstrom et al. 2008b). These results suggested that river-channel restoration efforts to extend high-quality habitat downstream could reduce downstream displacement of eggs and larvae and increase the range of adults (Hoagstrom et al. 2008a, 2008b). To test this hypothesis, a reach of the Pecos River channel at Bitter Lake National Wildlife Refuge was selected as a potential restoration site (see map in this case study). The reach lies within the occupied portion of the Pecos River and in the transition zone from high-quality, unchannelized habitat upstream to low-quality, channelized habitat downstream (USBR 2009).

The central restoration objective of this effort was to divert a reach of the Pecos River that flowed through a channelized portion of the stream into an unchannelized meander bend (USBR 2009). Nonnative saltcedar (*Tamarisk* spp.) was also removed. This allowed the river to meander and expand laterally, forming more natural habitat conditions similar to those present in unchannelized upstream reaches. Restoration was completed in 2011, and a follow-up study was conducted in 2012 and 2013, but dewatering of the river channel by upstream diversions and drought confounded this study (Mecham 2015). At the beginning of the study (June 2012), the Pecos bluntnose shiner was more abundant at the restoration site than in any other reach (see graphs in this case study), including in an unchannelized reach with high-quality habitat just upstream of the restoration site and within the Bitter Lake National Wildlife Refuge, and another site farther upstream in the designated upper critical habitat. This suggested the restored reach was highly suitable for the species when streamflow was persistent, possibly even offering higher-quality habitat than the reach in the designated critical habitat. How-

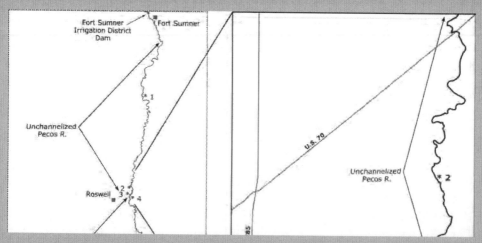

Idealized map showing the location of the Bitter Lake National Wildlife Refuge restoration site (the star on the left map) with the specific Refuge locations along the Pecos River delineated on the right.

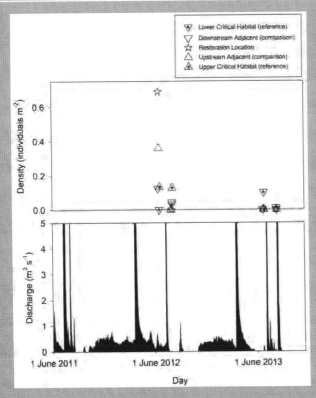

Top: The density of the Pecos bluntnose shiner (*Notropis simus pecosensis*) at study sites in the restored river channel, adjacent areas upstream (unchannelized) and downstream (channelized), and more distant areas in the upper critical habitat (unchannelized) and lower critical habitat (channelized). Surveys were conducted in June and August of 2012 and 2013 (details in Mecham 2015) (Chris Hoagstrom). Bottom: Mean daily discharge in the Pecos River near Acme, New Mexico, during the study period in the vicinity of the restoration project (from USGS stream gauge 083860000; https://waterdata.usgs.gov/nwis/inventory/?site_no=0838 6000&agency_cd=USGS). Note the extended periods of streamflow intermittence (zero discharge) in the summer of 2012 and 2013, prior to sampling in August 2012 and June and August 2013 (Hoagstrom et al. 2008a).

ever, streamflow became intermittent between June and August 2012, and the shiner disappeared from the restoration site and became rare elsewhere. Intermittent flow occurred again in June and August 2013, and the species remained rare overall and absent from the restored reach. Although restoration was partly aimed at improving retention of eggs and larvae, no larval or juvenile Pecos bluntnose shiners were collected from the restored reach or from any other study site, due to streamflow intermittence (Mecham 2015).

## Lessons Learned

The higher abundance of the Pecos bluntnose shiner in the restored river channel in June 2012 compared to other locations provides preliminary evidence that river-channel restoration benefits the species. However, habitat conditions rely on streamflow (Hoagstrom et al. 2008a), so channel restoration alone could not maintain the species when streamflow became intermittent.

***Additional insights.*** Subsequent studies have improved our knowledge about the summer spawning season of the species (Archdeacon et al. 2015) and showed that juvenile Pecos bluntnose shiners dispersed upstream from the channelized habitat and joined an adult population in a high-quality, unchannelized habitat (Chase et al. 2015). Because upstream dispersal is impossible when streamflow is intermittent, this further demonstrates the importance of continuous streamflow to the species. As reproduction and recruitment occur during summer, when the risk of flow interruption is highest, ensuring adequate streamflow, especially during the summer months, is a much-needed conservation action.

***The bottom line.*** The Pecos bluntnose shiner is sensitive to streamflow. Although channel restoration is beneficial, limited results should be expected unless improved streamflow management accompanies such initiatives. Additional studies should be conducted when streamflow is more substantial and persistent to document habitat improvements associated with the restoration effort. If adequate streamflow can be maintained, river-channel restoration could be an important component in a multifaceted effort for the shiner's conservation.

## CASE STUDY 6.9
Reintroduction of the Rio Grande Silvery Minnow in the Rio Grande Basin

Gary Garrett, Aimee Roberson, and Jeffery Bennett

### Introduction
In 2000, the USFWS, in collaboration with participating Game and Fish Departments, Reclamation, and other agencies, began a program to actively reintroduce experimental populations of the Rio Grande silvery minnow to selected reaches of the Rio Grande where the fish was found historically. To provide expertise in support of this effort, the Rio Grande Silvery Minnow Recovery Team (Recovery Plan) was created, which was a diverse team that consisted of fish scientists, hydrologists, conservationists, and water managers from a variety of federal and state agencies, institutes, and organizations. Along with other members of the Recovery Team, one of our central responsibilities was to develop the Rio Grande Silvery Minnow Recovery Plan, which described the when, where, and how of reintroducing the experimental population of the Rio Grande silvery minnow.

### Methods and Results
Over a period of one-and-a-half years, the Recovery Team conducted a reach-by-reach biological, biophysical, and chemical assessment of the Rio Grande and Pecos River Basins to identify sites where the reintroduction of the Rio Grande silvery minnow would most likely result in a self-sustaining population. The assessment considered (1) the reasons for the species' extirpation from the selected reach, (2) the presence of other native fish species that reproduce in a similar manner, (3) habitat conditions (including susceptibility to river drying and the presence of diversion structures), and (4) the presence of congeners (i.e., other species of the genus *Hybognathus*) (USFWS 2010).

Based on assessment results and other factors, the Recovery Plan identified three areas within the historical range of the minnow where the reestablishment of populations were thought to have the greatest potential for success, including the binational reach of the Rio Grande through Big Bend (USFWS 2010).

Over 2.7 million of the Rio Grande silvery minnow were reintroduced along the Big Bend reach of the river from 2008 to 2016. Stocking rates varied from as many as 510,000 in 2009 to as few as 70,000 in 2014. Although the monitoring program in 2010 documented the presence of Rio Grande silvery minnow eggs from two of the five initial release sites (Jason Remshardt, pers. comm.), survival to the adult stage has not been confirmed. Adults were often caught in the general vicinity of stocking sites in ensuing weeks, but there is no evidence of populations becoming established and self-sustaining.

Despite reintroduction efforts, experts believe the long-term viability of the fish will remain uncertain due to issues related to the river's hydrology and habitat—a concern with current river regulation and management and not the fish. Litigation as well as voluntary cooperation has resulted in some increase in flows, and several habitat restoration projects have been completed along the reaches in New Mexico and the international border between Mexico and the United States (see Case Study 5.5).

### Lessons Learned
- *Public Outreach.* Educating the public on the value of the reintroduction program and making it clear that the daily activities of water users and landowners would not be affected was critical to the pro-

gram being accepted and supported by entities who are often wary of regulatory restrictions associated with endangered species. An even greater public outreach program could have garnered more support for the reintroduction program as well as access to additional stocking locations.

The Rio Grande Silvery Minnow being released into the Rio Grande/Bravo along the U.S.-Mexico border in 2008. Photo by Mike Bender, USFWS.

- *Federal Designation.* The reintroduced Rio Grande silvery minnow was designated as nonessential and experimental under section 10(j) of the Endangered Species Act, which positively affected the overall outcome of the reintroduction program by reducing federal regulations and ensuring acceptance by landowners.

- *Stocking Considerations.* About 2.7 million Rio Grande silvery minnows were released over a period of eight years, which is really not a significant amount when it comes to minnows. There may be a minimum stocking threshold below which reproductive success significantly declines. One of the challenges is that the Rio Grande silvery minnow is that it is short-lived and they need to be able to find one other along long segments of river. Stocking of more fish in more locations (including the Lower Pecos River) would have probably improved results.

- *Complementary Riparian and Aquatic Restoration Actions.* Reintroducing the Rio Grande silvery minnow in areas where complementary stream restoration actions are underway is also important. For example, along the Big Bend reach of the Rio Grande, binational efforts to control invasive riparian plant species (saltcedar and giant cane) seem to be fostering processes that are establishing a more desirable and complex channel morphology that is also conducive to the long-term viability of the Rio Grande silvery minnow. Further monitoring and study are needed to understand the correlation among the management of invasive riparian plants, channel morphology, and the minnow, and greater coordination and collaboration between native fish reintroduction programs and other types of stream restoration efforts should be a priority.

- *Streamflow Management.* The long-term viability of the Rio Grande silvery minnow in the Rio Grande Basin hinges on streamflow. Flow in the basin is presently managed solely to meet human needs, with little consideration for environmental conditions. Both human and environmental objectives must be considered and, indeed, are often complementary. For example, improving the management of streamflow along the Big Bend segment of the Rio Grande to flush sediment and establish backwater, side-channel, and near-shore habitat during appropriate times of the year would initiate spawning and therefore help both the minnow and riverside towns. The Rio Grande silvery minnow would benefit through heightened spawning and improved habitat. Riverside towns would gain enhanced flood protection because of increased channel capacity. Such human-environmental win-win opportunities need to become more of a priority in the years to come, particularly as water scarcity increasingly becomes more of an issue.

New Mexico and Texas (Propst 1999). Today, the last remaining population—the Pecos blunt-nose shiner (subspecies *N. simus pecosensis*)—is restricted to about 300 km of the Pecos River in New Mexico (Hatch et al. 1985; Bestgen and Platania 1990) and is found at depths between 3 and 51 cm, where the substrate is shifting sand with a relatively wide river channel that exceeds 25 m in width (Bestgen and Platania 1990; Hoagstrom et al. 2008a). Dewatering that causes river drying is the greatest threat to this last remaining population. However, declines in base flow are also a threat because adults primarily use habitats with current velocities above 20 cm/s (Hoagstrom et al. 2008a).

The bluntnose shiner was listed as threatened by the U.S. federal government in 1987, which stimulated studies to better understand its status and vulnerability, as well as the measures needed to protect the remaining populations (Brooks et al. 1991; Propst 1999; Robertson 1997). River fragmentation by dams and reservoirs combined with habitat degradation of remaining riverine reaches have driven population declines and extirpations (Hoagstrom et al. 2008a, 2008b). As a result, recovery of the species will require establishing appropriate streamflow and maintaining suitable habitat conditions (USFWS 1987; 1992).

## The Rio Grande Silvery Minnow

The Rio Grande silvery minnow (*Hybognathus amarus*) is a small North American fish and one of seven North American members of the genus *Hybognathus*. Listed as endangered in the United States in 1994, the species is currently found in less than 5 percent of its historic natural habitat, which used to range from the Rio Grande Basin near Española, New Mexico, to near the Gulf of Mexico (USDOI 1994), and in the Pecos River from Santa Rosa, New Mexico, to its confluence with the Rio Grande. According to the USFWS, the Rio Grande silvery minnow is one of the most endangered fish in North America (USFWS 2010). The habitat of the minnow includes reaches characterized by low-flow velocities and silted bottoms (USFWS 2015).

The river segments where the Rio Grande silvery minnow was historically found were wide, braided, sand-bedded systems. They offered diverse riparian and aquatic habitat with extensive, active floodplains composed of numerous secondary channels, floodplain lakes, and marshes with significant woody debris. The riverine ecosystems of the Rio Grande Basin have changed dramatically in the last 150 years owing to a variety of human actions—such as the construction of dams, diversions, and levees, and the overallocation of water supplies—that have changed the hydrology of the river and fundamentally altered processes of sediment and solute transport as well as species interactions, producing a river system characterized by an overall loss of physical complexity and habitat diversity (Alo and Turner 2005). Flow regimes to which the minnow is adapted no longer occur, and high-magnitude spring flows that cue spawning are less frequent. These changes, coupled with deteriorated water quality, are the most likely reasons why the Rio Grande silvery minnow has disappeared from most of its historical range.

In 1994, the USFWS concluded that the Rio Grande silvery minnow should be listed as endangered because of the extremely limited habitat it currently occupies, its declining abundance, and because this species can be expected to become extinct in the foreseeable future if threats to the species and its habitat go unchecked (USDOI 1994).

# REFERENCES

Agouridis, C. T., S. R. Workman, C. W. Warner, and G. D. Jennings. 2005. "Livestock Grazing Management Impacts on Stream Water Quality: A Review." *Journal of the American Water Resources Association* 41:591–606.

Albertson, L. K., L. E. Koenig, B. L. Lewis, S. C. Zeug, L. R. Harrison, and B. J. Cardinale. 2013. "How Does Restored Habitat for Chinook Salmon (*Oncorhynchus tshawytscha*) in the Merced River in California Compare with Other Chinook Streams?" *River Research and Applications* 29:469–482.

Alò, D., and T. E. Turner. 2005. "Effects of Habitat Fragmentation on Effective Population Size in the Endangered Rio Grande Silvery Minnow." *Conservation Biology* 19:1138–48.

Anderson, B. W. 1995. "Salt Cedar, Revegetation and Riparian Ecosystems in the Southwest." Symposium Proceedings. California Exotic Pest Plant Council.

Anderson, B. W. 2015. *Experiences in Revegetation on the Lower Colorado River.* Bulletin of the Revegetation and Wildlife Management Center 8. Blythe, California.

Anderson, B. W. 2017. *Death of an Ecosystem: Murder and Autopsy of the Colorado River Riparian Ecosystem.* Bulletin of the Revegetation and Wildlife Center 9–10, Blythe, California.

Anderson, B. W., and S. A. Laymon. 1988. "Creating Habitat for the Yellow-Billed Cuckoo (*Coccyzus americana*)." In *Proceedings of the California Riparian Systems Conference: Protection, Management, and Restoration for the 1990s*; September 22–24, 1988, Davis, California. General Technical Report PSW-110., Berkeley, Calif.: U.S. Department of Agriculture, Pacific Southwest Forest and Range Experiment Station.

Anderson, B. W., and R. D. Ohmart. 1982. *Revegetation and Wildlife Enhancement Along the Lower Colorado River.* U.S. Dept. of the Interior, Bureau of Reclamation. Contract No. 7-07-30-V0009. https://www.usbr.gov/library/reclamationpubs.html.

Anderson, B. W., P. E. Russell, and R. D. Ohmart. 2004. *Riparian Revegetation: An Account of Two Decades of Experience in the Arid Southwest.* AVVAR Books, Blythe, California.

Archdeacon, T. P., Blocker, S. D., Davenport, S. R., and K. R. Henderson. 2015. "Seasonal Variation in Reproductive Condition of the Pecos Bluntnose Shiner (*Notropis simus pecosensis*)." *Western North American Naturalist* 75:271–80.

Armour, C. L., D. A. Duff, and W. Elmore. 1991. "The Effects of Livestock Grazing on Riparian and Stream Ecosystems." *Fisheries* 16:7–11.

Attard, C. R. M., L. M. Möller, M. Sasaki, M. P. Hammer, C. M. Bice, C. J. Brauer, D. C. Carvalho, J. O. Harris, and L. B. Beheregaray. 2016. "A Novel Holistic Framework for Genetic-Based Captive-Breeding and Reintroduction Programs." *Conservation Biology* 30:1060–69.

Australian Society for Fish Biology. 2004. "Conservation Status of Australian Fishes 2004." *Australian Society for Fish Biology Newsletter* 34 (2): 59–64.

Bailey, J. K., J. A. Schweitzer, and T. G. Whitham. 2001. "Salt Cedar Negatively Affects Biodiversity of Aquatic Macroinvertebrates." *Wetlands* 21:442–47.

Bark, R. H., D. E. Osgood, B. G. Colby, G. Katz, and J. Stromberg. 2009. "Habitat Preservation and Restoration: Do Homebuyers Have Preferences for Quality Habitat?" *Ecological Economics* 68: 1465–75.

Bark-Hodgins, R., and B. G. Colby. 2006. "An Economic Assessment of the Sonoran Desert Conservation Plan." *Natural Resources Journal* 46:709–25.

Barranco, A. 2001. "Invasive Species Summary Project: Saltcedar (*Tamarix ramosissima*)." Columbia .edu. November 11, 2001. http://www.columbia.edu/itc/cerc/danoff-burg/invasion_bio/inv_spp _summ/Tamarix_ramosissima.html.

Barrows, C. 1998. "The Case for Wholesale Removal." *Restoration and Management Notes* 16:135–39.

Bay, R. F. 2013. "Revegetation After Tamarisk Removal." In Tamarix: *A Case Study of Ecological Change in the American West*, edited by A. A. Sher and M. F. Quigley, 426–40. New York: Oxford University Press.

Bay, R. F., and A. A. Sher. 2008. "Success of Active Revegetation After Tamarisk Removal in Riparian Ecosystems of the Southwestern United States: A Quantitative Assessment of Past Restoration Projects." *Restoration Ecology* 16:113–28.

Bean, D., and T. Dudley. 2018. "A Synoptic Review of Tamarix Biocontrol in North America: Tracking Success in the Midst of Controversy." *BioControl* 63:361–76. http://doi.org/10.1007/s10526-018-9880-x.

Bell, G. P. 1997. "Ecology and Management of *Arundo donax*, and Approaches to Riparian Habitat Restoration in Southern California." In *Plant Invasions: Studies from North America and Europe*, edited by J. H. Brock et al., 103–13. New York: Barnes and Noble.

Belsky, A. J., A. Matzke, and S. Uselman. 1999. "Survey of Livestock Influences on Stream and Riparian Ecosystems in the Western United States." *Journal of Soil and Water Conservation* 54:419–31.

Bertrand, L. J., and M. LaLonde. 1985. "In Vitro Propagation and Nodulation by Frankia of Actinorhizal Russian Olive (*Elaeagnus angustifolia* L.)." *Plant and Soil* 87:143–52.

Bestgen, K. R., and S. P. Platania. 1990. "Extirpation of *Notropis simus simus* (Cope) and *Notropis orca* Woolman (Pisces: Cyprinidae) from the Rio Grande in New Mexico, with Notes on Their Life History." *Occasional Papers of the Museum of Southwestern Biology* 6:1–8.

Bice, C., N. Whiterod, and B. Zampatti. 2014. *The Critical Fish Habitat Project: Assessment of the Success of Reintroductions of Threatened Fish Species in the Coorong, Lower Lakes and Murray Mouth Region, 2011–2014.* Adelaide: SARDI Aquatic Sciences.

Birken, A. S., and D. J. Cooper. 2006. "Processes of Tamarix Invasion and Floodplain Development Along the Lower Green River, Utah." *Ecological Applications* 16:1103–20.

Bledsoe, B. P. 1999. "Specific Stream Power as an Indicator of Channel Pattern, Stability, and Response to Urbanization." PhD diss., Colorado State University.

Bourne, K. L. 2007. "The Effect of the Santa Cruz River Riparian Corridor on Single Family Home Prices Using the Hedonic Pricing Method." Master's thesis, University of Arizona.

Briggs, M. K. 1996. *Riparian Ecosystem Recovery in Arid Lands: Strategies and References.* Tucson: University of Arizona Press.

Briggs, M. K., and M. L. Flores. 2003. "Small-Scale Restoration Project Promotes Regional Restoration in the Colorado River Delta: A Testament to the Power of Tangible Restoration at the Community Level." *Southwest Hydrology* 2:24–26.

Brock, J. H. 1998. Invasion, Ecology and Management of *Elaeagnus angustifolia* (Russian Olive) in the Southwestern U.S.A. In *Plant Invasions: Ecological Mechanisms and Human Responses*, edited by U. Starfinger et al., 372. Leiden: Backhuys.

Brooks, A. P., P. C. Gehrke, J. D. Jansen, and T. B. Abbe. 2004. "Experimental Reintroduction of Woody Debris on the Williams River, NSW: Geomorphic and Ecological Responses." *River Research and Applications* 20:513–36.

Brooks, J. E., S. P. Platania, and D. L. Propst. 1991. *Effects of Pecos River Reservoir Operation on the Distribution and Status of Pecos Bluntnose Shiner* (Notropis simus pecosensis): *Preliminary Findings.*

Report to the U.S. Fish and Wildlife Service and U.S. Bureau of Reclamation, Albuquerque, New Mexico.

Brown, C. R. 1990. "Avian Use of Native and Exotic Riparian Habitats on the Snake River, Idaho." Master's thesis, Colorado State University.

Brown, R. W., and M. C. Amacher. 1999. *Selecting Plant Species for Ecological Restoration: A Perspective for Land Managers.* USDA Forest Service Proceedings RMRS-P-8 (https://www.fs.usda.gov/rmrs /search).

Buisson, L., W. Thuiller, S. Lek, P. Lim, and G. Grenouillet. 2008. "Climate Change Hastens the Turnover of Stream Fish Assemblages." *Global Change Biology* 14:2232–48.

Bunting, D., S. A. Kurc, and M. R. Grabau. 2011. "Using Existing Agricultural Infrastructure for Restoration Practices: Factors Influencing Successful Establishment of *Populus fremontii* over *Tamarix ramosissima*." *Journal of Arid Environments* 75:851–60.

Burnham, M. 2004. "Front Lines of Battle Against Invaders Increasingly Local." LandLetters. Accessed February 13, 2004. http://www.eenews.net/Landletter/Backissues/02120404.htm.

Campbell, A., and J. Dawes. 2004. *Encyclopedia of Underwater Life.* Oxford: Oxford University Press.

Caplan, T. 2002. "Controlling Russian Olives Within Cottonwood Gallery Forests Along the Middle Rio Grande Floodplain (New Mexico)." *Ecological Restoration* 20:138–39.

Carman, J. G., and J. D. Brotherson. 1982. Comparisons of Sites Infested and Not Infested with Saltcedar (*Tamarix pentandra*) and Russian Olive (*Elaeagnus angustifolia*). *Weed Science* 30:360–64.

Chamier, J., K. Schachtschneider, D. C. le Maitre, P. J. Ashton, and B. W. van Wilgen. 2012. "Impacts of Invasive Alien Plants on Water Quality, with Particular Emphasis on South Africa." *Water SA* 38 (2). https:/doi.org/10.4314/wsa.v38i2.19.

Chase, N. M., C. A. Caldwell, S. A. Carleton, W. R. Gould, and J. A. Hobbs. 2015. "Movement Patterns and Dispersal Potential of Pecos Bluntnose Shiner *(Notropis simus pecosensis)* Revealed Using Otolith Microchemistry." *Canadian Journal of Fisheries and Aquatic Sciences* 72:1575–83.

Christensen, E. M. 1963. "Naturalization of Russian Olive *(Elaeagnus angustifolia)* in Utah." *American Midland Naturalist* 70:133–37.

Claire, E., and R. Storch 1983. "Streamside Management and Livestock Grazing: An Objective Look at the Situation." In *Workshop on Livestock and Wildlife-Fisheries Relationships in the Great Basin*, edited by J. Menke, 111–28. Berkeley, Calif.: U.S. Forest Service.

Clark, L. B., A. L. Henry, R. Lave, N. F. Sayre, E. González, and A. A. Sher. 2019. "Successful Information Exchange Between Restoration Science and Practice." *Restoration Ecology* 27 (6): 1–10. http://doi.org /10.1111/rec.12979.

Cleverly, J. R. 2013. "Water Use by *Tamarix*." In Tamarix: *A Case Study of Ecological Change in the American West*, edited by A. A. Sher and M. F. Quigley, 85–98. New York: Oxford University Press.

Cohn, J. P. 2005. "Tiff over Tamarisk: Can a Nuisance Be Nice, Too?" *Bioscience* 55:648–54.

Cole, D. N., and P. B. Landres. 1995. "Indirect Effects of Recreation on Wildlife." *Wildlife and Recreationists: Coexistence Through Management and Research*, edited by R. L. Knight and K. J. Gutzwiller, 183–202. Washington, D.C.: Island Press.

Contreras-Balderas, S., P. Almada-Villela, M. de L. Lozano-Vilano, and M. E. Garcia-Ramırez. 2003. "Freshwater Fish at Risk or Extinct in Mexico." *Reviews in Fish Biology and Fisheries* 12:241–51.

Conteras-Balderas, S., and M. de L. Lozano-Vilano. 1994. "Water, Endangered Fishes, and Development Perspectives of Arid Lands of Mexico." *Conservation Biology* 8:379–87.

Contreras-Balderas, S., G. Ruizcampos, J. Schmitter-Soto, E. Díaz-Pardo, T. Contreras-MacBeath, M. Medina-Soto, et al. 2008. "Freshwater Fishes and Water Status in Mexico: A Country-Wide Appraisal." *Aquatic Ecosystem Health and Management* 11:246–56. https:/doi.org/10.1080/14634 980802319986.

Contreras-Macbeath, T. 2014. "An Analysis of the Spatial Distribution of Freshwater Fishes of Mexico, Their Conservation Status, and the Development of a Conservation Strategy for Species with Imminent Risk of Extinction Based on Contemporary Theories and Practices." PhD diss., Manchester Metropolitan University.

Contreras-MacBeath, T., T. H. Mejia, and R. Carrillo. 1998. "Negative Impact on the Aquatic Ecosystems of the State of Morelos, Mexico, from Introduced Aquarium and Other Commercial Fish." *Aquarium Sciences and Conservation* 2:67–78.

Cremer, K. W. 2003. "Introduced Willows Can Become Invasive Pests in Australia." *Biodiversity* 4 (4): 17–24. https:/doi.org/10.1080/14888386.2003.9712705.

Crook, D., B. Gillanders, A. Sanger, A. Munro, D. J. O'Mahony, S. Woodcock, S. Thurstan, and L. Baumgartner. 2010. "Outcomes of Native Fish Stocking in Rivers of the Southern Murray–Darling Basin." In *Native Fish Forum 2010* 28. Canberra: Murray–Darling Basin Authority.

Crook, D., D. J. O'Mahony, B. M. Gillanders, A. R. Munro, A. C. Sanger, S. Thurstan, S. and L. J. Baumgartner. 2016. "Contribution of Stocked Fish to Riverine Populations of Golden Perch *(Macquaria ambigua)* in the Murray–Darling Basin, Australia." *Marine and Freshwater Research* 67: 1401–9.

Dalton, M. G., B. E. Huntsman, and K. Bradbury. 1991. "Acquisition and Interpretation of Water-Level Data." In *Practical handbook of Groundwater Monitoring*, edited by D. M. Nielson, 367–96. Chelsea, Mich.: Lewis.

Dean, D. J., and J. C. Schmidt. 2011. "The Role of Feedback Mechanisms in Historic Channel Changes of the Lower Rio Grande in the Big Bend Region." *Geomorphology* 126:333–49.

DeCant, J. P. 2008. "Russian Olive, *Elaeagnus angustifolia*, Alters Patterns in Soil Nitrogen Pools Along the Rio Grande River, New Mexico, USA." *Wetlands* 28:896–904.

Dello Russo, G. 2013. "Tamarisk Management at Bosque del Apache National Wildlife Refuge: A Resource Manager's Perspective." In Tamarix: *A Case Study of Ecological Change in the American West*, edited by A. A. Sher and M. F. Quigley, chapter 21. New York: Oxford University Press.

DiTomaso, J. 1998. "Impact, Biology, and Ecology of Saltcedar (*Tamarix* spp.) in the Southwestern United States." *Weed Technology* 12:326–36.

DiTomaso, J., G. B. Kyser, S. R. Oneto, R. G. Wilson, S. B. Orloff, L. W. Anderson, S. D. Wright, J. A. Roncoroni et al. 2013. *Weed Control in Natural Areas in the Western United States*. Davis, Calif.: Weed Research and Information Center, University of California.

Doody, T. M., P. L. Nagler, E. P. Glenn, G. W. Moore, K. Morino, K. R. Hultine, and R. G. Benyon. 2011. "Potential for Water Salvage by Removal of Non-native Woody Vegetation from Dryland River Systems." *Hydrological Processes* 25:4117–31.

Dreesen, D., J. Harrington, T. Subirge, P. Stewart, and G. Fenchel. 2002. *Riparian Restoration in the Southwest: Species Selection, Propagation, Planting Methods, and Case Studies*. In *National Proceedings Forest and Conservation Nursery Associations—1999, 2000, and 2001*, edited by R. K. Dumroese, L. E. Riley, and T. D. Landis, 253–72. Proceedings RMRS-P-24. Ogden, Utah: USDA Forest Service, Rocky Mountain Research Station.

Dudley, R. K., and S. P. Platania. 2007. "Flow Regulation and Fragmentation Imperil Pelagic-Spawning Riverine Fishes." *Ecological Applications* 17:2074–86.

Duncan, K. W., and K. C. McDaniel. 1998. "Saltcedar (*Tamarix* spp.) Management with Imazapyr." *Weed Technology* 12:337–44.

Ebner, B. C., D. L. Morgan, A. Kerezsy, S. Hardie, S. J. Beatty, J. E. Seymour, J. A. Donaldson, et al. 2016. "Enhancing Conservation of Australian Freshwater Ecosystems: Identification of Freshwater Flagship Fishes and Relevant Target Audiences." *Fish and Fisheries* 17:1134–51.

Edwards, R. J., L. C. Clark, and K. G. Beck. 2014. "Russian Olive *(Elaeagnus angustifolia)* Dispersal by European Starlings *(Sturnus vulgaris)*." *Invasive Plant Science and Management* 7:425–31.

Elliott, J. G., A. C. Gellis, and S. B. Aby. 1999. "Evolution of Arroyos: Incised Channels of the Southwestern United States." In *Incised River Channels: Processes, Forms, Engineering and Management*, edited by S. E. Darby and A. Simon, 153–85. New York: Wiley.

Ellis, L. M. 1995. "Bird Use of Saltcedar and Cottonwood Vegetation in the Middle Rio Grande Valley of New Mexico, U.S.A." *Journal of Arid Environments* 30:339–49.

Elmore, W., and R. L. Beschta. 1987. "Riparian Areas: Perceptions in Management." *Rangelands* 9:260–65.

Engel-Wilson, R. W., and R. D. Ohmart. 1978. *Assessment of Vegetation and Terrestrial Vertebrates Along the Rio Grande Between Fort Quitman, TX and Haciendita, TX*. El Paso, Tx: International Boundary and Water Commission.

Environmental Protection and Biodiversity Conservation Act. 1999. http://www.environment.gov.au/cgi -bin/sprat/public/publicthreatenedlist.pl#fishes_critically_endangered.

EPA. 2017. "Invasive Non-native Species." https://www.epa.gov/watershedacademy/invasive-non-native -species.

Eubanks, E. C. 2004. *Riparian Restoration*. U.S. Department of Agriculture, Forest Service, August 2004. http://www.remarkableriparian.org/pdfs/pubs/TR_1737-22.pdf.

FISRWG. 1998. *Stream Corridor Restoration: Principles, Processes, and Practices*. By the Federal Interagency Stream Restoration Working Group (FISRWG)(15 Federal agencies of the US government). GPO Item No. 0120-A; SuDocs No. A 57. 6/2:EN 3/PT. 653. Washington, D.C.: Environmental Protection Agency.

Fleischner, T. L. 1994. "Ecological Costs of Livestock Grazing in Western North America." *Conservation Biology* 8:629–44.

Flores, R. 1990. *Hidrogeología de la Comarca Lagunera*. Universidad Juárez del Estado Durango: Gomez Palacio, DGO.

Forbes, J., R. J. Watts, W. A. Robinson, L. J. Baumgartner, P. McGuffie, L. M. Cameron, and D. A. Crook. 2015. "Assessment of Stocking Effectiveness for Murray Cod *(Maccullochella peelii)* and Golden Perch *(Macquaria ambigua)* in Rivers and Impoundments of South-Eastern Australia." *Marine and Freshwater Research* 67 (10): 1410–19.

Forest Ecosystem Management Team (FEMAT). 1993. *Forest Ecosystem Management: An Ecological, Economic, and Social Assessment*. Report of the Interagency Working Group from the President's Forest Conference. Portland, Ore.: U.S. Department of Agriculture, Forest Service, Region 6.

Franzreb, K. 1989. *Ecology and Conservation of the Endangered Least Bell's Vireo*. U.S. Fish and Wildlife Service, Biological Report 89. https://digitalmedia.fws.gov/digital/collection/document.

Friedman, J. M., G. T. Auble, P. B. Shafroth, M. L. Scott, M. F. Merigliano, M. D. Freehling, and E. R. Griffith. 2005. "Dominance of Non-native Riparian Trees in Western USA." *Biological Invasions* 7:747–51.

Frissell, C. A. 1993. "Topology of Extinction and Endangerment of Native Fishes in the Pacific Northwest and California (USA)." *Conservation Biology* 7:342–54.

Garrett, G. P., and R. J. Edwards. 2004. "Changes in Fish Populations in the Lower Canyons of the Rio Grande." In *Proceedings of the Sixth Symposium on the Natural Resources of the Chihuahuan Desert Region, October 14–17*, edited by C. A. Hoyt and J. Karges, 396–408. Fort Davis, Tex.: Chihuahuan Desert Research Institute. http://cdri.org/publications/proceedings-of-the-symposium-on-the-natural-resources-of-the-chihuahuandesert-region/.

Geocaching. 2018. "Sedge Mouth Meanders." https://www.geocaching.com/geocache/GC6RNK3_sedge-mouth-meanders?guid=44aacbcc-e856-45a5-8d89-08cb311a14c7.

GeoDimensions Pty Ltd. 2006. *Two Rivers Traffic Management Plan: A Strategy for Sharing Melbourne's Rivers and Bays.* Prepared for Parks Victoria, December, 2006. Accessed March 22, 2011. http://www.parkweb.vic.gov.au/resources07/07_2046.pdf.

Giessow, J., J. Casanova, R. Leclerc, G. Fleming, and J. Giessow. 2011. *Arundo donax (Giant Reed): Distribution and Impact Report.* Submitted to State Water Resources Control Board, California by the California Invasive Plant Council. http://www.cal-ipc.org/ip/research/arundo/index.php.

Gilligan, D., and C. Schiller. 2003. *Downstream Transport of Larval and Juvenile Fish in the Murray River.* Final Report for the National Resources Management Strategy, Project No. NRMS R7019. Sydney: New South Wales Fisheries.

Glenn, E. P., and P. L. Nagler. 2005. "Comparative Ecophysiology of *Tamarix ramosissima* and Native Trees in Western US Riparian Zones." *Journal of Arid Environments* 61:419–46.

González, E., Sher, A., Anderson, R., Bay, R., Bean, D., Bissonnete, G., B. Bourgeois, et al. 2017a. "Vegetation Response to Invasive Tamarix Control in Southwestern U.S. Rivers: A Collaborative Study Including 416 sites." *Ecological Applications* 27 (6): 1789–804.

González, E., Sher, A. A., Anderson, R. M., Bay, R. F., Bean, D. W., Bissonnete, D. Cooper, et al. 2017b. "Secondary Invasions of Noxious Weeds Associated with Control of Invasive Tamarix are Frequent, Idiosyncratic and Persistent." *Biological Conservation* 213:106–14. http://doi.org/10.1016/j.biocon.2017.06.043.

González, E., M. R. Felipe-Lucia, B. Bourgeois, B. Boz, C. Nilsson, G. Palmer, and A. A. Sher. 2017c. "Integrative Conservation of Riparian Zones." *Biological Conservation* http://doi.org/10.1016/j.biocon.2016.10.035.

González, E., V. Martínez-Fernández, P. B. Shafroth, A. A. Sher, A. L. Henry, V. Garófano-Gómez, and D. Corenblit. 2018. "Regeneration of Salicaceae Riparian Forests in the Northern Hemisphere: A New Framework and Management Tool." *Journal of Environmental Management* 218:374–87.

González, E., P. B. Shafroth, S. R. Lee, et al. 2020. "Riparian Plant Communities Remain Stable in Response to a Second Cycle of Tamarix Biocontrol Defoliation." *Wetlands.* https://doi.org/10.1007/s13157-020-01381-7.

Goolsby, J. A. 2017. "Biological Control of *Arundo donax*, an Invasive Weed of the Rio Grande Basin." Paper presentation. USDA-ARS Plains Area Edinburg (Moore Airbase), Texas. http://riograndewaterplan.org/downloads/presentations/07-12-2017/Item%205.pdf.

Goolsby, J. A., and P. J. Moran. 2009. "Host Range of *Tetramesa romana* Walker (Hymenoptera: Eurytomidae), a Potential Biological Control of Giant Reed, *Arundo donax* L. in North America." *Biological Control* 49:160–68.

Goolsby, J. A., P. J. Moran, A. E. Racelis, K. R. Summy, M. M. Jimenez, R. D. Lacewell, A. Perez de Leon, and A. A. Kirk. 2015. "Impact of the Biological Control Agent *Tetramesa romana* (Hymenoptera:

Eurytomidae) on *Arundo donax* (Poaceae: Arundinoideae) Along the Rio Grande River in Texas." *Biocontrol Science and Technology* 26 (1). https:/doi.org/10.1080/09583157.2015.1074980.

Grabau, M. R., D. Bangle, M. A. Milczarek, L. A. Hovland, and B. Raulston. 2013. "Irrigation Monitoring for Floodplain Riparian Restoration on the Lower Colorado River." In *Proceedings of 5th National Conference on Ecosystem Restoration, July 2013, Schaumberg, Ill.* https://www.researchgate.net/publication/299595843.

Grady, K. C., S. M. Ferrier, T. E. Kolb, S. C. Hart, G. J. Allan, and T. G. Whitham. 2011. "Genetic Variation in Productivity of Foundation Riparian Species at the Edge of Their Distribution: Implications for Restoration and Assisted Migration in a Warming Climate." *Global Change Biology* 17:3724–35. http://doi.org/10.1111/j.1365-2486.2011.02524.x.

Greening Australia Victoria. 2003. *Revegetation Techniques: A Guide for Establishing Native Vegetation in Victoria.* http://tamariskcoalition.org/sites/default/files/resource-center-documents/Australian_Revegetation_Techniques_PartB.pdf.

Growns, I., I. Wooden, and C. Schiller. 2004. "Use of Instream Habitat by Trout Cod *Maccullochella macquariensis* (Cuvier) in the Murrumbidgee River." *Pacific Conservation Biology* 10:261–65.

Gumm, J. M., J. L. Snekser, and M. Itzkowitz. 2008. "Conservation and Conflict Between Endangered Desert Fish." *Biology Letters (Royal Society Publishing)* 4:655–58.

Hammer, M. P., C. M. Bice, A. Hall, A. Frears, A. Watt, N. S. Whiterod, L. B. Beheregaray, J. O. Harris, and B. Zampatti. 2013. "Freshwater Fish Conservation in the Face of Critical Water Shortages in the Southern Murray–Darling Basin, Australia." *Marine and Freshwater Research* 64:807–21.

Hammer, M. P., T. S. Goodman, M. Adams, L. F. Faulks, P. J. Unmack, N. S. Whiterod, and K. F. Walker. 2015. "Regional Extinction, Rediscovery and Rescue of a Freshwater Fish from a Highly Modified Environment: The Need for Rapid Response." *Biological Conservation* 192:91–100.

Hardie, S. A., J. E. Jackson, L. A. Barmuta, and R. W. G. White. 2006. "Status of Galaxiid Fishes in Tasmania, Australia: Conservation, Listings, Threats, and Management Issues." *Aquatic Conservation* 16:235–50.

Hart, B. T., P. Bailey, R. Edwards, K. Hortle, K. James, A. McMahon, C. Meredith, and K. Swadling. 1990. "Effects of Salinity on River, Stream and Wetland Ecosystems in Victoria, Australia." *Water Research* 24:1103–17.

Harvey, M. D., and C. D. Watson. 1986. "Fluvial Process and Morphological Thresholds in Incised Channel Restoration." *Water Resources Bulletin* 2:359–68.

Hatler, W., and C. Hart. 2009. "Water Loss and Salvage in Saltcedar *(Tamarix)* Stands on the Pecos River, Texas." *Invasive Plant Science and Management* 2:309–317.

Heath, R. C. 2004. "Basic Groundwater Hydrology." Tenth printing, revised. Water-Supply Paper 2220.

Herrera, A. M., and T. L. Dudley. 2003. "Reduction of Riparian Arthropod Abundance and Diversity as a Consequence of Giant Reed (*Arundo donax*) Invasion." *Biological Invasions* 5:167–77.

Hoagstrom, C. W., J. E. Brooks, and S. R. Davenport. 2008a. "Recent Habitat Association and the Historical Decline of *Notropis simus pecosensis*." *River Research and Applications* 24:789–803.

Hoagstrom, C. W., J. E. Brooks, and S. R. Davenport. 2008b. "Spatiotemporal Population Trends of *Notropis simus pecosensis* in Relation to Habitat Conditions and the Annual Flow Regime of the Pecos River, 1992–2005." *Copeia* 1:5–15.

Hoagstrom, C. W., J. E. Brooks, and S. R. Davenport. 2011. "A Large-Scale Conservation Perspective Considering Endemic Fishes of the North American Plains." *Biological Conservation* 144: 21–34.

Hoffman, B. D., and L. M. Broadhurst. 2016. "The Economic Cost of Managing Invasive Species in Australia." *NeoBiota* 31:1–18.

Hohlt, J. C., B. J. Racher, J. B. Bryan, R. B. Mitchell, and C. Britton. 2002. "Saltcedar Response to Prescribed Burning in New Mexico." In *Research Highlights—2002: Range, Wildlife, and Fisheries Management.* Vol. 33. Edited by G. R. Wilde and L. M. Smith, 25. Lubbock: Texas Tech University, College of Agricultural Sciences and Natural Resources.

Howell, T. D., A. H. Arthington, B. J. Pusey, A. P. Brooks, B. Creese, and J. Chaseling. 2012. "Responses of Fish to Experimental Introduction of Structural Woody Habitat in Riffles and Pools." *Restoration Ecology* 20:43–55.

Hultine, K. R., T. L. Dudley, D. G. Koepke, D. W. Bean, E. P. Glenn, and A. M. Lambert. 2015. "Patterns of Herbivory-Induced Mortality of a Dominant Non-native Tree/Shrub (*Tamarix* spp.) in a Southwestern US Watershed." *Biological Invasions* 17:1729–42. https://doi.org/10.1007/s10530-014-0829-4.

Humphries, P. 2005. "Spawning Time and Early Life History of Murray Cod, *Maccullochella peelii peelii* (Mitchell) in an Australian River." *Environmental Biology of Fishes* 72 (4): 393–407.

Hupp, C. R., and W. R. Osterkamp. 1996. "Riparian Revegetation and Fluvial Geomorphic Processes." *Geomorphology* 14:277–95.

Ingram, B. A., B. Hayes, and M. L. Rourke. 2011. "Impacts of Stock Enhancement Strategies on the Effective Population Size of Murray Cod, *Maccullochella peelii*, a Threatened Australian Fish." *Fisheries Management and Ecology* 18:467–81.

Itami, R. 2008. "Level of Sustainable Activity: Moving Visitor Simulation from Description to Management for an Urban Waterway in Australia." In *Monitoring, Simulation and Management of Visitor Landscapes*, edited by H. R. Gimblett and H. Skov-Peterson, 349–70. Tucson: University of Arizona Press.

Itami, R., R. Gimblett, and A. Poe. 2017a. "Level of Sustainable Activity: A Framework for Integrating Stakeholders into the Simulation Modeling and Management of Mixed-Use Waterways." In *Participatory Modeling in Environmental Decision-Making: Methods, Tools, and Applications*, edited by S. Gray and R. Jordan, 211–39. New York: Springer.

Itami, R., R. Gimblett, and A. Poe. 2017b. "Level of Sustainable Activity in Prince William Sound: Defining and Managing Quality of Experience and Capacity in Wilderness Waterways." In *Sustaining Wildlands: Integrating Science and Community in Prince William Sound*, edited by A. Poe and H. R. Gimblett, 206–41. Tucson, Arizona: University of Arizona Press.

Iverson, M. 1998. *Effects of* Arundo donax *on Water Resources.* Trabuco Canyon: California Exotic Pest Plant Council.

Jelks, H. L., S. J. Walsh, N. M. Burkhead, S. Contreras-Balderas, E. D'az-Pardo, D. A. Hendrickson, J. Lyons, et al. 2008. "Conservation Status of Imperiled North American Freshwater and Diadromous Fishes." *Fisheries* 33:372–407.

Jolly, I. D., G. R. Walker, and P. J. Thorburn. 1993. "Salt Accumulation in Semi-arid Floodplain Soils with Implications for Forest Health." *Journal of Hydrology* 150:589–614.

Jones, C. 2016. *Soil Nutrient Fundamentals, Cover Crops, and Leaching.* http://landresources.montana.edu/soilfertility/documents/PDF/pres/SoilFundCCLeachGTriangleJan2016.pdf.

Katz, G. L., and P. B. Shafroth. 2003. "Biology, Ecology and Management of *Elaeagnus angustifolia* L (Russian olive) in Western North America." *Wetlands* 23:763–77.

Kauffman, J., N. Beschta, N. Otting, and D. Lytjen. 1997. "An Ecological Perspective of Riparian and Stream Restoration in the Western United States." *Fisheries* 22:12–24.

Kauffman, J., and W. C. Krueger. 1984. "Livestock Impacts on Riparian Ecosystems and Streamside Management Implications: A Review." *Journal of Range Management* 37:430–83.

Keller, D. L., B. G. Laub, P. Birdsey, and D. J. Dean. 2014. "Effects of Flooding and Tamarisk Removal on Habitat for Sensitive Fish Species in the San Rafael River, Utah: Implications for Fish Habitat Enhancement and Future Restoration Efforts." *Environmental Management* 54:465–78.

Kemmis, D. 1990. *Community and the Politics of Place.* Norman: University of Oklahoma Press.

Kennard, M. J., B. J. Pusey, J. D. Olden, S. J. Mackay, J. Stein, and N. Marsh. 2010. "Classification of Natural Flow Regimes in Australia to Support Environmental Flow Management." *Freshwater Biology* 55:171–93.

Kennedy, T. A., J. C. Finlay, and S. E. Hobbie. 2005. "Eradication of Invasive *Tamarix ramosissima* Along a Desert Stream Increases Native Fish Density." *Ecological Applications* 15:2072–83.

Keough, H. L., and D. J. Blahna. 2006. "Achieving Integrative, Collaborative Ecosystem Management." *Conservation Biology* 20:1373–82.

Kiernan, J. D., P. B. Moyle, and P. K. Crain. 2012. "Restoring Native Fish Assemblages to a Regulated California Stream Using the Natural Flow Regime Concept." *Ecological Applications* 22:1472–82.

King, A. J., Z. Tonkin, and J. Mahoney. 2009. "Environmental Flow Enhances Native Fish Spawning and Recruitment in the Murray River, Australia." *River Research and Applications* 25:1205–18.

Knopf, F. L., and T. E. Olson. 1984. "Naturalization of Russian-Olive: Implications to Rocky Mountain Wildlife." *Wildlife Society Bulletin* 12:289–98.

Lambert, A. M., T. L. Dudley, and K. Saltonstall. 2010. "Ecology and Impacts of the Large-Statured Invasive Grasses *Arundo donax* and *Phragmites australis* in North America." *Invasive Plant Science and Management* 3:489–94.

Lee, D., J. Sedell, B. Rieman, R. Thurow, and J. Williams. 1997. "Broad-Scale Assessment of Aquatic Species and Habitats." In *An Assessment of Ecosystem Components in the Interior Columbia Basin and Portions of the Klamath and Great Basins*, edited by T. M. Quigley and S. J. Arbelbide, chapter 4. General Technical Report PNW-GTR-405. Portland, Ore.: US Department of Agriculture, Forest Service, Pacific Northwest Research Station.

Lesica, P., and S. Miles. 2001. "Natural History and Invasion of Russian Olive Along Eastern Montana Rivers." *Western North American Naturalist* 61:1–10.

Lewins, R. 2001. *Consensus Building and Natural Resource Management: A Review.* University of Portsmouth, UK: CEMARE, Department of Economics.

Lintermans, M., S. Rowland, J. Koehn, G. Butler, B. Simpson, and I. Wooden. 2005. "The Status, Threats and Management of Freshwater Cod Species *Maccullochella* spp. in Australia." Management of Murray Cod in the Murray-Darling Basin: Statement, Recommendations and Supporting Papers.

Los Lunas Plant Materials Center (LLPMC). 2005. *The Pole Cutting Solution: Guidelines for Planting Dormant Pole Cuttings in Riparian Areas of the Southwest.* Los Lunas, N.M.: USDA-NRCS Plant Materials Center. www.nm.nrcs.usda.gov/news/publications/pole-cutting-solution.pdf.

Lucas, R. W., T. T. Baker, M. K. Wood, C. D. Allison, and D. M. Vanleeuwen. 2004. "Riparian Vegetation Response to Different Intensities and Seasons of Grazing." *Journal of Range Management* 57:466–74.

Lyon, J. P., C. Todd, S. J. Nicol, A. MacDonald, D. Stoessel, B. A. Ingram, R. J. Barker, and C. J. A. Bradshaw. 2012. "Reintroduction Success of Threatened Australian Trout Cod *(Maccullochella macquariensis)* Based on Growth and Reproduction." *Marine and Freshwater Research* 63:598–605.

Marks, J. C., G. A. Haden, M. O'Neill, M., and C. Pace. 2010. "Effects of Flow Restoration and Exotic Species Removal on Recovery of Native Fish: Lessons from a Dam Decommissioning." *Restoration Ecology* 18:934–43.

Mayden, R. L. 2004. "Biodiversity of Mexican Trout (Teleostei: Salmonidae: *Oncorhynchus*): Recent Findings, Conservation Concerns, and Management Recommendations." In *Homenaje al Doctor Andrés Reséndez Medina, un ictiólogico mexicano*, edited by M. de Lourdes Lozano Vilano and A. J. Contreras Balderas, 268–82. Monterrey, MX: Universidad Autónoma de Nuevo León.

McDaniel, K., T. Caplan, and J. Taylor. 2002. *Control of Russian Olive and Saltcedar Resprouts with Early and Late Summer Herbicide Applications.* Western Society of Weed Sciences 2002 Research Progress Report, Salt Lake City, Utah. March 14–17, 2002.

McDaniel, K., and K. W. Duncan. 2000. Summary paper written for the Governor's Task Force overseeing New Mexico's Phreatophyte Management Program.

McDaniel, K., and J. P. Taylor. 2003. "Saltcedar Recovery After Herbicide-Burn and Mechanical Clearing Practices." *Journal of Range Management* 56:439–45.

McGinley, M., and J. E. Duffy. 2012. "Invasive Species." In *Encyclopedia of Earth*, edited by J. C. Cutler. Washington, D.C.: Environmental Information Coalition, National Council for Science and the Environment.

MDBC. 2004. *Native Fish Strategy for the Murray-Darling Basin 2003–2013.* Canberra: Murray-Darling Basin Commission.

Mecham, D. J. 2015. "The Effects of Channelization and Channel Restoration on Aquatic Habitat and Biota of the Pecos River, New Mexico." Master's thesis, South Dakota State University.

Meehan, M. A., K. K. Sedivec, and E. S. DeKeyser. 2011. "Grazing Riparian Ecosystems: Grazing Intensity." North Dakota State University Extension Service Fact Sheet.

Meitzen, K. M., J. N. Phillips, T. Perkins, A. Manning, and J. P. Julian. 2017. "Catastrophic Flood Disturbance and a Community's Response to Plant Resilience in the Heart of the Texas Hill Country." *Geomorphology* 305:20–32. https://doi.org/10.1016/j.geomorph.2017.09.009.

Mendez-Estrella, R., J. Romo-Leon, A. Castellanos, F. Gandarilla-Aizpuro, and K. Hartfield. 2016. "Analyzing Landscape Trends on Agriculture, Introduced Exotic Grasslands and Riparian Ecosystems in Arid Regions of Mexico." *Remote Sensing* 8:664.

Merritt, D. M., and P. B. Shafroth. 2012. "Edaphic, Salinity, and Stand Structural Trends in Chronosequences of Native and Non-native Dominated Riparian Forests Along the Colorado River, USA." *Biological Invasions* 14:2665–85.

Meyer, J. L., M. J. Sale, P. J. Mulholland, and N. L. Poff. 1999. "Impacts of Climate Change on Aquatic Ecosystem Functioning and Health." *Journal of the American Water Resources Association* 35:1373–86.

Miller, R. R., W. L. Mickley, and S. M. Norris. 2005. *Freshwater Fishes of Mexico.* Chicago: University of Chicago Press.

Minckley, W. L., and J. E. Deacon. 1991. *Battle Against Extinction: Native Fish Management in the American West.* Tucson: University of Arizona Press.

Minckley, W. L., and M. E. Douglas. 1991. "Discovery and Extinction of Western Fishes: A Blink of the Eye in Geologic Time." In *Battle Against Extinction*, edited by W. L. Minckley and J. E. Deacon, 7–17. Tucson: University of Arizona Press.

Minckley, W. L., and P. C. Marsh. 2009. *Inland Fishes of the Greater Southwest: Chronicle of a Vanishing Biota.* Tucson: University of Arizona Press.

Minckley, W. L., P. C. Marsh, J. E. Deacon, T. E. Dowling, P. W. Hedrick, W. J. Matthews, and G. Mueller. 2003. "A Conservation Plan for Native Fishes of the Lower Colorado River." *BioScience* 53:219–34.

Moore, G. W., F. Li, L. Kui, and J. B. West. 2016. "Flood Water Legacy as a Persistent Source for Riparian Vegetation During Prolonged Drought: An Isotopic Study of *Arundo donax* on the Rio Grande." *Ecohydrology* 9:909–17.

Mossop, K. D., M. Adams, P. J. Unmack, K. L. Smith Date, B. B. M. Wong, and D. G. Chapple. 2015. "Dispersal in the Desert: Ephemeral Water Drives Connectivity and Phylogeography of an Arid-Adapted Fish." *Journal of Biogeography* 42:2374–88.

Moyle, P. B., and J. E. Williams. 1990. "Biodiversity Loss in the Temperate Zone: Decline of the Native Fish Fauna of California." *Conservation Biology* 4: 275–84.

Moyle, P. B., R. M. Yoshiyama, and R. A. Knapp. 1996. *Status of Fish and Fisheries*. Sierra Nevada Ecosystem Project: Final Report to Congress. Vol. 2, 953–73. University of California, Davis: Centers for Water and Wildland Resources.

Mueller, G. A., and P. C. Marsh. 2002. *Lost, a Desert River and its Native Fishes: A Historical Perspective of the Lower Colorado River*. (No. USGS/BRD/ITR—2002—0010). Fort Collins, Colo.: Geological Survey.

Nagler, P. L., U. Nguyen, H. L. Bateman, C. J. Jarchow, E. P. Glenn, W. J. Waugh, and C. van Riper III. 2017. "Northern Tamarisk Beetle (*Diorhabda carinulata*) and Tamarisk (*Tamarix* spp): Interactions in the Colorado River Basin." *Restoration Ecology* 26:348–59.

Nagler, P., and E. P. Glenn. 2013a. "*Tamarix*." In Tamarix: *A Case Study of Ecological Change in the American West*, edited by A. A. Sher and M. F. Quigley, chapter 5. New York: Oxford University Press.

Nagler, P., and E. P. Glenn. 2013b. "*Tamarix* and *Diorhabda* Leaf Beetle Interactions: Implications for *Tamarix* Water Use and Riparian Habitat." *Journal of the American Water Resources Association* 49:534–48.

Nagler, P., O. Hinojosa-Huerta, E. Glenn, J. Garcia-Hernandez, R. Romo, C. Curtis, A. Huete, and S. Nelson. 2005. "Regeneration of Native Trees in the Presence of Saltcedar in the Colorado River Delta, Mexico." *Conservation Biology* 19:1842–52.

Nardini, A. 2008. "Key Issues and Challenges in Decision Making Processes to Implement River Restoration." 4th ECRR International Conference on River Restoration, San Servolo, Venice, Italy, June, 16–21.

Neill, W. M. 1988. *Control of Tamarisk at Desert Springs*. N.p.: Desert Protective Council.

Northwest Power Planning Council (NWPPC). 2000. "Columbia River Basin Fish and Wildlife Program: A Multi-species Approach for Decision Making." Northwest Power Planning Council Document 2000–19, Portland, Ore. https://webarchive.library.unt.edu/eot2008/20081207214629/http://www .nwcouncil.org/library/2000/2000-14.htm.

Office of Technology Assessment (OTA). 1993. *Harmful Non-indigenous Species in the United States*. U.S. Congress. OTA-F-565. Washington, D.C.: U.S. Government Printing Office.

Paice, R. L., J. M. Chambers, and B. J. Robson. 2016. "Outcomes of Submerged Macrophyte Restoration in a Shallow Impounded, Eutrophic River." *Hydrobiologia* 778:179–92.

Parks Victoria. 2006. "Two Rivers Management Plan: A Strategy for Sharing Melbourne's Rivers and Bays." Prepared by Robert Itami GeoDimensions Pty Ltd 16 Tullyvallin Crescent Sorrento Victoria 3943 ABN 65 095 849 443. https://cals.arizona.edu/~gimblett/TwoRiversTrafficManagementPlan_20 -11-06_final.pdf.

Pearce, M. C., and D. G. Smith. 2003. "Saltcedar: Distribution, Abundance, and Dispersal Mechanisms, Northern Montana, USA." *Wetlands* 23:215–28.

Perdue, R. E. 1958. "*Arundo donax*: Source of Musical Reeds and Industrial Cellulose." *Economic Botany* 12:368–404.

Pickup, G., and R. F. Warner. 1976. "Effects of Hydrologic Regime on Magnitude and Frequency of Dominant Discharge." *Journal of Hydrology* 29:51–75.

Pimentel, D., L. Lach, R. Zuniga, and D. Morrison. 2005. "Update on the Environmental and Economic Costs Associated with Alien-Invasive Species in the United States." *Ecological Economics* 52:273–88.

Pimentel, D., S. McNair, S. Janecka, J. Wightman, C. Simmonds, C. O'Connell, E. Wong, et al. 2001. "Economic and Environmental Threats of Alien Plant, Animal and Microbe Invasions." *Agriculture, Ecosystems and Environment* 84:1–20.

Pino del Carpio, A., A. Villarroya, A. H. Arino, J. Puig, and R. Miranda. 2011. "Communication Gaps in Knowledge of Freshwater Fish Biodiversity: Implications for the Management and Conservation of Mexican Biosphere Reserves." *Journal of Fish Biology* 79:1563–91.

Platania, S. P., and C. S. Altenbach. 1998. "Reproductive Strategies and Egg Types of Seven Rio Grande Basin Cyprinids." *Copeia* 3:559–69.

Platts, W. S. 1979. "Livestock Grazing and Riparian/Stream Ecosystems." In *Proceedings on Forum-Grazing and Riparian/Stream Ecosystems*, 39–45. Trout Unlimited, Inc. https://catalog.hathitrust.org/Record/007472000/Cite.

Poe, A., and R. Gimblett, eds. 2017. *Sustaining Wildlands: Integrating Science and Community in Prince William Sound*. Tucson: University of Arizona Press.

Pollard, D. A., B. A. Ingram, J. H. Harris, and L. F. Reynolds. 1990. "Threatened Fishes in Australia: An Overview." *Journal of Fish Biology* 37:67–78.

Polunin, O., and A. Huxley. 1987. *Flowers of the Mediterranean*. London: Hogarth Press.

Pope, L., I. Rutherfurd, P. Price, and S. Lovett. 2006. "Controlling Willows Along Australian Rivers." River Management Technical Guideline No. 6, Land & Water Canberra, Australia.

Pratchett, M. S., L. K. Bay, P. C. Gehrke, J. D. Koehn, K. Osborne, R. L. Pressey, H. P. A. Sweatman, and D. Wachenfeld. 2010. "Contribution of Climate Change to Degradation and Loss of Critical Fish Habitats in Australian Marine and Freshwater Environments." *Marine and Freshwater Research* 62 (9): 1062–081.

Presidential Executive Order 13751. 2016. "Safeguarding the Nation from the Impacts of Invasive Species." The White House, Office of the Press Secretary, Washington D.C. https://obamawhitehouse.archives.gov/the-press-office/2016/12/05/executive-order-safeguarding-nation-impacts-invasive-species.

Propst, D. L. 1999. *Threatened and Endangered Fishes of New Mexico*. Technical Report 1. Albuquerque, N.M.: New Mexico Department of Game and Fish.

Propst, D. L., J. E. Williams, K. R. Bestgen, and C. W. Hoagstrom, eds. 2020. *Standing Between Life and Extinction: Ethics and Ecology of Conserving Aquatic Species in the American Southwest*. Chicago: University of Chicago Press.

Quigley, T. M., and S. J. Arbelbide, eds. 1997. *An Assessment of Ecosystem Components in the Interior Columbia Basin and Portions of the Klamath and Great Basins*. General Technical Report PNW-GTR-405. Portland, Ore: U.S. Forest Service, Pacific Northwest Research Station.

Rahel, F. J. 2000. "Homogenization of Fish Faunas Across the United States." *Science* 288:854–56.

Reichenbacher, F. W. 1984. "Ecology and Evaluation of Southwestern Riparian Plant Communities." *Desert Plants* 6:15–22.

Ricciardi, A., and J. B. Rasmussen. 1999. "Extinction Rates of North American Freshwater Fauna." *Conservation Biology* 13:1220–22.

Rieman, B., D. Lee, D. Burns, R. Gresswell, M. Young, R. Stowell, J. Rinne, and P. Howell. 2003. "Status of Native Fishes in the Western United States and Issues for Fire and Fuels Management." *Forest Ecology and Management* 178:197–211.

Rivers of Carbon. 2012. "What Is the Problem with Willows?" https://riversofcarbon.org.au/resources /willows-willow-management/.

RIVRLab. 2012. "Riparian Invasion Research Laboratory." http://rivrlab.msi.ucsb.edu/.

Robertson, L. 1997. "Water Operations on the Pecos River, New Mexico and the Pecos Bluntnose Shiner, a Federally Listed Minnow." In *Competing Interests in Water Resources: Searching for Consensus*, edited by H. W. Greydanus and S. S. Anderson, 407–21. Washington, D.C.: U.S. Committee on Irrigation and Drainage.

Romanov, A. M., J. Hardy, S. C. Zeug, and B. J. Cardinale. 2012. "Abundance, Size Structure, and Growth Rates of Sacramento Pikeminnow *(Ptychocheilus grandis)* Following a Large-Scale Stream Channel Restoration in California." *Journal of Freshwater Ecology* 27:495–505.

Rowland, S. J. 2013. "Hatchery Production for Conservation and Stock Enhancement: The Case of Australian Freshwater Fish." In *Advances in Aquaculture Hatchery Technology*, 557–95. Sawston, UK: Woodhead.

Ruiz-Campos G., F. Camarena-Rosales, A. Varela-Romero, S. Sanchez-Gonzales, J. de la Rosa-Velez. 2003. "Morphometric Variation of Wild Trout Populations from Northwestern Mexico (Pisces: Salmonidae)." *Reviews in Fish Biology and Fisheries* 13 (1): 91–110.

Schumm, S. A. 1984. "Patterns of Alluvial Rivers." *Annual Review of Earth and Planetary Sciences* 13:5–27.

Shafroth, P. B., G. T. Auble, and M. L. Scott. 1994. "Germination and Establishment of the Native Plains Cottonwood (*Populus deltoides* Marshall subsp. *monilifera*) and the Exotic Russian Olive (*Elaeagnus angustifolia* L.)." *Conservation Biology* 9:1169–75.

Shafroth, P. B., J. R. Cleverly, T. L. Dudley, J. P. Taylor, C. Van Riper III, E. P. Weeks, and J. N. Stuart. 2005. "Control of *Tamarix* in the Western United States: Implications for Water Salvage, Wildlife Use, and Riparian Restoration." *Environmental Management* 35:231–46.

Shafroth, P. B., J. M. Friedman, and L. S. Ischinger. 1995. "Effects of Salinity on Establishment of *Populus fremontii* (Cottonwood) and *Tamarix ramosissima* (Saltcedar) in Southwestern United States." *Great Basin Naturalist* 55:58–65.

Sheng, Z., A. Mcdonald, W. Hatler, C. Hart, and J. Villalobos. 2014. *Quantity and Fate of Water Salvage as a Result of Saltcedar Control on the Pecos River in Texas*. College Station: Texas A&M. https://oak trust.library.tamu.edu/handle/1969.1/6077.

Sher, A. A., and D. L. Marshall. 2003. "Seedling Competition Between *Populus deltoides* Native (Salicaceae) and Exotic *Tamarix ramosissima* (Tamaricaceae) Across Water Regimes and Substrate Types." *American Journal of Botany* 90:413–22.

Sher, A. A., D. L. Marshall, and J. P. Taylor. 2002. "Establishment Patterns of Native *Populus* and *Salix* in the Presence of Invasive Nonnative *Tamarix*." *Ecological Applications* 12:760–72.

Sher, A. A., and M. F. Quigley. 2013. Tamarix: *A Case Study of Ecological Change in the American West*. New York: Oxford University Press.

Simon, A., and S. E. Darby. 1997. "Process-Form Interactions in Unstable Sand-Bed River Channels: A Numerical Modeling Approach." *Geomorphology* 21:85–106.

Simpson, R. R., and A. J. Mapleston. 2002. "Movements and Habitat Use by the Endangered Australian Freshwater Mary River Cod, *Maccullochella peelii mariensis*." *Environmental Biology of Fishes* 65:401–10.

Skidmore, P. B. 2017. "Riparian Restoration in the Colorado River Basin." In *Case studies of Riparian and Watershed Restoration in the Southwestern United States: Principles, Challenges, and Successes*, edited by B. E. Ralston and D. A. Sarr, 73–76. U.S. Geological Survey Open-File Report 2017–1091. https://pubs.er.usgs.gov/.

Smithsonian Marine Station at Fort Pierce (SMSFP). 2007. "Non-native and Invasive Species." https://www.sms.si.edu/irlspec/Nonnatives.htm.

Sogge, M. K., E. H. Paxton, and A. A. Tudor. 2006. "Saltcedar and Southwestern Willow Flycatchers: Lessons from Long-Term Studies in Central Arizona." In *Monitoring Science and Technology Symposium: Unifying Knowledge for Sustainability in the Western Hemisphere*, edited by C. Aguirre-Bravo et al., 238–41. Proceedings RMRS-P-42CD. Fort Collins, Colo.: U.S. Department of Agriculture, Forest Service, Rocky Mountain Research Station.

Sogge, M. K., S. J. Sferra, and E. H. Paxton. 2008. "*Tamarix* as Habitat for Birds: Implications for Riparian Restoration in the Southwestern United States." *Restoration Ecology* 16:146–54.

Soto-Galera, E., D. Pardo, E. López López, and J. Lyons. 1999. "Fish as Indicators of Environmental Quality in the Rio Lerma Basin, México." *Aquatic Ecosystem Health Management* 1:267–76.

Stannard, M., D. Ogle, L. Holzworth, J. Scianna, and E. Sunleaf. 2002. *History, Biology, Ecology, Suppression and Revegetation of Russian-Olive Sites* (Elaeagnus angustifolia L.). Plant Materials No. 47, Technical Notes. Boise, Id.: USDA-National Resources Conservation Service.

Stella, J. C., M. K. Hayden, J. J. Battles, H. Piégay, S. Dufour, and A. K. Fremier. 2011. "The Role of Abandoned Channels as Refugia for Sustaining Pioneer Riparian Forest Ecosystems." *Ecosystems* 14:776–90.

Stoleson, S. H., and D. M. Finch. 2001. "Breeding Bird Use of and Nesting Success in Exotic Russian Olive in New Mexico." *Wilson Bulletin* 113:452–55.

Stromberg, J. C., V. B. Beauchamp, M. D. Dixon, S. J. Lite, and C. Paradzick. 2007. "Importance of Low-Flow and High-Flow Characteristics to Restoration of Riparian Vegetation Along Rivers in Arid Southwestern United States." *Freshwater Biology* 52:651–79.

Stromberg, J. C., M. K. Chew, P. L. Nagler, and E. P. Glenn. 2009a. "Changing Perceptions of Change: The Role of Scientists in Tamarix and River Management." *Restoration Ecology* 17 (2): 177–86.

Stromberg, J. C., S. J. Lite, M. D. Dixon, and R. L. Tiller. 2009b. "Riparian Vegetation: Pattern and Process." In *Ecology and Conservation of the San Pedro River*, edited by J. C. Stromberg and B. Tellman. Tucson: University of Arizona Press.

Susskind, L. 1999. "A Short Guide to Consensus Building." In *The Consensus Building Handbook: A Comprehensive Guide to Reaching Agreement*, edited by L. Susskind, S. McKearnan, and J. Thomas-Larmer, 3–57. London: Sage.

Swenson, E. A., and C. L. Mullins. 1985. "Revegetating Riparian Trees in Southwestern Floodplains." In *Riparian Ecosystems and Their Management: Reconciling Conflicting Uses*, edited by R. R. Johnson et al., 135–39. First North American Riparian Conference, April 16–18, 1985, Tucson, Arizona. General Technical Report RM-120. Fort Collins, Colo.: U.S. Department of Agriculture, Forest Service, Rocky Mountain Forest and Range Experiment Station.

Tamarisk Coalition. 2009. *Colorado River Basin Tamarisk and Russian Olive Assessment.* http://www.tamariskcoalition.org/sites/default/files/files/TRO_Assessment_FINAL%2012-09.pdf.

Tamarisk Coalition. 2014. "Suggested Methodologies for Cottonwood Pole Willow Whip and Longstem Plantings." Grand Junction, Colo.

Tamarisk Coalition. 2017. 2017 Tamarisk Coalition Workshop. Proceedings and Notes. Grand Junction, Colo.

Taylor, J. P. 1987. *Imazapyr (Arsenal) Use and Performance on Saltcedar (*Tamarix pentandra*), Willow (*Salix and Baccharis *spp.*) and Phragmites (*Phragmites communis*) at the Bosque del Apache NWR.* Albuquerque, N.M.: U.S. Fish and Wildlife Service.

Taylor, J. P., and K. C. McDaniel. 1998. "Restoration of Saltcedar (*Tamarix* sp.)-Infested Floodplains on the Bosque del Apache National Wildlife Refuge." *Weed Technology* 12:345–52.

Taylor, J. P., and K. C. McDaniel. 2004. "Revegetation Strategies After Saltcedar (*Tamarix* spp.) Control in Headwater, Transitional, and Depositional Watersheds." *Weed Technology* 18:1278–82.

Trombulka, S. C., and C. A. Frissell. 2000. "Review of Ecological Effects of Roads on Terrestrial and Aquatic Communities." *Conservation Biology* 14:18–30.

Trout Cod Recovery Team. 2004. *Recovery Plan for Trout Cod*, Maccullochella macquariensis *(Cuvier, 1829):2005–2009*. Heidelberg, Victoria: Department of Sustainability and Environment.

USACE. 2017. "USACE Regulatory Program and Permits." http://www.usace.army.mil/Missions/Civil -Works/Regulatory-Program-and-Permits/Nationwide-Permits/.

USBR. 2009. *Pecos River Channel Restoration at the Bitter Lake National Wildlife Refuge, Chaves County, New Mexico, Environmental Assessment.* Albuquerque, N.M.: U.S. Bureau of Reclamation, Albuquerque Area Office.

USBR. 2013. *Quality of Water Colorado River Basin.* U.S. Department of Interior, Bureau of Reclamation, Upper Colorado Region, Progress Report No. 24. https://www.usbr.gov/uc/progact/salinity/pdfs /PR24final.pdf.

USDA. 1995. *Decision Notice/Decision Record, FONSI, Ecological Analysis and Appendices for the Inland Fish Strategy (INFISH), Interim Strategies for Managing Fish-Producing Watersheds in Eastern Oregon and Washington, Idaho, Western Montana and Portions of Nevada.* Missoula, Mont.: U.S. Forest Service, Region 1.

USDA. 2010. *Field Guide for Managing Saltcedar.* U.S. Department of Agriculture. U.S. Forest Service. Southwestern Region. TP-R3-16-2. https://www.fs.usda.gov/Internet/FSE_DOCUMENTS/stelprdb 5180537.pdf.

USDA. 2014. *Field Guide for Managing Russian Olive in the Southwest.* https://www.fs.usda.gov/Internet /FSE_DOCUMENTS/stelprdb5410126.pdf.

USDA/USDOI. 1995. *Decision Notice / Decision Record, FONSI, Ecological Analysis and Appendices for the Interim Strategy for Managing Anadromous Fish-Producing Watersheds in Eastern Oregon and Washington, Idaho, and Portions of California (PACFISH).* Portland, Ore: U.S. Forest Service, Region 6.

USDOI. 1994. *Endangered and Threatened Wildlife and Plants: Final Rule to List the Rio Grande Silvery Minnow as an Endangered Species.* Federal Register 59:36988–36995. Washington, D.C.: Office of the Federal Register, National Archives and Records Administration.

USFWS. 1987. *Endangered and Threatened Wildlife and Plants*: Notropis simus pecosensis *(Pecos Bluntnose Shiner).* Final Rule. Federal Register 52:5295–5303. Washington, D.C.: Office of the Federal Register, National Archives and Records Administration.

USFWS. 1992. *Pecos Bluntnose Shiner Recovery Plan.* Albuquerque, N.M.: U.S. Fish and Wildlife Service, Region 2.

USFWS. 2010. *Rio Grande Silvery Minnow (*Hybognathus amarus*) Recovery Plan, First Revision.* Albuquerque, N.M.

USFWS. 2015. *U.S. Fish and Wildlife Service: Strategic Plan for the U.S. Fish and Wildlife Service Fish and Aquatic Conservation Program: FY2016–2020.* https://www.fws.gov/fisheries/pdf_files/FAC_Strategy Plan_2016-2020.pdf.

Valdez-Moreno, M., N. V. Ivanova, M. Elías-Gutiérrez, S. Contreras-Balderas, and P. D. N. Hebert. 2009. "Probing Diversity in Freshwater Fishes from Mexico and Guatemala with DNA Barcodes." *Fish Biology* 74 (2): 377–402.

Vandersande, M. W., E. P. Glenn, and J. L. Walworth. 2001. "Tolerance of Five Riparian Plants from the Lower Colorado River to Salinity, Drought and Inundation." *Journal of Arid Environments* 49:147–59.

Van Haveren, B. P., and W. L. Jackson. 1986. "Concepts in Stream Riparian Rehabilitation." In *Proceedings of Wildlife Management Institute 51st North American Wildlife and Natural Resources Conference*, 1–18. March 21–26, Reno, Nevada.

van Riper, C., K. L. Paxton, C. O'Brien, P. B. Shafroth, and L. J. McGrath. 2008. "Rethinking Avian Response to *Tamarix* on the Lower Colorado River: A Threshold Hypothesis." *Restoration Ecology* 16:155–67.

Van Wilgen, B. W., A. Khan and C. Marrais. 2011. "Changing Perspectives on Managing Biological Invasions: Insights from South Africa and the Working for Water Programme." In *Fifty Years of Invasion Ecology: The Legacy of Charles Elton*, edited by D. M. Richardson, 377–93. Oxford: Wiley-Blackwell.

Wager, R., and P. Jackson. 1993. *The Action Plan for Australian Freshwater Fishes. Environment Australia, Department of Agriculture, Water and the Environment.* https://www.environment.gov.au/about-us /publications.

Walker, K. F., F. Sheldon, and J. T. Packridge. 1995. "A Perspective on Dryland River Ecosystems." *Regulated Rivers: Research and Management* 11:85–104.

Walters, M. A., R. O. Teskey, and T. M. Hinckley. 1980. *Impact of Water Level Changes on Woody Riparian and Wetland Communities.* Vol. 2. Eastern Energy and Land Use Team, National Water Resources Analysis Group, Office of Biological Services. Kearneysville, W.Va.: U.S. Fish and Wildlife Service.

Watson, C. C., D. S. Biedenham, and B. P. Bledsoe. 2002. "Use of Incised Channel Evolution Models in Understanding Rehabilitation Alternatives." *Journal of the American Water Resources Association* 38:151–60.

Weber, M. A., T. Meixner, and J. C. Stromberg. 2016. "Valuing Instream-Related Services of Wastewater." *Ecosystem Services* 21:59–71.

Wedderburn, S., M. P. Hammer, C. M. Bice, L. N. Lloyd, N. S. Whiterod, and B. P. Zampatti. 2017. "Flow Regulation Simplifies a Lowland Fish Assemblage in the Lower River Murray, South Australia." *Transactions of the Royal Society of South Australia* 141:169–92.

Wedderburn, S., N. S. Whiterod, and D. C. Gwinn. 2019. *Determining the Status of Yarra Pygmy Perch in the Murray-Darling Basin.* Report to the Murray-Darling Basin Authority and the Commonwealth Environmental Water Office. Adelaide: University of Adelaide and Aquasave-NGT.

Williams, J. D., J. E. Johnson, D. A. Hendrickson, S. Contreras-Balderas, M. Navarro-Mendoza, D. E. McAllister, and J. E. Deacon. 1989. "Fishes of North American Endangered, Threatened, or of Special Concern." *Bulletin of the American Fisheries Society* 14 (6): 2–20.

Winegar, H. H. 1977. "Camp Creek Channel Fencing—Plant, Wildlife and Soil and Water Response." *Rangeman's Journal* 4:10–12.

Worf, G. L., and J. S. Stewart. 1999. *Russian Olive Disorder: Phomopsis Canker.* Madison: University of Wisconsin System Board of Regents and University of Wisconsin-Extension, Cooperative Extension.

Ye, Q., and M. Hammer. 2009. "Fishes." In *Natural History of the Riverland and Murraylands*, edited by J. T. Jennings, 334–52. Adelaide: Royal Society of South Australia Inc.

York, John. 1990. Personal communication. Soil Conservation Service. Phoenix, Arizona.

Yu, T., F. Qi, J. Si, X. Zhang, C. Zhao. 2017. "*Tamarix ramosissima* Stand Evapotranspiration and Its Association with Hydroclimatic Factors in an Arid Region in Northwest China." *Journal of Arid Environments* 138:18–26.

Zambrano, L., M. Perrow, C. Macias-García, and V. Aguirre-Hidalgo. 1999. "Impact of Introduced Carp *(Cyprinuscarpio)* in Subtropical Shallow Ponds in Central Mexico." *Journal of Aquatic Ecosystem Stress and Recovery* 6:281–88.

Zavaleta, E. 2000. "Valuing Ecosystem Services Lost to Tamarix Invasion in the United States." In *Invasive Species in a Changing World*, edited by H. A. Mooney and R. J. Dobbs, 261–300. Washington, D.C.: Island.

Zavaleta, E., R. J. Hobbs, and H. A. Mooney. 2001. "Viewing Invasive Species Removal in a Whole-Ecosystem Context." *Trends in Ecology and Evolution* 16:454–59.

Zohary, M. 1962. *Plant Life of Palestine*. New York: Ronald Press.

Zohary, M., and A. J. Willis. 1992. *The Vegetation of Egypt*. London: Chapman & Hall.

Zouhar, K. 2005. "*Elaeagnus angustifolia*." In *Fire Effects Information System*. U.S. Department of Agriculture, Forest Service, Rocky Mountain Research Station, Fire Sciences Laboratory (Producer). http://www.fs.fed.us/database/feis/.

# 7

# Monitoring the Results of Stream Corridor Restoration

*Daniel Bunting, Andrew M. Barton, Brooke Bushman, Barry Chernoff,*
*Kelon Crawford, David J. Dean, Eduardo González, Jeanmarie Haney,*
*Osvel Hinojosa-Huerta, Helen M. Poulos, Jeff Renfrow, Holly Richter,*
*Carlos Alberto Sifuentes Lugo, Juliet Stromberg, Dale Turner, Kevin Urbanczyk,*
*and Mark K. Briggs*

## INTRODUCTION

Often overlooked and underfunded, ecological monitoring is an essential component of stream restoration work. It helps practitioners to identify successful restoration practices, detect ineffective ones, and adjust their adaptive-management activities to improve efficacy (Bernhardt and Palmer 2011). Monitoring, along with research and modeling, are the three legs of the scientific stool that support ecosystem restoration and management. Monitoring tells us what is happening, research tells us why and how it is happening, and modeling provides insights about what can happen under different management alternatives.

Specific and applicable monitoring strategies can vary with the type of stream corridor restoration project that is being conducted and the entity that is funding it. Funding entities may require *implementation monitoring* (also referred to as *construction monitoring* [Hutto and Belote 2013]) to assess whether a project is implemented as designed and if maintenance (such as follow-up treatments or repairs) is required (Roni et al. 2002). Implementation monitoring may require regular visits by project personnel to ensure that installed treatments are functioning as intended. Such monitoring is crucial for identifying and executing maintenance and repair needs that can save a project early in its inception and improve the overall odds of success.

Another type of monitoring that restoration practitioners may need to perform is *permit monitoring* (also referred to as *performance monitoring* [The National Academies of Sciences, Engineering, and Medicine 2017]), which is often driven by the need to meet regulatory compliance. Permit monitoring typically uses standardized criteria to assess project performance, often producing a report card that is submitted to the overseeing agency, client, or funder. The USACE, for example, has a compensatory mitigation program that requires permittees to meet performance standards during a five-year monitoring phase before the permittee can receive full credit for a completed mitigation project.

The focus of this chapter is instead on *outcome* or *effectiveness monitoring* (we will use simply "monitoring" from this point forward), which is a question-driven, science-based approach that tracks and documents the success of your stream corridor restoration project over time (Block et al. 2001). For stream restoration projects, the central question that monitoring addresses is, "How successful was the restoration project in meeting its stated objectives?" Monitoring quantifies how well restoration tactics are meeting restoration objectives, providing a solid foundation for you to make management decisions based on results (i.e., adaptive management). Depending on the objectives of a stream restoration project, monitoring can focus on evaluating a variety of factors, including the following:

- *Populations of specific species.* This includes the species central to the objectives of the restoration effort, but potentially other species as well, including rare or invasive species that pose threats to restoration and management goals;
- *Socioeconomic factors.* These are particularly relevant for community-based stream restoration efforts as well as efforts that attempt to quantify value of ecosystem services provided by a restored stream;
- *Long-term trends.* Identifying the long-term trends in biophysical and chemical conditions and processes along the stream being restored (as well as trends related to control sites or streams) is critical to understanding the effectiveness of restoration actions and serves as a basis for adaptive management. This sort of monitoring can include such factors and processes as streamflow, shallow groundwater, soil chemistry, and water quality. These may not be directly associated with the objectives of your stream restoration effort, yet they will have a strong bearing on how your restoration objectives are realized. For example, streamflow and shallow groundwater conditions will be relevant to riparian-revegetation projects, and water-quality conditions will be relevant to efforts to reintroduce native fish.

Unfortunately, the results of many stream restoration projects are not monitored, which is a tragic loss to restoration ecology. Without monitoring, it is impossible to learn from and adapt management activities based on past restoration experiences and increases the likelihood that past mistakes or ineffective tactics will be repeated (Bernhardt et al. 2005; Nilsson et al. 2016). Monitoring is often neglected for several reasons; high on the list is when resources needed for monitoring are not secured or are omitted in the budgeting process (Holmes et al. 2005; Bernhardt and Palmer 2007). Other restoration projects that begin with funding for monitoring end up using those funds to cover cost overruns during the project's implementation phase.

Perhaps more regrettable is when time and money are spent on poorly designed monitoring efforts that do not allow a thoughtful and rigorous evaluation of project results. This is often because collected data are (1) not directly relevant to the goals or objectives of the restoration project (Bernhardt and Palmer 2007), (2) not collected over a sufficient period or in a manner that does not allow a statistically rigorous evaluation of results (Bernhardt and Palmer 2007; Poulos et al. 2014; Poulos and Chernoff 2016; Nilsson et al. 2016), or (3) based too strongly on subjective public opinion or visual appeal (Palmer et al. 2007). An additional challenge, even for well-designed monitoring efforts, is the lack of a strategy for disseminating monitoring results, which can greatly limit contributions to the pool of restoration knowledge.

As part of this introduction on monitoring, we want to stress that monitoring should be initiated prior to the implementation of restoration tactics. The monitoring chapter is placed after chapter 6—implementation of restoration tactics, owing to the emphasis of monitoring on evaluating how well restoration tactics realize progress toward stated restoration goals and objectives. Yet this requires monitoring to be set up and conducted prior to the implementation of restoration tactics to allow a before and after comparison of conditions. We stress this point again as part of the "Ten Considerations for a Robust Monitoring Program" (section immediately below), but want to emphasize this upfront for practitioners who may be skimming through the guidebook chapters as part of developing an initial timeline for rolling out the major actions of their stream restoration program. With the above point in mind, the flow of your stream corridor restoration program would follow figure 1.1, with the adjustment that monitoring protocols would be finalized and implemented prior to the implementation of restoration tactics and then implemented again after to evaluate restoration progress.

Methods for monitoring stream corridor biophysical conditions are well-documented, and there are numerous resources that cover monitoring design, data collection and analysis, and interpretation of results. In this chapter we will not walk you through time-worn step-by-step methodologies but instead will focus on (1) summarizing important criteria and considerations for developing a robust monitoring program, (2) providing numerous references on monitoring methodologies associated with different types of stream corridor restoration efforts (so you have the necessary background to develop your own monitoring program), and (3) presenting case studies that highlight lessons learned from efforts that monitor riparian vegetation, native fish, shallow groundwater, and streamflow.

One quick aside before wading deeper into the monitoring waters. We want to make sure to highlight the importance of maintenance actions. Maintenance actions are usually undertaken at the scale of the on-the-ground restoration actions that have been implemented and typically aim to reduce impacts that could otherwise deleteriously influence the effectiveness of restoration tactics. Examples of maintenance actions include repairing herbivory or stock fences (to prevent wildlife or cattle from impacting a newly planted riparian site), repairing gabions, checking dams and the like following a large storm, keeping invasive species at bay while native species are being established, checking on irrigation (as when supplemental water is being used to establish riparian revegetation), managing recreational use, and maintaining signage.

Two points to keep in mind regarding maintaining your restoration site. First, conducting maintenance actions (planning them and allocating the resources to conduct them) can make the difference between the success and failure of your restoration project. This is particularly true the first few years after restoration tactics have been completed, when newly planted vegetation or recently introduced fish species are most vulnerable and less resilient to a variety of undesirable impacts. For example, in the arid parts of the world where green leafy vegetation is at a premium, a newly planted riparian site can be very tempting to many species. A single opening in an unmaintained stock fence may be all it takes for a persistent cow, goat, or deer to gain access; it is amazing what damage one cow can do. Second, involving local citizens to bolster maintenance efforts and vigilance can improve the trajectory of a project while instilling local buy in. Unless practitioners reside near their target stream, maintenance actions can be difficult to conduct on a routine basis, which makes it even more important to involve local citizens. For further insights on involving the local community in stream restoration, please look to both the community-based sections of chapter 6 and chapter 8.

# TEN CONSIDERATIONS FOR A
# ROBUST MONITORING PROGRAM

As noted above, the central goal of monitoring associated with stream restoration is to provide the data and information to evaluate progress toward realizing restoration objectives. Below, we highlight our top ten considerations for developing a robust monitoring program. Although all ten are relevant regardless of budget, funding availability will play a role in determining how well you will be able to incorporate these considerations as part of your monitoring plan. Some of the considerations (like one, two, and three) should be incorporated no matter the state of your resources, while others may have to wait until more resources become available. Also, keep in mind that the considerations summarized below are not all or nothing. They are inherently flexible. If you have crumbs for a budget, a shallow dive into these considerations may be all you can do. If you have the whole cake, a deeper dive may be possible.

## Consideration One: Develop Clear, Detailed, and Quantifiable Stream Restoration Objectives

Answering the question, "How well did the restoration results meet the project objectives?" requires having stream restoration objectives that are clear, sufficiently detailed, and quantifiable. Sound and realistic restoration objectives for your stream restoration project provide the foundation for robust monitoring. This topic is discussed in chapter 2, but we revisit it here in the context of monitoring. Figure 7.1 provides sample scenarios to depict how and why quantifying objectives is important.

Vague restoration objectives (such as "cottonwoods will be established along Rip Log Creek") cannot be meaningfully quantified nor do they provide a means to describe the success of your stream restoration project. If 1 percent of the planted cottonwoods survive, would that be considered successful? Or, is the threshold of success 50 percent or more? Beyond a simple percentage of survival, what extent and distribution of cottonwoods is expected, and over what time frame? There is simply no way to gauge progress toward stream restoration objectives that are inadequately defined.

In contrast, consider efforts along the RGB to reintroduce the Rio Grande silvery minnow (Case Study 6.9). The USFWS and partners spent considerable time at the onset of the reintroduction program developing sound and quantifiable restoration objectives. Measures of success of this program include establishing fish densities as measured in October (to reflect the status of the minnow population following the critical summer low-flow period). In this case, success was defined as October densities of one or more fish per 100 m$^2$ of surface flow at designated monitoring sites for ten years of a fifteen-year period (approximately a recovered fish population density), with at least one fish per 100 m$^2$ of flow for five or fewer years (approximately a self-sustaining population density) and at least 0.3 fish per 100 m$^2$ for two or fewer years (a population density requiring hatchery management and population augmentation)

### Why quantification of site conditions and restoration objectives and results is so important

**Scenario 1:** Site baseline conditions need to be measured prior to implementing restoration tactics to quantify the effectiveness of different restoration treatments (T1, T2, and T3).

**Scenario 2:** Quantifying conditions at a control site provides the basis for distinguishing the impacts of restoration tactics on site conditions from impacts brought about by natural processes that can affect bottomland conditions throughout a stream system (e.g., distinguishing impacts of drought or a flood on site conditions versus those due to the implementation of restoration tactics).

**Scenario 3:** Quantifying conditions at a reference site will provide a quantifiable target to which outcomes of restoration tactics on site conditions can be compared. Although restoration actions may appear to be effective, identifying a reference site will provide practitioners opportunities to quantifiably evaluate the effectiveness of restoration tactics on a range of success criteria that might include key hydroecological thresholds, native vegetation cover, diversity of key species of wildlife, etc.

**Scenario 4:** The desired future condition of a restoration site will often be described by multiple restoration objectives (e.g., objectives that concern streamflow, channel morphology, riparian vegetation, wildlife use, etc.) that need to be quantified to allow adequate evaluation of progress toward their realization. For example, restoration objectives dealing with streamflow can be put forward as mean streamflow needed during the dry season to maintain populations of threatened native fish species, objectives concerning channel morphology can be described by desired width/depth ratio, riparian vegetation by desired cover of key native plants, etc.. Please see Chapter Two for more information on concepts discussed in all of the above scenarios.

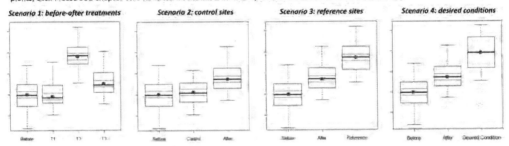

**FIGURE 7.1** Use of box plots to explore sample scenarios. Example scenarios show how a good study design can quantify restoration objectives and measure treatment success. Here, box plots are used to depict how the impacts of restoration actions can be isolated to measure success. Box plots, in general, provide a snapshot of distributed data (i.e., dot = mean with confidence intervals; bold line = median; box = 50 percent data surrounding the mean; whiskers = next 49.3 percent within the whiskers; outliers = remaining 0.7 percent outside of whiskers).

(USFWS 2016). These specific population-density objectives provide the basis for monitoring to quantify progress toward specific recovery objectives in a manner that allows thoughtful future adaptation. For example, if monitoring results in one part of the RGB basin show little promise of meeting objectives, project biologists would have the data and information to focus future reintroduction efforts (as well as possibly complementary on-the-ground restoration efforts) in parts of the watershed where the silvery minnow population is responding more positively.

## Consideration Two: Include Monitoring in the Planning Phase of Your Project, and Set Aside Funds for It

Monitoring cannot be thought of as a dispensable or supplemental element whose inclusion is dependent on what budget remains after restoration tactics are implemented. Rather, it is an essential element for answering the fundamental question about the overall success of the stream-restoration project: How well did the restoration project realize its objectives? Discussing the monitoring needs of your project in its earliest planning phase can help secure a monitoring budget along with other project elements. Up-front discussions with the client, funder, or lead agency during the planning or proposal stage that stress the importance of monitoring will improve opportunities for the inclusion of monitoring in the project's final scope and budget. In many cases, monitoring requires a long-term approach to allow a sufficient period for obtaining reliable results. For example, planted tree cohorts may be evaluated for survival after planting in the short-term (six to twelve months after planting); however, establishment that leads to natural seed production may take longer (three to five years), and other attributes (e.g., the use of the planted vegetation as habitat by key wildlife species, or impacts of restoration on sedimentation or carbon sequestration) longer still to manifest themselves (Bissels et al. 2004; Lennox et al. 2011; González et al. 2017). Likewise, aquatic-ecosystem monitoring, which is often more expensive than terrestrial monitoring, may require several years to evaluate a range of metrics such as water quality, macroinvertebrate assemblages (the organisms of various species that co-occur and interact in the area), and native-fish recovery and reproduction after introduction (Moerke et al. 2004; Smith et al. 2011; Pierce et al. 2015). Therefore, articulating the needs of ecological monitoring to the funder from the planning phase will help solidify project objectives and specify how they will be monitored in the long term.

A frequently cited rule of thumb is that at least 20 percent of the total restoration project budget should be set aside in support of monitoring activities (Greipsson 2012). Given that funding for stream restoration efforts is often geared toward implementation, you may need to push the funding entity to support monitoring actions. The bottom line is that entities who are funding stream restoration projects should be interested in monitoring to provide a cost-benefit justification for the results achieved with their money. Don't take "no" for an answer. Monitoring is beneficial to those who are doing the on-the-ground implementation and funders alike. Also, many funding entities work within a fiscal year budget and have only a specific amount of money available for their priority projects. When monitoring is prohibitively expensive for the funder, a good option may be to include long-term monitoring during the next fiscal year. With some influence and assurance, the funder will understand and justify the need for finding additional money for monitoring, even if it ultimately comes from a different bucket, phase, or fiscal year.

Regardless, if your budget is minimal, it is understood that there will only be so much you will be able to do. But do not push monitoring aside. Monitoring is inherently flexible. When minimal resources are available, monitoring can be more qualitative, possibly based in large part on a descriptive summary and repeat photography. When ample resources are available, data collection can be undertaken to quantify results in a statistically robust way that provides a strong foundation for formulating an adaptive response. At the end of the day, you need information that allows insight into how well you achieved your stream restoration objectives.

## Consideration Three: Initiate Monitoring Prior to Implementing Restoration Tactics

Care should be taken to ensure that the sampling begins prior to the implementation of restoration tactics (Blossey 1999, McDonald et al. 2016). You cannot answer how well your restoration project accomplished its objective(s) without collecting site data that characterize conditions prior to restoration; a starting point is required to allow a comparison between pre- and posttreatment conditions. Whether you are planting obligate riparian trees, reintroducing native fish, trying to reestablish desirable channel morphological conditions, or something else, understanding the ecological conditions prior to initiating restoration actions is essential for evaluating results (Rumm et al. 2016; England and Wilkes 2018).

While often challenging to obtain with available resources, multiple years of preimplementation data will give insight into the natural or seasonal variability in environmental site conditions prior to treatment implementation. Depending on the restoration objective and the parameters being measured (i.e., the performance indicators), this step may be crucial for determining whether your treatments are responsible for the changes observed after restoration is completed. While some metrics may be evaluated in a single field campaign (e.g., the extent and distribution of riparian trees prior to restoration), others may require multiple campaigns (e.g., annual plant diversity, numbers of wildlife) to capture variation between seasons or years (e.g., climatic variability, appropriate baseflows) (Bhattacharjee et al. 2006; Gurnell et al. 2006; Darrah and van Riper 2018). Regardless of the parameters being monitored, multiple years of data collection will provide more statistical power and increase the validity of inferring cause-and-effect relationships between restoration treatments and postproject conditions.

An additional benefit to monitoring site conditions for several years prior to implementing restoration tactics is that such pretreatment data and information can complement the data and information gleaned from assessing the biophysical conditions of your stream corridor and its watershed (see chapter 3). This will strengthen the information base needed to make informed decisions on such important preimplementation tasks as site prioritization (such as the need to change location of a site to an area characterized by biophysical conditions more favorable to realizing restoration objectives), altering restoration objectives (e.g., planting drought-tolerant plants in place of obligate riparian plants when drying trends are indicated), or modifying restoration objectives to be more realistic as a result of better understanding site conditions.

## Consideration Four: Establish Control Sites

To learn as much as we can from management actions, restoration should be viewed as a scientific experiment (Brudvig 2011). Monitoring biophysical conditions at the restoration sites as well as at control sites—sites with similar biophysical conditions to the sites being restored except without restoration intervention—provides a strong basis for understanding the effectiveness of restoration (González et al. 2015; McDonald et al. 2016). A control site serves as a covariate to account for natural variability and, in essence, provides a pathway for you to directly understand the effects of the restoration action or treatment (tree planting, shrub removal, streamflow modification, etc.) from other sources of variation (such as temperature extremes, flood events, or legacy effects of past disturbances) that may influence conditions at both the restoration sites and control sites. Control sites should be established along the same stream as the restoration project or in a nearby tributary of similar biophysical and climatic conditions. The key is to set up control sites in a manner that isolates, as best as possible, the natural variables that may impact both the control and treated site equally. The variables that are monitored between treated sites and control sites, the protocols used to monitor them, and monitoring schedules all need to be identical. The difference between a well-matched control site and the restored site, as it proceeds into the future after restoration, is then the isolated effect of restoration (The National Academies of Sciences, Engineering, and Medicine 2017) (see figure 7.1). If possible, multiple control sites should be used to reflect the inherent variability between different sites as well as to increase statistical power (McDonald et al. 2016).

## Consideration Five: Provide the Environmental Context for Restoration Results

When funds are sufficient, a secondary goal should be to provide practitioners with an environmental context for evaluating stream restoration results. Conditions of streamflow and shallow groundwater, the extent and distribution of invasive species, water quality, soil chemistry, disturbance regimes (particularly flooding), nutrient cycling, and other parameters can have a significant impact on the results of your stream corridor restoration project. Monitoring that focuses only on quantifying changes in restoration targets will not provide practitioners sufficient context for understanding restoration results in a manner conducive for effective adapted management (González et al. 2015).

For example, if the objective of your project is to reestablish native riparian plants, monitoring should consider both the plants themselves and the variables that influence their survival, potentially including monitoring of weather conditions (should be a commonality for all types of stream restoration projects), streamflow, water quality, and depth-to-groundwater, among others. Why did our riparian revegetation project experience low survival rates? Well, there may have been a strong precipitation event and flooding (based on assessment of weather data and streamflow data). And/or maybe there was a drought (weather data and depth to shallow groundwater data). This does not imply that, as a restoration practitioner, you need to take on all monitoring tasks on your own. Other institutions, organizations, or agencies may already be conducting monitoring on climate, streamflow, water quality, or groundwater elevations, for example, in your watershed (see appendices A and B for a list of organizations that collect water and climate data), and practitioners should take advantage of natural resource monitoring that is taking place in their watershed.

## Consideration Six: Monitor Ecosystem Services

Whenever possible, the paradigm shift of restoration ecology: from reestablishing pristine ecosystems to recovering natural processes compatible with socioeconomic interests of humans, must be reflected in the evaluation of restoration outcomes as well (Palmer et al. 2014). Ecosystem services are all environmental benefits provided by biodiversity and ecosystems to humans (Millennium Ecosystem Assessment 2005), and according to the most recent and widely used classification (The Common International Classification of Ecosystem Services, CICES v5.1, Haines-Young and Potschin 2018), includes three categories: provisioning (e.g., food, fresh water, and raw materials), regulating (e.g., climate regulation, pollution remediation), and cultural (e.g., aesthetics, intellectual and spiritual stimulation). Defining the ecosystem services that may be enhanced by the realization of the central restoration goals of your stream restoration project, and monitoring them, is important in this regard. The point here to consider is that, although improving streamflow and the diversity of aquatic habitat along your stream to benefit native fish may be the central goal of your restoration effort (just to offer an example), the realization of that goal may also provide a variety of ecosystem services. Improved habitat for native fish may translate to greater visitation by people or other ecosystem services that can be quantified by assessing willingness to pay (Zhao et al. 2013), hedonic property value analysis (Lewis et al. 2008), cost-benefit analysis (Acuña et al. 2013), or other means.

Monitoring the impact of your stream restoration project on ecosystem services may be pie in the sky if your total monitoring budget is $2,000. And the ecosystem service picture can become complicated very quickly. The increased visitation to your restoration site alluded to in the above example may reach a point where the impacts of increased visitation compromises the hydroecologic restoration improvements made by your restoration project. Regardless, it is important to be informed about such complexities. Also, keep in mind the long-term nature of stream restoration as well as the potential benefits of having data and information that quantify the potential ecosystem service value of your stream restoration project and where those tipping points might be. What may be out of reach today could be within reach tomorrow. One option is to begin with straightforward and cheap strategies for monitoring the ecosystem services provided by your stream restoration project (e.g., questionnaires and/ or tallying visitation to your site by tourists) and expand on them in the future when you have more resources. As food for thought, here are a few resources that may help to get you going down the ecosystem service road: https://www.sustainingwildlands.com/; Bagstad et al. 2012; Peters et al. 2013; Wang et al. 2014; Gimblett et al. 2017.

## Consideration Seven: Involve Citizens

Given appropriate training, citizens can fill important roles in monitoring the results of restored ecosystems (Dickinson et al. 2010; Edwards et al. 2018). As practitioners involved in the initial phases of a stream restoration project go on to other endeavors, local citizens may be the only people who have consistent access to restoration sites and can continue monitoring stream conditions in a reliable fashion. Citizens who become actively involved in monitoring become citizen scientists as they gain awareness, understanding, and proficiency in the restoration project itself, as well as in the scientific concepts and skills required to collect, organize, and analyze monitoring data. The hands-on nature of most stream monitoring efforts provides an educational experience and often exposes people to species, places, and topics they might not otherwise encounter. It is also noteworthy that involving citizens in monitoring can fulfill public outreach and broader impact requirements for many research grants, including those awarded by the National Science Foundation (Silvertown 2009). You can include citizens in your stream restoration effort or monitoring program by making a presentation at a planned community event, talking with community leaders, or reaching out to local nonprofit organizations, universities, and high schools (see also chapter 8) (Meitzen et al. 2018).

## Consideration Eight: Consider the Use of Remote Sensing and Geographical Information System Technologies as Part of the Monitoring Program

Remote sensing is the science of obtaining information about objects or areas from a distance (as contrasted with on-the-ground monitoring procedures), usually from an aircraft (including drone platforms) or satellites. Among many other sources, NOAA provides a good introduc-

tion and overview of remote sensing on their remote sensing webpage: https://oceanservice
.noaa.gov/facts/remotesensing.html. Dramatic advances in remote sensing have made many
applications available to stream restoration practitioners, even those with modest budgets. For
example, for a recent project at Cuatro Ciénegas, Mexico, a drone platform and all complementary hardware and software were purchased for less than US $5,000. We are strong advo-
cates of using remote sensing technology as part of a stream restoration monitoring program.
There is no fork in the road, where practitioners have to select between on-the-ground and
remote sensing methodologies. Rather, a complementary strategy where both approaches are
used can be very effective, providing the means to broaden the spatial and temporal scales of a
monitoring program as well as the diversity of the data that are collected (Dufour et al. 2018).
(See "Use of Remote Sensing and Geographic Information Technologies in Monitoring" later
in this chapter for more on this topic.)

## Consideration Nine: Don't Reinvent the Wheel. Use Field-Tested, Proven Monitoring Methods

Whether your monitoring program is focused on riparian plant communities, birds, fish,
aquatic invertebrates, channel morphology, streamflow, groundwater, other parameters, or
a combination thereof, it is a safe bet that applicable monitoring methodologies have been
developed and are well documented. A wide range of existing and proven monitoring methodologies are available for guiding such activities at your site (see table 7.1). Take advantage of
these and other past work to select and adapt an approach that best aligns with the setting, conditions, and objectives of your stream restoration effort. There is no need to start from scratch.

## Consideration Ten: Know That Help Is Available

Developing a thoughtful, robust monitoring program for your stream restoration project can
be complicated. Unless you are a monitoring expert, a statistician, or just super-smart, you
and the grand majority of the rest of us will need guidance. If you find parts of this chapter
daunting, simply skip to the next section, but do not feel disheartened and close the book. Help
can be secured from many quarters. As noted above, make sure to take advantage of the well-
documented monitoring efforts that have come before you, and seek out monitoring experts
when needed. Discussing your restoration project and monitoring plan with practitioners who
are or have been associated with a restoration project that is similar to yours can be extremely
valuable. Reach out to a statistician at a nearby university whose input can help you immensely
to design your monitoring program or outline the most suitable analytical approach.

# MAIN STEPS FOR DEVELOPING YOUR MONITORING PROGRAM

First let us familiarize ourselves with a few commonly used monitoring terms. A *management
unit* is a geographically bounded area that is identified for practical purposes and targeted for
management. In the context of stream restoration, it is the bottomland environment along a
reach or stream segment where restoration is occurring and/or where control sites are being

**TABLE 7.1** Selected references that detail methodologies for monitoring riparian plants, freshwater fish, aquatic macroinvertebrates, channel morphology, and streamflow

| Selected references | Description or use |
|---|---|
| **Riparian plants** | |
| Coles-Ritchie et al. 2004 | Summarizes and compares several techniques for evaluating riparian vegetation conditions |
| Nagler et al. 2004, 2018 | Uses normalized difference vegetation indices (NDVI) to quantify aerial coverage of selected riparian plants at large scales |
| Palmquist et al. 2018, 2019 | Describes the design and implementation of long-term monitoring protocols |
| Poulos and Barton 2018 | Summarizes step-by-step vegetation-sampling protocols |
| Stromberg et al. 2006 | Uses riparian vegetation indicators to gauge the impact of changes in surface water and groundwater |
| Winward 2000 | Reviews three sampling methods for monitoring vegetation resources in riparian ecosystems |
| https://irma.nps.gov/DataStore/DownloadFile/604463 | Summarizes monitoring protocols for big rivers in park units in the Northern Colorado Plateau Network, how to establish a sentinel site and set up vegetation transects |
| **Freshwater fish** | |
| Gurtin et al. 2003 | Summarizes the classification and description of aquatic habitat, monitoring of fish populations (particularly backwater habitat), sampling, and analytical methods |
| Laporte et al. 2014 | Assesses the impact of key environmental variables (e.g., streamflow velocity) on vulnerable fish |
| Mercado-Silva et al. 2002 | Assesses fish-based index of biotic integrity for streams in central Mexico |
| Moring et al. 2014 | Assesses fish assemblage composition and mesohabitat during different flow regimes |
| Poulos and Chernoff 2016 | Uses mulitivariate time-series methods to assess effects of ecosystem restoration actions on fish community interactions |
| van Zyll de Jong and Cowx 2016 | Assesses salmonid populations and habitat |
| **Benthic invertebrates** | |
| Bug Lab–BLM/USU National Aquatic Monitoring Center https://www.usu.edu/buglab/MonitoringResources/MonitoringProtocols/#item=26 | Provides clear, accurate, and timely information on aquatic resource assessments for resource managers and the public |
| Cuffney et al. 1993 | Provides guidance on site, reach, and habitat selection and methods and equipment for qualitative and quantitative multihabitat sampling |
| Jones 2011 | Summarizes step-by-step field and postfield procedures for sampling invertebrates |

(*continued*)

**TABLE 7.1** (*continued*)

| Selected references | Description or use |
|---|---|
| Leps et al. 2016 | Provides a standardized benthic sampling method for understanding the response of benthic community to stream restoration efforts |
| Haase et al. 2004 | Summarizes standardized and practical protocols to sort and sample macroinvertebrates |
| **Channel morphology** | |
| Chapter 3 of this guidebook | Provides an overview and references on channel morphology and surveying |
| Harrelson et al. 1994 | Guides for on evaluating physical characteristics of a stream (including channel morphology and streamflow) |
| Legleiter et al. 2004 | Uses remote sensing to evaluate channel morphology and in-stream habitat |
| Shroder 2013 | Uses digital photography/imagery, videography, and historical photography to detect and measure geomorphological change |
| **Streamflow** | |
| Chapter 3 of this guidebook | Provides specific guidance accompanied by many references for analyzing water and sediment flow |
| Dobriyal et al. 2017 | Reviews several methods for monitoring streamflow, including methods categorized as timed-volume, velocity-area, formed-constriction, and noncontact |
| Davies et al. 2010 | Assesses river health in the Murray-Darling Basin |
| **Shallow groundwater** | |
| Baxter et al. 2003 | Uses minipiezometers to measure vertical hydraulic gradient, hydraulic conductivity, and specific discharge in gravel and cobble streams |
| **Water biochemistry** | |
| Hotzel and Croome 1999 | Covers all aspects of phytoplankton monitoring |
| Hem 1989 | Although not on monitoring per se, this text is a must-have on the study and interpretation of natural water chemistry |
| Schmitt et al. 2004 | Summarizes monitoring of contaminants and their impacts on fish in large rivers |
| **Other** | |
| Legge et al. 2018 | Reviews monitoring of threatened and endangered species |

established. *Ecological units* are the specific sites—within a management unit—along the stream reach or segment where monitoring will be conducted. Typically, these are sites where restoration actions have taken place or where control sites have been identified. They are often selected based on a mix of environmental, management, and sociopolitical factors, including land ownership, land use and management history, access, vegetation, topography, channel morphology, and soil characteristics. Stratification within ecological units often is necessary to allow comparison between different elements of an ecological unit that are distinctly similar (e.g., riparian plant characteristics on floodplain surfaces in one ecological unit as compared to floodplain riparian plant characteristics at another; pool fish diversity in one ecological unit compared to pool fish diversity in another). Monitoring is then carried out over time within each monitoring unit by sampling points that are randomly or systematically distributed. A *sample point* is a single sample of data collected at a specific location on the ground. Multiple sample points within a monitoring unit type are called *replicates*. Monitoring at each sample point is then implemented using a consistent sampling method (plots, sampling frames, transects, etc.) that is repeated at multiple sample points, or replicates. Please refer to the glossary for other monitoring terms that you may not be familiar with.

The main steps for developing a sound and robust monitoring program are summarized below. To the extent possible, all steps should be addressed before initiating restoration tactics, with the aim of capturing prerestoration conditions as a basis for assessing the effectiveness of restoration tactics in meeting restoration objectives.

## Step 1: Define Monitoring Objectives

Monitoring objectives need to be clearly defined and reflect the objectives of your stream restoration effort. Just as the development of sound and realistic objectives is the foundation for the entire stream restoration program, clearly defined monitoring objectives form the foundation for a strong monitoring program and provide the basis for question-driven monitoring that allow statistical analysis of monitoring results (Lindenmayer and Likens 2010). If the objective of your stream restoration project is to reestablish native, obligate riparian trees, monitoring objectives will need to assess, in one form or another, the establishment and growth of the vegetation that was planted, as well as (potentially) such plant community characteristics as plant density, diversity, and vertical structure complexity, among others. If the restoration objective is to reestablish native fish, the objective of monitoring will focus on population numbers and evidence of fecundity of the desired fish species. In addition, don't forget about including in your monitoring objective the measurement of parameters that will have a strong bearing on restoration outcomes (like monitoring water quality in the context of restoring native fish to your system).

## Step 2: Develop Monitoring Methodologies

This is a big step that covers a broad swath of thematic territory. It includes identifying the parameters that will be measured, the location of monitoring sites, sampling methods, and schedule. Before getting into these "mini-steps" that constitute a stream monitoring program, we encourage you to keep in mind that monitoring methods need to be each of the following:

1. *Realistic*—As you develop monitoring methods, continue to ask yourself if your plan is realistic in terms of cost, personnel, materials, and supplies that are typically available. Putting forward the most robust monitoring design in the world will be useless if it cannot be implemented. Regarding personnel, you will need to have, find, or train your monitoring team to carry out

the monitoring protocols. If your monitoring program centers on assessing changes in chan-nel morphology, you will need personnel who can survey (e.g., trained in the use of real-time kinematic (RTK), total station, and other survey equipment). If your monitoring program is assessing riparian vegetation response, having personnel who can identify plants will be essen-tial, and so on.

2. *Replicable*—Well-designed monitoring programs employ replicable methods and have clear analysis pathways that are tied to the questions of interest. Monitoring protocols should be documented clearly and in sufficient detail to allow others to repeat the steps. Keep in mind that, ideally, monitoring should continue well into the future (possibly over a decade, depend-ing on restoration goals and objectives) and may often require different people to conduct the same measurements many years later. Therefore, emphasis should be placed on clearly doc-umenting protocols to limit subjectivity and allow replicable consistency when training new personnel.

3. *Flexible*—Understanding that the desire is to conduct monitoring well into the future, the over-all monitoring framework needs to be sufficiently flexible to effectively respond and adapt to changes in the availability of resources, natural resource conditions, and management. Try to avoid a monitoring framework that is all or nothing. Acknowledge there may be lean years when carrying out the entire monitoring program may be extremely difficult. For years of funding scarcity, the monitoring design should describe protocols that can be carried out inexpensively (e.g., retaking photo points), and identify which parts of the monitoring design are essential and which could potentially be skipped during years when resources are at a minimum.

4. *Consistent over time*—Monitoring methods should be designed for the long term and should not be significantly altered midstream (Sutherland 2006). Although monitoring needs to be flexible and adapt to project needs, basic monitoring protocol decisions (such as bird point counts, transects counts, plant cover, or plant-density measurements) should remain consis-tent over time.

Six components are often considered essential to monitoring methodologies (Elzinga et al. 1998; Elzinga 2001; Poulos and Barton 2018): (1) identification of the geographic area and sites to be monitored, (2) selection of parameters or indicators to be monitored, (3) selection of the statistical approach, (4) stratification of monitoring sites into ecological units, (5) determina-tion of the number of sampling points and randomization, and (6) finalization of sampling protocols, including frequency and the timing of sampling.

## Step 2A: Identify the Geographic Area and Sites to Be Monitored

For stream restoration projects that are at the reach scale or greater (i.e., not confined to one site), monitoring sites along the stream reach will need to be selected. That is, depending on the scale of your stream restoration effort, resources will often not allow you to monitor the entire stream reach or segment, thus necessitating the selection of specific places along a reach where monitoring will take place or be concentrated. Palmquist and colleagues (2018) provided step-by-step instructions on how to determine an appropriate number of monitoring sites as well as how to randomize their placement along a particular stream reach. Random-izing the location of monitoring sites along a stream is often stratified based on the location and timing of restoration actions, land use and management history, geography, vegetation, and geomorphology characteristics. In addition, there are practical considerations that practitioners will need to pay attention to. For example, for monitoring efforts along the RGB through the remote Boquillas Canyon (see Case Study 7.1), geomorphology characteristics as well as access issues had the strongest bearing on determining the location as well as the number of monitoring sites that could be realistically monitored. For this effort, it was considered important to monitor the riparian plant community's response to invasive plant management on a mix of point bars and longitudinal bars; this meant that a mix of both types of channel morphological surfaces were needed as part of the monitoring design. Conversely, keep in mind that there may be sites along the reach or segment of stream that is being restored that should not be considered for monitoring; these may include archaeological sites, sites where human activities are occurring, sites where different management actions took place or will take place in the future, and sites that are not representative of the reach. In the parlance of monitoring, such sites are called "exclusionary," while the remaining parts of the stream reach where monitoring can be conducted are called "inclusionary." Identifying the parts of your stream reach that will be excluded and then randomizing where monitoring sites will be located along the remainder of the reach is an effective approach.

## Step 2B: Select the Indicators or Parameters to Be Measured

The parameters or indicators that will be measured should relate directly to the objectives of your monitoring program. For stream restoration efforts that focus on riparian plants, common monitoring parameters include percent survival of planted vegetation, plant diversity, composition, percent cover, canopy structure, and density. Depending on the objectives of the restoration project, other important parameters might include stem size, age distribution of tree species, leaf area, vertical structure (cover or leaf area along height strata), abundance within functional types, and plant vigor, health, or condition of particular species. Per Step 2, above, be realistic in selecting the parameters that need to measured. A long laundry list of parameters to be measured will be more costly than a reduced, targeted list. Why measure percent cover of annual grasses if your monitoring objectives focus on trees?

For riparian revegetation projects, permanently tagged or marked trees can be remeasured to obtain survival rates, growth rates, and qualitative measures of vigor (Elzinga 2001; Bunting et al. 2013). For example, in binational riparian revegetation efforts within small tributaries of the RGB along the U.S.-Mexico border (see the riparian revegetation discussion in chapter 6), the near-term monitoring focus is on the vitality (dead or alive) and growth of planted vegetation, while the focus of long-term monitoring is on measuring parameters associated with the riparian plant community, including plant density, diversity, vertical structure complexity, among other parameters.

For stream restoration efforts focused on native fish, parameters might include a variety of population matrices, including the numbers and densities of the target species, overall species diversity, biomass, or age diversity, among others. For efforts along the Rio Grande / Río Bravo to reintroduce the Rio Grande silvery minnow (see Case Study 6.9), the USFWS found it impractical to estimate changes in population numbers of the species as a whole due to numerous challenges associated with sampling: the species is relatively small and lives in flowing water, spawning success is highly variable, there is large spatial and temporal variability of the fish population throughout the river system, and fish losses can be due to a variety of factors that are not well-known and whose impact on the silvery minnow's population can vary with time and space. Due to these challenges, the USFWS opted to monitor silvery minnow densities (the number of fish per area of surface flow sampled) to reflect mortalities for all age classes of the silvery minnow that occurred during the course of the year.

More robust monitoring protocols also measure the physical and chemical parameters that have a strong bearing on the success of restoration (assuming that the restoration objective is biologically grounded). In the case of monitoring the Rio Grande silvery min- now along the Rio Grande Basin, these additional parameters include the measurement of streamflow, physical habitat properties (stream width, depth, habitat type, substrate), and a variety of water-quality parameters, including temperature, dissolved oxygen, and salinity (Edwards and Garrett 2013).

## Step 2C: Select a Statistical Approach

Stream practitioners should consider stream restoration as a science experiment whose cen- tral question focuses on determining whether restoration results occurred by random chance or if the restoration treatment was responsible for the changes that occurred. Adequately answering this question requires an objective statistical approach. Statistics! Just the word may make you reach for the Pepto-Bismol. We introduce this topic with the understanding that many of us do not have the knowledge, background, or desire to develop a strong statis- tical framework for monitoring. Always keep in mind that help may be just a phone call away. Reach out to a local university to see if a statistician would be amenable to work with you on developing tests, methods, and approaches. As Sir Ronald Aylmer Fisher (1938)—a historical genius statistician—once said, "To consult the statistician after an experiment is finished is often merely to ask him to conduct a post mortem examination. He can perhaps say what the experiment died of."

The overview of statistical considerations given below is brief, but technical, and can be jargony. If you find it rough going, please feel free to skip to the next section as long as you have someone on your team who can help. The bottom line: developing a strong statistical frame- work for your monitoring program is essential, but assistance is likely available if statistics is not your cup of tea. If you want to brush up on statistics yourself, here are a few references:

- *A Primer of Ecological Statistics* (second edition) (Gotelli and Ellison 2013)
- *Numerical Ecology* (Legendre and Legendre 2012)
- *Practical Statistics for Field Biology* (Fowler et al. 2013)

Developing a statistically valid monitoring approach requires three important actions: (1) stratifying monitoring sites into ecological units (see B.4 below), (2) determining the num- ber of sampling points needed, and (3) randomizing the placement of sampling points, or plots (see B.5). As noted previously, the inclusion of control sites in your monitoring design will greatly strengthen statistical power.

The types of statistical tests and analyses to be conducted need to be identified in advance of the monitoring program implementation. Our ability to detect statistically significant change is determined by several components, including the random and inherent variability (i.e., natural variability from season to season or reach to reach regardless of the treatment) of the response variable, the scale of treatment, and the number of samples. Statistics speak in numbers with a degree of confidence and, as such, are valuable for communicating restoration results to funders, stakeholders, and policymakers. Numerous statistical tests (such as ANOVA, Chi-square, unpaired/paired t-tests, and linear mixed-effects models) are available for evaluating various components of your restoration project.

One option commonly used in stream restoration to assess the significance of results is hypothesis testing. A hypothesis is created to test whether the results of a selected restoration treatment is statistically significant (i.e., the results are due to something other than chance alone). If so, then an inference or conclusion can be drawn that the changes, positive or negative, are due to the treatment that was implemented. Indeed, creating and testing hypotheses forms the cornerstone of the scientific method, whose four basic steps include (1) making an observation that describes a problem, (2) creating a hypothesis, (3) testing the hypothesis, and (4) drawing conclusions that allow us to adapt restoration tactics to improve success (adaptive management) (see figure 7.2).

For stream restoration objectives that are focused on improving aquatic habitat for native fish, testable hypotheses might pertain to the expansion of backwater habitat as a means to increase the number of fish or minimum streamflow required in the low-flow period of the year. A well-developed hypothesis will further describe the restoration tactics employed to achieve the objectives (e.g., removing a levee to provide backwater habitat for key fish species, establishing environmental reservoir storage to support minimum flow requirements during the low-flow period, or riparian tree planting to provide habitat for key bird species). A great example of hypothesis testing associated with reconnecting old meanders (albeit, one that comes from a nonarid region) is Lorenz et al. 2016, who hypothesized that only a

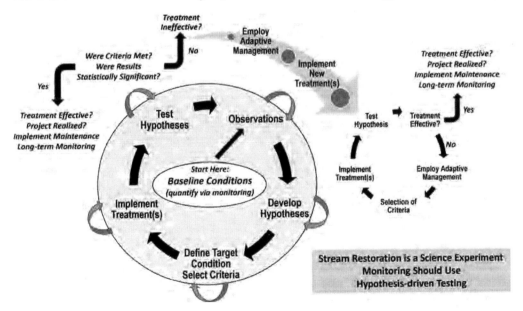

**FIGURE 7.2** A hypothesis-driven approach guides monitoring from design to project completion. Hypotheses are developed and tested, and results quantify levels of success or failure, which are needed for management decisions. Quantified outcomes tell whether a project has been realized or whether adaptive management should be employed to improve the trajectory.

fully reconnected meander would allow recolonization of rheophilic (i.e., living in fast-moving waters) macroinvertebrates.

A reliable statistical framework will also allow practitioners to potentially address deeper cause-and-effect questions. For example, "Is riparian plant diversity in this ecosystem limited by a lack of ecosystem disturbance?" leading, perhaps, to assessing riparian plant diversity before and after flooding.

As you discuss the development of your monitoring design with a statistician, a few statistical terms are important to become familiar with. A "null hypothesis" is the hypothesis that there is no significant difference between specified populations; any observed difference is due to chance or sampling error. In the example mentioned above on connecting old meanders, the null hypothesis would be that connecting old meanders would result in no difference in recolonization of rheophilic macroinvertebrates. "Confidence level" refers to the percentage of all possible samples that can be expected to include the true population parameter. For example, a 95 percent confidence interval means that if the same population is sampled (e.g., density of a key fish species or percent cover of a particular plant species) on numerous occasions, the resulting intervals will represent the true population parameter in 95 percent of the cases.

The Greek letter α, or alpha, is the confidence level and describes the probability of rejecting the null hypothesis when the null hypothesis is actually true (see the discussion of the null hypothesis in Step 2C). In statistical parlance, this is referred to as making a *Type I error* (also known as a "false positive" finding). *Type II errors* also exist, which is failing to reject a false null hypothesis (also known as a "false negative"). Alpha is typically expressed as a proportion via the equation $(1-\alpha)$, where alpha equals .05 if the confidence level is 95 percent $(1-0.95)$, indicating that there is a 5 percent risk of rejecting the null hypothesis when it is true. Another point to keep in mind is that field studies (i.e., monitoring) can warrant a higher alpha value than is typically used in scientific experiments. Changing an alpha value from 0.05 to 0.10 (i.e., a reduction from 95 percent to 90 percent confidence), for example, increases the likelihood of rejecting a null hypothesis and detecting a difference. By lowering Type I error and increasing Type II error, you err on the side of detecting a difference when a difference exists, versus not detecting a difference when one does not exist. This may be reasonable in situations with small sample sizes and when managers need recommendations on understanding if restoration activities are affecting species or processes.

While statistical approaches may seem daunting to some practitioners, the set-up, implementation, and analysis of statistics is often much easier than understanding the inherent jargon. As noted above, we highly recommend involving statisticians and/or natural resource scientists when you first begin to develop your monitoring program to ensure that your design will provide statistically meaningful results. There are also free statistical software resources online that can be taken advantage of, including R Statistical Language: http://www.R-project .org; a t-test calculator: http://www.socscistatistics.com/tests/ttestdependent/Default.aspx; and Invivostat: http://invivostat.co.uk. Considering these and other resources as part of developing your overall monitoring design can make the difference between a well-intended but ineffective monitoring design and one that is statistically powerful and provides meaningful insights to assessing stream restoration results and formulating effective adaptive-management responses.

## Step 2D: Stratify Monitoring Sites into Meaningful Ecological Units

Sampling natural resource conditions as part of a monitoring program is typically done in one of four ways: (1) completely randomized sampling, (2) stratified random sampling, (3) systematic sampling, or (4) subjective sample placement. Please see introductory texts on statistics for detail on each of these sampling designs (e.g., Gotelli and Ellison 2013). Each method has advantages and disadvantages. A stratified random design is often required when conditions (e.g., soils, water availability, vegetation communities) vary across your restoration site, necessitating that the site be stratified into similar ecological units (allowing more of an apple-to-apple comparison). Given the diversity of microhabitats and morphological surfaces in stream ecosystems, some stratification of the monitoring sites into relatively homogeneous ecological units is typically warranted no matter what the monitoring objectives are or the parameters that are being measured. Once the monitoring site has been stratified, sampling plot locations are distributed randomly within each of the stratified management units (see Case Study 7.1, below). This approach ensures that sampling plots are distributed among management units with a statistically rigorous sampling intensity, allowing managers to assess responses to management or environmental pressures at multiple scales, from individual monitoring units to the entire study area. For the monitoring of aquatic habitat, monitoring sites can be stratified into riffle, pool, and run, or in any of a variety of ways depending on objectives (e.g., Orzetti et al. 2010; Ogren and Huckins 2015: Gazendam et al. 2016).

When monitoring riparian plant communities, stratification based on morphological surfaces of the bottomland environment is often an effective approach. The riparian bottomland environment can be divided into such morphological surfaces as active floodplains (surfaces that are inundated every year), secondary floodplains (higher elevated surfaces that are inundated only by higher-magnitude flow events like once every five years or greater), and higher terraces (inundated with extremely low frequency, from every fifty to one hundred years). These distinctions provide a sampling design that allows vegetation plots to be randomly distributed within each surface. This approach allows you to compare the vegetation communities on these surfaces between different sites or grouped together, to characterize how the riparian plant community on various channel surfaces have changed over time (two examples include González et al. 2017 as well as the Riparian Monitoring Case Study along the RGB in Big Bend [Case Study 7.1]).

## Step 2E: Determine the Number of Sampling Points and Randomize Their Placement

One of the key questions that you need to ask yourself as part of developing your monitoring design is, How many plots, transects, or measurement points do I need to adequately assess the significance of change over time? This concept is generally called "sampling depth." If not enough data are collected, your capacity to make strong inferences from the results of data analysis will be limited. There are many ways to determine the adequate sampling depth for a meaningful statistical analysis.

Two approaches that we have used to determine monitoring sampling intensity are highlighted below.

***Approach One: Using Your Own Data to Determine Sampling Depth.*** Let's say you are concerned about an invasive riparian plant species along your stream, and you wonder

**TABLE 7.2** Example of monitoring percent cover for an invasive riparian plant species along four transects over two years

| Transects | Year 1 | Year 2 | Difference |
|---|---|---|---|
| Transect 1 | 17.3 | 19.8 | 2.5 |
| Transect 2 | 19.4 | 23.6 | 4.2 |
| Transect 3 | 18.0 | 20.1 | 2.1 |
| Transect 4 | 21.9 | 23.5 | 1.6 |
| Mean | 19.2 | 21.8 | 2.6 |
| Standard Deviation | 2.0 | 2.1 | 1.1 |

whether it increased from year 1 to year 2. To quantify change, you measure percent cover along four vegetation transects that you have established along a key part of your stream in year 1 and then resampled them in year 2. Table 7.2 shows your data, with means and standard deviations calculated.

A paired t-test is commonly used to assess the significance of change and can be done in Microsoft Excel using the data analysis feature. In this example, we do a paired t-test to determine whether the change in mean percent cover of the invasive plant from year 1 to year 2 was significant, yielding: $t = -4.6$, df = 3, $P < 0.01$. What this means is that there is a statistically significant increase in the invasive plant at a probability level of 1 percent. In other words, the chance that the conclusion that the invasive plant has increased from year 1 to year 2 is incorrect is less than 1 percent. This is an important conclusion as it may spur you and your colleagues to take management action.

But what if your data were different, and you did not find a statistically significant difference? Maybe you simply had not set up enough transects to actually detect a difference. That would be important to know. How can you tell? Fortunately, it is easy to calculate how many transects you will need, and whether or not you actually sampled enough, by using the following formula:

$$((SDdiff)^2 * (Z_{alpha} + Z_{beta})^2) / (\text{Year 1 mean} * \% \text{ detection})^2$$

Using the data above once again, we first make the assumption that we need to be able to detect a 10 percent change. The standard deviation of the difference between year 1 and year 2 (i.e., 1.1) is the key, and we call that $SD_{diff}$. The values for the Z-coefficients are explained in the footnote.* Then, we simply use the year 1 mean of 19.2 and a 10 percent detection level of 0.1 to complete the calculation:

---

* The actual formula is Number Transects Needed = $((SD_{diff})^2 * (Z_{alpha} + Z_{beta})^2) / (\text{Year 1 mean} * \% \text{ detection})^2$

$SD_{diff}$ = Standard deviation of the differences between paired samples (see equation and examples below).

$Z_{alpha}$ = Z-coefficient for the false-change (Type I) error rate. We assume the reasonable rate of 10 percent for concluding that there's a difference when there's not. That number would be 1.64.

$Z_{beta}$ = Z-coefficient for the missed-change (Type II) error rate. We assume the reasonable rate of 10 percent for concluding that there's no difference when there really is one. That number would be 1.28.

Number Transects Needed = $((1.1^2 * (1.64 + 1.28)^2) / (19.2 * 0.1)^2) = 2.8$ transects

This equation tells us that, given the data collected so far, four transects are sufficient to detect a 10 percent difference. This is a general approach to figuring out how many samples you need in order to detect change over time, given your actual data. This approach can also be used for figuring out how many samples are enough for detecting differences between transects, plots, or point samples subject to a management treatment, such as prescribed fire vs. control transects. Similar analyses can be carried out for before vs. after a treatment combined with control vs. treatment, but the calculations are more complicated than those shown. An online paired t-test calculator is available here for this purpose: http://www.socscistatistics .com/tests/ttestdependent/Default.aspx.

*Approach Two: Using a Species-Area Curve.* An alternative approach to determining the number of transects to use in monitoring is to use a species-area curve (often used more when monitoring biology [particularly number of species] versus monitoring of chemical or physical parameters). The basis for this approach is that as you sample more and more area, the number of *new* species encountered declines until nearly all species have been included. Once you have reached that point, it is reasonable to assume that you have adequately sampled the ecological variation in the area in question.

With this as a basis, you can use a simple iterative sampling approach where all constituent species are documented along the first transect, with new species added along the second, continuing with sampling along additional transects until few or no new species are being added with additional transects (i.e., this is known as reaching the asymptote, or the point where the curve levels off, and when you have reached an adequate sampling depth). Hypothetical data are provided in Table 7.3 and graphed in Figure 7.3. Notice that only one new species is added by sampling transect 9 and no new ones are added with transect 10. The safe assumption is that 10 transects are sufficient for capturing most of the ecological variation in this tract of land.

This species-area curve approach, in fact, would be a reasonable approach to use for determining the number of transects to use for any management unit, whether it encompasses a large or small area. If adding more transects does not add to the number of species, you've probably included most of the ecological variation in the cover type. One advantage of this approach compared to that described previously is that adequacy of sampling can be determined within one field season. Carrying out the calculations described previously after two years of monitoring would, then, be an effective check on this simpler method.

## Step 2F: Finalize Sampling Protocols

In addition to the above, monitoring methodologies need to describe the how (sampling methods) and when (timing and frequency of sampling) of monitoring. As already noted, regardless of the specific parameters being measured, it is highly likely that numerous standard monitoring protocols are available and well-documented. For the great majority of stream restoration

---

MDC = Minimum detectable change size. If you want to detect a 10 percent change from the 19.2 percent found in year 1 when you sampled in year 2, then that would be $19.2 * 0.1 = 1.92$.

Note that other values for the two Z coefficients and MDC could be chosen.

TABLE 7.3 Hypothetical species-area data

| Transect | New Species | Cumulative Species |
|---|---|---|
| 1 | 12 | 12 |
| 2 | 8 | 20 |
| 3 | 6 | 26 |
| 4 | 6 | 32 |
| 5 | 4 | 36 |
| 6 | 4 | 40 |
| 7 | 3 | 43 |
| 8 | 2 | 45 |
| 9 | 1 | 46 |
| 10 | 0 | 46 |

*Note*: Plants are sampled over ten transects in a sequential manner, starting with transect 1. New species are recorded for each newly sampled transect, as is the cumulative species number. Area (defined by the number of transects sampled) and total species are graphed in figure 7.3, revealing the species-area curve for this management site.

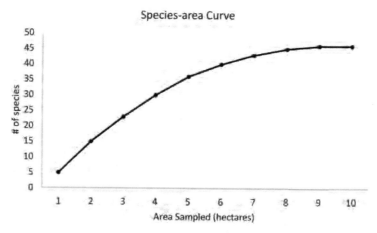

FIGURE 7.3 Example of the species-area curve relationship where the asymptote in the number of species is reached at nine transects (as described in the example in the text). The area sampled increases with the number of transects conducted. Graph by Helen M. Poulos.

projects, the challenge will not be developing monitoring protocols from scratch but selecting the most appropriate methodology from the plethora of options available (see table 7.1).

Regarding the "when" of monitoring, remeasuring the same variables at the same locations over meaningful time intervals is an effective way to quantify change over time. Monitoring methods should also describe the frequency and timing of data collection over the long term. Permanent plots or transects can be revisited seasonally or annually, for instance, to determine how presence, abundance, diversity, and other parameters have changed over time. Designing a sampling plan that sufficiently captures ecological variability, given the limitations of time and resources, is challenging. How much sampling needs to be carried out over time to effectively gauge progress toward restoration objectives? This is the central question. On the one hand, more sampling is always better in terms of the ability to detect change. On the other, monitoring more units or at greater frequencies than needed wastes time and money (Morandi et al. 2014).

Setting regular time intervals (e.g., quarterly or annually) for collecting data may be the most appropriate option for some restoration objectives. For restoration objectives that are biologically orientated, data collection may need to be correlated with a specific season that, once identified and monitoring has begun, should not vary temporally to minimize biases in data collection due to seasonal variation in the abundance of organisms. For example, measurements of vegetation cover percentage, if conducted in both dry and wet seasons or in both spring and fall, will likely give results driven more by seasonal differences in plant cover than differences due to restoration actions.

For other restoration projects, monitoring that follows unpredictable events such as floods or fire may be appropriate because biophysical changes to the restored site are likely to occur from them (Densmore and Karle 2009). For stream restoration projects that have a central objective of altering the channel morphology, monitoring will probably reveal little change during years of drought (when flow is marginal and flooding nonexistent). However, for projects with such restoration objectives, it may be effective to define flow thresholds that would trigger monitoring (for example, a magnitude or duration of flow that practitioners believe would foster changes in channel morphology).

One final thought on monitoring protocols, once you get out in the field and begin collecting data, make sure to record the positions of permanent plots or transects with the most precise geographical positioning system (GPS) equipment available. Also, mark the points with iron bars or any kind of mark that can be found (sometimes years later) even if conditions can dramatically change. It is astonishing how much site conditions can change and how difficult it can be to find a plot or transect even with excellent GPS equipment. Having a metal detector in the field helps in situations where metal pins, rebar, and the like are used and may become buried.

## Step 3: Finalize a Data-Management Approach

One of the great challenges for long-term monitoring efforts is for collected data to be documented in a manner that allows end users to assess its quality and usability (Lovett et al. 2007, McEachern and Sutter 2010). Ensuring that collected data and the information the data provide are available for future data collection and analysis requires detailed and comprehensive management procedures (Fancy and Bennetts 2012). To help address this important need, a data-management section should be included in the monitoring design to provide guidance that will ensure that collected data are complete, of desired quality, available for analysis and sharing, and archived for future use. Key topics that need to be addressed under the banner of data management include roles and responsibilities, data completeness and quality, development of metadata, archiving data, and evaluating existing data from other sources (Sutter et al. 2015). Whether you record data by hand on datasheets and enter them into digital spreadsheets or analysis software upon returning from the field, or you collect data digitally in the field (by a GPS unit or tablet application, for example) or by a data acquisition system (such as a field-based datalogger), the final organization of all data must be quality-assured (how fit the data are to meet their purpose) and quality-controlled (how the quality of data are maintained over time). This also will allow you to determine whether data are missing and what key gaps need to be addressed.

Key considerations for effectively managing datasets associated with monitoring projects are well-documented (Reichman et al. 2011; Fancy and Bennetts 2012; Sergeant et al. 2012; Ruegg et al. 2014; Sutter et al. 2015; Poulos and Barton 2018). As practitioners develop their own strategies for organizing, storing, and managing monitoring data, a key consideration is to develop a graded approach that matches the scale of data management to the needs of the organization and the complexity of the project (Sutter et al. 2015).

## Step 4: Develop a Data Analysis Strategy

Identifying the methods that will be used to analyze the data that have been collected as part of your monitoring program is an essential component of the monitoring design and needs to be done up front (i.e., before sampling is initiated). That stated, we acknowledge that a discussion of analytical techniques can elicit different reactions depending on the audience, ranging from glee and excitement (maybe more from the researcher/statistician camp) to a fight-or-flight response (perhaps the practitioner camp). If you are in the latter group, we encourage a superficial read if only to familiarize yourself with some of the analytical strategies that are available. We are not trying to turn readers who do not have the background or desire into statisticians. Remember that help is available! In addition, a discussion of analytical methods comes with many terms specific to this topic. Please see the glossary for definitions.

Statistical analyses can be conducted on *univariate* data (such as the average density of a planted species) or *multivariate* data (with dependent variables, such as a matrix of species densities; see Roni et al. 2002 for more detail). In a recent review of 169 articles assessing the success of restored riparian vegetation, González et al. (2015) found that experimental designs were generally poor, and the analytical techniques used were rather simplistic. Thirty studies did not implement any statistical analysis. Of those that did, univariate analyses were the most common (133 out of the 169 articles); multivariate analyses were used less frequently (46 out of 169 articles). No matter which analytical technique is used, the steps outlined above (defining objectives, developing methodologies, and managing data) must be completed first to produce clean datasets that can be used in data analysis. Situations with before-after-control-impact (BACI) type designs (Stewart-Oaten et al. 1986), for instance, can employ repeated measures, linear mixed-effects models (Pinheiro and Bates 2000), or blocked Analysis of Variance (ANOVA), to name a few methods (Morandi et al. 2014; McDonald et al. 2016).

When comparing before and after data collected for the same site, a paired t-test can be employed, which tests for a difference between each set of pairs. For example, you could compare species abundance in sites before and following restoration activity. In situations where several years of data are collected, it may be more appropriate to use a repeated measures analysis. Repeated measures, used when data are collected at the same sites over time, can help account for variation caused by the passage of time or year-to-year variation (see any of the statistical references that we cited earlier).

A *blocked design* can be used when sections of stream reaches (or blocks, in the case of riparian research) are established, containing areas with a control site and one or more experimental sites (Bateman et al. 2008; Cooper and Anderson 2012; Schlatter et al. 2017). For example, Bateman et al. (2008) established four blocks along the Middle Rio Grande in New Mexico to monitor reptiles and amphibians in sites where nonnative plants were removed while other sites were assigned to sections that had been treated with prescribed fire or revegetated with native shrubs. Each block contained one control and three experimental sites with varying degrees of nonnative plant removal and replanting. Examples of block design used to test the impact of managed flow on regeneration of riparian trees include Cooper and Andersen 2012 and Zamora-Arroyo 2017.

Linear *mixed-effects models* provide a more comprehensive and flexible framework for examining the temporal effects of stream restoration activities on vegetation, fishes, and invertebrate taxa over multiple time-steps (more than two years). They can also be used to account for designs with different and nested spatial levels. The giant cane removal case study (Case Study 7.1) reviews a statistical analytical approach to evaluate changes in the cover of individual plant taxa and plant types (grasses, forbs, and woody plants) over time with respect to repeated management activities.

*Multivariate analysis* is another useful tool for detecting trends in the structure of datasets, testing for differences in multivariate attributes of groups of variables and for classification (Legendre and Legendre 2012). Such analyses are appropriate when a researcher wants to compare several dependent variables. Analyses such as principal component analysis (PCA), nonmetric multidimensional scaling (nMDS), or factor analysis can be used to reduce the number of variables used. Simplifying the model to have fewer variables is often preferable.

Transient taxa and differential temporal use by organisms are key characteristics of ecosystems (e.g., the arrival of golden perch [*Macquaria ambigua*] in tributaries of the Murray River [Koster et al. 2014]), just as fluctuations in abiotic components (such as rain and temperature) also vary within those same systems. Quantifying these fluctuations and their interactions provides a framework for assessing the effectiveness and impacts of river-management activities on fundamental ecosystem characteristics.

This can be done with new temporal multivariate modeling, which allows environmental managers to establish success metrics for evaluating restoration outcomes in the context of species interactions (Hampton et al. 2013; Poulos and Chernoff 2016). Temporal shifts in species' interactions can provide insights into ecosystem responses to environmental perturbations, such as dam removal. Temporal fluctuations in ecosystem components (such as the types and abundances of organisms) are integral parts of river system dynamics, and such variations may shift one or more components of an ecosystem without compromising the resilience in the ecosystem. For example, in rocky stream communities, there is usually a synchronous hatching of important aquatic invertebrates, such as a species of mayfly; the temporary absence of that species of mayfly beforehand from that aquatic community does not change the fundamental characteristics or biotic interactions within the river reach.

Similarity percentage (SIMPER [Hammer et al. 2001]) and analysis of similarity (ANOSIM [Clarke 1993]) can also be used to identify differences in a biological community over time or by treatment type following a stream restoration project. Such methods compute the percentage contribution of each species to the dissimilarities between all pairs of sampling units (i.e., sampling intervals, treatments, etc.). Poulos et al. (2014) provide an example of how such analyses can be employed to evaluate the effects of dam removal on fish assemblages.

**TABLE 7.4** A summary of selected methods and instruments for measuring riparian vegetation

| Selected methods and instruments for monitoring riparian vegetation — Methods | Descriptions and uses | Plant presence, abundance, composition, richness, diversity | Vegetation and ground cover | Plant, tree, or stem density | Canopy cover, foliar density, or volume | Tree diameters (diameter breast height and basal diameter) | Plant height, vegetation structure | Plant production, biomass, vigor |
|---|---|---|---|---|---|---|---|---|
| Point intercept, transects | Simple line transect used to measure all plant cover, typically by species, including overlapping vegetation | • | • | | | | | |
| Line-point intercept, transects (Herrick et al. 2005) | Simple line transect using a single point to document overstory, understory, and ground cover at defined intervals (e.g., every 1 m) | • | • | | | | | |
| Step-point, transects (Evans and Love 1957; BLM 1996) | Rapid field technique using a defined interval (e.g., 5 steps) to survey vegetation at the toe of a boot, or using a rod attached to a frame, generally fitted with fixed nails to determine cover at defined points | • | • | • | | | | |
| Quadrats, sampling frames (Braun-Blanquet 1932, BLM 1996; Coulloudon et al. 1999) | Provides a subsample area with known dimensions, can be gridded or delineated | • | • | • | | | | |

| Method | Description | | | | | |
|---|---|---|---|---|---|---|
| Relevés (Braun-Blanquet 1932, MDNR 2013); Daubenmire frames | Provides a small survey or sampling area (e.g., 1 × 1 m or 50 × 20 cm Daubenmire) to obtain cover, typically using the median within a range of cover estimates (e.g., 25–50% = 37.5%) | | | | • | • | • |
| Nested plots (Barnett and Stohlgren 2003) | Smaller sampling areas within the larger sampling area provide different zones to measure different plant forms (e.g., herbs vs. trees) | • | • | • | • | • | • |
| Belt transects (Hill et al. 2005) | Large, linear plots with defined intervals (e.g., 100 × 10 m with 10 m intervals) to provide replicates or analyze metrics extending away from a target area, such as a stream | | • | | | | • |
| Point-centered quarter (Cottam and Curtis 1956; Mueller-Dombois and Ellenberg 2002) | Plotless technique to estimate density using a quadrant extending from a sample point, the distance is measured to each plant closest to the center point within each quadrant | | • | | • | | |

(continued)

TABLE 7.4 (*continued*)

## Selected methods and instruments for monitoring riparian vegetation

| Methods | Descriptions and uses | Plant presence, abundance, composition, richness, diversity | Vegetation and ground cover | Plant, tree, or stem density | Canopy cover, foliar density, or volume | Tree diameters (diameter breast height and basal diameter) | Plant height, vegetation structure | Plant production, biomass, vigor |
|---|---|---|---|---|---|---|---|---|
| Vertical line intercept | Documents cover foliar density or volume in the vertical plane (e.g., live foliage observed within each decimeter [10 cm] extending into a tree canopy and within an imaginary distance from a stadia rod [e.g., 1 dc]) | • | | | • | | • | |
| Harvest, wet/dry sampling (BLM 1996) | Sampling live vegetation by harvesting either whole plants or live tissues; generally involves weighing live 'green' biomass in situ, drying, and then reweighing dry biomass | | | | | | | • |
| **Instruments** | **Definitions, descriptions, examples** | | | | | | | |
| Lidar (Farid et al. 1996) | Light detection And ranging, ground And aerial (fixed-wing aircraft or helicopter) | • | • | | | | • | |

| Method | Description | | | | | |
|---|---|---|---|---|---|---|
| Satellite imagery, aerial photography | For example, Landsat, MODIS, Google Earth; use of infrared spectrum for vegetation indices | • | • | | | • |
| Spherical densiometer (Strickler 1959) | Convex, concave, fish-eye lens | | | • | | • |
| Photography, ground | repeat RGB, phenology, hemispherical photos for canopy analysis | • | | • | | • |
| DBH tape | Diameter at breast height (i.e., 1.3 m above ground surface) or basal diameters, also circumferences | | | | • | |
| Biltmore stick | Diameter at breast height and with hypsometer to estimate tree height | | • | • | | |
| Clinometer, compass | Obtain geometry (e.g., angles) to compute tree height | | • | | | |
| Rangefinder | Uses infrared pulse laser to quickly obtain distances from fixed point | | • | | | |

341

# CASE STUDY 7.1
Monitoring the Effectiveness of Managing Giant Cane Along the Rio Grande/ Río Bravo in Big Bend

Helen M. Poulos, Joe Sirotnak, Jeff Renfrow, Javier Ochoa-Espinoza, and Mark K. Briggs

## Background
This case study builds upon the Case Study 6.5 regarding the management of giant cane (*Arundo donax*) along the binational reach of the Rio Grande/Bravo River (RBG) in Big Bend (see maps in Case Studies 6.5 and 7.8). Over the last fifteen years along the binational reach of the RGB in Big Bend, two exotic invasive riparian plant species—salt cedar (*Tamarix* spp.) and giant cane (*Arundo donax*)—have been the focus of adaptive vegetation management along the river by federal and state agencies on both sides of the U.S.-Mexico border. Between 2004 and 2009, salt cedar was a management target, given the widespread dominance of this invasive species along river channel margins. The combined impact of intensive treatment by hand (manual cutting and spot treatment with herbicide) of tens of thousands of salt cedar stems, the release of the salt cedar leaf beetle in 2008, and a high magnitude reset scouring flood event in late 2008 has greatly reduced the extent and distribution of salt cedar along the Big Bend reach.

However, as the extent and distribution of salt cedar decreased, the abundance of giant cane surged (though no correlation between the decrease in salt cedar and decline of giant cane has been proven). Several aspects of the dramatic and rapid increase in giant cane along the Big Bend reach of the RGB instigated a shift from managing salt cedar during the 2004–9 time period, to managing giant cane beginning in 2010 to today (these aspects are reviewed in Case Study 6.5).

In this case study, we provide an example of how riparian vegetation monitoring data collected during a fourteen-year period using two different monitoring protocols—one centered on monitoring salt cedar and an updated monitoring protocol focused on monitoring the response to managing giant cane—can be analyzed to provide useful guidance both for informing past management effectiveness and for guiding future management actions. Changing data collection protocols midstream is far from ideal and, indeed, flies in the face of fundamental recommendations on monitoring made in this very book that emphasizes the importance of not changing protocols once a monitoring plan is in place. That stated, changing monitoring protocols in response to shifting management objectives is an all-too-common real-world occurrence, and the adaptive management process itself requires that managers continually respond to the rapidly changing biological community. In this situation, when management priorities shifted to controlling giant cane, monitoring objectives and data collection protocols also changed to accommodate the physiological differences, establishment patterns, and response to treatment between the two invasive species.

## Monitoring Objectives and Methods
The data collection protocols to assess both salt cedar management (2000–2009) and giant cane management emphasize plot- and transect-based plant species cover to evaluate how repeated management affected the prevalence of these exotic species as well as the woody and herbaceous taxa response following treatment at local (i.e., site) and landscape (i.e., entire reach) scales. The main difference between the two data collection techniques was the number and size of the plots used and the manner that they were located in the RGB's active bottomland environment. Beginning in 2004, monitoring the results of salt cedar management centered on surveying small numbers of permanently

Personnel on the Big Bend binational team survey vegetation along the RGB using randomly positioned 1 × 1 m plots (PVC square). The position of all plots is entered using an RTK and/or total station (left). Vegetation conditions are typically not as open as indicated in the left photograph. More typical (at least for our work along the RGB) is what is depicted on the right. Using higher numbers of small monitoring plots (versus small numbers of larger plots) has not only statistical advantages, but also practical ones, particularly regarding placement in the field and measurement. Photos by Mark Briggs.

established long, narrow plots (2 × 20 m) that spread across the slope contour of active floodplain surfaces (emphasis was on remeasuring the same plot [same piece of ground] during each monitoring campaign), and line-intercept methods were used to estimate plant cover. Although these plots were located only on active channel surfaces (surfaces inundated at least once every two years), their long length meant that they covered a variety of aggrading and degrading surfaces within the active floodplain environment (e.g., aggrading and degrading floodplain surfaces and aggrading and degrading low terrace surfaces).

When management attention shifted from salt cedar to giant cane in 2010, three conclusions were made regarding data collection protocols. First, a greater number of smaller plots were needed to quantify the effectiveness of treatment of giant cane, in particular to answer questions related both to the effectiveness of treatment (e.g., Did the combination of prescribed burning followed four to six weeks later by herbicide application produce high giant cane mortality?) and the overall response of the riparian vegetation community to treatment (e.g., How quickly did giant cane reestablish following treatment?). Second, given the establishment of giant cane throughout the river's bottomland environment (on primary and secondary floodplain surfaces) as well as quasi-abandoned floodplain surfaces ("abandoned" as a result of aggradation processes that increased the elevation of these surfaces well above that of surfaces that are routinely inundated), relating vegetative response following treatment to specific morphological surface was added as a monitoring objective to better understand how vegetation response following treatment differed by morphological surface. To address this objective, each monitoring site was stratified based on the channel morphological surfaces encountered (from primary floodplain to secondary to quasi-abandoned surfaces). Transects perpendicular to the river extending from quasi-abandoned surfaces to primary floodplain surfaces were installed with vegetation plots randomly placed along the transect by morphological surface. Third, based on our experience monitoring salt cedar, we concluded that relocating and measuring the same plot every year is prohibitively challenging owing to the dramatic and rapid aggregational and degradational processes that were occurring along the river. The revised vegetation monitoring protocols follow so-called Big River Protocols developed by the National Park Service NPS (https://irma.nps.gov/Data Store/DownloadFile/604465).

*(continued)*

Percent cover using the 2 × 20 m plots was based on line-intercept (using the longitudinal leeward side of the plot as the transect), while percent cover using 1 × 1 m plots was estimated visually after calibrating with cardboard cutouts that reflect different percent plant cover.

Both data collection procedures estimated cover for both woody and herbaceous (i.e., graminoids and forbs) taxa. Individual plants were identified to the species level, and they were also grouped by life-form (woody, graminoid, or forb). Five monitoring sites were established along a 32 km reach of the RGB in Boquillas Canyon to evaluate the effectiveness of salt cedar treatment, and five monitoring sites were established along a 35 km (22 mi) reach of the RGB upstream of Boquillas Canyon and two monitoring sites along a 10 km (6.2 mi) stretch of the RGB in Black Gap. Monitoring sites were located in areas where invasive-plant-management actions were conducted and in areas that were accessible and had appropriate premanagement conditions of channel morphology (a mix of channel morphological surfaces) and riparian vegetation (high densities of salt cedar or giant cane prior to treatment).

The central objectives of data analysis are to quantify (1) changes in the vegetation complex over the entire monitoring and management time-series, (2) the effectiveness of salt cedar and giant cane eradication efforts as two separate management targets, and (3) the response of the riparian vegetation community by channel morphological surface. We used paired t-tests and nonmetric multidimensional scaling to evaluate temporal changes in the plant community cover in response to giant cane removal.

### Results

Giant cane management at the two treated sites resulted in significant decreases in *A. donax* cover over the pretreatment to posttreatment interval according to the paired t-test ($P = 0.00001$; t stat = $-4.85$). Total woody plant cover increased ($P = 0.0005$; t stat = $-3.48$), graminoid cover decreased ($P = 0.00001$; t stat = $-3.85$), and forb cover increased ($P = 0.00001$; t-stat = 7.13) significantly in response to the treatments relative to the untreated sites.

The riparian vegetation community experienced significant shifts in species composition over the time-series, according to the community-scale nMDS results, and especially in the years following *A. donax* removal. The lack of overlap in the 95 percent confidence ellipses of the ordination space in the years 2014 and 2015 of the nMDS highlight this trend. Changes were most notable in the latter years of the study, demonstrating the importance of continued monitoring at the site, even in the years following the river restoration treatments. This overall shift in species composition resulted from changes in plant species dominance over the time-series, which was characterized by a shift away from *A. donax* and other nonnative grass cover toward a posttreatment species matrix comprising native shrub taxa including seepwillow and desert willow (i.e. *Baccharis neglecta*, *Baccharis salicifolia*, *Bahia absinthifolia*, and *Chilopsis linearis*) (see figure on the next page of this case study).

The analysis of the vegetation data collected as part of this monitoring effort indicate that (1) the use of prescribed fire and herbicide treatment dramatically reduces the extent and distribution of giant cane after one to two treatments, (2) giant cane does not quickly reestablish following treatment, (3) such single-species management actions can promote rapid reestablishment of other riparian plants that result in cascading community- and landscape-scale effects on the greater Boquillas Canyon riparian plant community.

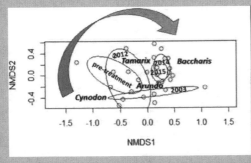

Nonmetric multidimensional scaling of plots in species space over the study year plotted with 95 percent confidence ellipses for each year of the study. The arrow depicts the overall temporal trend in the data by demonstrating that management of *Arundo* and *Tamarix* have been successful in restoring a mix of native taxa to the riparian system. In particular, *Baccharis* spp.—a genus of native early successional woody plants (photo on right)—have reestablished and regained dominance in plant cover over the latter years of the study after repeated treatment of *Arundo* and *Tamarix*. Figure by Helen M. Poulos.

## Lessons Learned

Key lessons learned from monitoring the riparian vegetation response to managing invasive species along the RGB, include the following:

- Collecting pretreatment data is essential for evaluating the effectiveness of management and for developing an adaptive management response.
- Long-term and continued vegetation monitoring is paramount, especially to understanding the long-term consequences and sustainability of managing invasive species (i.e., to address the key question, Will the extent and distribution of target invasive species rapidly reestablish to premanagement levels?).
- Univariate and multivariate statistical analyses provide robust analytical tools for evaluating shifts in the plant community in response to management actions. Practitioners can take advantage of freely available open-source software that provides a wide range of analytical tools to managers at no cost (e.g., R is a freely available language and environment for statistical computing and graphics: https://cran.r-project.org/). Code and sample datasets are included in the supplementary materials so that managers can access these resources for adaptation to their own datasets.
- Data sets that were collected using different methods and sampling intensities can still be used for post hoc time-series analysis, provided that the appropriate statistical analysis is used.
- Attempts to use strict, monumented (staked) repeatedly measured plots or transects is difficult in dynamic fluvial riparian systems. The most time-consuming task in attempting to repeat-measure specific transects was often finding buried or scoured stakes. Often, stakes were never relocated. Therefore, in rapidly changing fluvial systems, randomly locating plots or transects within a defined area as part of each monitoring campaign is a more time-efficient and overall effective approach.
- Keep data collection methods simple and goals attainable with small crews. Simple methodologies that can be done quickly, inexpensively, and without specialized equipment are more likely to yield useful long-term data sets.

Statistical software can run into the hundreds of dollars and may require annual license updates. Analytical software is open source and available for free download. The open-source statistical program R Foundation for Statistical Computing (R) will run on a variety of platforms and can be downloaded online (www.r-project.org), but requires that users have familiarity with writing codes. Past 3.x (https://folk.uio.no/ohammer/past/) is another freely available multivariate statistical package for evaluating temporal shifts in community composition. Specific to wildlife studies, Program MARK (http://www.phidot.org/software/mark /docs/book/) and Presence (www.mbr-pwrc.usgs.gov/software/presence.html) are freeware to analyze capture-mark-recapture data and are used by fishery biologists, herpetologists, ornithologists, and mammologists. Other open source software programs can be found by visiting www.statsci.org/free.html.

## Step 5: Identify Pathways for Sharing and Disseminating Monitoring Results

Finding appropriate avenues to disseminate monitoring information is as important as collecting and analyzing the data. Many monitoring projects fail to analyze and report interim data or broadly distribute data when the benchmark analysis has been completed. Systematic reporting is necessary to provide periodic updates to interested parties and promote peer review. By presenting monitoring information at annual meetings and conferences attended by agency managers and the public, you can outline the importance of monitoring and ensure future continuity.

Monitoring data should be routinely archived for future analysis and for assessments that are developed into reports. While some datasets and results will be proprietary, practitioners and managers must understand the importance of sharing data with the broader restoration community. As the relevance and importance of stream restoration continues to grow, we must keep sight of the bigger picture: connecting small- and large-scale projects to benefit the regional landscape. Technology is available to efficiently and creatively disseminate data both widely and transparently. Online platforms such as Dryad (www.datadryad.org) have made scientific data readily accessible and freely reusable. Federal institutions such as the U.S. Geological Survey now upload their data in public repositories such as ScienceBase (www .sciencebase.gov).

## MONITORING RESULTS OF STREAM RESTORATION: EXAMPLES FROM THE FIELD

Case studies presented here include examples of monitoring of riparian ecosystems, birds, fish, shallow groundwater, and streamflow. The focus is on the process that practitioners used to develop their monitoring programs and some of the key lessons learned in designing and implementing them. To the extent possible, references are provided for you to access more details on the sampling and analytical protocols that were used.

### Monitoring Riparian Plant Communities

Methods for monitoring riparian vegetation resources are well-documented and come in a variety of formats that involve sampling a variety of parameters (see table 7.4 and Case Study 7.1).

*Plant cover* can be estimated using point or line-point transects, quadrats (squares or rectangles of land marked off for the study of flora and fauna), or sampling frames; *density* can be derived from belt-transects, quadrats, or sampling frames, or by dimensionless methods such as relevés (Minnesota Department of Natural Resources 2013) or point-centered quarter (Cottam and Curtis 1956). The point-centered quarter method also can be used to measure composition and size-class distribution of stands of woody plants (Mueller-Dombois and Ellenberg 2002). Line-point intercept (Herrick et al. 2005) is another rapid, accurate method for quantifying vegetation and soil surface conditions. By adding sampling at different vertical layers, from the ground to the tallest vegetation, information on vegetation structure can also be easily captured. When measuring ground cover in a quadrat, the use of cover classes (such as Braun-Blanquet) can reduce sampling time (Bullock 2006). Stem diameter and basal area of woody species typically are measured using a Biltmore stick or diameter tape, both of which measure the diameter at breast height (DBH, 1.3 m / 4.3 ft.) or by basal diameter (e.g., 10–15 cm / 4–6 in. above ground surface), which is often more appropriate for measuring tree species in riparian zones, which tend to be low-branching. To measure *canopy cover or density*, a spherical densiometer (Strickler 1959) or hemispherical photos are often used. The vertical line intercept method also is used to characterize the vertical structure of vegetation. In addition, data describing vegetation height can be extracted from LiDAR (also see remote sensing, below) (Farid et al. 2006), which can provide detailed measurements of topography, bathymetry, and vertical canopy structure. A nested plot design (e.g., Whittaker plots) can accommodate sampling plants of different size and growth form (i.e., herbs, shrubs, trees) (Barnett and Stohlgren 2003). One also can use qualitative techniques such as relevés to rapidly assess the abundance and species present (Mueller-Dombois and Ellenberg 2002; Bullock 2006).

For many semiarid streams, the total number of plant species along channel margins can number in the hundreds, with herbaceous species often outnumbering the woody species. Classifying plants into functional types (e.g., upland plants versus facultative or obligate wetland plants) can help to reduce the number of classifications required (Merritt et al. 2010). Species can also be grouped according to known physical and physiological traits (such as rooting depth, tolerance of anoxia or drought, and canopy height at maturity) rather than by taxonomy or association (occurrence in the same community in the field). Such trait-based associations may be useful in predicting shifts in entire functional groups of species in response to changes in environmental characteristics, such as those caused by management, human activities, or climate change (Lavorel and Garnier 2002; Violle et al. 2007; Sterk et al. 2016). An example might be a shift from phreatophytic (referring to species requiring a constant groundwater source) or mesic vegetation to xeric, drought-tolerant vegetation as a result of groundwater levels dropping below a threshold. Several studies are underway that use traits such as rooting depth, leaf size, plant height, and seed size to classify plants into functional types that share common adaptations to environmental conditions (Stromberg and Merritt 2016; Bejarano et al. 2018; Aguiar et al. 2018).

## Monitoring Birds in the Context of Stream Restoration

Bird surveys, including plot counts, point counts, transects, automated bird calls, capture-mark-recapture, nest surveys, fledgling surveys, and so on, can collectively document species presence, richness, diversity, and abundance (Ralph et al. 1995; Bibby et al. 2000; Hinojosa-

## CASE STUDY 7.2

Long-Term Monitoring of Marsh Birds in the Colorado River and Cienega de Santa Clara, Baja California, and Sonora, Mexico

Osvel Hinojosa-Huerta

Despite having lost roughly 80 percent of historical wetlands, the Colorado River Delta remains one of the most important wetlands in the Sonoran Desert, supporting nearly 101,000 ha (nearly 250,000 acres) of riparian woodlands, wetlands, brackish marshes, and large estuarine ecosystems. These crucial ecosystems provide habitat to over 360 species of migratory and resident birds, including marsh birds such as Yuma Ridgway's Rail (*Rallus obsoletus yumanensis*), as well as stopover habitat for many waterbirds and neotropical migrants.

As part of the Northwest Water and Wetlands Conservation Program, Pronatura Noroeste began a long-term monitoring program in 1999 to better understand the relationship of birds to changes in surface water, groundwater, and vegetative conditions as a result of restoration actions, including efforts to eradicate nonnative riparian plants and plant native riparian species, as well as the response of the Delta's bird community to the 2014 pulse flow (see Case Study 5.5).

Monitoring was conducted in eight parts of the Colorado River Delta that are identified by experts as being of particular importance to marsh and other key bird species. Standardized protocols using call response broadcast surveys were conducted twice a year at 556 points over a period of seventeen years (1999–2016). Monitoring results indicate that the Ciénega de Santa Clara supports the largest concentration of marsh birds in the Delta, including an estimated 6,304 Yuma Ridway's Rails, with a stable population, as this wetland and the flows that maintain it have been protected over this period of time. Along the Hardy River, the population of Yuma Ridgway's Rails has increased exponentially since 2009, at an average annual rate of 25 percent, in response to increased instream flows into the area. In general, results of monitoring indicated that the abundance of birds in restored areas increased on average 28 percent and diversity increased by about 53 percent, demonstrating improvements to ecosystem health following restoration activity.

### Lessons Learned

- Long-term monitoring (i.e., a decade or more) is essential to a robust adaptive management program to guide restoration actions and land and water management for the benefit of native species.

Huerta et al. 2008). In general, line transects may be better for lower density, active species in fairly even habitats, while point counts are better for quiet, cryptic species or for censusing larger numbers of species (Bibby et al. 2000). Surveying litter, organic material, and micro- and macroinvertebrates is also common for monitoring habitat health, as some avian and mammalian communities require these elements for refuge, diet, and nutrition.

## Monitoring Aquatic Macroinvertebrates, Riparian Invertebrates, and Fish

Macroinvertebrates have long been used as indicators to assess water quality. Particular species or families can be used to assess broader system conditions. In general, healthy streams

The elusive Yuma Ridgway Rail can often be heard before seen. Photo by Francisco Zamora.

- Monitoring of wetland and riparian birds is particularly helpful to gauge success of stream restoration efforts because they respond to habitat changes at varying scales (KBO and USFS Pacific Southwest Research Station 2013). The methods we used are effective and cost-effective to implement.
- In this case, the results of our long-term monitoring underscore the resiliency of the Colorado River Delta's wetland, riparian, and estuary ecosystems, holding promise that future environmental flow efforts and on-the-ground restoration can continue to improve the Delta's overall environmental condition. However, such conservation actions can only occur with continued binational attention to supporting on-the-ground restoration efforts and to securing water for key wetland and riparian areas.
- A diversity of public-private partnerships is critical in maintaining monitoring efforts over the long term. Though spearheaded by Pronatura Noroeste, this long-term monitoring effort was supported by numerous partners, including the National Fish and Wildlife Foundation, CONANP, the Sonoran Institute, the Sonoran Joint Venture, and the Walton Family Foundation.

support a high diversity and high number of macroinvertebrate taxa, whereas an unhealthy stream may have less diversity and relatively higher numbers of pollution-tolerant species. By analyzing the taxonomic assemblage of macroinvertebrate communities, experts have come up with biotic indices to indirectly monitor stream health. This is possible because particular species, and in some cases families and genera, are sensitive to pollution, whereas others are not. The presence of those groups that are pollution-sensitive indicates health in a stream. The most common macroinvertebrate biotic indices used include the Benthic-Index of Biological Integrity (B-IBI) (Kerans and Karr 1994) and the Hilsenhoff Biotic Index (http://cfb.unh.edu /StreamKey/html/index.html); these indices require considerable expertise to use. Approaches that identify organisms at the family level also are available, allowing rapid stream evaluation

# CASE STUDY 7.3
## Monitoring the Effects of Small Dam Removal on Fish Community Composition: Removal of the Zemko Dam Along the Eightmile River, Connecticut (U.S.)

Helen M. Poulos and Barry Chernoff

The objective of this monitoring study was to quantify changes in fish community composition in response to the removal of a small dam known as the Zemko Dam, along the Eightmile River in Salem, Connecticut (see map in this case study). Zemko Dam was a small dam where water flowed over a ledge and through a controlled opening that regulated the flow of water out of the reservoir and powered a gristmill and a sawmill during different periods of its 260-year existence. Thousands of small dams like this have been constructed throughout the United States, Mexico, and Australia. Many are more than 200 years old and show signs of neglect and abandonment. Resource managers and local communities are increasingly interested in small dam removal as a strategy for dealing with these degraded structures to reduce the risk of flooding from dam failure and to restore the ecological integrity of aquatic ecosystems (Poff et al. 1997; Hart and Poff 2002; Hogg et al. 2013; Hogg et al. 2015). Yet the ecological impacts of dam removal on river systems, particularly regarding understanding the responses of fish species composition to dam management, remains poorly understood. In this case study, we highlight the key impacts of the Zemko Dam removal on fish assemblage composition as an example of a multiyear monitoring program that evaluated the effects of river restoration on fish assemblage structure. For a detailed account, please see Poulos et al. (2014).

## Methods

We sampled fish abundance over a six-year period (2005–2010) to quantify changes in fish abundance prior to dam removal, during drawdown, and for three years following dam removal. Fish population dynamics were examined above the dam, below the dam, and at two reference sites by indicator species analysis, linear mixed effects models, nonmetric multidimensional scaling, and analysis of similarity. We inventoried the fishes during the growing season between May and October from 2005 to 2010 above the dam site, below the dam site, and at the two reference sites. Fishes were collected for a thirty-minute period that covered 75 m of stream during each of the sampling dates with a Smith-Root LR-24 backpack electroshocker and 0.16 cm stretch mesh nets.

The Eightmile River is characterized by riffle-pool-run morphology. Riffles are relatively shallow portions where loose bedrock, boulders, or cobbles break the surface, except

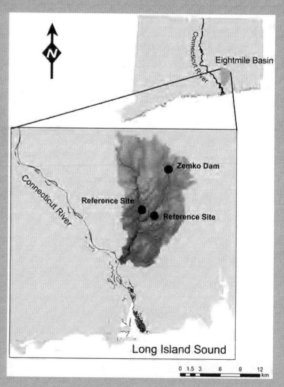

Map of the Eight-Mile River watershed, Zemko Dam, and study sites. Map by Helen M. Poulos.

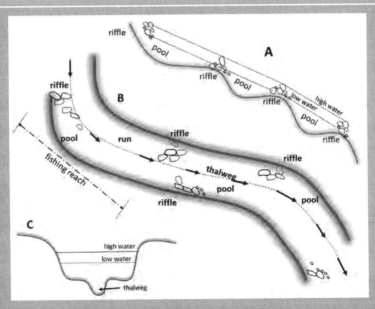

Schematic of riffle-pool-run morphology. Diagram by Helen M. Poulos.

during the highest waters. Given the pitch of the stream, the increased kinetic energy of the water going over a riffle scours the substrate, forming pools at the downstream border of the riffle. Beyond the pool, the elevation of the stream bottom rises toward the next riffle, often forming a relatively shallow, flat section, termed a run (marked B in the diagram in this case study). The thalweg, or talweg, is the lowest section of the stream bottom (marked C in the diagram in this case study), forming the channel with the fastest flowing water.

We sampled fish along what is known as a "fishing reach," which is a standard unit of fish census that extends from the head of a riffle to the head of the adjacent riffle. For a study such as this one of population dynamics, it is important that we sampled several fishing reaches to serve as replicates. It is also critical to block the stream at the ends of the fishing reach and perform a three-pass study, removing fishes in each pass, to determine the effectiveness of the sampling effort. To sample aquatic invertebrates in the stream a combination of techniques was used, including Surber-samples in the riffles, a drift-catch net in the runs, and kick-nets to sample invertebrates in leaf or other debris. An alternative technique is to leave novel substrate for colonization such as rock-bags or leaf-bags.

Fish relative abundance was calculated by species for each sampling interval at each site to make direct comparisons among sample sites over time. Relative abundance was calculated using count data for each species at each sampling interval. This value was expressed as the percentage abundance of each species relative to the total fish abundance for a sampling interval. We used indicator species analysis (Dufrene and Legendre 1997) and PC-Ord Software (McCune and Mefford 2011) to identify key indicator species across all years for the reference, above-dam, and below-dam sites. The relative abundances of each species' were analyzed by site over time (2005–10) via a mixed model procedure. We used R (R Development Core Team 2017) and the lme4 (De Boeck et al. 2011), languageR (Baayen 2007), and multilevel (Bliese 2006) packages to perform linear mixed-effects analyses of the relationships among a species' relative abundance, year, and sample site.

*(continued)*

## CASE STUDY 7.3
(*continued*)

### Results

We observed significant shifts in fish relative abundance over time in response to dam removal. Changes in fish species composition were variable and they occurred within one year of the draw-down. A complete shift from lentic to lotic fishes failed to occur within three years after the dam was removed. However, we did observe increases in fluvial and transition (i.e. pool head, pool tail, or run) specialist fishes both upstream and downstream from the former dam site. While the reference communities recovered from the effects of dam removal, the sites above and below the previous dam were still in flux and had not resembled the community structure of the reference communities (Poulos and Chernoff 2016).

The Zemko Dam removal case study serves as an example of the importance of ecological monitoring in river restoration because such data allow managers to assess the success of restoration projects with objective metrics. Evaluating temporal changes in fish assemblages in response to environmental management provides an important context for assessing how aquatic, marine, or terrestrial adaptive management activities influence community structure and key biotic interactions among taxa.

### Lessons Learned

- Although not a lesson that is directly connected to monitoring, we want to highlight that our results demonstrate the importance of dam removal for restoring river connectivity for fish movement. Removal of small dams should be strongly considered as a priority in areas where restoration of hydrological conditions will promote conservation of native fishes.
- Long-term monitoring is essential to understanding shifts in fish community structure following dam removal. In this study, results showed that the dam sites continue to remain in a state of reorganization for years after the dam removal event, and the inclusion of two reference sites also revealed that neither of the two dam sites displayed evidence that a shift toward the community configuration of the reference site had occurred within the three years following the dam removal. Stream recovery can take decades or even centuries, given that many of these impoundments have been in place for over a century (Doyle et al. 2005; Catalano et al. 2007; Maloney et al. 2008). Only by monitoring over an

but with less statistical power. They typically consist of using a dipnet to sample riffles and other aquatic habitat where macroinvertebrates are found. Sampling results are generally compared to a reference value (i.e., an index score averaged across properly functioning, healthy streams). Another approach is to sample upstream and downstream from a known pollution source. These results could further be compared to reaches where a restoration treatment has been undertaken to improve stream habitat for macroinvertebrates or fish, or across the course of multiple seasons (Purcell et al. 2002; Parkyn et al. 2003; Larned et al. 2006; Selvakumar et al. 2010; Giling et al. 2015).

Invertebrate surveys of riparian systems also can yield valuable monitoring data, as when assessing beetle damage to vegetation or invertebrate recovery important for bird species. For example, Nelson and Wydoski (2008) used a riparian biotic index to look at butterfly community composition in sites along the Arkansas River (U.S.) and identified a gradient of butterfly

- extended time period will we be able to understand the long-term consequences of dam removal as well as other stream management actions.
- Analyzing the data using the multivariate autoregressive 1 (MAR 1) framework (see Poulos and Chernoff 2016) was critical to our understanding of the outcomes of dam removal. From the analyses we were able to model not only the change in community structure over time but also changes in community dynamics and measure the degree to which the sites were resilient or reactive over time (Poulos and Chernoff). These are key because community interaction dynamics tell us about population growth of species in relation to the population density of other species over the monitoring period. For example, sampling data collected just below the dam showed that as the population density (number of fish per square meter) of Fallfish (*Semotilus corporalis*) increased, the population-growth rate of Tesselated Darters (*Etheostoma olmstedi*) decreased, which is indicative of the predator-prey relationship. However, at the reference sites, the population-growth rate of Tesselated Darters was independent of the density of Fallfish. Monitoring the ecology of interspecific dynamics is key to assessing ecosystem function to ensure that the ecosystem is functioning within expected parameters. A species list and the change of species over time are important monitoring results, but only part of what is needed to assess restoration results. We need to also be able to demonstrate that the ecological components of the environment are sufficient to maintain the targeted species in the long term.
- It is important to collect data that provide insights into resilience and reactivity. Resilience metrics measure both the return of the community to its initial configuration and the return of variability after a perturbation, in this case dam removal. Reactivity describes the degree to which perturbation pushes communities toward reorganization into new species structures and ecologies. Many restoration projects such as dam removal have reorganization in mind but rarely use objective metrics in an adaptive way to evaluate their success.
- The monitoring effort was relatively inexpensive and replications were relatively easy. Both of these characteristics were key to extending the monitoring effort long after the monitoring budget was exhausted. Managers should favor targeted and inexpensive monitoring programs that have a realistic chance of being conducted for many years over expensive monitoring programs, even if the expensive efforts allow the measurement of additional parameters with greater accuracy. Little will be gained if an expensive monitoring program can be only be done once due to budget constraints.

composition ranging from sites where tamarisk (a nonnative species) had been removed, sites that had a blend of cottonwood/willow (both native species) and tamarisk, and sites dominated by native vegetation. They found that tamarisk-dominated areas did not have higher riparian biotic index values (reflecting richness of species and the presence of particular species) after treatment. Ants are another example of terrestrial invertebrates that have been used as ecological indicators of restoration progress in riparian zones (Gollan et al. 2011).

In the confines of a book on stream restoration, there are several broad, well-documented topics of which we can only scratch the surface. Monitoring of freshwater fishes is certainly one of them. The citations we provide below are only the surface of a deep well of literature on monitoring and sampling of inland, freshwater fish populations of Australia, Mexico, and the United States. Although just the surface, we hope they are helpful to readers who are taking their first dive into this important topic.

- Inventory, monitoring, and sampling methods: Bonar et al. (2009); De la Maza Benignos et al. (2017); Espinosa-Perez y Salinas Rodriguez (2014); Grainger et al. (2013); Pidgeon (2004); Sostoa et al. (2005); Taylor et al. (2017);* Thomsen et al. (2012);† Van Haverbeke (2013)
- International collaboration: Bonar et al. (2017)
- Trophic ecology: Behn and Baxter (2019)
- Websites: Freshwater species and ecosystems, State of the Environment, Australia (https://soe.environment.gov.au/theme/biodiversity/topic/2016/freshwater-species-and-ecosystems); Desert Fish Council (https://www.desertfishes.org/); Desert Landscape Conservation Center (https://desertlcc.org/resource/rarest-fish-world-desert-fishes-and-their-response-changing-climate); FISHBIO (https://fishbio.com/); Native Fish Conservation Network (https://native fishconservation.org/); the North American Freshwater Migratory Fish Database (U.S., Canada, and Mexico) (https://habitat.fisheries.org/the-north-american-freshwater-migratory-fish -database-a-database-to-inform-habitat-conservation-efforts-of-freshwater-fish-species-in -the-united-states-canada-and-mexico/)

The fish monitoring case study that we highlight above is from the Eightmile River, Connecticut (U.S.). Although the Eightmile River is not located in a "water deficient" region, the study is included in the guidebook to highlight its long-term nature and utility of the sampling and analytical methods for evaluating the impacts of dam removal on fish assemblages; methods that can be adapted and applied to dryland streams. See also monitoring methods associated with protecting the Pecos Bluntnose Shiner (Case Study 6.8) and reintroducing the Rio Grande Silvery Minnow (Case Study 6.9).

## Monitoring Shallow Groundwater

Evaluation and monitoring of stream restoration actions, as well as the restoration objective itself, often focuses on surface waters and ecosystems, neglecting the connection of the river to the groundwater below it and in the adjacent floodplain (Boulton 2007; Hester and Gooseff 2010; Lehr et al. 2015). Such a myopic focus on surface water conditions ignores the substantial evidence that the disconnection of shallow groundwater from surface water is one of the most important causes of aquatic and riparian ecosystem deterioration (Schumm 1999; Parola and Hansen 2011). In many parts of the world, valley bottoms have been buried in sediment due to increased upland soil erosion as a result of deforestation and other land-clearing actions, farming, severe wildfires, and overgrazing, among other disturbances. Subsequent human actions to convey valley water and sediment downstream (e.g., by channel dredging, removal of channel debris, channelization, and channel reallocation) have promoted channel incision into the new alluvial deposits, resulting in floodplain surfaces (and the roots of floodplain vegetation) that are disconnected from the shallow aquifer. In some regions, this disconnection is exacerbated by the overpumping of shallow groundwater, which lowers the water table even further. In arid and semiarid regions of the

---

* Use of acoustic telemetry.

† Analyzed environmental DNA to monitor endangered freshwater biodiversity.

# CASE STUDY 7.4

Groundwater Monitoring for River Restoration: Upper San Pedro River Case Study

Jeanmarie Haney, Brooke Bushman, and Holly Richter

## Background

Groundwater recharge via constructed infiltration ponds and injection wells has been utilized for many decades to store unused surface water underground for future human use. In southeast Arizona, a new and novel application of this technology is being used to keep a desert river flowing despite growing groundwater demands for municipal and residential use. This approach could have a wide application in similar arid-region river basins.

The Cochise Conservation and Recharge Network (CCRN) is a collaborative partnership formed in 2015 with the goal of implementing projects to sustain both a vibrant local economy and a flowing desert river in southeast Arizona. The CCRN is using outcomes from many years of collaboration-driven hydrological and ecological studies to implement a series of near-channel aquifer recharge projects using constructed infrastructure such as infiltration ponds and injection wells. The network consists of eight sites and/or projects, encompassing 2,567 ha (6,344 acres) along approximately 40 km (25 mi.) of the San Pedro River north from the international border (see map in this case study). Based on groundwater modeling, the hope is that CCRN's work will reduce depletion of shallow groundwater elevations from pumping and foster groundwater conditions that will protect the river's surface flow in the long-term (Richter et al 2014; Lacher et al. 2014).

## Goals for Monitoring

Monitoring the effectiveness of the recharge projects, both individually and as a network along the length of the river, is essential for determining if goals are being met and what adaptations may be needed. Monitoring consists of three main components: verification of specific water-budget and groundwater-model parameters at the regional scale (per Gungle et al. 2016); wet/dry mapping of the river in late June (the hottest and driest time of the year), so that year-to-year variability in surface water trends at the river reach can be compared via spatial patterns of wetted length (Turner and Richter 2011) (see also wet/dry mapping monitoring in Case Study 7.5); and groundwater monitoring to track changes in groundwater elevations from the prerecharge baseline. This case study focuses on groundwater monitoring at project sites. Standard methods were followed (e.g., Barcelona et al. 1985; Cunningham and Schalk 2011).

Monitoring of groundwater elevations provide the data needed to track the effectiveness of groundwater recharge. The data are used to track aquifer response and to refine the operation of recharge facilities at each location and to continually refine the groundwater model to improve the accuracy of predicted systemic responses at the regional scale. Longer-term success of the recharge facility network in sustaining the shallow groundwater levels and baseflows of the San Pedro River is tested by the iterative use of refined groundwater modeling, utilizing groundwater-level monitoring data, USGS stream-gauge data for baseflow periods, and spatial surface water trends from wet/dry mapping.

## Monitoring Design and Results

Each recharge site is unique in the timing and circumstances under which it was conceived and is operated. Despite the variations in facility operations and monitoring by the group's various members,

*(continued)*

## CASE STUDY 7.4
(*continued*)

the CCRN has been able to integrate the monitoring results at various temporal and spatial scales, providing both details on individual sites and high-level regional metrics over the long term.

Water levels in six monitor wells at the City of Sierra Vista's effluent recharge facility have been measured monthly since April 2002; data analysis in 2016 indicated an average water-level rise of 3.3 meters (10.9 feet) compared to the baseline (July 2002). Regional water-level mapping indicates an extensive groundwater mound has been created due to recharge at the facility (Schmerge et al. 2009). At Cochise County's pilot stormwater recharge facility at Palominas, monitoring wells were installed when the facility was constructed, adjacent to the recharge basins along the length of the facility and up-gradient. The wells were instrumented for continuous data collection prior to the first runoff event, and prerunoff data were used to establish a baseline. After two years of facility operation, the overall average water-level rise compared to the baseline was 0.2 meters (0.68 feet). At two planned recharge sites, existing wells are being monitored to establish baseline conditions. Continuous-monitoring equipment was installed where feasible; quarterly monitoring is conducted in areas where permanent monitoring equipment could not be installed. At each new location where recharge facilities are constructed, new monitoring wells will be sited, designed, and constructed to align with the goals of the recharge facility.

### Lessons Learned
Determining how groundwater elevations respond to artificial recharge via constructed infrastructure across a network of sites has been challenging, but rewarding. Each recharge facility has unique circumstances and timelines defining its construction, operation, and reporting requirements. Monitoring and reporting must account for sites that recharge water from continuous (effluent) or sporadic (stormwater) sources, for example. A single facility may serve multiple purposes, such as flood control and recharge. Our four main lessons learned:

- *Monitor baseline groundwater conditions* prior to the construction of recharge facilities. This step is essential. Ideally, baseline monitoring should be conducted several years prior to interventions and can be begun by measuring elevations in available wells near project sites. Budgeting and planning for installation of monitoring wells specific to project needs—with installation following facility design but prior to facility construction—should be part of the project.
- *Identify cost-saving opportunities* to allow monitoring efforts to continue in the long term. In this effort, three cost-saving opportunities presented themselves: (1) coordination of data downloads, analyses, and report-writing from individual recharge sites reduced costs of contractor mobilization and increased contractor efficiency; (2) identifying one entity as the "sponsor" of the monitoring program—the entity that does the budgeting, contracting, and record keeping, and maintains timelines—yielded additional savings by fostering a high level of coordination for long-term monitoring; and (3) partnering with operators of any existing operational recharge facilities in the area, to incorporate their groundwater-level data, allows broader regional analysis with little increased cost.
- *Maintain a technical work group of individual site partners* to review network monitoring results, including annual groundwater-level change, and compare those results with baseflow and continuous river-flow trends as reported by other entities in the watershed; this enables adaptation as additional

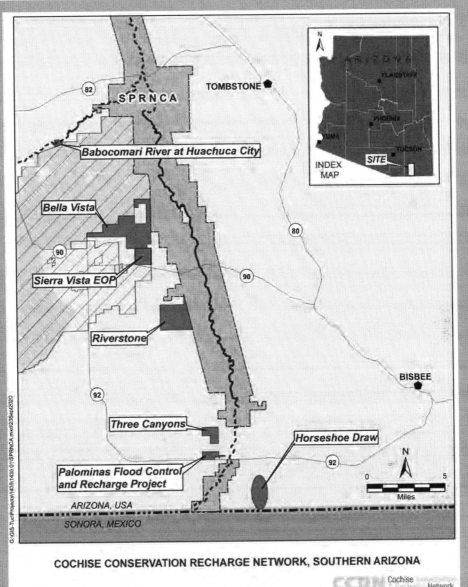

**COCHISE CONSERVATION RECHARGE NETWORK, SOUTHERN ARIZONA**

The San Pedro Riparian National Conservation Area is managed by the U.S. Bureau of Land Management and contains some of the best remaining riparian habitat in the southwestern United States, providing irreplaceable habitat for migrating birds and numerous other species. Map from The Nature Conservancy.

lessons are learned. For example, using results from monitoring to update groundwater and surface water models to improve the certainty of modeling simulations can influence the monitoring design, location, and operation of new facilities.

- *Develop templates for precise and regular communication to decision makers of key monitoring outcomes* and progress toward goals; this keeps decision makers engaged and raises the probability that funding will continue to meet the long-term needs of the monitoring program.

world, such hydromorphological changes have been the death knell for organisms that rely on shallow groundwater for their survival.

Stream corridor restoration objectives are consequently increasingly focused on reestablishing the floodplain/groundwater connection by implementing actions that will reestablish the former elevation of the incised channel bed, reduce pumping of shallow groundwater, and/or recharge water into shallow aquifers that are hydrologically connected to streamflow. To evaluate the effectiveness of such restoration actions, the monitoring of shallow groundwater elevations is necessary.

If you have the resources to install your own monitoring wells, you would have the advantage of placing them at locations and in patterns most advantageous to providing an accurate portrayal of groundwater conditions. For example, installing wells perpendicular to the stream channel will provide information on how groundwater levels change with distance from the main channel as well as whether the stream reach that you are working on is a gaining or losing reach with respect to streamflow. Such information can be extremely valuable for designing riparian revegetation strategies where knowing how the depth to saturated soil varies at potential planting sites allows ecosystem managers to identify sites where conditions are most likely to support desirable species (see "Estimating Groundwater/Surface Water Exchange" in Chapter 6).

One of the challenges is that groundwater elevations can vary significantly over time—over the course of years and even during the course of a single season. Knowing how groundwater conditions vary throughout the year—and ideally year-to-year—is important because large seasonal groundwater fluctuations can significantly affect the results of your stream restoration project. Therefore, the more often the water-table depth is measured and the longer the monitoring period, the more accurately the actual water-table fluctuation can be determined. If funding or resources limit your ability to monitor seasonal fluctuations over the course of several years, then investigating groundwater depth during the hot, dry season prior to planting is still valuable before finalizing treatments. This will provide useful information on how rapidly the saturated zone drops during the hot, dry season when ET is the greatest. (ET is the combination of water taken up and transpired by plants and the water directly evaporated from the stream, soil, and living and dead plant surfaces.) If the saturated zone maintains itself at thresholds appropriate for your realizing your restoration objective (e.g., at depths below surface elevations suitable for establishing desirable plant species for a riparian revegetation effort) during the height of the hot, dry season, it is a sign that groundwater conditions are suitable for realizing your stream restoration objectives.

There are a variety of sources of information on groundwater concepts, methods for measuring groundwater, and ways to develop an accurate portrayal of groundwater conditions from measurements, including Todd and Mays 2004; Heath 2004; Briseño-Ruiz et al. 2011; the groundwater information website from the University of California, Davis (http://groundwater.ucdavis.edu); and Gidahatari's "Implementación de redes de monitoreo de agua subterránea" web resource (http://gidahatari.com/ih-es/implementacion-redes-monitoreo-aguas-subterraneas). Information is also readily available on how to select sites, construct and install monitoring wells, and use automatic water-level monitoring equipment (along with how often to measure), and other aspects of monitoring shallow groundwater conditions (Taylor and Alley 2001; Baxter et al. 2003; HDR 2014). Lightweight and inexpensive "minipiezometers" also can be used to conduct numerous measurements in remote locations (Baxter et al. 2003). The number and locations of monitoring points required to give a realistic picture of groundwater depth and fluctuation depend on several factors, including the size, topography, geology, and soil characteristics of the site. Small sites characterized by relatively homogeneous hydrogeological conditions will require fewer monitoring points than larger, more complex sites (Dalton et al. 1991).

In situations where budget constraints and other limitations (securing needed permits, access, or topographical conditions, etc.) preclude the installation of monitoring wells, you may need to rely on existing wells in and near the target restoration reach of the stream. This is the approach taken by The Nature Conservancy along the San Pedro River, which flows out of northern Sonora, Mexico, into southeast Arizona, U.S. (see Case Study 7.4).

## Monitoring Streamflow

Chapter 3 reviews and provides references for measuring streamflow and the types of analysis available to describe hydrological conditions and trends. Appendix A summarizes key agencies in Australia, Mexico, and the United States that maintain, analyze, and interpret streamflow gauge data. However, although streamflow gauges are not always located along the streams being restored, monitoring streamflow is nevertheless essential for all stream corridor restoration projects. Since budgets are typically more insufficient than ample, approaches for monitoring streamflow need to be developed that do not require the installation and maintenance of expensive equipment. Two projects described below tackle this challenge in different ways. Along the San Pedro River (Case Study 7.5), practitioners are implementing "wet/dry mapping," which measures the extent of surface water during critical times of the year along this intermittent stream. On the Rio Grande / Río Bravo, which constitutes the international U.S.-Mexico border, practitioners are using relatively inexpensive and readily available hardware and software to monitor flow stage (Case Study 7.6).

## Monitoring Channel Morphology

Repeated surveys of channel morphological features along alluvial-dominated channel reaches can provide the data needed to

- assess overall channel stability, as indicated by temporal changes in channel gradient, channel width, width-depth ratios, alluvial material, etc.;
- develop stage-discharge models and other types of flow topographical-based modeling;
- map the extent and distribution of bottomland morphological surfaces and quality of habitat; and
- provide the context for understanding biological change, including such biotic parameters as the extent and distribution of riparian vegetation, quality of aquatic habitat, benthic characteristics, and so on.

Monitoring of channel morphology is often conducted at the reach scale, with collected data at a given site or a range of sites considered to be representative of a length of river that

# CASE STUDY 7.5
Citizen Science: Monitoring Perennial Reaches with Wet/Dry Mapping, San Pedro River, Arizona

Brooke Bushman and Dale Turner

## Background

Initiated in 1999, wet/dry mapping of the binational San Pedro River is a collaborative, citizen-science volunteer monitoring program. Begun as a joint effort by the BLM and The Nature Conservancy, this monitoring program annually maps the spatial extent of surface water in the San Pedro River channel. What was initially an effort to map the 80 km (50 mi.) length of the BLM's San Pedro Riparian National Conservation Area, the mapping program has grown to include 480 km (300 mi.) of the river and its tributaries, from the headwaters in Sonora, Mexico, to the Gila River confluence in Arizona. Collecting water presence data was the impetus, but equally important was engaging more than a hundred people each year to connect with the river. Volunteers include agency staff, nonprofit groups, private landowners, and the public, all of whom are essential partners in protecting this desert river.

## Goals for Monitoring

Beyond providing volunteer participants a firsthand experience that we hope leads to river stewardship, the data from wet/dry mapping have been useful in many settings. Collected data are analyzed by agency scientists in conjunction with other hydrological datasets to assess groundwater conditions in the watershed and to identify suitable reaches to stock native fish. Results of wet/dry mapping also help identify conservation priorities by providing the basis for identifying important river or tributary reaches where perennial flows could be supported through land-protection efforts, and where groundwater-replenishment projects could support flows and riparian habitat. Over time, progressively more of the river and its tributaries have been mapped, with our current total amounting to about 83 percent of the total river length. The goal is to annually map as much of the river as possible, by engaging and requesting access permission to map the patchwork of private and publicly owned reaches of the river.

## Methods (Monitoring Design)

Wet/dry mapping is conducted every year on and around the third Saturday in June to coincide with the lowest flow before the expected start of the summer rainy season. Teams of citizen scientists travel along the San Pedro River and its tributaries using GPS to mark the locations where water is present. They are trained for personal safety and consistency in data collection prior to mapping each year and assigned to river reaches approximately five miles in length. At dawn, teams assemble at predetermined starting locations, equipped with essential supplies to stay cool and hydrated on one of the hottest summer days. Surveyors record the beginning and end points of surface water occurrence, using paper data forms and consumer-grade GPS units. Most teams walk their designated survey reaches, but longer dry portions have been surveyed on horseback or all-terrain vehicles. To allow the mapping to be completed in one day, the full river is broken into five sections of approximately 32 to 80 km (20 to 50 mi.). Survey team leaders are assigned to coordinate the training, equipment, and personnel needs to cover each of the monitoring sections. After mapping, teams deliver the GPS units and data sheets to the section leaders. Several Nature Conservancy staff coordinate the section

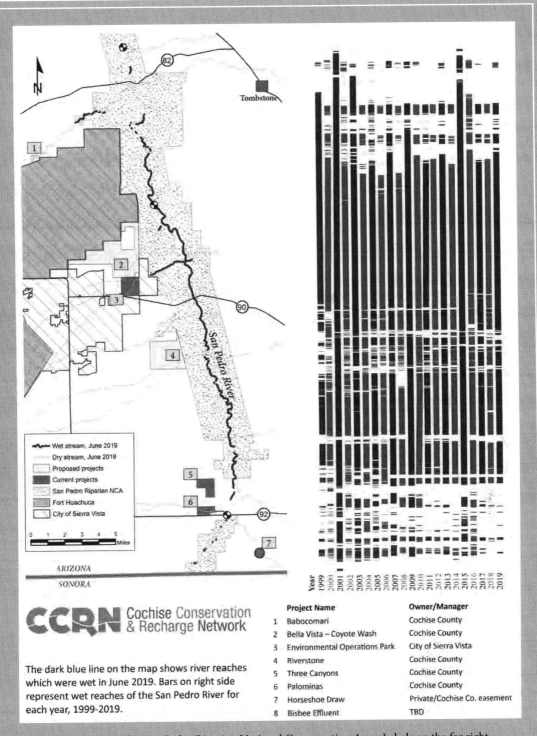

**CCRN** Cochise Conservation & Recharge Network

The dark blue line on the map shows river reaches which were wet in June 2019. Bars on right side represent wet reaches of the San Pedro River for each year, 1999-2019.

| | Project Name | Owner/Manager |
|---|---|---|
| 1 | Babocomari | Cochise County |
| 2 | Bella Vista – Coyote Wash | Cochise County |
| 3 | Environmental Operations Park | City of Sierra Vista |
| 4 | Riverstone | Cochise County |
| 5 | Three Canyons | Cochise County |
| 6 | Palominas | Cochise County |
| 7 | Horseshoe Draw | Private/Cochise Co. easement |
| 8 | Bisbee Effluent | TBD |

Wet/dry results from the San Pedro Riparian National Conservation Area. Labels on the far right identify the ten analysis segments, each covering 8.1 km (5 miles). Five of the locations shown in the figure that are part of the Cochise Conservation and Recharge Network were acquired for groundwater protection or recharge projects. Map and graph by The Nature Conservancy.

*(continued)*

361

leaders to ensure consistent training, perform data-quality control after mapping, and deliver the data to The Nature Conservancy for analysis and creation of the maps.

End-point coordinates are then imported to a Geographic Information System (ArcGIS, Environmental Systems Research Institute) and transferred to the closest points on a linear representation of the river. By using linear referencing, wet reaches and overall surveyed reaches are converted to tables with start and stop points along the line. The resulting wetted lengths for various reaches are analyzed for trends using the nonparametric Mann-Kendall test. More detailed methods are provided in Turner and Richter (2011) and on a website (http://azconservation.org/projects/water/wet_dry_mapping).

## Results

Wet/dry mapping is a simple method that provides a whole-river snapshot of surface water conditions, with results allowing the identification of reaches with truly persistent surface water in a groundwater-dependent river basin (see graph with map in this case study). The wet/dry monitoring program complements both data from streamflow gauges (which provide continuous values for a few discrete sampling points) and a shallow groundwater monitoring program that makes use of a limited number of sampling wells (see also "Groundwater Monitoring for River Restoration: Upper San Pedro River Case Study"). More importantly, data collected over the years reveal trends in the presence of surface water. Although results on their own do not identify the drivers behind surface water changes, consistent trends suggest correlations with areas of groundwater extraction. In addition, results reveal the benefits of conservation actions that are reducing water use and increasing recharge near the river.

References describing the collection and analysis of wet/dry mapping data to inform our understanding of streamflow and groundwater conditions are becoming more numerous (Lacher et al. 2014; Gungle et al. 2016; Allen et al. 2019). For our work on the San Pedro, we have found that at least five years of data are needed to show a statistically significant trend.

Wet/dry maps of the San Pedro River and its tributaries from 2007 to the present are posted online and publicly available at www.azconservation.org. Beyond the maps themselves, no data (such as GPS points of water location) are made available from privately owned lands, to respect landowner concerns.

## Lessons Learned

- *Use a Consistent Survey Date.* Conducting wet/dry mapping at approximately the same date every year strengthens confidence in results. This date was identified using historical data describing the onset of southern Arizona's summer rainy season, the North American monsoon. The onset of the monsoon

has a consistent character and behavior (Brierley et al. 2010). The attributes to be measured must be relevant to the flow regime for the stream being measured as well as the objectives of the monitoring program itself (Fryirs 2003; Fryirs et al. 2008; Fryirs and Brierley 2009). There are a variety of resources that provide detailed information on data-collection procedures and placement of transect lines for measuring cross-sectional profiles. Here are some selected resources:

Volunteers with CONANP, which partners with TNC on wet/dry monitoring of this binational basin, come together to map the headwaters of the San Pedro River in Sonora, Mexico. Photo by Isaias Ochoa Gutierrez, CONANP.

season has, however, fluctuated over the years of the project. If the monsoon start date becomes consistently earlier or later, we will carefully reevaluate the mapping date.

- *Start Small.* At the onset, wet/dry mapping focused on a relatively small segment of the San Pedro River and was only scaled up as other partners committed resources to the mapping.

- *Keep It Simple.* Maintaining our focus on documenting only water presence and other clearly defined and limited qualitative information has been essential to the successful use and coordination of citizen scientists. Because of the sheer number of people mobilized on this day, the temptation exists to complicate the procedure by collecting additional data. Our mantra is "Keep it simple and focused."

- *Enhance River Stewardship.* Section leaders annually consider the current river stewardship issues and strategically recruit volunteers such as legislative aides, municipal staff, and conservation groups, whose perspectives may be enhanced by experiencing the river. Team composition can be equally important for relationship-building among key decision makers. Engaging local conservation groups as section leaders also provides a role for them in establishing the science that guides their work.

- Big River protocols developed by the National Park Service: https://irma.nps.gov/DataStore/DownloadFile/604465; https://irma.nps.gov/DataStore/DownloadFile/604463 (for standard operating procedures).

- Selected manuals and books: Harrelson et al. 1994; Gordon et al. 2004; Fryirs and Brierley 2012.

- Selected papers that include detail on survey methods: Osterkamp and Hedman 1982; Wheaton et al. 2010; Erwin et al. 2012; Dean and Schmidt 2013; Bangen et al. 2014.

## CASE STUDY 7.6
Real-Time Measurement of Streamflow Using Inexpensive and Readily Available
Equipment: A Monitoring Example Along the Rio Grande / Río Bravo, U.S.-Mexico
Border Region

Kelon Crawford and Jeff Renfrow

The Rio Grande / Río Bravo (RGB) through Black Gap is a gaining reach owing to discharge from numerous springs (see maps in Case Studies 6.5 and 7.8). This reach is also dominated by extensive and dense stands of giant cane, which have contributed to channel narrowing processes that have buried prime aquatic and riparian habitat (see Case Study 6.5). Discharge from many of the springs along this reach passes through substantial recently deposited alluvium (typically deposited on a mix of floodplain surfaces that are dominated by giant cane) before flowing into and contributing to RGB flow. One question that our binational team has in connection with our efforts to manage giant cane is if eradication of giant cane could bolster streamflow, particularly during the low-flow season when even a minor increase in flow could be significant to a variety of native flora and fauna—the theory being that spring discharge to the river could be enhanced owing to decreased evapotranspiration losses after giant cane has been eradicated. To answer this question, regular measurement of RGB streamflow is required upstream and downstream of the Black Gap reach before and after treating giant cane. Given the lack of streamflow gauges in suitable locations and the need to frequently measure flow (something that could not be accomplished manually by field personnel given the remoteness of the site), the only way to accomplish this task is to install our own instrumentation.

Although our monitoring budget was thin, we were pleasantly surprised to find inexpensive data loggers that allowed continuous measurement of stage based on readings of water pressure (i.e., pressure transducers) (see photos of transducer installation in this case study). The equation to determine water level from pressure data is given by $h = (P - P_r) / (rho * g)$, where $h$ is water level, $P$ is the pressure measured by the sensor in the stream, $Pr$ is the pressure measured by the reference sensor (atmospheric pressure), $rho$ is the density of the water in the stream (mass/volume), and $g$ is the acceleration due to gravity ($9.81$ m/s$^2$ or $32.17$ ft/s$^2$).

For our work, we used three of HOBO U20 water-level loggers. Two data loggers are installed along the Black Gap reach of the RGB upstream and downstream of where giant cane was treated at locations that (1) have bedrock exposed on banks, (2) offer protection from strong current, and (3) allow loggers to be placed deep enough to remain submerged during low flow. The third data logger is used as a barometric reference and installed in a protected site away from the river that was no more than 15 km (10 miles) from the river loggers and at a similar elevation (see Google Earth image in this case study). The data loggers are US$495 each, are 14 cm (5.5 in.) long and were installed using hose clamps and other readily accessible hardware. The data loggers record data every fifteen

In general, keep in mind that there are a variety of social and biophysical considerations that may affect the exact location for the ultimate placement of a particular transect, including

- *access*;
- *tributary confluences*—establishing cross-channel transects immediately upstream or down-stream of tributary confluences may complicate data interpretation;

Installation of HOBO transducers along the RGB at Black Gap was accomplished by fixing bolts in exposed bedrock next to the river in a manner that allowed allow easy access and removal in order to download data (as long as flow was not substantial) (left). HOBO units were installed in 12-gauge half-slotted metal (right). Photos by Kelon Crawford.

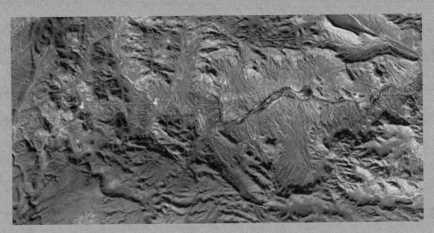

Google Earth image of the RGB at Black Gap, showing placement of the HOBO data loggers. The remote and rugged nature of this area necessitates the use of reliable, automatic data loggers. The distance between the upstream data logger and downstream is just over 24 km.

minutes and we visit the site once every three to six months to download the data. In our case, the loggers have to be removed from the bracket in order to download data, which is a quick and easy process but requires low water to access the mounting brackets. Labor costs include installation of the instruments and conventional current meter measurement of flow at selected stages to develop a rating curve at sites where the HOBOs are located.

- *presence of infrastructure* such as roads or buildings that may affect local biophysical conditions;
- *streamside biota*—grouping transects based on the classification of streamside vegetation communities may be important depending on the objectives of the project; and
- *homogeneity of the channel morphology*—it is usually valuable to establish transects along reaches of similar geological and morphological characteristics.

Surveying Channel Morphology Along the Rio Grande/ Río Bravo in Big Bend

David Dean

The RGB in the Big Bend region is subject to rapid geomorphic change consisting of channel narrowing during years of low peak flow, and channel widening during rare, large, long-duration floods. Additional background on various aspects of the binational restoration response along the RGB in Big Bend is covered in Case Studies 6.5 and 6.9. For the binational team working in this region, quantifying the rate and magnitude of infilling during years of channel narrowing, and the degree of channel widening during rare, large floods, is imperative for understanding how changes in geomorphic form may be affecting instream and riparian habitat, as well as for quantifying environmental flow needs to maintain desirable channel morphology and identifying appropriate management actions to minimize or even reverse undesirable geomorphic changes.

Below, we review several aspects of surveying channel morphology as reflected by our experiences along the RGB in Big Bend.

## Where to Survey and How Many Sites and Transects

In general, monitoring channel morphology consists of establishing permanent channel cross sections that are repeatedly surveyed at time intervals of interest. The survey itself consists of measuring x,y,z points across the transects. Although cross-section surveys can focus in all hydraulic environments (e.g., pools, riffles, runs, meander bends), make sure you and your team narrow surveying to locations that will allow you to address questions central to your monitoring objectives. If infilling of pools is of interest, establishing cross sections within pool habitats will be important. If quantifying channel change as a result of management actions is of interest, make sure surveying is conducted not only where the management action has been conducted but in reference sites as well to allow comparison between sites where management actions did and did not take place.

Along the RGB in Big Bend, the two main objectives of monitoring are to understand (1) channel narrowing and widening processes, and (2) potential impacts of management actions (specifically, the eradication of *Arundo donax* [see Case Study 6.5] on channel morphology). To address both of these objectives, our team established three monitoring reaches where giant cane has been managed and reference sites where management of giant cane is off limits. These three reaches range in length from 2.5 to 6 km long, with sixteen cross sections in the shortest reach, and thirty-three cross sections in the longest reach. At each site, cross sections' locations were determined by channel morphological complexity, sinuosity, and management actions, among other considerations.

Ideally, channel cross sections should go from one side of the stream channel to the other. For our work along the RGB, we tried to survey during low-flow periods when wading the channel was possible. If you are working along a large perennial river, you may need to use a small boat/raft that allows deeper readings to be made. If you are only concerned with changes to the banks and flood-plain, you may not need to survey the entire channel cross section and can survey from transect end points toward the channel, stopping when water is waist deep. If you have a healthy budget and/or are working with an agency that has access to more sophisticated equipment, echo/depth sounding may be possible (e.g., USGS: https://pubs.usgs.gov/fs/2018/3021/fs20183021.pdf). Such options may be costly, but lower cost commercial options generally used by fisherman exist, even though some sacrifices in measurement accuracy may occur (e.g., Fish Finders: https://humminbird.johnsonout doors.com/fish-finders).

## Equipment Needs

Survey cross sections do not need to be costly and can be done with readily available survey equipment, ranging from simple levels and tape measures to total stations and high-resolution GPS receivers that take advantage of navigation satellites (e.g., Global Navigation Satellite System [GNSS]). Along the RGB in Big Bend, surveying is done with both total stations and RTK GPS units. Total stations are used at monitoring sites in narrow canyons with poor GPS satellite coverage. Understanding the level of accuracy required for your monitoring effort is key to identifying the survey equipment most appropriate for your stream corridor project. Does quantifying geomorphic change along your stream corridor demand centimeter precision, or are you measuring gross channel change, such as bank erosion, that can be measured in a less precise manner? If the former, measurement of a greater number of points with more expensive and accurate equipment is likely warranted. If the latter, you may be able to get away with measuring fewer points using less expensive equipment. For example, if you're simply measuring bank retreat, engineers' levels or tapes may be sufficient. The bottom line: Don't spend limited monitoring resources on renting or purchasing expensive survey equipment that is not needed. Money saved on equipment may allow another round of data collection and/or allow you to survey an additional site.

## When and How Often to Survey

Along reaches of the RGB where giant cane management actions are planned, it is important to conduct the first survey before or as soon as possible after the management action is conducted. We have found it much easier to wait to do the first cross-channel survey until after the prescribed burn has been conducted (prescribed burning is one of two main steps used to treat giant cane along this reach). Not only is surveying through dense cane not a lot of fun but establishing line of site and straight transect lines can be almost impossible. An effective prescribed burn addresses these issues. As long as the survey is done soon after the prescribed burn has been completed, it will adequately define pretreatment conditions. The time interval for repeating the surveys should be defined up front. For our work in the RGB, we repeat surveys at least once every three years and/or after occurrence of a flow event of sufficient magnitude and duration to do channel work.

## Survey Setup and Measurement

Below, we summarize several considerations for setting up and conducting your survey based on our experiences along the RGB.

1. Transect locations: For sites or reaches of interest, establish cross sections that allow the main hydraulic controls (i.e., features that control the water-surface slope and hydraulic energy of the stream) of the target site or reach to be captured. For example, if your reach includes such hydraulic controls as riffles or bars, runs and pools, make sure to place cross sections upstream, within, and downstream of these features. Cross sections should also be placed above, in the middle, and downstream from sharp bends, as well as in straight runs. In general, it is best to have a higher concentration of cross sections in the parts of the stream that are "complex," and fewer cross sections in simpler reaches.

(*continued*)

2. Instrument station setup: If using a permanent benchmark or feature as an instrument location (e.g., total station, RTK GPS base station), make sure that the instrument location is established in areas that are unlikely to be eroded or buried by riverine sediment and that allow clear line of sight throughout the monitoring site (and thus the potential of surveying multiple transects without moving the instrument). If establishing a base station location for RTK GPS surveys where there are no permanent georeferenced benchmarks, raw static GPS data logged to the instrument can be post processed to obtain an accurate position using the NGS Online Positioning User Service (OPUS). If a total station or other manual instrument is being used, keep in mind that it may take hours for the surveyor to conduct the survey, so make sure that the instrument is set up in a location that not only allows as broad a view as possible of the monitoring site, but also provides an easy and safe space to stand on (in canyon country, such sites can be a challenge to find). Also, you may need to survey in a local coordinate system if real-world coordinates are unknown. If surveying in a local coordinate system, coordinates of 6000, 5000, 1000 are often chosen as the local easting, northing, and elevation, respectively.

3. Benchmarks: Make sure to establish permanent stable benchmarks at each end of survey transects as well as where the base station is set up. If possible, locations of both cross-channel endpoints as well as the base station locations should be located outside the active bottomland environment and preferably on bedrock where benchmark locations can be permanently marked via etching or with a nail. If no stable bedrock exists in the study area, posts and/or rebar may be used; however, these may be interfered with by livestock, or may be subject to heave/sinking, thereby increasing uncertainty in measurements over time.

4. Transect setup: Regardless of the survey equipment you are using, setting up a tagline between the cross-section endpoints can help the survey team easily follow the transect line; this is likely necessary if using an engineer's level or total station. If using an RTK GPS, some surveying software allows the surveyor to "stake out" a line using the coordinates of the two endpoints, thus negating the need for a tagline. Note that the endpoints of each cross section must be surveyed before using this line stakeout option. If surveying with an engineer's level, a tape measure will need to be used to measure horizontal distance (either placed along the ground surface or clipped to a tagline such that distances can be recorded at each elevation measurement), and it is best to set up the instrument on one of the cross-section endpoints to allow a direct line of site to the next endpoint.

5. Measurement: Point density of the survey should be adequate to define breaks in slope. Thus for surfaces that rapidly change in elevation, such as channel banks, levees, terraces, and so on, points may be as dense as a few points per meter. In areas with little topographical variation, survey density can be less, as long as the surface is adequately defined by the survey; point density may be as small as 1 point every ~3 meters.

## Data Analysis

1. Basic graphing: For each cross section, calculate the straight-line distance between the left and right endpoints of each surveyed cross section, as well as its elevation (remember, left/right is always orientated looking downstream). Plot each cross section with the distance from the left endpoint on the x-axis, and elevation on the y-axis. Surveys of the same cross section conducted at different time periods can be laid on top of one another to examine geomorphic change (see graph in this case study).

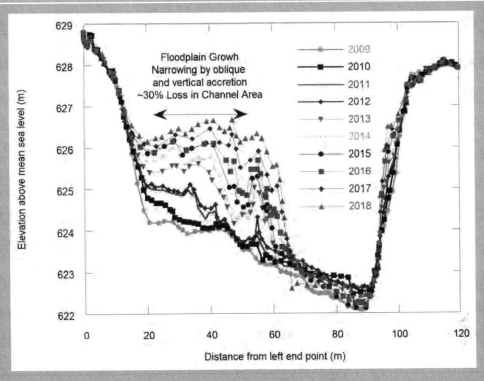

Cross-sectional changes at one of several monitoring sites along the Rio Grande/Bravo in Big Bend National Park. Note that the Rio Grande/Bravo is an extremely dynamic river with substantial geomorphic change every year. Geomorphic change along most rivers are likely to be much less than shown here. Graph by David Dean.

All of the above (as well as much of the initial analyses below) can be done in Microsoft Excel or other spreadsheet/graphing programs.

2. Initial analyses: Calculate the minimum channel-bed elevation of each cross section to quantify the degree of channel-bed aggradation/degradation. Calculate channel width if easily defined. In rivers that experience substantial geomorphic change, and thus have a suite of channel-margin/floodplain surfaces, identifying the boundaries of the channel for width calculations can be subjective. In such cases, calculations of channel area between the two endpoints may be a better metric for analyzing geomorphic changes. When calculating changes in cross-section area, all endpoints must be manually corrected such that the same endpoint coordinates are used every year. This ensures that all calculated changes in channel area are because of geomorphic change, and not uncertainty/variability in the endpoint coordinates. Generally, the endpoint coordinates obtained in the first year of surveys are maintained and used for the cross-sectional area analyses of each successive year. If aggradation occurs at the endpoints, the new elevation of the endpoints should be used for that year, and successive years. On the Rio Grande, many geomorphic surfaces exist that are inundated every year, thereby making determinations of channel width a subjective measure that confounds analyses of channel width. Thus cross-section area is the primary attribute used to analyze geomorphic change; cross-section area includes changes in both width and bed elevation (see graph in this case study as an example).

(*continued*)

## CASE STUDY 7.7
(*continued*)

### Other Considerations

1. Define a specific set of surveying codes to characterize features of interest (e.g., TOB = top of bank, LEP = left end point). See Standard Operating Procedure (SOP) #8 in Big Rivers Monitoring Protocol for Park Units in the Northern Colorado Plateau Network (Big Rivers Monitoring Protocol) for examples (https://irma.nps.gov/DataStore/Reference/Profile/2254835).

2. Consider making facies maps to characterize habitats and their change over time. See SOP #14 in Big Rivers Monitoring Protocol for Park Units in the Northern Colorado Plateau Network for examples (https://irma.nps.gov/DataStore/DownloadFile/604463).

3. If the aim of surveying is to acquire the necessary topographical data to build hydraulic models to investigate inundation extent, or water-surface elevations of floods, make sure cross sections are placed close enough to fully characterize the hydraulic environment. For example, for a riffle/pool sequence, cross sections should be positioned just upstream of the riffle, in the middle of the riffle, just downstream of the riffle, in the pool, and at the downstream end of the pool. Hydraulic models can be calibrated using surveyed high-water marks or by installing transducers to obtain water-surface elevations at the downstream end of your reach (Dean and Schmidt 2013; SOP #7 in Big Rivers Monitoring Protocol [https://irma.nps.gov/DataStore/DownloadFile/604463]).

4. If surveying with a total station or an RTK GPS, complete digital elevation models of bars/floodplains, channel features can be made. These digital elevation models can be compared between successive years, and volumetric calculations of erosion and deposition can be determined (see Wheaton et al. 2010; Erwin et al. 2012; Bangen et al. 2014; Big Rivers Monitoring protocol).

Longitudinal profile surveys are typically done along the thalweg of the river (the centerline of the channel), which is obviously much easier to do during low-flow periods. For cross-channel transects, make certain to install endpoint monuments at both ends of the transect outside the influence of one hundred–year flow events. The channel cross-sections should encompass the active floodplain surfaces and as much of the surrounding terrace surfaces as possible, given limitations of funding, landownership, and land use. For wide channels on large river systems (i.e., where distances may make it difficult to establish monuments outside the active channel on every transect), surveyors could tie the survey end points into at least one benchmark outside the one hundred–year surface. Other considerations are reviewed from experiences monitoring channel morphology along the reach of the Rio Grande/Bravo that forms the international border between the United States and Mexico (see Case Study 7.7).

## Use of Remote Sensing and Geographic Information Technologies in Monitoring

Remote sensing, the science of obtaining information without making physical contact with the target, includes data collected by numerous devices ranging from ground-based cameras to aerial and satellite platforms. Perhaps the foundation of remote sensing and the evolution of its applications arose from the combined use of artificial satellites and a GPS, a system that uses multiple navigational satellites to allow users to accurately locate a point on the earth. Over the last twenty-five years, the advancements in remote sensing technologies and Geographical Information Systems—or Science—(GIS) have been utilized extensively to assess natural resource conditions worldwide. GIS makes spatial assessments of data and can bring multiple types of data (including data collected via airborne and satellite remote sensing platforms) into the same analysis scenario to help guide environmental and natural resource management in all regions in the world.

Remote sensing data can be collected using a variety of platforms. Aerial photography takes advantage of the bird's-eye view by capturing images of the landscape from above the canopy at varying distances. Cameras can be attached to fixed-wing aircraft or remote-controlled drones (i.e., unmanned aerial vehicles) to produce quality images of small- to medium-scale stream reaches (Cruzan et al. 2016). Regardless of the application, deploying airborne sensors with subsequent calibration to ground control points allows for myriad monitoring opportunities at very high resolution over large areas. While there are benefits and disadvantages to remote sensing versus pedestrian surveys, the ability to cover large areas makes remote sensing a desirable strategy, particularly when pedestrian surveys are impractical due to coverage or access issues.

Satellite imagery has a distinct advantage because it is used to collect relevant data of high spectral, spatial, radiometric, and temporal resolution. Remote sensing satellites (such as the Landsat Program) complement global positioning satellites by providing a variety of data that can be used to assess weather conditions, bathymetry, vegetation, biomass, soil moisture, and thermal conditions, among other things. Recent efforts have combined cameras with infrared sensors to document physiological traits of plants. Multispectral cameras have the capability to monitor normalized difference change indices; these include several vegetation indices (VIs) such as Normalized Difference VI (NDVI), Soil Adjusted VI (SAVI), and the Enhanced VI (EVI), which are indicators of plant health and condition (Bannari et al. 1995; Nagler et al. 2012). Other indices include the Normalized Difference Water Index (NDWI) and Normalized Difference Burn Index (NDBI), which can be used to monitor physical changes such as water-surface area and recent burn footprints, respectively. Visible and infrared cameras also can be attached to many platforms, including monitoring towers.

LiDAR is another remote sensing method that emits intense, focused beams of light and measures the time it takes for the reflections to be detected by the sensor (NOAA Coastal Services Center 2012). The ability of LiDAR to measure variable distances from the target to the source at very high densities takes advantage of first return and ground signals to differentiate between ground surface and other structures (trees, buildings, etc.). LiDAR can be used to differentiate tree type (sometimes species), estimate biomass, and determine leaf area by surveying targeted locations during leaf-on and leaf-off seasons. It can detect subtle topographical changes and can be used to determine changes in river terraces and banks as well as channel geomorphological change. LiDAR can be used on the ground or on an aircraft. The disadvantage of LiDAR is that it requires substantial memory storage and specific software to analyze huge datasets.

Aerial photography and satellite imagery are, perhaps, the most valuable remote sensing products for monitoring restoration outcomes because many scenes have been imaged and archived throughout the years and are often accessible from open sources online. Google Earth, for instance, now has historical imagery available for viewing. While some satellite remote sensing analyses can be prohibitively expensive, imagery from the Moderate-Resolution Imaging Spectroradiometer (MODIS) and Landsat can be acquired for free (through the USGS Landsat Global Archive, for example). Most aerial and satellite images have spatial resolution that is high enough for vegetation to be distinguishable from water, bare soil, and human perturbations (e.g., asphalt/dirt roads, buildings). Common vegetation indices that can be measured remotely include leaf area index, the fraction of photosynthetically active radiation (fAPAR) absorbed, and plant water use, which are all valuable measures of plant health and condition. Computer imaging software can be used to convert aerial pictures to digital images (Wolf and Dewitt 2000), which allows distinct groups of vegetation or other land covers to be manually digitized and subsequently classified. A summary of remote sensing products and their spatial and temporal resolutions, costs, and utility are outlined in table 7.5.

Whether collected by aerial or satellite platforms, remote sensing data should be incorporated with other biophysical data that reflect the target region's topography, climate, soils, plant communities, political boundaries, land use, and other factors. To accomplish this, ground verification and calibration of the collected data is critical, ensuring that the interpreted elements of the remote sensing data conform to ground conditions. This is usually accomplished by dividing the area of interest into quadrants or transects so that locations on the ground can be accurately identified and georeferenced with the help of GPS technologies. Note that most imagery acquired from formal open-source programs such as MODIS or Landsat will already be georeferenced and preprocessed to a certain quality control level. With advances in technology, equipment, and tools, the use of remote sensing technologies in your monitoring program has become a cost-effective and realistic option for many stream practitioners. As open public sources for data sharing become more popular, more high-resolution imagery will become available (see http://publiclaboratory.org for landscape monitoring using balloons and kites, for example). Data can be collected and analyzed at temporal and spatial scales that vary with the techniques employed. Several local agencies and governments (i.e., city, county, state) acquire high-resolution products for use in local planning, many of which are viewable, if not downloadable, online. It could be valuable to seek out local entities to inquire if data products are available for free or at low cost for your particular study area.

Change detection for GIS is an approach that identifies differences between the state of land features or land cover over the same spatial extent by observing them at different times (Campbell and Wynne 2011). Changes can be detected manually by an observer or analyst or determined statistically using computer software. Imaging software such as ArcGIS can be used to digitize vector images for computer analysis. This process changes the analog image to a raster image, which is a grid of pixels (i.e., picture elements) that can be investigated using computer software. After different land covers are identified and classified, measurement tools in software programs can measure the total area occupied by each class. It is important to choose images taken during the same time of year to account for changes due to seasonal variability. In addition, natural variations from year-to-year can be analyzed and accounted for by acquiring yearly images for additional change-detection analyses. In the simplest sense, this approach will allow the user to monitor such changes as vegetation area after a restoration

**TABLE 7.5** Summary of selected tools/resources that are used to acquire remote sensing data

| | Spatial resolution (m) | Pan spatial resolution (m) | Multispectral resolution (main bands)[a] | Temporal resolution (days) | Availability | Overall costs[b] (USD) | Applications |
|---|---|---|---|---|---|---|---|
| Camera | 12–16 mp | N/A | R, G, B | user defined | N/A | 50–500 | Portable digital camera, time-lapse |
| Phenocam | 5+ mp | N/A | R, G, B | user defined | N/A | 250–5000 | Fixed platform, motion or time-lapse |
| Gigapixel | 5 mp | N/A | R, G, B | user defined | N/A | 30,000+ | Fixed, time-lapse, ecosystem scale |
| MODIS-Terra/Aqua | 250–1,000 | x | R, G, B, IR, TIR | 8 to 16 | 1999–present | free | Change detection, NDVI/EVI, land surface temp, landscape scale |
| Landsat (5,7,8)[c] | 30 | 15 | R, G, B, NIR, TIR | 16 | 1984–present | free | Change detection, VIs, river-reach scale |
| Sentinel-2 | 10–20 | | R, G, B, Re, NIR, NIRn, SWIR | 5–10 | 2015–present | free | Change detection, VIs, river-reach scale |
| IKONOS | 3.2 | 0.82 | R, G, B, IR | <3 | 1999–2014 | 250–1,000s | Change detection, VIs, species scale |
| Quickbird | 2.44 | 0.61–0.65 | R, G, B, IR | 3.5 | 2001–2014 | 350–1,000s | Change detection, VIs, species scale |
| WorldView-3 | 1.24 | 0.31 | R, G, B, NIR, SWIR | <1 | 2014–present | 350–1,000s | Change detection, VIs, species scale |
| Geoeye-1 | 1.65 | 0.41 | R, G, B, IR | | 2008–present | 350–1,000s | Change detection, VIs, species scale |
| LiDAR | 1–20 cm (v); 20–2000 cm (h) | N/A | N/A | N/A | N/A | free–100,000s | Elevation, contours, canopy cover, wetland delineation, inundation mapping, geomorphic change |

[a] Red (R), green (G), and blue (B) bands represent human vision and general camera products, whereas infrared (IR, [i.e.; near (N), shortwave (SW), and thermal (T)]) are wavelengths outside of human detection limits, but useful for vegetation health analyses and cloud/aerosol masking.

[b] Costs are highly variable, dependent on desired specifications. Overall costs for cameras include estimates for setup and installation (e.g., platforms, batteries, solar panels, remote/internet, software). For satellite product acquisition, minimum range in cost accounts for minimum area required for purchasing scenes from vendors. Prices do not include analysis software. Satellite imagery minimum costs are based on cost/scene and minimum area required per order.

[c] Landsat 5 (TM, available 1984–2013); Landsat 7 (ETM+, available 2003–present); Landsat 8 (OLI, available 2013–present). Landsat 9 (OLI-2, availability unknown), a copy of Landsat 8, is set to launch 2020.

# CASE STUDY 7.8
Assessing Channel Geomorphic Change Along the Big Bend Reach of the
Rio Grande / Río Bravo Using Survey Data Collected from Different Time Periods
Using Different Equipment and Techniques

Kevin Urbanczyk

## Introduction
Recent studies on the geomorphology of sand and gravel bars along the RGB in Big Bend National Park exemplify how a combination of survey technologies (in this case GIS, GPS, LiDAR, and Total Station) can be applied to assess and address an environmental problem (Urbanczyk and Bennett 2012). The project is designed to track geomorphic change in a river system that is in sediment surplus. Along the Big Bend reach of the RGB, the combined impacts of river impoundment, water diversion, the spread of invasive species, and a host of upland disturbances that increased tributary sediment input into the river have transformed this once wide, shallow, meandering channel of the Rio Grande that had little vegetation into a narrow and deep channel that is dominated by invasive vegetation (giant cane [*Arundo donax*] and tamarisk [*Tamarix ramosissima*]). In particular, giant cane has become established in dense stands along channel margins and is contributing to channel narrowing by stabilizing recently deposited sediment and promoting sediment deposition during periodic flooding (Dean and Schmidt 2013). Channel-narrowing processes have buried high-quality riparian and aquatic habitat, making the reversal of these processes a priority for a binational team of agencies, institutes, NGOs, and riverside communities that are working in this region (see Case Study 6.5).

## Study Objective and Location
Between 2011 and 2019, multiple state and federal agencies, institutions, and NGOs embarked on a concerted effort to eradicate giant cane along channel margins on both sides of the river in the Big Bend region (i.e., on both the U.S. and Mexican sides of the RGB) (see Case Study 6.5). One of the objectives of this binational management effort was to determine if eradication of giant cane would make underlying alluvium more vulnerable to scour and evacuation by prevailing discharges. In other words, could management of giant cane help stem, and potentially even reverse, channel narrowing processes? Given this management objective, monitoring of channel morphology conditions along reaches where management of giant cane took place became a priority. However, limited resources for monitoring precipitated an opportunistic monitoring response that involved multiple players using different survey equipment and techniques. Although far from ideal, such a monitoring scenario where different groups are involved over the long term using survey equipment that is available to them is probably more common than not. The objective of this pilot study was to identify an approach that allowed accurate quantification of channel morphological change using survey data collected with different survey equipment and techniques. If such could be done in a satisfactory manner, the approach would allow an assessment of channel change throughout the Big Bend reach before, during and after giant cane management.

## Methods and Results
Surveys of the sand bars at the entrance to Boquillas Canyon (see map and photo in this case study) began in 2004 and involved using total stations to measure topographical variation across three cross

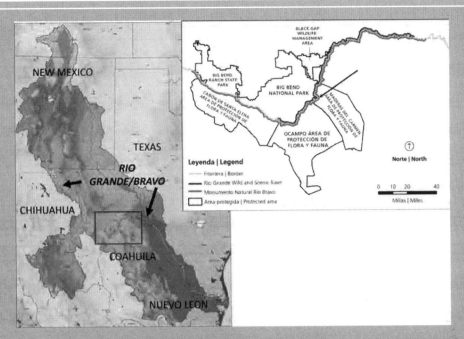

Map showing the location of this pilot study along the RGB at the entrance to Boquillas Canyon in Big Bend National Park just downstream of the boundary between the protected areas of Ocampo and Maderas del Carmen, Mexico.

sections of the sandbars. In late September to early October 2008, a high-magnitude flood (of a statistical recurrence interval of around fifteen to twenty-five years) scoured much of the vegetation and fine sediments from the sandbar. Postflood topographical surveys of the sandbar were conducted in the fall of 2011 and again in the fall of 2013 using both a Total Station and RTK* GPS unit. To add to the postflood dataset, an aerial LiDAR survey of the entire Rio Grande Basin reach in Big Bend National Park, which included all of Boquillas Canyon, including the entrance sandbars, was conducted in the summer of 2012.

The 2011, 2012, and 2013 datasets were processed into three-dimensional models and analyzed using the geomorphic change detection (GCD) process described by Wheaton et al. (2010), which compares digital elevation models (DEMs, which are raster maps that depict elevation) from different periods and calculates a DEM of Difference (DOD), which is a measurement of volumetric change where surface elevations have increased or decreased. The technique also accounts for individual DEM uncertainties and propagates the uncertainty errors through to the resultant DOD in the form of thresholds that exclude undetectable changes. Methods to extrapolate volumetric change using the 2004 cross-section dataset are described in Urbanczyk and Bennett 2012 and provide the basis for assessing nearly a decade of geomorphic change at this location.

To complete the GCD analysis, DEMs for the survey data collected in 2011 and 2013 had to be constructed, which involved delineating polygon areas of interest that are covered by the data for each survey campaign. The field point data were processed into UTM coordinates and merged into a single GIS layer that was converted to a triangular irregular network (TIN)—TINs are a means to represent surface morphology digitally—and then converted to a DEM. The 2012 aerial LiDAR data were already

*(continued)*

processed into a DEM and, therefore, already suitable for the GCD analysis. The final product of this analysis is the DOD (DEM of Difference). This is a map that shows in detail the degree of topographic change over time in the surveyed area. Among other considerations, such quantification of channel morphologic change allows thoughtful assessment and selection of restoration sites with conditions suitable to the realization of restoration objectives.

## Take Home Points and Lessons Learned

1. *Addressing an Important Monitoring Need.* Although not ideal, data collection as part of a long-term monitoring program is often done ad hoc by different entities who are taking advantage of window opportunities and resource availability. In this light, methods that allow robust analysis of datasets collected using different sampling methodologies can be extremely useful to managers and other decision makers, providing the potential to quantify long-term change that may not have been previously possible.

2. *Analysis is Technical, So Help May Be Required.* The analytical methods summarized here are technical and potentially necessitate the participation of someone proficient in the analysis of survey data. That said, the need to analyze datasets collected using different survey techniques is a real-life challenge, one that can likely attract the attention of a professor and students (or other experts) who live near you. Indeed, my participation in this study began and continues with strong participation of my students and river managers.

Channel morphologic change at the entrance of Boquillas Canyon can be dramatic and is typical of the boom and bust nature of many streams of dryland regions. Being able to quantify morphologic change from survey data collected over time using different ethodologies can help practitioners identify sites where restoration funding can be most effectively spent. Photograph by Kevin Urbanczyk.

3. *On-the-Ground Surveys Can Be Done Relatively Cheaply.* Although the 2012 aerial LiDAR campaign described in this case study was expensive (costing roughly $75,000 to provide data for over 166 km [100 miles] of the Rio Grande Basin through Big Bend), the on-the-ground surveys were conducted relatively cheaply, using survey equipment that is not prohibitively expensive or can be rented, making such surveys a realistic possibility for stream practitioners with thin budgets.

*RTK positioning takes advantage of global navigation satellite data and allows centimeter survey accuracy (with the level of accuracy affected by the number of satellites the unit is communicating with that, in turn, can be influenced by both the timing of survey activities as well as on-the-ground obstructions [like high canyon walls]).

treatment or a flood, wet/dry areas of a river or stream, or the extent of sandbar aggradation or sediment deposition.

Other software such as ERDAS Imagine or ENVI image analysis can be used to analyze changes in vegetation indices (NDVI, EVI, etc.). For example, an NDVI image from 2010 can be subtracted from 2000 to yield a new image whereby positive output values indicate vegetation increase and negative values indicate vegetation loss. Remote sensing has been used to monitor the extent of tamarisk leaf defoliation after the introduction of the tamarisk leaf beetle in 2007 along the Dolores River (Dennison et al. 2009; Nagler et al. 2011).

The use of GIS and remote sensing technologies in natural resource monitoring is rapidly evolving and there are many good references that readers can pursue for a deeper dive (Woodward et al. 2018). In addition to the above citations, please consider the following resources, which were selected based more on methodologies that were used and not geography (i.e., some are from nonarid environments, but methods can be adapted for use in other locations):

- Affordable and rapid assessment approaches in stream restoration (Hubbart et al. 2017; Evans et al. 2020);
- Using high-resolution multispectral airborne remote sensing to map riparian vegetation along the Middle Rio Grande (Akasheh et al. 2008);
- Integration of remote sensing and field data to assess setback levees (Konrad et al. 2008);
- Assessing the accuracy of remote sensing data (Congalton and Green 2009);
- Using high-resolution remote sensing data to estimate channel width and transient channel storage zones (Bingham et al. 2012);
- Using ground, aerial, and satellite photography to quantify geomorphic change (Legleiter et al. 2004; Shroder 2013);
- Overview of availability and use of GIS, GPS, and satellite data (Urbanczyk 2017).

Stream practitioners will often be faced with the challenge of using ground and remote sensing data collected using different techniques over time to evaluate biophysical changes along their stream. Case Study 7.8 shows one strategy to overcome this challenge to assess geomorphological change along the Rio Grande/Bravo of the southwestern United States and northern Mexico.

# REFERENCES

Acuña, V., J. R. Díez, L. Flores, M. Meleason, and A. Elosegi. 2013. "Does It Make Economic Sense to Restore Rivers for Their Ecosystem Services?" *Journal of Applied Ecology* 50:988–97.

Aguiar, F. C., P. Segurado, M. J. Martins, M. D. Bejarano, C. Nilsson, M. M. Portela, and D. M. Merritt. 2018. "The Abundance and Distribution of Guilds of Riparian Woody Plants Change in Response to Land Use and Flow Regulation." *Journal of Applied Ecology* 55 (5): 2227–40. https:/doi.org/10.1111/1365-2664.13110.

Akasheh, O. Z., C. M. U. Neale, and H. Jayanthi. 2008. "Detailed Mapping of Riparian Vegetation in the Middle Rio Grande River Using High Resolution Multi-spectral Airborne Remote Sensing." *Journal of Arid Environments* 72 (9): 1734–44.

Baayen, R. H. 2007. the languageR package. https://cran.r-project.org/web/packages/languageR /languageR.pdf.

Bagstad, K. J., D. Semmens, R. Winthrop, D. Jaworski, and J. Larson. 2012. *Ecosystem Services Valuation to Support Decisionmaking on Public Lands—A Case Study of the San Pedro River Watershed, Arizona*. U.S. Geological Survey Scientific Investigations Report 2012–5251.

Bangen, S. G., J. M. Wheaton, N. Bouwes, B. Bouwes, and C. Jordan. 2014. "A Methodological Intercomparison of Topographic Survey Techniques for Characterizing Wadeable Streams and Rivers." *Geomorphology* 206:343–61.

Bannari, A., D. Morin, F. Bonn, and A. R. Huete. 1995. "A Review of Vegetation Indices." *Remote Sensing Reviews* 13:95–120.

Barcelona, M. J., J. P. Gibb, J. A. Helfrich, E. E. Garske. 1985. *Practical Guide for Ground-Water Sampling. Illinois State Water Survey, SWA Contract Report 374*. November 1985.

Barnett, D. T., and T. J. Stohlgren. 2003. "A Nested-Intensity Design for Surveying Plant Diversity." *Biodiversity and Conservation* 12:255–78. https://doi.org/10.1023/A:1021939010065.

Bateman, H. L., M. J. Harner, and A. Chung-MacCoubrey. 2008. "Abundance and Reproduction of Toads (*Bufo*) Along a Regulated River in the Southwestern United States: Importance of Flooding in Riparian Ecosystems." *Journal of Arid Environments* 72:1613–19.

Baxter, C., F. R. Hauer, and W. W. Woessner. 2003. "Measuring Groundwater–Stream Water Exchange: New Techniques for Installing Minipiezometers and Estimating Hydraulic Conductivity." *Transactions of the American Fisheries Society* 132:493–502.

Behn, K. E., and C. V. Baxter. 2019. "The Trophic Ecology of a Desert Riverfish Assemblage: Influence of Seasonand Hydrologic Variability." *Ecosphere* 10 (1). https://doi.org/10.1002/ecs2.2583.

Bejarano, M. D., C. Nilsson, and F. C. Aguiar. 2018. "Riparian Plant Guilds Become Simpler and Most Likely Fewer Following Flow Regulation." *Journal of Applied Ecology* 55 (1):365–76.

Bernhardt, E., M. A. Palmer, J. D. Allan, G. Alexander, K. Barnas, S. Brooks, J. Carr, et al. 2005. "Synthesizing U.S. River Restoration Efforts." *Science* 205:636–37.

Bernhardt, E., and M. Palmer. 2007. "Restoring Streams in an Urbanizing World." *Freshwater Biology* 52:738–51.

Bernhardt, E., and M. Palmer. 2011. "River Restoration: The Fuzzy Logic of Repairing Reaches to Reverse Catchment Scale Degradation." *Ecological Applications* 21 (6): 1926–31. https://doi.org/10.1890/10-1574.1.

Bhattacharjee, J., J. P. Taylor, and L. M. Smith. 2006. "Controlled Flooding and Staged Drawdown for Restoration of Native Cottonwoods in the Middle Rio Grande Valley, New Mexico, USA." *Wetlands* 26 (3): 691–702.

Bibby, C. J., N. D. Burgess, D. A. Hill, and S. H. Mustoe. 2000. *Bird Census Techniques*. 2nd ed. Elsevier Press.

Bingham, Q. G., B. T. Neilson, C. M. U. Neale, and M. B. Cardenas. 2012. "Application of High-Resolution, Remotely Sensed Data for Transient Storage Modeling Parameter Estimation." *Water Resources Research* 48:1–15.

Bissels, S., N. Hölzel, T. W. Donath, and A. Otte. 2004. "Evaluation of Restoration Success in Alluvial Grasslands Under Contrasting Flooding Regimes." *Biological Conservation* 118 (5): 641–50.

Bliese, P. 2006. *Multilevel Modeling in R (2. 2): A Brief Introduction to R, the Multilevel Package and the Nlme Package*. Washington D.C.: Walter Reed Army Institute of Research.

BLM. 1996. *Sampling Vegetation Attributes*. Interagency Technical Reference. BLM National Applied Resource Sciences Center. BLM/RS/ST-96/002+1730. Supersedes BLM Technical Reference 4400–

4, Trend Studies, May 1995, Bureau of Land Management. https://www.blm.gov/learn/blm-library/agency-publications/technical-references.

Block, W. M., A. B. Franklin, J. P. Ward Jr., J. L. Ganey, and G. C. White. 2001. "Design and Implementation of Monitoring Studies to Evaluate the Success of Ecological Restoration on Wildlife." *Restoration Ecology* 9:293–303.

Blossy, B. 1999. "Before, During and After: The Need for Long-Term Monitoring in Invasive Plant Species Management." *Biological Invasions* 1:301–11.

Bonar, S. A., W. A. Hubert, and D. W. Willis. 2009. *Standard Methods for Sampling North American Freshwater Fishes.* American Fisheries Society. http://www.openarchives.org/OAI/2.0/oai_dc/; http://www.openarchives.org/OAI/2.0/oai_dc.xsd.

Bonar, S. A., N. Mercado-Silva, W. A. Hubert, T. D. Beard Jr., D. Goran, J. Kubečka, B. D. S. Graeb, et al. 2017. "Standard Methods for Sampling Freshwater Fishes: Opportunities for International Collaboration." *Fisheries* 42 (3): 150–56.

Boulton, A. J. 2007. "Hyporheic Rehabilitation in Rivers: Restoring Vertical Connectivity." *Freshwater Biology* 52:632–50.

Braun-Blanquet, J. 1932. *Plant Sociology.* Translated by G. D. Fuller and H. S. Conrad. New York: McGraw-Hill.

Brierley, G., R. Helen Reid, K. Fryirs, and N. Trahan. 2010. "What Are We Monitoring and Why? Using Geomorphic Principles to Frame Eco-hydrological Assessments of River Condition." *Science of the Total Environment* 408:2025–33.

Brudvig, L. A. 2011. "The Restoration of Biodiversity: Where Has Research Been and Where Does It Need to Go?" *American Journal of Botany* 98:549–58. https:/doi.org/10.3732/ajb.1000285.

Bullock, J. M. 2006. "Plants." In *Ecological Census Techniques.* 2nd Ed, edited by W.J. Sutherland, 186–Cambridge: Cambridge University Press.

Bunting, D. B., S. Kurc, and M. Grabau. 2013. "Long-Term Vegetation Dynamics After High-Density Seedling Establishment: Implications for Riparian Restoration and Management." *River Research and Applications* 29 (9): 1119–30.

Campbell, J. B., and R. H. Wynne. 2011. "Change Detection" In *Introduction to Remote Sensing,* edited by J. B. Campbell and R. H. Wynne, 445–64. 5th ed. https://www.guilford.com/books/Introduction-to-Remote-Sensing/Campbell-Wynne/9781609181765/contents.

Catalano, M. J., M. A. Bozek, and T. D. Pellett. 2007. "Effects of Dam Removal on Fish Assemblage Structure and Spatial Distributions in the Baraboo River, Wisconsin." *North American Journal of Fisheries Management* 27:519–30.

Clarke, K. R. 1993. "Non-parametric Multivariate Analyses of Changes in Community Structure." *Australian Journal of Ecology* 18:117–43.

Coles-Ritchie, M. C., R. C. Henderson, E. K. Archer, and C. Kennedy. 2004. *Repeatability of Riparian Vegetation Sampling Methods: How Useful Are These Techniques for BroadScale, Long-Term Monitoring?* USDA Forest Service—General Technical Report RMRS-GTR. https://www.fs.usda.gov/rmrs/publications/series/general-technical-reports?field_citation_value_op=word&field_citation_value=&field_abstract_value_op=word&field_abstract_value=GTR+RM-245.

Congalton, R. G., and K. Green. 2009. *Assessing the Accuracy of Remotely Sensed Data: Principles and Practices.* 2nd ed. Boca Raton, Fla.: CRC Press Taylor & Francis Group. https://doi.org/10.1016/j.jag.2009.07.002.

Cooper, D. J., and D. C. Andersen. 2012. "Novel Plant Communities Limit the Effects of a Managed Flood to Restore Riparian Forests Along a Large Regulated River." *River Research and Applications* 28 (2): 204–15.

Cottam, G., and J. T. Curtis. 1956. "The Use of Distance Measure in Phytosociological Sampling." *Ecology* 37:451–60.

Coulloudon, B., K. Eshelman, J. Gianola, N. Habich, L. Hughes, C. Johnson, M. Pellant, et al. 1999. *Sampling Vegetation Attributes*. Technical Reference 1734–4. Denver, Colo.: Bureau of Land Management.

Cruzan, M. B., B. G. Weinstein, M. R. Grasty, B. F. Kohrn, E. C. Hendrickson, T. M. Arredondo, and P. G. Thompson. 2016. "Small Unmanned Aerial Vehicles (Micro-UAVs, Drones) in Plant Ecology." *Journal of Applied Plant Science* 4 (9). https://doi.org/10.3732/apps.1600041.

Cuffney, T. F., M. E. Gurtz, and M. R. Meador. 1993. *Methods for Collecting Benthic Invertebrate Samples as Part of the National Water-Quality Assessment Program*. Open-File Report 93–406. Raleigh, N.C.: U.S. Geological Survey.

Cunningham, W. L., and C. W. Schalk, comps. 2011. *Groundwater Technical Procedures of the U.S. Geological Survey: U.S. Geological Survey Techniques and Methods 1–A1* https://pubs.er.usgs.gov/.

Dalton, M. G., B. E. Huntsman, and K. Bradbury. 1991. "Acquisition and Interpretation of Water-Level Data." In *Practical Handbook of Ground-Water Monitoring*, edited by D. M. Nielsen, 367–97. Chelsea, Mich.: Lewis.

Darrah, A. J., and C. van Riper. 2018. "Riparian Bird Density Decline in Response to Biocontrol of Tamarix from Riparian Ecosystems Along the Dolores River in SW Colorado, USA." *Biological Invasions* 20 (3): 709–20.

Davies, P. E., J. H. Harris, T. J. Hillman, and K. F. Walker. 2010. "The Sustainable Rivers Audit: Assessing River Ecosystem Health in the Murray–Darling Basin, Australia." *Marine and Freshwater Research* 61:764–77.

Dean, D. J., and J. C. Schmidt. 2013. "The Geomorphic Effectiveness of a Large Flood on the Rio Grande in the Big Bend Region: Insights on Geomorphic Controls and Post-flood Geomorphic Response" *Geomorphology* 201:183–98.

De Boeck, P., M. Bakker, R. Zwitser, M. Nivard, A. Hofman, F. Tuerlinckx, and I. Partchev. 2011. "The Estimation of Item Response Models with the Lmer Function from the Lme4 Package in R." *Journal of Statistical Software* 39 (12): 1–28.

De la Maza Benignos, M., C. Aguilera, L. Eguiarte, O. Leal-Nares, S. Luna, R. Mendoza, D. Condado, J. Montemayor Leal, and V. Souza Saldivar. 2017. *Cuatro Ciénegas y su estado de conservación a través de sus peces*. Pronatura Noreste, A.C. Report can be obtained from CONANP, Área de Protección de Cuatro Cienegas, Cuatro Cienegas, Coahuila, Mexico.

Dennison, P. E., P. L. Nagler, K. R. Hultine, E. P. Glenn, and J. R. Ehleringer. 2009. "Remote Monitoring of Tamarisk Defoliation and Evapotranspiration Following Saltcedar Leaf Beetle Attack." *Remote Sensing of Environment* 113:1462–72.

Dickinson, J., B. Zuckerberg, D. Bonter. 2010. "Citizen Science as an Ecological Research Tool: Challenges and Benefits." *Annual Review of Ecology, Evolution, and Systematics* 41:149–72.

Dobriyal, P., R. Badola, C. Tuboi, and S. A. Hussain. 2017. "A Review of Methods for Monitoring Streamflow for Sustainable Water Resource Management." *Applied Water Science* 7:2617–28.

Densmore, R. V., and K. F. Karle. 2009. "Flood Effects on an Alaskan Stream Restoration Project: The Value of Long-Term Monitoring." *Journal of the American Water Resources Association* 45 (6): 1424–33.

Doyle, M. W., E. H. Stanley, C. H. Orr, A. R. Selle, S. A. Sethi, S. A., and J. M. Harbor. 2005. "Stream Ecosystem Response to Small Dam Removal: Lessons from the Heartland." *Geomorphology* 71:227–44.

Dufour, S., P. M. Rodríguez-González, and M. Laslier. 2018. "Tracing the Scientific Trajectory of Riparian Vegetation Studies: Main Topics, Approaches and Needs in a Globally Changing World." *Science of the Total Environment* 653:1168–85.

Dufrene, M., and P. Legendre. 1997. "Species Assemblages and Indicator Species: The Need for a Flexible Asymmetrical Approach." *Ecological Monographs* 67 (3): 345–66.

Edwards, P. M., G. Shaloum, and D. Bedell. 2018. "A Unique Role for Citizen Science in Ecological Restoration: A Case Study in Streams." *Restoration Ecology* 26 (1): 29–35.

Edwards, R. J., and G. P. Garrett. 2013. *"Biological Monitoring of the Repatriation Efforts for the Endangered Rio Grande Silvery Minnow (*Hybognathus amarus*) in Texas."* Austin: Texas Parks and Wildlife Department.

Elzinga, C. L. 2001. "Results of 15 Years of Monitoring in the Coram Research Natural Area, Montana." Unpublished report. Forestry Sciences Laboratory, Missoula, Mont.

Elzinga, C. L., J. W. Salzer, and J. W. Willoughby. 1998. *Measuring and Monitoring Plant Populations.* Denver, Colo: U.S. Department of the Interior, Bureau of Land Management, National Applied Resource Sciences Center.

England, J., and M. A. Wilkes. 2018. "Does River Restoration Work? Taxonomic and Functional Trajectories at Two Restoration Schemes." *Science of the Total Environment* 618:961–70.

Erwin, S. O., J. C. Schmidt, J. M. Wheaton, and P. R. Wilcock. 2012. "Closing a Sediment Budget for a Reconfigured Reach of the Provo River, Utah, United States." *Water Resources Research* 48 (10). https://doi.org/10.1029/2011WR011035.

Espinosa-Pérez y Salinas Rodríguez. 2014. *Protocolo de muestreo de peces en aguas continentales para la aplicación de la Norma de Caudal Ecológico.* Programa Nacional de Reservas de Agua. Mexico City: CONAGUA.

Evans, A. D., K. H. Gardner, S. Greenwood, and B. Pruitt. 2020. *Exploring the Utility of Small Unmanned Aerial System (sUAS) Products in Remote Visual Stream Ecological Assessment.* https://doi.org/10.1111/rec.13228.

Evans, R. A., and R. M. Love. 1957. "The Step-Point Method of Sampling: A Practical Tool in Range Research." *Journal of Range Management* 10:208–12.

Fancy, S. G., and R. E. Bennetts. 2012. "Institutionalizing an Effective Long-Term Monitoring Program in the U.S. National Park Service." In *Design and Analysis of Long-term Ecological Monitoring Studies,* edited by R. A. Gitzen et al., 481–97. Cambridge: Cambridge University Press.

Farid, A. D., C. Goodrich, and S. Sorooshian. 2006. "Using Airborne Lidar to Discern Age Classes of Cottonwood Trees in a Riparian Area." *Western Journal of Applied Forestry* 21:149–58.

Fisher, R. A. 1938. Presidential Address to the First Indian Statistical Congress, 1938 https://www.gwern.net/docs/statistics/decision/1938-fisher.pdf.

Fowler, J., L. Cohen, and P. Jarvis. 2013. *Practical Statistics for Field Biology.* John Wiley.

Fryirs, K. 2003. "Guiding Principles for Assessing Geomorphic River Condition: Applications of a Framework in the Bega Catchment, South Coast, New South Wales, Australia." *Catena* 53:17–52.

Fryirs, K., A. Arthington, and J. Grove. 2008. "Principles of River Condition Assessment." In *River Futures: An Integrative Scientific Approach to River Repair,* edited by G. J. Brierley and K. A. Fryirs, 100–24. Washington D.C.: Island Press.

Fryirs, K., and G. J. Brierley 2009. "Naturalness and Place in River Rehabilitation." *Ecology and Society* 14 (1): 20. http://www.ecologyandsociety.org/vol14/iss1/art20.

Fryirs, K., and G. J. Brierley. 2012. *Geomorphic Analysis of River Systems: An Approach to Reading the Landscape.* Wiley Online Library. https://onlinelibrary.wiley.com/doi/book/10.1002/9781118305454.

Gazendam, E., B. Gharabaghi, J. D. Ackerman, and H. Whiteley. 2016. "Integrative Neural Networks Models for Stream Assessment in Restoration Projects." *Journal of Hydrology* 536:339–50.

Giling, D. P., R. Mac Nally, and R. M. Thompson. 2015. "How Sensitive Are Invertebrates to Riparian-Zone Replanting in Stream Ecosystems?" *Marine and Freshwater Research* 67 (10). http://dx.doi.org /10.1071/MF14360.

Gimblett, R., C. A. Scott, and M. Hammersley. 2017. "Rewilding, Sacred Spaces, and 'Outstanding Remarkable Values.'" *International Journal of Wilderness* 23 (2). https://ijw.org/dam-removal-white -salmon-river/.

Gollan, J. R., L. L. De Bruyn, N. Reid, D. Smith, and L. Wilkie. 2011. "Can Ants Be Used as Ecological Indicators of Restoration Progress in Dynamic Environments? A Case Study in a Revegetated Riparian Zone." *Ecological Indicators* 11 (6): 1517–25.

González, E., A. Masip, E. Tabacchi, and M. Poulin. 2017. "Strategies to Restore Floodplain Vegetation After Abandonment of Human Activities." *Restoration Ecology* 25:82–91.

González, E., A. A. Sher, E. Tabacchi, A. Masip, and M. Poulin. 2015. "Restoration of Riparian Vegetation: A Global Review of Implementation and Evaluation Approaches in the International, Peer-Reviewed Literature." *Journal of Environmental Management* 158:85–94.

Gordon, N. D., T. A. McMahon, B. L. Finlayson, C. J. Gippel, and R. J. Nathan. 2004. *Stream Hydrology: An Introduction for Ecologists.* 2nd ed. Chicheste, UK: John Wiley.

Gotelli, N. J., and A. M. Ellison. 2013. *A Primer of Ecological Statistics.* Oxford: Oxford University Press.

Grainger, N., J. Goodman, and D. West. 2013. *Introduction to Monitoring Freshwater Fish.* Version 1.1. Department of Conservation, New Zealand. https://www.doc.govt.nz/Documents/science -and-technical/inventory-monitoring/im-toolbox-freshwater-fish/im-toolbox-freshwater-fish -introduction-to-monitoring-freshwater-fish.pdf.

Greipsson, S. 2012. *Restoration Ecology.* Burlington, Mass.: Jones and Bartlett Learning.

Gungle, B., J. B. Callegary, N. V. Paretti, J. R. Kennedy, C. J. Eastoe, D. S. Turner, J. E. Dickinson, L. R. Levick, Z. P. and Sugg. 2016. *Hydrological Conditions and Evaluation of Sustainable Groundwater Use in the Sierra Vista Subwatershed, Upper San Pedro Basin, Southeastern Arizona.* Version1.1, October 2016. U.S. Geological Survey Scientific Investigations Report 2016–5114. http://dx.doi.org /10.3133/sir20165114.

Gurnell, A., A. J. Boitsidis, K. Thompson, and N. J. Clifford. 2006. "Seed Bank, Seed Dispersal and Vegetation Cover: Colonization Along a Newly-Created River Channel." *Journal of Vegetation Science* 17 (5): 665–74.

Gurtin, S. D., J. E. Slaughter IV, S. J. Sampson, and R. H. Bradford. 2003. "Use of Existing and Reconnected Backwater Habitats by Hatchery-Reared Adult Razorback Suckers: A Predictive Model for the Imperial Division, Lower Colorado River." *Transactions of the American Fisheries Society* 132:1125–37.

Haase, P., S. Lohse, S. Pauls, K. Schindehütte, and A. Sundermann. 2004. "Assessing Streams in Germany with Benthic Invertebrates: Development of a Practical Standardized Protocol for Macroinvertebrate Sampling and Sorting." *Limnologica* 34:349–65.

Haines-Young, R., and M. B. Potschin. 2018. "Common International Classification of Ecosystem Services (CICES) V5.1 and Guidance on the Application of the Revised Structure." www.cices.eu.

Hammer, Ø. D., A. Harper, and P. D. Ryan. 2001. "PAST: Paleontological Statistics Software Package for Education and Data Analysis." *Palaeontologia Electronica* 4:9.

Hampton, S. E., E. E. Holmes, L. P. Scheef, M. D. Scheuerell, S. L. Katz, D. E. Pendleton, and E. J. Ward. 2013. "Quantifying Effects of Abiotic and Biotic Drivers on Community Dynamics with Multivariate Autoregressive (MAR) Models." *Ecology* 94:2663–69.

Harrelson, C. C., C. L. Rawlins, and J. P. Potyondy. 1994. *Stream Channel Reference Sites: An Illustrated Guide to Field Technique.* U.S. Department of Agriculture, Forest Service, Rocky Mountain Forest and Range Experiment Station. GTR RM-245. https://www.fs.usda.gov/rmrs/publications/series/general -technical-reports.

Hart, D. D., and N. L. Poff. 2002. "A Special Section on Dam Removal and River Restoration." *BioScience* 52:653–55.

HDR. 2014. *Groundwater Monitoring Well Installation and Groundwater Level Monitoring Report: Rio Grande Canalization Project Restoration Sites.* Final report prepared by HDR EOC Spring Branch, Texas for the International Boundary & Water Commission, United States & Mexico, U.S. Section, El Paso, Texas. https://www.ibwc.gov/EMD/reports_studies.html.

Hem, J. D. 1989. *Study and Interpretation of the Chemical Characteristics of Natural Water.* 3rd printing. U.S. Geological Survey Water-Supply Paper 2254. https://pubs.er.usgs.gov/.

Herrera, A. M., and T. L. Dudley. 2003. "Reduction of Riparian Arthropod Abundance and Diversity as a Consequence of Giant Reed *(Arundo donax)* Invasion." *Biological Invasions* 5:167–77.

Herrick, J. E., J. W. Van Zee, K. M. Havstad, L. M. Burkett, and W. G. Whitford. 2005. Monitoring Manual for Grassland, Shrubland and Savanna Ecosystems. Volume I: *Quick Start.* Volume II: *Design, Supplementary Methods and Interpretation.* Las Cruces, N.M.: USDA-ARS Jornada Experimental Range. https://www.fs.usda.gov/Internet/FSE_DOCUMENTS/stelprdb5172121.pdf.

Hester, E. T., M. N. Gooseff. 2010. "Moving Beyond the Banks: Hyporheic Restoration Is Fundamental to Restoring Ecological Services and Functions of Streams." *Environmental Science Technologies* 44:1521–25.

Hill, D. A., M. Fasham, G. Tucker, M. Shewry, and P. Shaw. 2005. *Handbook of Biodiversity Methods: Survey, Evaluation and Monitoring,* 219–22. Cambridge: Cambridge University Press.

Hinojosa-Huerta, O., H. Iturribarría-Rojas, E. Zamora, A. Calvo-Fonseca. 2008. "Densities, Species Richness and Habitat Relationships of the Avian Community in the Colorado River, Mexico." *Studies in Avian Biology* 37:74–82.

Hogg, R., S. M. Coghlan Jr, and J. Zydlewski. 2013. "Anadromous Sea Lampreys Recolonize a Maine Coastal River Tributary After Dam Removal." *Transactions of the American Fish Society* 142:1381–94.

Hogg, R., S. M. Coghlan Jr, J. Zydlewski, and C. Gardner. 2015. "Fish Community Response to a Small-Stream Dam Removal in a Maine Coastal River Tributary." *Transactions of the American Fish Society* 144:467–79.

Holmes, P. M., D. M. Richardson, K. J. Esler, E. T. F. Witkowski, and S. Fourie. 2005. "A Decision-Making Framework for Restoring Riparian Zones Degraded by Invasive Alien Plants in South Africa." *South African Journal of Science* 101:553–64.

Hotzel, G., and R. Croome. 1999. "A Phytoplankton Methods Manual for Australian Freshwaters. Land and Water Resources Research and Development Corporation." Occasional Paper 22/99, GPO Box 2182, Canberra ACT 2601.

Hubbart J, E. Kellner, P. Kinder, K. Stephan. 2017. "Challenges in Aquatic Physical Habitat Assessment: Improving Conservation and Restoration Decisions for Contemporary Watersheds." *Challenges* 8:31.

Hutto, R. L., and R. T. Belote. 2013. "Distinguishing Four Types of Monitoring Based on the Questions They Address." *Forest Ecology and Management* 289:183–89.

Ives, A. R., B. Dennis, K. L. Cottingham, and S. R. Carpenter. 2003. "Estimating Community Stability and Ecological Interactions from Time-Series Data." *Ecological Monographs* 73 (2): 301–30.

Jones, N. E. 2011. *Benthic Sampling in Natural and Regulated Rivers: Sampling Methodologies for Ontario's Flowing Waters*. Ontario Ministry of Natural Resources, Aquatic Research and Development Section, River and Stream Ecology Lab, Aquatic Research Series 2011–05, Ontario, Canada. https://www.ontario.ca/page/benthic-sampling-natural-and-regulated-rivers.

KBO and USFS Pacific Southwest Research Station. 2013. *Bird Monitoring as an Aid to Riparian Restoration: Findings from the Trinity River in Northwestern California*. Rep. No. KBO2013–0012. Ashland, Ore: Klamath Bird Observatory.

Kerans, B. L., and J. R. Karr. 1994. "A Benthic Index of Biotic Integrity (B-IBI) for Rivers of the Tennessee Valley." *Ecological Applications* 4 (4): 768–85.

Konrad, C. P., R. W. Black, F. Voss, C. M. U. Neale. 2008. "Integrating Remotely Acquired and Field Data to Assess Effects of Setback Levees on Riparian and Aquatic Habitats in Glacial-Melt Water Rivers." *River Research and Applications* 24:355–72.

Koster, W. M., D. R. Dawson, J. R. Morrongiello, and D. A. Crook. 2014. "Spawning Season Movements of Macquarie Perch *(Macquaria australasica)* in the Yarra River, Victoria." *Australian Journal of Zoology* 61 (5): 386–94.

Lacher, L. J., D. S. Turner, B. Gungle, B. M. Bushman, and H. E. Richter. 2014. "Application of Hydrologic Tools and Monitoring to Support Managed Aquifer Recharge Decision Making in the Upper San Pedro River, Arizona, USA." *Water* 6:3495–527.

Lambert, A. M., T. L. Dudley, and K. Saltonstall. 2010. "Ecology and Impacts of the Large-Statured Invasive Grasses *Arundo donax* and *Phragmites australis* in North America." *Invasive Plant Science and Management* 3:489–94.

Laporte, M., A. Bertolo, P. Berrebi, and P. Magnan. 2014. "Detecting Anthropogenic Effects on a Vulnerable Species, the Freshwater Blenny *(Salaria fluviatilis)*: The Importance of Considering Key Ecological Variables." *Ecological Indicators* 36:386–91.

Larned, S. T., A. M. Suren, M. Flanagan, B. J. F. Biggs, and T. Riis. 2006. "Macrophytes in Urban Stream Rehabilitation: Establishment, Ecological Effects, and Public Perception." *Restoration Ecology* 14 (3): 429–40.

Lavorel, S., and E. Garnier. 2002. "Predicting Changes in Community Composition and Ecosystem Functioning from Plant Traits: Revisiting the Holy Grail." *Functional Ecology* 16 (5): 545–56. http://doi.org/10.1046/j.1365-2435.2002.00664.x.

Legendre, P., and L. Legendre. 2012. *Numerical Ecology*. Vol. 24. 3rd ed. Elsevier.

Legge, S., D. B. Lindenmayer, N. M. Robinson, B. C. Scheele, D. M. Southwell, and B. A. Wintle, eds. 2018. *Monitoring Threatened Species and Ecological Communities*. Clayton, Victoria: CSIRO.

Legleiter, C. J., D. A. Roberts, W. A. Marcus, and M. A. Fonstad. 2004. "Passive Optical Remote Sensing of River Channel Morphology and In-Stream Habitat: Physical Basis and Feasibility." *Remote Sensing of Environment* 93:493–510.

Lehr, C., F. Pöschke, J. Lewandowski, and G. Lischeid. 2015. "A Novel Method to Evaluate the Effect of a Stream Restoration on the Spatial Pattern of Hydraulic Connection of Stream and Groundwater." *Journal of Hydrology* 527:394–401.

Lennox, M. S., D. J. Lewis, R. D. Jackson, J. Harper, S. Larson, and K. W. Tate. 2011. "Development of Vegetation and Aquatic Habitat in Restored Riparian Sites of California's North Coast Rangelands." *Restoration Ecology* 19 (2): 225–33.

Leps, M., A. Sundermann, J. D. Tonkin, A. W. Lorenz, and P. Haase. 2016. "Time Is No Healer: Increasing Restoration Age Does Not Lead to Improved Benthic Invertebrate Communities in Restored River Reaches." *Science of the Total Environment* 557–58:722–32.

Lewis, L. Y., C. Bohlen, and S. Wilson. 2008. "Dams, Dam Removal, and River Restoration: A Hedonic Property Value Analysis." *Contemporary Economic Policy* 26 (2): 175–86.

Lindenmayer, D., and G. E. Likens. 2010. "The Science and Application of Ecological Monitoring." *Biological Conservation* 143:1317–28. https:/doi.org/10.1016/j.biocon.2010.02.013.

Lorenz, S., M. Leszinski, and D. Graeber. 2016. "Meander Reconnection Method Determines Restoration Success for Macroinvertebrate Communities in a German Lowland River." *International Review of Hydrobiology* 101 (3–4): 123–31.

Lovett, G. M., D. A. Burns, C. T. Driscoll, J. C. Jenkins, M. J. Mitchell, L. Rustad, J. B. Shanley, G. E. Likens, and R. Haeuber. 2007. "Who Needs Environmental Monitoring?" *Frontiers in Ecology and the Environment* 5:253–60.

Maloney, K. O., H. Dodd, S. E. Butler, and D. H. Wahl. 2008. "Changes in Macroinvertebrate and Fish Assemblages in a Medium-Sized River Following a Breach of a Low-Head Dam." *Freshwater Biology* 53:1055–68.

McCune, B., and M. J. Mefford. 2011. PC-ORD: *Multivariate Analysis of Ecological Data*. Gleneden Beach, Ore.: MJM Software Designs.

McDonald, T., C. D. Gann, J. Jonson, and K. W. Dixon. 2016. *International Standards for the Practice of Ecological Restoration—Including Principles and Key Concepts*. Washington, D.C.: Society for Ecological Restoration.

McEachern, K., and R. Sutter. 2010. "Assessment of Eleven Years of Rare Plant Monitoring Data from the San Diego Multiple Species Conservation Plan." Draft final report for Contract 08W3CA5001030 between the U.S. Geological Survey and the San Diego Area Governments.

Meitzen, K. M., J. N. Phillips, T. Perkins, A. Manning, A. and J. P. Julian. 2018. "Catastrophic Flood Disturbance and a Community's Response to Plant Resilience in the Heart of the Texas Hill Country." *Geomorphology* 305:20–32.

Mercado-Silva, N., J. D. Lyons, G. Sagado-Maldonado, and M. Medina Nava. 2002. "Validation of a Fish-Based Index of Biotic Integrity for Streams and Rivers of Central Mexico." *Reviews in Fish Biology and Fisheries* 12 (2–3): 179–91.

Merritt, D. M., M. L. Scott, N. L. Poff, G. T. Auble, and D. A. Lytle. 2010. "Theory, Methods and Tools for Determining Environmental Flows for Riparian Vegetation: Riparian Vegetation-Flow Response Guilds." *Freshwater Biology* 55:206–25.

Millennium Ecosystem Assessment. 2005. "Ecosystems and Human Well-Being: Biodiversity Synthesis." Washington, D.C.: World Resources Institute.

Minnesota Department of Natural Resources (MDNR). 2013. *A Handbook for Collecting Vegetation Plot Data in Minnesota: The Relevé Method*. 2nd ed. Minnesota Biological Survey, Minnesota Natural Heritage and Nongame Research Program, and Ecological Land Classification Program. Biological Report 92. St. Paul: Minnesota Department of Natural Resources. https://files.dnr.state.mn.us/eco/mcbs/releve/releve_singlepage.pdf.

Moerke, A. H., K. J. Gerard, J. A. Latimore, R. A. Hellenthal, and G. A. Lamberti. 2004. "Restoration of an Indiana, USA, Stream: Bridging the Gap Between Basic and Applied Lotic Ecology." *Journal of the North American Benthological Society* 23 (3): 647–60.

Morandi, B., H. Piégay, N. Lamouroux, and L. Vaudor. 2014. "How Is Success or Failure in River Restoration Projects Evaluated? Feedback from French Restoration Projects." *Journal of Environmental Management* 137:178–88.

Moring, J. B., C. L. Braun, and D. K. Pearson. 2014. *Mesohabitats, Fish Assemblage Composition, and Mesohabitat Use of the Rio Grande Silvery Minnow over a Range of Seasonal Flow Regime in the Rio Grande/Rio Bravo del Norte, in and Near Big Bend National Park, Texas, 2010–2011*. U.S. Geological Survey Scientific Investigations Report 5210.

Mueller-Dombois, D., and H. Ellenberg. 2002. *Aims and Methods of Vegetation Ecology*. Caldwell, N.J.: Blackburn Press.

Nagler, P., T. Brown, K. R. Hultine, C. van Riper III, D. W. Bean, P. E. Dennison, and R. S. Murray. 2012. "Regional Scale Impacts of *Tamarix* Leaf Beetles *(Diorhabda carinulata)* on the Water Availability of Western U.S. Rivers as Determined by Multi-scale Remote Sensing Methods." *Remote Sensing of Environment* 118:227–40. http://www.sciencedirect.com/science/article/pii/S0034425711004068.

Nagler, P., E. P. Glenn, C. S. Jarnevich, and P. B. Shafroth. 2011. "Distribution and Abundance of Saltcedar and Russian Olive in the Western United States." *Critical Reviews in Plant Sciences* 30:508–23.

Nagler, P., E. P. Glenn, T. L. Thompson, and A. Huete. 2004. "Leaf Area Index and Normalized Difference Vegetation Index as Predictors of Canopy Characteristics and Light Interception by Riparian Species on the Lower Colorado River." *Agriculture and Forest Meteorology* 125:1–17.

Nagler, P., H. Hinojosa-Huerta, E. Glenn, J. E. Garcia-Hernandez, R. Romo, C. Curtis, A. Huete, and S. Nelson. 2005. "Regeneration of Native Trees in the Presence of Invasive Saltcedar in the Delta of the Colorado River, Mexico." *Conservation Biology* 19:1842–52.

Nagler, P. L., C. Jarchow, J. Christopher, E. P. Glenn. 2018. "Remote Sensing Vegetation Index Methods to Evaluate Changes in Greenness and Evapotranspirationin Riparian Vegetation in Response to the Minute 319 Environmental Pulse Flow to Mexico." *Proceedings of the International Association of Hydrological Sciences* 380:45–54.

The National Academies of Sciences, Engineering, and Medicine (NASEM). 2017. *Effective Monitoring to Evaluate Ecological Restoration in the Gulf of Mexico*. Washington, D.C.: National Academies Press. https://doi.org/10.17226/23476.

Nelson, S. M., and R. Wydoski. 2008. "Riparian Butterfly *(Papilionoidea and Hesperioidea)* Assemblages Associated with *Tamarix*-Dominated, Native Vegetation-Dominated, and *Tamarix* Removal Sites Along the Arkansas River, Colorado, USA." *Restoration Ecology* 16:168–79.

Nilsson, C., A. L. Aradottir, D. Hagen, G. Halldórsson, K. Høegh, R. J. Mitchell, K. Raulund-Rasmussen, K. Svavarsdóttir, A. Tolvanen and S. D. Wilson. 2016. "Evaluating the Process of Ecological Restoration." *Ecology and Society* 21 (1): 41.

NOAA Coastal Services Center. 2012. *Lidar 101: An Introduction to Lidar Technology, Data, and Applications*. Revised ed. Charleston, S.C.: NOAA Coastal Services Center.

Ogren, S. A., and C. J. Huckins. 2015. Culvert Replacements: Improvement of Stream Biotic Integrity? *Restoration Ecology* 23 (6): 821–28.

Orzetti, L. L., R. C. Jones, and R. F. Murphy. 2010. "Stream Condition in Piedmont Streams with Restored Riparian Buffers in the Chesapeake Bay Watershed 1." *Journal of the American Water Resources Association* 46 (3): 473–85.

Osterkamp, W. R., and E. R. Hedman. 1982. "Perennial-Streamflow Characteristics Related to Channel Geometry and Sediment in the Missouri River Basin." U.S. Geological Survey Professional Paper 1242.

Palmer, M., D. J. Allan, J. Meyer, and E. Bernhardt. 2007. "River Restoration in the Twenty-First Century: Data and Experiential Future Efforts." *Restoration Ecology* 15:472–81. https://doi.org/10.1111/j.1526-100X.2007.00243.x.

Palmer, M., K. L. Hondula, and B. J. Koch. 2014. "Ecological Restoration of Streams and Rivers: Shifting Strategies and Shifting Goals." *Annual Review of Ecology, Evolution, and Systematics* 45:247–69.

Palmquist, E. C., B. E. Ralston, D. A. Sarr, and T. C. Johnson. 2018. *Monitoring Riparian-Vegetation Composition and Cover Along the Colorado River Downstream of Glen Canyon Dam, Arizona.* U.S. Geological Survey Techniques and Methods, bk. 2, chap. A14. https://doi.org/10.3133/tm2A14.

Palmquist, E. C., S. A. Sterner, and B. E. Ralston. 2019. "A Comparison of Riparian Vegetation Sampling Methods Along a Large, Regulated River." *River Research and Applications* 35 (6): 759–67.

Parkyn, S. M., R. J. Davies-Colley, N. J. Halliday, K. J. Costley, and G. F. Croker. 2003. "Planted Riparian Buffer Zones in New Zealand: Do They Live Up to Expectations?" *Restoration Ecology* 11 (4): 436–47.

Parola, A. C., and C. Hansen. 2011. "Reestablishing Groundwater and Surface Water Connections in Stream Restoration." *Sustain* 24:2–7.

Peters, D. P. C., B. T. Bestelmeyer, K. M. Havstad, A. Rango, S. R. Archer, A. C. Comrie, H. R. Gimblett, et al. 2013. "Desertification of Rangelands." In *Climate Vulnerability: Understanding and Addressing Threats to Essential Resources*, 239–58. Cambridge, Mass.: Academic Press.

Pidgeon, B. 2004. *A Review of Options for Monitoring Freshwater Fish Biodiversity in the Darwin Harbour Catchment.* Final Report for Water Monitoring Branch, Natural Resources Management Division, Department of Infrastructure, Planning & Environment. Palmerston, NT, Australia.

Pierce, R., C. Podner, and L. Jones. 2015. "Long-Term Increases in Trout Abundance Following Channel Reconstruction, Instream Wood Placement, and Livestock Removal from a Spring Creek in the Blackfoot Basin, Montana." *Transactions of the American Fisheries Society* 144 (1): 184–95.

Pinheiro, J. C., and D. M. Bates. 2000. *Mixed-Effects models in S and S-PLUS.* New York: Springer.

Poff, N. L., J. D. Allan, M. B. Bain, J. R. Karr, K. L. Prestegaard, B. D. Richter, R. E. Sparks, and J. C. Stromberg. 1997. "The Natural Flow Regime." *Bioscience* 47 (11): 769–84.

Poulos, H. M., and A. M. Barton. 2018. "Vegetation Sampling Protocol for Xeric Habitats of the Northeast." Prepared for Northeast Regional Conservation Needs Program.

Poulos, H. M., and B. Chernoff. 2016. "Effects of Dam Removal on Fish Community Interactions and Stability in the Eightmile River System, Connecticut, USA." *Environmental Management* 59:249–63. https://doi.org/10.1007/s00267-016-0794-z.

Poulos, H. M., K. E. Miller, M. L. Kraczkowski, A. W. Welchel, R. Heineman, and B. Chernoff. 2014. "Fish Assemblage Response to a Small Dam Removal in the Eightmile River System, Connecticut, USA." *Environmental Management* 54:1090–101.

Purcell, A. H., C. Friedrich, and V. H. Resh. 2002. "An Assessment of a Small Urban Stream Restoration Project in Northern California." *Restoration Ecology* 10 (4): 685–94.

Ralph, C. J., S. Droege, and J. R. Sauer. 1995. "Managing and Monitoring Birds Using Point Counts: Standards and Applications," In *Monitoring Bird Populations by Point Counts*, edited by C. J. Ralph et al., 161–68. USDA Forest Service, Pacific Southwest Research Station, General Technical Report PSW-GTR-149.

R Development Core Team. 2017. *A Language and Environment for Statistical Computing*. Vienna, Austria: R Foundation for Statistical Computing.

Reichman, O. J., M. B. Jones, and M. P. Schildhauer. 2011. "Challenges and Opportunities of Open Data in Ecology." *Science* 331:703–5.

Richter, H. E., B. Gungle, L. J. Lacher, D. S. Turner, and B. M. Bushman. 2014. "Development of a Shared Vision for Groundwater Management to Protect and Sustain Baseflows of the Upper San Pedro River, Arizona, USA." *Water* 6:2519–38. https://doi.org/10.3390/w6082519.

Roni P., T. J. Beechie, R. E. Bilby, F. E. Leonetti, M. M. Pollock, and G. R. Pess. 2002. "A Review of Stream Restoration Techniques and a Hierarchical Strategy for Prioritizing Restoration in Pacific Northwest Watersheds." *North American Journal of Fisheries Management* 22:1–20.

Ruegg, J., C. Gries, B. Bond-Lamberty, G. J. Bowen, B. S. Felzer, N. E. McIntyre, P. A. Soranno, K. L. Vanderbilt, and K. C. Weathers. 2014. "Completing the Data Life Cycle: Using Information Management in Macrosystems Ecology Research." *Frontiers in Ecology and the Environment* 12:24–30.

Rumm, A., F. Foeckler, O. Deichner, M. Scholz, and M. Gerisch. 2016. "Dyke-Slotting Initiated Rapid Recovery of Habitat Specialists in Floodplain Mollusc Assemblages of the Elbe River, Germany." *Hydrobiologia* 771 (1): 151–63.

Schlatter, K. J., M. R. Grabau, P. B. Shafroth, and F. Zamora-Arroyo. 2017. "Integrating Active Restoration with Environmental Flows to Improve Native Riparian Tree Establishment in the Colorado River Delta." *Ecological Engineering* 106:661–74.

Schmerge, D. S., F. Corkhill, and S. Flora. 2009. *Water-Level Conditions in the Upper San Pedro Basin, Arizona, 2006*. ADWR Water Level Change Map Series Report No. 3. https://new.azwater.gov/sites/default/files/WLCMSReportNo.3_UpperSanPedro.pdf.

Schmitt, C. J., G. M. Dethloff, J. E. Hinck, T. M. Bartish, V. S. Blazer, J. J. Coyle, N. D. Denslow, and D. E. Tillitt. 2004. *Biomonitoring of Environmental Status and Trends (BEST) Program: Environmental Contaminants and Their Effects on the Fish in the Rio Grande Basin*. Scientific Investigations Report 5108. Columbia, Mo.: Columbia Environmental Research Center.

Schumm, S. A. 1999. "Causes and Controls of Channel Incision." In *Incised River Channels*, edited by S. E. Darby and A. Simon, 2033. Chichester, UK: John Wiley.

Selvakumar, A., T. P. O'Connor, and S. D. Struck. 2010. "Role of Stream Restoration on Improving Benthic Macroinvertebrates and In-Stream Water Quality in an Urban Watershed: Case Study." *Journal of Environmental Engineering* 136 (1): 127–39.

Sergeant, C. J., B. J. Moynahan, and W. F. Johnson. 2012. "Practical Advice for Implementing Long-Term Ecosystem Monitoring." *Journal of Applied Ecology* 49:969–73.

Shroder, J. F. 2013. "Ground, Aerial, and Satellite Photography for Geomorphology and Geomorphic Change." In *Treatise on Geomorphology*, edited by J. F. Shroder and M. P. Bishop, 25–42. San Diego, Calif.: Academic Press.

Silvertown, J. 2009. "A New Dawn for Citizen Science." *Trends in Ecological Evolution* 24 (9): 467–71.

Smith, J. G., C. C. Brandt, and S. W. Christensen. 2011. "Long-Term Benthic Macroinvertebrate Community Monitoring to Assess Pollution Abatement Effectiveness." *Environmental Management* 47 (6): 1077–95.

Sostoa, A., D. García de Jalón, E. Garcia-Berthou. 2005. "Protocolos de muestreo y análisis para Ictiofauna." In *Metodología para el establecimiento el Estado Ecológico según la Directiva Marco del Agua*, 22–32. Spain: Ministerio de Medio Ambiente.

Sterk, M., G. Gort, H. De Lange, W. Ozinga, M. Sanders, K. Van Looy, and A. Van Teeffelen. 2016. "Plant Trait Composition as an Indicator for the Ecological Memory of Rehabilitated Floodplains." *Basic and Applied Ecology* 17 (6): 479–88.

Stewart-Oaten, A., W. W. Murdoch, and K. R. Parker. 1986. "Environmental Impact Assessment: Pseudoreplication in Time?" *Ecology* 67:929–40.

Strickler, G. S. 1959. "Use of the Densiometer to Estimate Density of Forest Canopy on Permanent Sample Plots." PNW Old Series Research Notes No. 180, 1–5.

Stromberg, J. C. and D. M. Merritt. 2016. "Riparian Plant Guilds of Ephemeral, Intermittent and Perennial Rivers." *Freshwater Biology* 61(8):1259–75.

Stromberg, J. C., J. L. S. J. Lite, T. J. Rychener, L. R. Levick, M. D. Dixon, and J. M. Watts. 2006. "Status of the Riparian Ecosystem in the Upper San Pedro River, Arizona: Application of an Assessment Model." *Environmental Monitoring and Assessment* 115:145–73.

Sutter, R. D., S. B. Wainscott, J. R. Boetsch, C. G. Palmer, and D. J. Rugg. 2015. "Practical Guidance for Integrating Data Management into Long-Term Ecological Monitoring Projects." *Wildlife Society Bulletin* 39 (3): 451–63. https:/doi.org/10.1002/wsb.548.

Taylor, C. J., and W. M. Alley. 2001. *Ground-Water-Level Monitoring and the Importance of Long-Term Water-Level Data*. U.S. Geological Survey Circular 1217. https://pubs.er.usgs.gov/.

Taylor, M. D., R. C. Babcock, C. A. Simpfendorfer, and D. A. Crook. 2017. "Where Technology Meets Ecology: Acoustic Telemetry in Contemporary Australian Aquatic Research and Management." *Marine and Freshwater Research* 68 (8): 1397–402. https://doi.org/10.1071/MF17054.

Thomsen, P. F., J. Kielgast, L. L. Iversen, C. Wiuf, M. Rasmussen, M. Thomas, P. Gilbert, L. Orlando, and E. Willerslev. 2012. "Monitoring Endangered Freshwater Biodiversity Using Environmental DNA." *Molecular Ecology* 21:2565–73.

Turner, D. S., and H. E. Richter. 2011. "Wet/Dry Mapping: Using Citizen Scientists to Monitor the Extent of Perennial Surface Flow in Dryland Regions." *Environmental Management* 47:497–505.

Urbanczyk, K. M. 2017. "GIS, GPS, and Satellite Data." In *Sustainable Water Management and Technologies*, edited by D. H. Chen, 79–103. Boca Raton, Fla.: CRC Press.

Urbanczyk, K. M. and J. B. Bennett. 2012. *Geomorphology of Sand Bars in Boquillas Canyon, Big Bend National Park*. Geological Society of America, Abstracts with Programs, March 2012. Vol. 44. No. 1, p. 2. https://www.geosociety.org/GSA/Events/Search_Abstracts/GSA/Events/Abstracts.aspx?hkey=b496a8bd-a572-4af1-b935-ebaab4d4cc71.

USDA. 2012. The PLANTS Database. National Plant Data Team. Greensboro, N.C., USA. https://plants.sc.egov.usda.gov/java/.

USDA. 2016. *Final Biological and Conference Opinion for Bureau of Reclamation, Bureau of Indian Affairs, and Non-Federal Water Management and Maintenance Activities on the Middle Rio Grande, New Mexico*. USFWS Consultation No. 02ENNM00–2013-F-0033. Albuquerque, New Mexico. https://www.fws.gov/southwest/es/NewMexico/BO_MRG.cfm.

Van Haverbeke, D. R., D. M. Stone, L. J. Coggins Jr., and M. J. Pillow. 2013. "Long-Term Monitoring of an Endangered Desert Fish and Factors Influencing Population Dynamics." *Journal of Fish and Wildlife Management* 4 (1): 163–77. https:/doi.org/10.3996/082012-JFWM-071.

van Zyll de Jong, M., and I. G. Cowx. 2016. "Long-Term Response of Salmonid Populations to Habitat Restoration in a Boreal Forest Stream." *Ecological Engineering* 91:148–57.

Violle, C., M. L. Navas, D. Vile, E. Kazakou, C. Fortunel, I. Hummel, and E. Garnier. 2007. "Let the Concept of Trait Be Functional!" *Oikos* 116:882–92. https://doi.org/10.1111/j.0030-1299.2007.15559.x.

Wheaton, J. M., J. Brasington, S. E. Darby, and D. A. Sear. 2010. "Accounting for Uncertainty in DEMs from Repeat Topographic Surveys: Improved Sediment Budgets." *Earth Surface Processes and Landforms* 35:136–56.

Winward, A. H. 2000. *Monitoring the Vegetation Resources in Riparian Areas*. Gen. Tech. Rep. RMRS-GTR-47. Ogden, Utah: U.S. Department of Agriculture, Forest Service, Rocky Mountain Research Station.

Wolf, P. R., and B. A. Dewitt. 2000. *Elements of Photogrammetry with Applications in GIS*. New York: McGraw-Hill.

Woodward, B. D., P. H. Evangelista, N. E. Young, A. G. Vorster, A. M. West, S. L. Carroll, R. K. Girma, et al. 2018. "CO-RIP: A Riparian Vegetation and Corridor Extent Dataset for Colorado River Basin Streams and Rivers." *International Journal of Geo-Information* 7:397. https:/doi.org/10.3390/ijgi7100397.

Zamora-Arroyo, F. 2017. "Integrating Active Restoration with Environmental Flows to Improve Native Riparian Tree Establishment in the Colorado River Delta." *Ecological Engineering* 106:661–74.

Zamora-Arroyo, F., P. L. Nagler, M. K. Briggs, D. Radtke, H. Rodriquez, J. Garcia, C. Valdes, A. Huete, and E. P. Glenn. 2001. "Regeneration of Native Trees in Response to Flood Releases from the United States into the Delta of the Colorado River, Mexico." *Journal of Arid Environments* 49:49–64.

Zhao, J., Q. Liu, L. Lin, H. Lv, and Y. Wang. 2013. "Assessing the Comprehensive Restoration of an Urban River: An Integrated Application of Contingent Valuation in Shanghai, China." *Science of the Total Environment* 458:517–26.

# 8

# Going Long

## ENSURING THAT YOUR STREAM RESTORATION
## EFFORT CONTINUES TO GROW

*Mark K. Briggs, Collin Haffey, Laura McCarthy, Jon Radke, Karen J. Schlatter, Lindsay White, and Francisco Zamora*

We are now approaching the restoration finish line and it has been an honor to have joined you on this journey. Together, we have discussed lessons learned from past stream restoration efforts, developed restoration goals and objectives, evaluated stream and watershed conditions, considered climate change and environmental flow, implemented restoration tactics, and monitored results, while learning from numerous real-life stream restoration examples. Whether your restoration project was small in scale—perhaps a riparian revegetation project at a single site—or a multiyear restoration effort at the watershed scale, project completion marks an important milestone that deserves to be celebrated. You have completed a project whose honorable goal was to repair a damaged ecosystem—a goal that will benefit both people and wildlife for many years to come.

As you celebrate the completion of your restoration project, please don't walk away without a backward glance. The outcomes of far too many stream restoration projects were lost because the entities involved never looked back. We will never know whether these projects were successful, nor will we have an opportunity to learn from their results or mistakes. Even for successful projects (the trees that were planted grow to maturity, the exotic species that were eradicated do not return, the fish that were reintroduced survive), a greater opportunity will be lost if no effort is made to maximize the effort for greater impact. Streams will live on well after we have crossed the finish line. Different timescales are at play.

The salient point—one that needs to be particularly heard by the entities that fund stream restoration efforts—is that improving stream conditions takes time. It is likely that the stream that is being restored has been affected over centuries by a myriad of human impacts, driven by millions and sometimes billions of dollars: construction of large dams, water-diversion canals, flood control, and other large engineering projects along large rivers are not cheap. In the face of such impacts and change, a two-year grant can only do so much to budge the stream condition needle toward natural/wild/better, no matter how well restoration is executed. As noted previously, this is not to say that small restoration projects are meaningless. Quite the contrary. Such projects are the foundation from which future restoration efforts can be launched. The salient point, now that you have completed your restoration project, is don't pack up your field gear and head to another watershed. Stay where you are, take advantage of your momentum, and build on your results.

How well you will be able to do this will depend largely on the level of support you have from riverside communities, private and public funders, and the entities that manage or own the land where your stream is located. The collaborative end goal is a four-way partnership between communities and local citizens, scientists, managers, and practitioners; all four elements are needed for successful river restoration on a grand scale (Bernhardt et al. 2007; Gonzalez et al. 2017; Hawley 2018). As noted in chapter 6, establishing trust with local managers, landowners, and communities often takes considerable time. Local entities who may have been only partly supportive at the start of a restoration project and waited to see how the restoration experience played out, may have developed strong confidence in the restoration process and enthusiasm for more. Indeed, establishing a foundation of trust as part of the restoration program can be as important an outcome as realizing your biophysical objectives and is key to attracting additional funding to support additional restoration for greater impact (Ruiz-Mallén et al. 2015; Hawley 2018).

Therefore, as you finish your restoration project, we encourage you to go beyond the standard role of leader or boss and become a coordinator and facilitator for follow-on collaboration in pursuit of greater impacts. Our aim in this final chapter is to offer suggestions for capitalizing on momentum and attracting the support you will need to go long and achieve greater restoration impact.

## KEEP THE COMMUNICATION FLOWING

"Communication" covers a wide swath of thematic territory. It is critical that we—as river restoration practitioners, river managers, scientists, and people living and working on rivers—actively foster sociopolitical support and community participation in river restoration by emphasizing the win-win benefits of stream restoration to humans and wildlife. Here are a few suggestions under the communication banner from past (and ongoing) stream restoration projects that have enhanced collaboration and attracted additional funding, allowing initial restoration efforts to be expanded for greater long-term impact.

### Throw a Party

There are few better reasons to throw a party than to honor the shared achievement of completing a stream corridor restoration project. When projects are successful—even if successes are small in scale—take the time to publicly highlight them. Sharing news about restoration success fosters goodwill and can increase public, political, and funding support for longer-term efforts. Take the opportunity to celebrate the stream you worked to restore, the people with whom you worked, and the restoration project itself. As you make your invitation list, remember to include others who may not have been directly involved yet whose knowledge and support will be critical for both the long-term viability of the stream restoration effort you just completed and for expanding restoration efforts to other sites, reaches, or segments of your stream. These "others" could include public officials, current and potential future funders, local citizens, community leaders, scientists, businesses, journalists, water managers, and water users. In other words, illuminate your recently completed restoration project for local citizens (the ultimate stakeholders) and others who have a strong bearing on the funding future or expanded projects, managing the land and water of your stream, or setting priorities for future restoration and research. So add names to your invitation list, grab a grill and some beverages, and take the party to your stream (see figures 8.1 to 8.3).

## Make a Public Communication Splash

The completion of a stream restoration project is welcome news and provides an opportunity for making a public splash about your stream restoration project. Public communication includes a variety of approaches that restoration practitioners can use to engage the public in dialogue, such as public speaking events, newspaper and magazine articles, appearances on television news, and distributing information through social media. Welcome the help of expert communicators and invite them to your stream. Journalists can do the work for you

**FIGURE 8.1** At a public event celebrating the completion of a restoration project along the reach of the Santa Cruz River that passes through the San Xavier District of the Tohono O'odham Nation (near Tucson, Arizona), community elders offer blessings and support for bringing back the wetlands and riparian ecosystems of their youth. Photo by Mark Briggs.

**FIGURE 8.2** March 27, 2014, marked the day that the "pulse flow" release of water passed through Morelos Dam (bottom photo, R. Lester, Deakin Uni, Australia) on the Colorado River straddling the U.S.-Mexico border. The pulse flow allowed river water to reach the Sea of Cortez for the first time in fifteen years and was the first time in history that water was dedicated to environmental use in this transboundary watershed. Federal, state, and local governments, NGOs, community members, researchers, and river enthusiasts gathered at Morelos Dam to celebrate the occasion (top two photos).

to get word to the public about your stream restoration effort, particularly if they have well-established channels for distributing information and a readership ready to receive it (see figures 8.4 and 8.5).

There are also advantages to developing your own information materials about your project for wider distribution. This allows you the freedom to put the project's successes in the best light; articulate the benefits accruing to the stream, the ecosystem, and the human residents; acknowledge the important role of stakeholders in the effort; and point to what more can be done (see figures 8.6 and 8.7).

**FIGURE 8.3** In New South Wales, Australia, a planned flow along the Darling River traveled over 1,000 km and reached the historic town of Bourke. A celebration of the event was held at the historic wharf, involving representatives of First Nations, municipalities, and state and federal government agencies in recognition of the first time that a planned release from large storages in two river systems had flowed down tributaries through the Darling River system.

**FIGURE 8.4** Journalists from the *New York Times* were invited to a transboundary reach of the RGB to witness a prescribed burn of an invasive grass as an example of binational collaboration on river management (Fernandez 2016).

**FIGURE 8.5** A group of journalists from the Institute for Journalism and Natural Resources (U.S.) takes a tour of restoration sites in the Colorado River Delta. As a result of this trip, at least three articles were prepared and widely distributed, contributing substantially to the public's understanding of U.S. and Mexican collaborative work in the Delta and the challenges therein. Photo by Jennifer Pitt.

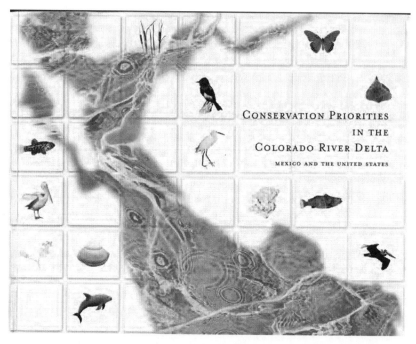

**FIGURE 8.6** This beautifully illustrated, 103-page booklet prepared by the Sonoran Institute, Environmental Defense Fund, University of Arizona, Centro de Investigación en Alimentación y Desarrollo, Pronatura Sonora, and World Wildlife Fund proved instrumental in fomenting public support for restoring the Colorado River Delta, Mexico (Zamora-Arroyo et al. 2005).

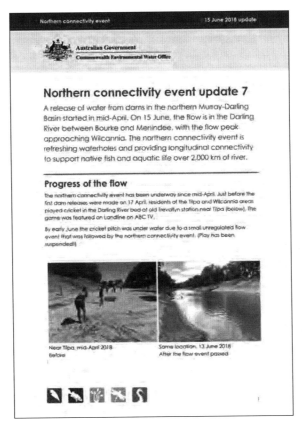

**FIGURE 8.7** Northern Connectivity Event Updates produced by the Commonwealth Environmental Water Office of the Australian Government summarize the movement of planned environmental flow events along the river system, initial results, monitoring that is underway, and community engagement activities (past and upcoming). These updates are amply illustrated with color photographs and graphics and written in nontechnical (but not overly simple) language for residents who are lacking information about the rivers that flow through their region.

## Take Advantage of Social Media

Social media is one of the best ways today to pique interest and keep people engaged in your restoration projects. A variety of social networks are available to distribute news about your restoration project or a specific event. Facebook, Twitter, YouTube, Instagram, and Tumblr are just a few of the social media platforms used worldwide to cheaply and effectively broadcast news. In the Colorado River Delta, for example, the Sonoran Institute uses its Facebook page to routinely secure the participation of hundreds of volunteers in its annual tree-planting event, drumming up interest and providing logistical details (see https://www .facebook.com/sonorainstitute/). After the event is finished, volunteers are recognized for their work through the posting of photos, videos, and stories from the event on Facebook and Instagram. Your younger staff members will be adept at most of these platforms, but there are good resources available to help even social-media-challenged practitioners use them and design a campaign (Hoechlin 2017; Solaris 2019).

## EXPANDING AND STRENGTHENING PARTNERSHIPS

Strong and diverse partnerships are needed to ensure that restoration gains are not reversed when a money source dries up, key politicians are replaced, community leaders move away, or managers change priorities. Put another way, stream restoration efforts built around a single funder, politician, or community leader are inherently vulnerable to impermanence. As you cross the restoration finish line, take advantage of the moment to not only solidify relations with current partners but also reach out and establish new partnerships that will improve likelihood for long-term success. Government agencies, NGOs, water users, industry, municipalities, businesses, local citizen groups, universities, and others may all be motivated to improve stream conditions for a variety of reasons, each bringing valid and critical perspectives to the table. The diversity of perspectives gained will get you to broadly supported decisions beneficial to the majority of stakeholders (Uiito and Duda 2002).

Over the last decade, novel public-private partnerships have demonstrated a variety of ways to successfully take stream corridor restoration efforts to the next level. Building diverse and long-lasting partnerships is often sparked by one or more individuals or entities who are motivated by a particular issue. In some cases, the spark comes from private entities who live and work in the area, as was the case at Banrock Station, Southern Australia (Case Study 8.1). In other cases, the spark can come from outside the region—as in the binational effort to restore the Colorado River Delta (Case Study 8.2), which was initiated by several NGOs.

Diverse public-private partnerships can sustain restoration efforts even for streams that are in complex socioeconomic and political landscapes. We are reminded of what a colleague of ours, Carlos Sifuentes (CONANP and long-time member of the binational team along the RGB) noted: "The Rio Grande/Bravo does not divide us, but brings our two countries together to work on a shared resource of mutual importance." Although opinions of citizens who live and work in such complicated landscapes can vary tremendously as to what it means to restore a stream and how restoration should be carried out, it is often the case that they will agree on the value of a stream that is in good condition versus a stream that is not. In this sense, streams (perhaps more than most ecosystem types) can be the bridge that allows diverse, multisectoral, public-private partnerships to form and work together on shared and mutually beneficial objectives (see figure 8.8).

The development of diverse public-private partnerships can stem from many issues concerning stream conditions originating from outside or inside the target watershed. In the Colorado River Delta, concerns on water quality expressed by local citizens proved to be the spark that led to a broad, binational response (see Case Study 8.2).

An unforeseen event (such as a major flood or drought) may also precipitate a synergistic public-private response to ameliorating the negative consequences when future similar events occur. Such is the case with the Rio Grande Water Fund, which was born out of the need to protect upper watershed forests that serve as source waters for the cities of Santa Fe and Albuquerque, New Mexico (see Case Study 8.3 as well as the location map in Case Study 5.6).

In our final example of a diverse public-private partnership, the impetus to help move binational efforts along the Rio Grande/Bravo forward came from well outside the region in the form of a global business whose own long-term environmental goals aligned well with the restoration objectives for the binational stretch of river (see Case Study 8.4).

# CASE STUDY 8.1
A Public-Private Partnership to Reclaim Wetlands at Banrock Station,
Southern Australia

Lindsay White

Since 1993, Banrock Station in South Australia has been owned by Accolade Wines, now one of the major wine producers in the world. That year, in parallel with the development of the vineyard, the company initiated the restoration of more than 1,500 ha (3,706 acres) of wetlands and adjoining woodlands. The effort began with the removal of sheep and cattle from the land; they had been the primary drivers of deteriorating conditions, causing or exacerbating the fragmentation of the wetlands, the loss of native flora from trampling and overgrazing, increased soil erosion, and deteriorating water quality. Banrock Station also became the first site in the State of South Australia to implement an active management of its wetlands through the reestablishment of a more natural hydrological regime, with a succession of wet and dry phases. In 2000, the Banrock Station Wine and Wetland Centre opened to the public, offering wetland walks as well as wine and al-fresco dining with views overlooking the river floodplain (see photo below). In 2002 the site was declared as one of Australia's Ramsar Sites—a wetland site of international importance—as designated under UNESCO's 1971 Ramsar Convention.

In 2007, with support from the state government, Banrock Station was able to relocate the vineyard irrigation pumps from the main lagoon to the river. It was then able to restore the natural dry phase to the wetlands in late summer and autumn, replicating a more natural hydrological regime of winter/spring floods and summer/autumn drying.

Wetlands as viewed from the deck of the Banrock Station Wine and Wetland Centre. Photo by Richard Mintern, CEWO.

In 2015, Banrock Station and Accolade Wines entered into a three-year partnership with the Commonwealth Environmental Water Holder (a federal agency that aims to restore environmental water throughout the commonwealth and provides a key mechanism for water reform in the Murray-Darling Basin) to deliver over two gigaliters (Gl, about 1,620 acre-feet) of water to augment regeneration of lignum (*Muehlenbeckia florulenta*), red gum (*Eucalyptus camaldulensis*), and black box (*Eucalyptus largiflorens*) and improve floodplain and aquatic habitat for such vulnerable species as the regent parrot and the southern bell frog (both species are listed as vulnerable by the Australian government). This partnership has been so successful that both parties are looking to extend it another five years.

Thanks to these measures, the wetland has now been returned to a healthy and functioning ecosystem. Introduced European carp have been removed and the soil structure improved, while native fauna and flora have returned. Reintroducing the dry phase in the seasonal cycle of the wetland has also saved about 1.15 Gl of precious river water from evaporation over two years. At the time of writing, Banrock Station had become the top birding destination in South Australia and the thirty-fifth most

(*continued*)

A celebration of Banrock staff and state and federal agencies showcase how successful and enduring public-private partnerships can produce important wins for people and the environment. Photo by Richard Mintern, CEWO.

popular in Australia as a whole, underscoring how unique public-private partnerships can lead to both significant conservation achievements and economic benefits from ecotourism.

### Lessons Learned

One of the great lessons learned from Benrock Station is that private entities and businesses can be the leaders and instigators of conservation and restoration efforts, providing the basis for collaborative, long-lasting public-private partnerships.

Public-private partnerships are not the only option for effective collaboration in stream restoration projects. Research consortiums and other collaborative information-sharing platforms can foster research, monitoring, and sharing of stream restoration results. Increased collaboration between restoration scientists and practitioners is needed now more than ever, with the "science-practice gap" frequently cited as a major factor limiting both the science and practice of restoration (Arlettaz et al. 2010; Dickens and Suding 2013). The completion of a restoration project provides a great opportunity to strengthen bonds between scientists and practitioners, while benefiting the restoration project itself (Clark et al. 2019). Collaborative, long-term monitoring of results can be promoted and furthered and the line between academia and practice blurred via the involvement of professors and students in real-life restoration experiments (Giardina et al. 2007; Palmer 2009; Cabin et al. 2010). The bottom line is that free and open information-sharing among all stakeholders provide avenues to explore

# CASE STUDY 8.2
Informal Community-Based Partnerships Seed Larger Public-Private Partnerships in the Colorado River Delta, Mexico

Karen J. Schlatter and Francisco Zamora

Collaborative partnerships involving community citizens, NGOs, agencies, and institutions have fostered much of the restoration progress that has been made in the Colorado River Delta of Mexico. For example, in the Hardy River—a tributary of the Colorado River that is fed by agricultural runoff and effluent from the Las Arenitas Wastewater Treatment Plant—concerns over water quality led to the creation of an informal community group called the Ecological Association of Hardy-Colorado River Users (AEURHYC). In 2006 the group partnered with NGOs and government entities to develop the Hardy River Management Plan, which opened up new collaborative mechanisms involving the Mexican state of Baja California, citizens' groups, and NGOs. Eventually, this effort led to a formal agreement with Baja California for the construction of a treatment wetland at Las Arenitas to improve the quality of water flowing into the Hardy River. The agreement dedicated 15.2 mm³ (12,330 acre-feet) of effluent flow to the Hardy River, the first agreement of its kind in Mexico. In return, participating groups committed to specific maintenance and monitoring actions to uphold the statutes of the agreement.

With that success, the Sonoran Institute (an NGO based in Arizona that has been working binationally in the Colorado River Delta for over fifteen years) initiated an "Adopt-the-River" program to involve

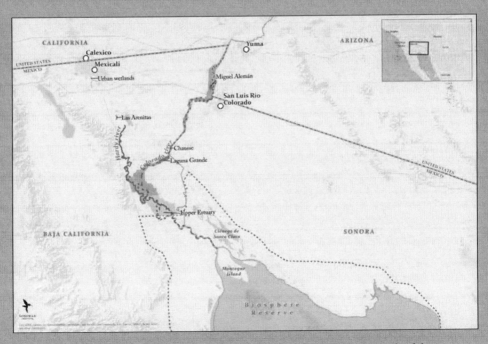

Map of the Colorado River Delta showing several priority restoration sites south of the international border, including the Colorado River, Hardy River, and Cienega de Santa Clara. Map by A. Melendez, Sonoran Institute.

*(continued)*

# CASE STUDY 8.2
*(continued)*

As part of the Adopt-the-River program, local community members and Sonoran Institute staff members plant hundreds of native riparian trees as part of a restoration demonstration site on the Hardy River, Baja California, Mexico (left). Results of revegetation efforts five years later (right). Photos © Sonoran Institute, Guadalupe Fonseca Molina.

local citizens in restoration work. Participating groups include agricultural and fishing cooperatives, irrigation districts, government agencies, businesses, and schools (Zamora-Arroyo et al. 2008). Since its inception, Adopt-the-River has engaged over fifteen thousand people and helped to restore over seven hundred acres of native riparian habitat.

These and other small-scale, community-based efforts fostered a broader binational response that included the establishment of the joint International Boundary Water Commission / Comisión Internacional de Límites y Aguas (IBWC/CILA) binational technical task force in 2000 that facilitated studies on Colorado River Delta ecology. It was this task force and the results of the studies that it supported that eventually led to the NGO-led report shown in figure 8.6 (Zamora-Arroyo et al. 2005), which serves as the Delta's official restoration guide (Zamora-Arroyo and Flessa 2009) and provided the foundation for the unprecedented release of environmental flow waters described in Case Study 5.5.

## Lessons Learned

- Informal small-scale collaboration can provide the foundation for and blossom into formal partnerships that allow restoration to be taken to the next level, and in some cases this is a necessary ingredient (Zamora-Arroyo and Flessa 2009). That is, seeing is believing, and small-scale informal restoration efforts can provide the near-term, tangible, proof-of-concept required for attracting the additional partnerships needed within a more formal collaborative framework (Zamora-Arroyo et al. 2008).
- It's not necessary to start with a documented agreement signed by involved parties before beginning your restoration project. If you have a restoration project that makes sense regarding its objectives, tactics, and support, move forward. Formalities can happen later.

**FIGURE 8.8** As part of binational efforts to restore the Rio Grande/Bravo along the border between Mexico and the United States, Big Bend National Park, Comision Nacional de Areas Naturales Protegidas, the World Wildlife Fund, and community leaders organized an all-women canoe trip consisting of community leaders from Mexican and U.S. riverside towns. The purpose of the canoe trip was to strengthen the involvement of riverside communities, and of women from these communities in particular, in restoration efforts along this critical binational river.

lessons learned from past experiences, build trust, and move projects forward (van der Zaag and Savenije 2000; Uiito and Duda 2002).

A word on international and transboundary partnerships: Language differences can be a challenge to effective information-sharing for transboundary stream restoration, so the translation of documents and involvement of the press on both sides of the border must be a priority. And no matter the location, practitioners should keep in mind several technical and institutional challenges that may need to be overcome as they develop shared information platforms. The technical challenges usually pertain to developing the platform itself, establishing portholes to allow multidirectional data transfer, and agreeing to and instituting processes for maintaining and updating the database. Institutional challenges, however, can be more formidable, given the need to secure formal approval for sharing data and information at the highest levels of participating organizations, agencies, and institutions. Addressing such challenges can take time.

For transboundary restoration efforts, binational commissions can enhance technical and scientific communication, help develop long-term research objectives, and provide a mechanism for effective data- and information-sharing. One formal collaborative binational research commission is the Research Coordination Network for the Colorado River Delta (RCN-CRD), a binational research team comprising experts in the fields of law and natural and social sciences. The RCN-CRD began as a year-long pilot operation to assess the consequences

The Rio Grande Water Fund: A Long-Term Public-Private Partnership That Protects Source Water at the Watershed Scale

Laura McCarthy and Collin Haffey

The Rio Grande Water Fund (RGWF) is a public-private partnership that includes five local county governments; federal actors such as the USDA National Forest Service, the Bureau of Land Management (BLM), the USFWS, the USGS; state-level counterparts; local community associations such

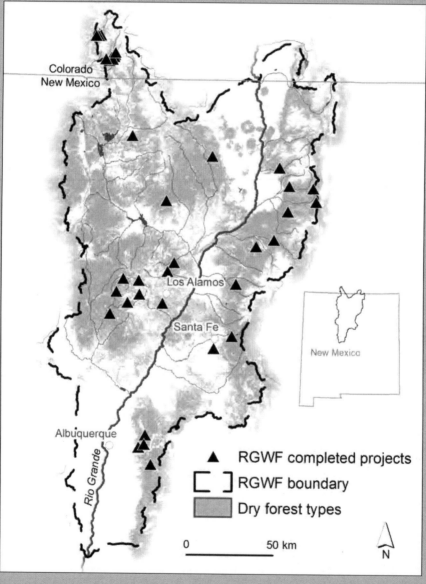

Map showing projects completed by RGWF through 2019. All projects fall within RGWF focal areas and have been strategically placed to maximize watershed protection. Map by Steve Bassett, The Nature Conservancy.

Fire managers for the Carson and Santa Fe National Forests gather around a map of fire risk to establish Potential Operational Delineations. Such opportunistic, multipartner planning is coordinated by the Rio Grande Water Fund and allows naturally ignited fires to be managed in a way that builds forest resiliency and reduces the likelihood of catastrophic fire. In 2019 alone, nearly twenty thousand acres were "treated" by naturally ignited fire events. Photo by Collin Haffey, The Nature Conservancy.

as the Chama Peak Land Alliance and Rocky Mountain Youth Corps; Hispanic land-grant and Native American tribal communities; and private sector interests such as the New Mexico Forest Industry Association and Business Water Task Force. Other water-service delivery and infrastructure actors at local and national scales such as water utilities, the flood control authority, and Army Corps of Engineers are also frequently engaged. Collectively, these partners represent the diverse set of land ownership and water users found in the fund's area who commit to working together to secure clean water for communities in the watershed and downstream.

The benefits of bringing together multiple stakeholders under the RGWF umbrella are numerous and have helped to:

- Coordinate projects by diverse stakeholders across jurisdictional boundaries to improve the efficiency and effectiveness of the effort;
- Leverage and multiply funding sources to expand work in other important parts of the watersheds; and
- Mobilize a collaborative, multipartner approach to protect watersheds and water supply in at-risk forests totaling nearly 688,000 ha (1.7 million acres), using inclusive priority-setting and coordinated capacity-building in forest management.

*(continued)*

## CASE STUDY 8.3
*(continued)*

### Lessons Learned

Several elements were critical to the success of this partnership:

- *Public/Private Partners in a Transboundary Project*—Public-private coalitions make working across jurisdictional boundaries possible. Working at the watershed scale often involves working on lands that are owned or managed by a diversity of entities. To effectively address such watershed-scale issues as the protection of source waters, a coalition of landowners and managers was essential. As of 2019, the RGWF coalition has implemented dozens of restoration and forest-thinning projects on over 55,000 ha (138,000 acres) of upper forest land owned or managed by six different entities as well as private lands in order to reduce potential for catastrophic fire.

- *Private Funding*—Private funding was key to securing longer-term federal and state support for the RGWF. While federal and state funding makes up most of RGWF contributions today, the first US$2 million came from private foundations and was critical to the fund's formation. Private funding was also used to leverage federal and state funding for greater impact, with the combined pool of funding supporting both restoration treatments and complementary activities such as planning, education, outreach, and monitoring. As of 2019, the RGWF leveraged $4.99 million of private funding to focus the investment of $48 million in federal, state, and local funds.

- *Identification of Responsibilities for Long-Term Success*—As partnerships mature, it is important to identify the main role that each partner has in moving the stream restoration project forward. For the RGWF, identifying fiscal and administrative responsibilities was particularly important. Participants extensively deliberated alternatives for structuring the RGWF during the coalition's startup phase, and the project's champions felt strongly that the state of New Mexico could not support the administrative costs of a new organization. Therefore, a structure was created whereby the Nature Conservancy administers the private and some federal investments, and the RGWF partners facilitate government-to-government transfer of funding for projects. The RGWF's executive committee (composed of a diversity of participatory stakeholders) coordinate about which projects in the focal areas will move forward. As a well-established global NGO, TNC brings a track record of fiscal accountability to reassure investors and serves as a trusted facilitator for collaborating partners.

of elevated salinity levels on the Ciénega de Santa Clara—the largest wetland of the Colorado River Delta. The team coordinated binational research efforts, shared information, and worked with government agencies to secure additional water flows to the cienega. Today, the network has expanded to involve natural scientists, social scientists, and legal scholars from both Mexico and the United States to understand how natural and human-caused variation in water supply affects the ecosystem of the Colorado River Delta (https://www.geo.arizona.edu/rcncrd/).

Although formal research commissions can offer a wide range of expertise and knowledge and provide a variety of benefits, their inherent institutional complexity can slow

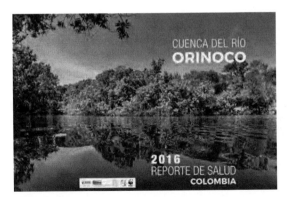

**FIGURE 8.9** This basin report card effort for the Orinoco River Basin gave the basin a grade of "B minus," highlighting issues relating to water quality and fragmentation. More important than the grade itself, the basin report card process brought together community leaders, ranchers, farmers, scientists, and conservationists from throughout the basin, fostering collaboration to address a wide range of conservation priorities.

decision-making and implementation of on-the-ground actions. Establishing informal relationships while simultaneously working to formalize others will give your project a desired amount of flexibility.

Basin report cards are another means to bring together the watershed community and share natural resource, socioeconomic, and political information. Basin report cards evaluate the elements and characteristics of a basin that stakeholders most value and assign grades on the health of those priority areas. The process is initiated via citizen workshops where shared values that depend on their system's health—including social and economic benefits—are identified. Methods to measure and track the status of these values over time are then identified. Using existing data—and finding creative solutions when necessary—stakeholders assign and share the grades. Through this collaborative process, stakeholders develop a common understanding of the basin's health and a shared vision for its future, empowering them to demand management of their freshwater resources to protect those shared values (see figure 8.9). Many basin report cards are a means not only of sharing information but of identifying research and monitoring priorities to allow for more robust data collection and scoring.

## STREAM RESTORATION PRIORITIES FOR THE NEXT GENERATION OF PRACTITIONERS

The development of this guidebook reflects the experiences of fifty-five authors from four countries who have contributed to stream restoration efforts as practitioners, scientists, funders, and community activists. As part of saying farewell to you, we wanted to take advantage of the authors who have contributed to this guidebook to identify a list of ongoing stream restoration challenges that we feel, if addressed, could greatly improve the effectiveness of stream corridor restoration. In contrast to the advice the guidebook offers to improve the effectiveness of stream restoration projects planned for the immediate future, this is a look further down the road, as the baton is passed to the next generation of stream restoration practitioners. We started the guidebook with an elite 8 list, why not end with one? Our hope is that the list will promote deliberation, conversation, and debate that will lead to answers and more effective stream restoration globally.

# CASE STUDY 84
## Businesses Can Be Strong Stream Restoration Partners

Jon Radtke

In 2018, the Coca-Cola Company (TCCC) announced it had achieved its 2020 global replenishment goal, along with its global bottling partners throughout the Coca-Cola system. This means the amount of water that was returned to communities and nature as part of the replenishment program equaled the amount of water used in Coca-Cola beverages and their production. With this achievement, Coca-Cola is the first Fortune 500 company to reach such an aggressive water replenishment target. From our perspective, the replenishment goal means that when a consumer drinks a Coca-Cola product, they can be confident that Coca-Cola and its bottling partners are committed to responsible water use today and in the future.

The Coca-Cola system has achieved its water replenishment goals through 248 community water partnership projects in 71 countries that focus on safe-water access, watershed protection, and water for productive use. In many cases, projects also provide access to sanitation and education, help improve local livelihoods, assist communities with adapting to climate change, improve water quality, enhance biodiversity, engage on policy, and build awareness on water issues. Only those projects that increased water availability for the environment and communities were counted as part of the replenishment goal. The methodology used to quantify replenishment benefits was developed in collaboration with the Nature Conservancy and LimnoTech (an environmental consulting firm) and peer-reviewed for technical accuracy.

## Lessons Learned

- For stream restoration practitioners, private companies (from small to big) can have interest in and be supportive of stream conservation/restoration efforts. TCCC's support of stream restoration efforts around the globe was spurred by its global replenishment goal. Other companies may have goals that are more locally driven, such as protecting source water supplies. Don't hesitate to reach out to companies in your watershed to see if they also have a replenishment goal and would be supportive of your stream restoration project.
- A partnership between a stream restoration effort and a business (whether local or global) will likely not be solidified with one phone call or one meeting. The prolonged interaction of interested parties will likely need to occur.
- In many cases, TCCC's support of the environmental and community projects that are part of its global replenishment portfolio have lasted over eight years. For example, TCCC's support of and participation

### How can we . . .

1. *. . . more effectively reach out to the public about the connections between water supply reliability and stream health?* In other words, how do we better help people see that healthy streams are critical to providing the quantity and quality of water that irrigated agriculture, municipalities (rural and urban communities), industry, and recreation interests need for their own long-term viability?

As part of its replenishment goal, TCCC supported restoration work along the binational stretch of the Rio Grande/Río Bravo. A portion of the binational partnership photographed here includes members from Big Bend National Park, Rio Grande Scientific Support Services, TCCC, WWF, Profauna, and Comisión Nacional de Áreas Naturales Protegidas. Photo by Audra Melton.

in the binational restoration work along the RGB (see photo in this case study) lasted over a decade. Not all public-private partnerships may be that long-lasting, but from our experience, such long-term and dedicated commitment is exactly what was needed to make sustainable improvement in stream conditions.

- TCCC personnel got to the field on numerous occasions to evaluate results of restoration and replenishment work firsthand and to plan next steps. Many of the binational team members commented that they had never previously experienced such direct and long-term involvement from a funder. Attention, funders! Getting to the field and seeing results firsthand is critical for effective evaluation and provides an opportunity to get to know restoration team members and plan next steps. Such shared experiences of funders, administrators, and on-the-ground restoration personnel is an important key to long-term success.

2. *. . . better demonstrate that restoring rivers makes economic sense?* Like it or not, we live in a world that is largely driven by money. We need to be better prepared to demonstrate to society that restoring rivers is not only desirable from an ecological standpoint but also economically. Although much progress has been made on this topic, more needs to be done to quantify the economic benefits of stream restoration (i.e., showing the value of ecosystem services that healthy streams provide to society) and use such information to develop innovative financing

mechanisms for restoration (e.g., California's water resources bonds that fund watershed restoration and protection).

3. . . . *maintain healthy streamflow and sediment transport regimes that maintain ecosystem processes, functions, and biotas, even in downsized or fragmented watersheds?* A key part of answering this question will be continue to improve on ways to collect, organize, analyze, and convey technical information to the public in a manner that will garner support for stream restoration and lead to long-term protection of the streamflow needed to maintain native stream ecosystems.

4. . . . *share the results of stream restoration more broadly to increase public participation and support of stream restoration?* We need to do a better job sharing real-time, real-world stories about native stream species and stream restoration projects with politicians, community leaders, water managers, business leaders, and others in the public realm. For example, in the Murray-Darling Basin, several native fish species have been fitted with acoustic tags to track them (e.g., students can get online to see where Max, the Murray Cod, is today), and migratory water birds have been fitted with satellite trackers on their legs and light solar panels on their backs. Initiatives such as these can help spark greater public interests as well as provide data. The more effective that practitioners are at piquing public interest in streams and their importance, the more local citizens will be involved and supportive of restoration efforts.

5. . . . *provide education and training to practitioners?* Most restoration practitioners are self-taught or rely on short courses provided primarily by practitioners. Professional certification restoration programs that include multidisciplinary training are needed. What can be done to elevate and promote restoration education programs? What role can academia play and which disciplines could take the lead in bridging the science-practice gap? How academia can provide practitioners timely information on how to restore stream ecosystems in the context of rapidly evolving climate change is an important example of a science-practice gap that will require future attention. Advances in technologies (e.g., communication platforms, remote sensing) are occurring rapidly and can certainly help to bridge science-practice gaps by fostering better planning, evaluation, and monitoring. But how can practitioners be better informed of these new technologies and learn how to effectively apply them to their own restoration projects?

   An important part of addressing this challenge is to provide practitioners avenues to more effectively learn from each other's experiences. The technical aspects of stream restoration are of great interest to restoration practitioners, and sharing this information is the heart of this guidebook. What other ways can stream restoration experiences be more effectively and quickly shared among practitioners throughout the world? A repository of cases, practices, and experiences that would allow us to learn from failures as much as from successes would be very valuable to the profession. Establishing websites and other information-sharing platforms that are devoted to this question could go a long way to improving the effectiveness of stream restoration globally. A great example of this is the Collaborative Conservation and Adaptation Strategy Toolbox (CCAST), which is an online library of adaptation case studies managed by the U.S. Fish and Wildlife Service and is highlighted in chapter 4.

6. . . . *better support long-term research/monitoring of stream restoration results?* Advancements in the design and planning of stream restoration that is resilient and successful in the long term can only be made if restoration results can be monitored well after project completion. Yet

monitoring is frequently underfunded. Support for long-term observations of key parameters requires the strengthening of national (or even regional) systems such as the National Science Foundation's Long Term Ecological Research Network (U.S.). In addition, closer ties between the restoration community, academia, and citizen science projects that involve the local public directly in monitoring and research will be important and mutually beneficial to restoration practitioners, universities, and the general public.

7. ... *balance the restoration of deteriorated streams with the importance of protecting streams that remain in good condition?* This question touches on two key issues: (1) coordinating and communicating within the conservation community (agencies, NGOs, institutions, businesses, funding entities, communities) regarding who is doing what and where in a particular watershed, and (2) identifying priorities that help guide overall restoration response. How do we take advantage of the inherent resiliency of stream ecosystems? Stream ecosystems often come back when drivers of deterioration are eliminated. The removal of unnecessary and unsafe dams is high on this list. Eliminating such drivers of stream decline may allow restoration to occur (with consummate benefits to both people and native species) without the need for additional restoration action. How can such opportunities be effectively identified and orchestrated?

8. ... *establish long-term funding mechanisms for stream restoration?* Long-term support of stream restoration actions is critical to long-term success. We need to go beyond the paradigm of securing grants that lead to one-and-out restoration projects to long-term synergistic projects that can dramatically move the restoration needle. Are there effective approaches for securing funding and support in the long term (i.e., a decade commitment versus two-year grants or funding cycles)? A study of stream restoration efforts lasting a decade or more could help identify funding sources, mechanisms, and novel partnerships that have successfully fostered long-term support. There are also existing models of long-term funding for conservation that potentially could be applied to stream restoration. One example is the U.S. Land and Water Conservation Fund, which relies on royalties from offshore drilling to fund conservation programs. There are also regulatory frameworks that disincentivize impacts and simultaneously fund restoration (or could be modified to do so). Examples in the United States include the Clean Water Act and Federal Emergency Management Agency, which have frameworks to collect impact fees to fund restoration, yet lack administration and enforcement.

■ ■ ■

We now leave you at your restoration site and say farewell. Our sincere hope is that this marks only the beginning of a long and fruitful relationship. Looking back at how stream restoration projects begin can often be humorous. For several of this guidebook's authors, our restoration projects began with a napkin sketch of our restoration project constructed with a few colleagues over lunch, and we have the photographs to prove it. Perhaps you have your own beginning story. Maybe that story led you to pick up this guidebook, or maybe this guidebook is your beginning story. Regardless of how it began, we truly do not want your interest to wane or your project to stall. Use this guidebook as a reference. Mark it up! Cross out text you aren't interested in or don't agree with, underscore what works for you. Inhale the lessons from the case studies and breathe them into your own stream restoration project.

And do not hesitate to reach out to us. The stream corridor restoration community is a tight-knit, gregarious, sometimes cantankerous and opinionated, but always welcoming community. If you weren't a member before you picked up this guidebook, you certainly are now. Give us a call or drop us an email and tell us about your project, what you've learned, or any questions or concerns you have (see author bios). If you have questions that we cannot answer, we likely will be able to steer you toward others who can. All the guidebook authors agree that there are few endeavors more rewarding than restoring a stream corridor. Please keep us by your side through the thick and thin of it. We look forward to continuing the restoration journey with you.

# REFERENCES

Arlettaz, R., M. Schaub, J. Fournier, T. S. Reichlin, A. Sierro, J. E. M. Watson, and V. Braunisch. 2010. "From Publications to Public Actions: When Conservation Biologists Bridge the Gap Between Research and Implementation." *Bioscience* 60:835–42.

Bernhardt, E. S., E. B. Sudduth, M. A. Palmer, J. D. Allan, J. L. Meyer, G. Alexander, J. Follastad-Shah, et al. 2007. "Restoring Rivers One Reach at a Time: Results from a Survey of U.S. River Restoration Practitioners." *Restoration Ecology* 15 (3): 482–93.

Cabin, R. J., A. Clewell, M. Ingram, T. McDonald, and V. Temperton. 2010. "Bridging Restoration Science and Practice: Results and Analysis of a Survey from the 2009 Society for Ecological Restoration International Meeting." *Restoration Ecology* 18:783–88.

Clark, L. B., A. L. Henry, R. Lave, N. F. Sayre, E. González, and A. A. Sher. 2019. "Successful Information Exchange Between Restoration Science and Practice." *Restoration Ecology* 27 (6): 1241–50.

Dickens, S. J. M., and K. N. Suding. 2013. "Spanning the Science-Practice Divide: Why Restoration Scientists Need to Be More Involved with Practice." *Ecological Restoration* 31 (2): 134–40.

Fernandez, M. 2016. "U.S.–Mexico Teamwork Where the Rio Grande Is But a Ribbon." *New York Times*, April 22, 2016. https://www.nytimes.com/2016/04/23/us/us-mexico-teamwork-where-the-rio-grande -is-but-a-ribbon.html?_r=1.

Giardina, C. P., C. M. Litton, J. M. Thaxton, S. Cordell, L. J. Hadway, and D. R. Sandquist. 2007. "Science Driven Restoration: A Candle in a Demon Haunted World? Response to Cabin (2007)." *Restoration Ecology* 15:171–76.

González, E., A. Masip, E. Tabacchi, and M. Poulin. 2017. "Strategies to Restore Floodplain Vegetation After Abandonment of Human Activities." *Restoration Ecology* 25:82–91.

Hawley, R. J. 2018. "Making Stream Restoration More Sustainable: A Geomorphically, Ecologically, and Socioeconomically Principled Approach to Bridge the Practice with the Science." *Bioscience* 68 (7): 517–28.

Hoechlin, N. 2017. *Mastering Business Social Media Marketing in Theory and Practice.* JNR Publishing Group. http://jnrpublishing.com/.

Palmer, M. A. 2009. "Reforming Watershed Restoration: Science in Need of Application and Applications in Need of Science." *Estuaries and Coasts* 32:1–17.

Ruiz-Mallén, I., C. Schunko, E. Corbera, M. Rös, and V. Reyes-García. 2018. "Meanings, Drivers, and Motivations for Community-Based Conservation in Latin America." *Ecology and Society* 20 (3): 33. http://dx.doi.org/10.5751/ES-07733-200333.

Solaris, J. 2019. "Social Media for Events (2019 edition): A Complete Guide to Marketing Your Events Using Social Media." https://www.eventmanagerblog.com/social-media-events.

Uiito, J., and A. Duda. 2002. "Management of Transboundary Water Resources: Lessons from International Cooperation for Conflict Prevention." *Geographical Journal* 168:365–78.

van der Zaag, P., and H. H. Savenije. 2000. "Towards Improved Management of Shared River Basins: Lessons from the Maseru Conference." *Water Policy* 2 (1–2): 47–63.

Zamora-Arroyo, F., and K. W. Flessa. 2009. "Nature's Fair Share: Finding and Allocating Water for the Colorado River Delta." In *Conservation of Shared Environments: Learning from the United States and Mexico*, edited by L. Lopez-Hoffman et al., 23–28. Vancouver: University of British Columbia Press.

Zamora-Arroyo, F., O. Hinojosa-Huerta, E. Santiago, E. Brott, and P. Culp. 2008. "Collaboration in Mexico: Renewed Hope for the Colorado River Delta." *Nevada Law Review* 8 (3):871–89.

Zamora-Arroyo, F., J. Pitt, S. Cornelius, E. Glenn, O. Hinojosa-Huerta, M. Moreno, J. García, P. Nagler, M. de la Garza, and I. Parra. 2005. *Conservation Priorities in the Colorado River Delta, Mexico and the United States.* Prepared by the Sonoran Institute, Environmental Defense, University of Arizona, Pronatura Noroeste Dirección de Conservación Sonora, Centro de Investigación en Alimentación y Desarrollo, and World Wildlife Fund—Gulf of California Program. Tucson, Ariz.: Sonoran Institute.

# APPENDIX A

## SELECTED SOURCES OF NATURAL RESOURCE DATA AND INFORMATION IN AUSTRALIA, MEXICO, AND THE UNITED STATES

## INTERNATIONAL

An updated, comprehensive, two-part treatise on stream gauging (fieldwork and computation of discharge) was published by the World Meteorological Organization (WMO 2010). This invaluable report consolidates techniques and procedures accumulated during many decades by agencies including the USGS, EPA, NRCS, IBWC, and USACE.

## AUSTRALIA

### Selected Federal Agencies

#### Department of Agriculture
The Department of Agriculture (formerly the Department of Fisheries and Forestry) has the dual roles of providing services to the agricultural, food, fisheries, and forest industries, and addressing the challenges of natural resource management. The *Caring for Our Country* initiative, an online repository of digital reports, provides open access to a digital archive of publicly funded information derived from Australian government investments in natural resource managements. The repository has print and electronic documents on or derived from biodiversity conservation activities, along with land-management, biophysical, social and economic research, and policy activities (http://www.daff.gov.au/publications).

#### Department of the Environment
The Department of the Environment implements Australian government policies on the environment, heritage, water, and climate actions. Its website supports a link dedicated to water publications and resources (http://www.environment.gov.au/topics/water/publications-and -resources).

## The Commonwealth Environmental Water Office

The Commonwealth Environmental Water Office (CEWO) delivers water for the environment to help rivers flow, keep native plants healthy, and support feeding and breeding of native animals, birds, fish, and frogs. CEWO provides a host of information and services that educate the public on the befits of healthy rivers to humans and native species. The CEWO used to sit under the Commonwealth Department of the Environment. https://www.environment.gov.au/water/cewo.

# MEXICO

## Selected Federal Agencies

A comprehensive guide to agencies in Mexico that collect natural resource information is available from the Centro de Estudios Jurídicos y Ambientales (http://www.ceja.org.mx/publicaciones.php). Responsibilities of selected natural resource agencies in Mexico are summarized below.

## Comisión Nacional del Agua (CONAGUA)

CONAGUA is an agency of the Ministry of the Environment and Natural Resources with mandates to administer and manage the water resources of Mexico in a manner beneficial to the people and ecosystems of the country (http://www.conagua.gob.mx/home.aspx). CONAGUA is responsible for maintaining stream gauges and collecting water-quality data. Streamflow and water-quality data can be requested at sina@conagua.gob.mx, and also from http://siga.cna.gob.mx/.

In 2008, CONAGUA compiled and summarized data collected from 1,186 monitoring sites across Mexico (CONAGUA 2010). In addition to water-quality data, this thorough summary provides statistics on national water use, water quality, annual summaries of the program, descriptions of strategic projects, records of numerous dams and reservoirs along streams of many different sizes, and agendas for future projects and research. Other sources of hydrological data from Mexico include the Ministry of Finance, the Federal Congress, state governments and state congresses, and the Ministry of Environment and Natural Resources.

## Comisión Nacional de Áreas Naturales Protegidas (CONANP)

CONANP is responsible for the management of parks, reserves, sanctuaries, monuments, and other protected areas in Mexico (https://www.conanp.gob.mx/anp/anp.php). Much of the agency is staffed by conservationists, engineers, and scientists, many of whom are associated with universities; published products are diverse and can be identified through its website.

## Instituto Nacional de Estadística, Geografía, e Informática (INEGI)

INEGI produces digital orthophotos and topographical and geological maps for most of Mexico that can be viewed from an internet map service accessed in a GIS from a web mapping service or purchased on CD (http://www.inegi.org.mx/).

## Comisión Nacional para el Conocimiento y Uso de la Biodiversidad (CONABIO)

CONABIO is an interdepartmental commission, created in 1992. Its mission is to promote, coordinate, support, and carry out activities aimed at increasing awareness of biodiversity and the benefits that conserving biodiversity can provide to society. CONABIO is an institution that generates information and manages the National Information System on Biodiversity. CONABIO also provides advice to various users and implements the national and global biodiversity-information networks, complying with international commitments on biodiversity entered into by Mexico, and carries out actions directed toward conservation and sustainable use of biodiversity in Mexico (http://www.conabio.gob.mx/).

Main data bases are at http://www.conabio.gob.mx/institucion/snib/doctos/acerca.html.

## Servicio Geológico Mexicano (SGM)

SGM is the Mexican Geological Service that produces geological and economic information for all interested in the mining and geological knowledge of Mexico; this agency also collects and organizes data and information associated with the environment, water, land use and land management, and geological hazards. Main data bases are at http://www.sgm.gob.mx/index .php?option=com_content&task=view&id=17&Itemid=30.

## Secretaría del Medio Ambiente y Recursos Naturales (SEMARNAT)

SEMARNAT is the Ministry of Environment and Natural Resources and is the federal agency in Mexico responsible for promoting the protection, restoration, and conservation of ecosystems and natural resources and environmental goods and services in Mexico to promote their use and sustainable development. Databases are integrated at the Sistema Nacional de Información Ambiental y de Recursos Naturales, a set of freely available statistical, cartographic, and documentary data that allows users to collect, organize, and disseminate information about the environment and natural resources. Of particular relevance to stream corridor restoration is the integration of information on inventories of natural resources and the monitoring of air, water, and soil quality, as well as ecological land records and programs that focus on environmental protection.

# UNITED STATES

## Federal Agencies

### U.S. Geological Survey (USGS)

Most stream gauges in the United States are maintained by the USGS; its surface water portal (http://waterdata.usgs.gov/nwis/sw) includes raw discharge data at fifteen- or thirty-minute intervals for each stream gauge, along with summary statistics (i.e., mean and peak-daily discharges and mean monthly and annual discharges). Accessible data are rigorously tested for accuracy. In some cases, previously processed data and reports summarizing hydrological analyses are available (Granado et al. 2017). The USGS also maintains several online toolkits with data charts and graphs, such as WaterWatch (http://waterwatch.usgs.gov/index.php), which can be automatically generated. At some USGS stream gauges sediment data also are actively collected or have been; these data can be accessed at http://waterdata.usgs.gov/nwis/qwdata.

The USGS publishes topographical and geological maps for many parts of the country (http://ngmdb.usgs.gov/), has a clearinghouse for geospatial data on its Cumulus Portal for Geospatial Data (http://cumulus.cr.usgs.gov/), and compiles and frequently updates geospatial data in its Seamless Data Warehouse (http://seamless.usgs.gov/index.php). Geospatial database programs such as ESRI ArcMap have built-in features that calculate all of the measures mentioned. Geospatial software uses digital elevation maps, which are digital representations of topographical maps available at http://ned.usgs.gov/.

The USGS also collects, publishes, and posts water-quality and sediment-discharge data from numerous gauge sites, and is the lead agency for developing sampling and data-analysis protocols (Wilde 2011). The data-collection methods standardly used by USGS water-quality personnel are peer reviewed, current, and published in the National Field Manual for the Collection of Water-Quality Data (http://pubs.water.usgs.gov/twri9A/).

Specialized information sources associated with the USGS include the National Water-Quality Assessment Program assessments of water-quality conditions, trends, and how they are affected by natural processes and human activities (http://water.usgs.gov/nawqa/), and Science in Your Watershed (http://water.usgs.gov/wsc/index.html), providing easy identification of your watershed and a summary of various physical data sets, data products, and reports.

## Environmental Protection Agency (EPA)

The EPA is the lead federal agency for establishing and enforcing water-quality standards, relying on states and tribes for local data collection, developing assessment reports, and establishing Total Maximum Daily Loads for impaired waters. A specialized source of EPA information is the EPA Watersheds website (http://water.epa.gov/type/watersheds/). Data on the EPA website relate largely to water quality, but the website also lists regulatory and citizen-based groups for each watershed and other pertinent links.

The EPA has an excellent tool for individual catchments at http://www.epa.gov/waters/tools/index.html. Its Surf Your Watershed website provides information on potential sources of pollution, land-use and management changes, and how to find the owners and managers of impoundments. (http://cfpub.epa.gov/surf/locate/index.cfm).

The Watershed Assessment of River Stability & Sediment Supply website (http://water.epa.gov/scitech/datait/tools/warsss/) is a comprehensive source for geomorphic assessment. This step-by-step approach quantifies geomorphic conditions in a channel reach and includes worksheets to guide the collection, organization, and analysis of data.

## Natural Resources Conservation Service (NRCS)

The USDA NRCS has a Rapid Watershed Assessments program that provides estimates of where conservation investments can be effectively used to satisfy landowners, conservation districts, and other community organizations and stakeholders within a watershed. Assessment information helps users to set priorities and identify actions that can reduce impacts of land-use change on biophysical conditions (http://www.nrcs.usda.gov/Internet/FSE_DOCUMENTS/stelprdb1042217.pdf).

## U.S. Department of Agriculture (USDA)

The USDA Aerial Photo Field Office has extensive archives of historical and current aerial photographs, many of which are available electronically (http://www.fsa.usda.gov/FSA/apfo app?area=apfohome&subject=landing&topic=landing).

## International Boundary and Water Commission (IBWC)

The IBWC is the federal agency that applies boundary and water treaties of the United States and Mexico and settles disputes that arise in their application. The IBWC is an international body composed of U.S. and Mexican sections, each headed by an engineer-commissioner appointed by the respective president. Each section is administered independently. The U.S. section of the IBWC is headquartered in El Paso, Texas, and operates under the foreign-policy guidance of the Department of State. The Mexican section is administered by the Mexican Ministry of Foreign Affairs and is headquartered in Ciudad Juarez, Chihuahua, Mexico (http://www.cila.gob.mx/).

The IBWC maintains one streamflow gauge on the Colorado River's northern international boundary, one at the international boundary of the Tijuana River, and fifty-five along the mainstem Rio Grande/Bravo and borderland tributaries (https://www.ibwc.gov/Water_Data /rtdata.htm).

## State and Local Agencies

### Arizona

The Arizona Department of Environmental Quality, established in response to concerns about groundwater quality, also administers programs focusing on land and air. The ADEQ website (http://www.azdeq.gov/environ/water/watershed/index.html) provides links to water-related information.

The Arizona Department of Water Resources collects and manages hydrological data, including groundwater inventories, a hydrology library, imaged records, and other useful information (http://www.azwater.gov/AzDWR/IT/DataCenter/DataCenter.htm).

### California

The California Data Exchange Center (CDEC) installs, maintains, and operates a hydrological data-collection network of automatic snow-reporting gauges and precipitation and stream-stage sensors for flood forecasting. The surface water link is accessed at http://cdec.water.ca .gov/index.html, and Northern California has data sites that can be found at the CDEC: http:// cdec.water.ca.gov/cgi-progs/rivfcast/SCALBUL. A statewide list of reservoir-storage volumes can be found at http://cdec.water.ca.gov/misc/monthly_res.html. Access to groundwater data is available at http://www.water.ca.gov/groundwater/casgem/online_system.cfm and http:// www.water.ca.gov/waterdatalibrary/.

### New Mexico

The New Mexico Environmental Department (http://www.nmenv.state.nm.us/) houses its Water Resources and Management Division (http://www.nmenv.state.nm.us/nav_water.html).

## *Texas*

The Texas Natural Resources Information System (TNRIS.org) provides extensive information for Texas, including aerial photographs, Landsat images, topographical maps, and hydrological datasets.

The Texas Commission on Environmental Quality (http://www.tceq.state.tx.us/) manages the Surface Water Quality Monitoring (SWQM) program (http://www.tceq.texas.gov/water quality/monitoring/index.html), which coordinates the collection of physical, chemical, and biological samples from numerous surface water sites.

## *Mexico-U.S. Border Region*

The Border Environment Cooperation Commission and the North American Development Bank are active along the U.S.-Mexico border region. They complement federal agencies of Mexico that study and protect the natural resources of the border region. Active also along the border corridor are The Nature Conservancy, the World Wildlife Fund, and the IBWC. The U.S. section of the IBWC and the CRP (Texas Clean Rivers Program), with the TCEQ, collect water-quality data in the Texas portion of the Rio Grande Basin (http://www.tceq.texas.gov /waterquality/clean-rivers).

# APPENDIX B

## MAJOR CLIMATE SERVICE PROVIDERS AND RESOURCES, BY COUNTRY

## AUSTRALIA

### Bureau of Meteorology (BOM)

BOM (http://www.bom.gov.au/) is the national weather, climate, and water agency for Australia. It provides observational data and regular forecasts (on daily to seasonal timescales) of temperature, precipitation, and streamflow for the entire nation, as well as by region. From the BOM "Water Information" page (http://www.bom.gov.au/water/), users can access a variety of information, from streamflow forecasts to water restrictions to water assessments. Included here are summaries of recent rainfall and streamflow and an interactive map that displays parameters, such as runoff and soil moisture, for the past three years. The website's "Climate and Past Weather" page (http://www.bom.gov.au/climate/) provides links to maps of recent and average conditions, reports and summaries, seasonal forecasts, observational data, data on climate extremes, and maps and time series that show recent climate trends for the country. From this page the user can also request weather-station data. BOM regularly collaborates with the Commonwealth Scientific and Industrial Research Organisation (CSIRO; see below) on various climate-related projects, such as the State of the Climate 2016.

### Commonwealth Scientific and Industrial Research Organisation (CSIRO)

CSIRO (https://www.csiro.au/) is a corporate entity of the Australian government that conducts research in a myriad of fields, disseminates scientific information, and facilitates the utilization of research results. Through various centers and projects, it works with practitioners and provides data useful for their needs. For example, CSIRO's Sustainable Yields projects (https://www.csiro.au/en/Research/LWF/Areas/Water-resources/Assessing-water-resources/Sustainable-yields) assess current and future water availability for Australia's major water systems, providing reports and maps. The CSIRO website also provides biennial State of the

Climate reports—a joint effort between BOM and CSIRO—which describe recent changes in Australia's climate, such as changes to temperature and precipitation extremes, and briefly describe future changes. The latest (2018) biennial report can be found at https://www.csiro.au/en/showcase/state-of-the-climate.

## Climate Change in Australia

A collaboration between BOM and CSIRO, Climate Change in Australia (https://www.climate changeinaustralia.gov.au/en/) is a website devoted to providing future climate projections specifically for resource managers in Australia. There are over ten tools available, each given a complexity rating based on the information provided and the level of climate science knowledge needed to fully understand and use the tool. The website provides extensive support and guidance for using the data and projection tools, and provides a decision tree that allows users to find information specific to their needs. Available tools include a climate analogs tool that displays locations where current climate approximates the future climate at a given location; a map explorer that displays projections of various parameters, such as rainfall and temperature and allows users to download data from eight different models; a tool that allows users to download future daily or monthly time-series data at a specific station; and a projections builder tool that allows users to generate application-specific projections for an impact assessment. The website also provides "cluster reports" that offer detailed regional climate information to assist regional decision makers in understanding climate projections and how they apply to their region.

## National Climate Change Adaptation Research Facility (NCCARF)

NCCARF (https://www.nccarf.edu.au/) is an initiative by the Australian federal government that supports stakeholders throughout the country as they prepare for and adapt to the risks of climate change. The program promotes an integrated approach to research and adaptation by bringing together researchers and decision makers at conferences, seminars, and workshops (the dates of which can be found on the website). The NCCARF website provides factsheets made in consultation with stakeholders that describe various impacts of climate change and has an adaptation library where users can search by location, topic, and resource type to find resources such as case studies, reports, and guidelines.

Researchers at Adaptation Research Networks (https://www.nccarf.edu.au/adaptation -networks), initiated by NCCARF and hosted by universities across Australia, conduct focused research on topics from human health to water resources and freshwater biodiversity. Each network provides contact information for climate change adaptation experts by area of expertise. The Natural Ecosystems Network, an adaptation network initiated by NCCARF and hosted by James Cook University, offers free memberships that enable members to access data and tools, be notified of trainings and workshops, and find relevant expert advice and collaborative contacts.

## AdaptNRM

AdaptNRM (http://adaptnrm.csiro.au/about-adaptnrm/) is a national initiative through Australia's Department of Environment that supports and guides natural resource managers through the process of climate adaptation planning. The project was completed in 2016, but the website will remain active indefinitely. It provides managers with five modules, from adaptation planning to shared learning, that deliver synthesized guidance with support from technical guides and datasets. The project's research team consists of scientists from CSIRO and NCCARF, but the outputs and initiatives were created with input from other national researchers and natural resource managers and practitioners to ensure that the information is useful. Among some of the resources provided on the website are an adaptation checklist, links and references for planning, decision-support tools, and a guide to managing weeds in a changing climate.

## CoastAdapt

CoastAdapt (http://coastadapt.com.au) is an information-delivery and decision-support framework. It focuses on Australia's coastal regions, the risks they face from climate change and sea-level rise, and how to respond to those risks. The site contains information and guidance to help people from all walks of life understand climate change and the responses available to manage the impacts. CoastAdapt also links users to climate change resources on the NCCARF and other websites that are relevant to Australia more broadly (https://coastadapt .com.au/it's-not-just-coast-practical-knowledge-adaptation-across-australia).

## National Environmental Science Programme (NESP) Research Hubs (Department of the Environment and Energy)

NESP (http://www.environment.gov.au/science/nesp) aims to support decision makers in their efforts to understand, manage, and conserve Australia's environment by providing information and research conducted through six research hubs, including the Earth Systems and Climate Change (ESCC) Hub. The ESCC Hub (http://nespclimate.com.au/), a partnership of CSIRO, BOM, and Australian universities, aims to improve understanding of past, current, and future climate and improve the utility of climate change information. The website outlines research projects, such as improving regional projections and managing climate extremes, and provides contact information for the scientists involved and affiliated publications. Researchers also post articles about their research on the news page of the website.

## State Government Agencies

Several of the state government agencies in Australia provide resources as well. Below are two examples.

AdaptNSW (http://climatechange.environment.nsw.gov.au/) is a regional resource for the state of New South Wales (NSW), produced by the NSW Department of Environment and Heritage. The website provides climate change information for NSW, including climate projections, impacts, adaptation information, and educational resources. In addition to an interactive map displaying climate projections for the whole state, the website provides projections for each of the state's planning regions, complete with a PDF snapshot and downloadable data and maps. Educational resources on the website include adaptation guidelines, research publications, videos, and links to other state, national, and international resources.

The government of Victoria (https://www.climatechange.vic.gov.au/) provides ample climate resources for Victoria, including emissions reports, projections, and resources for local government. The webpage dedicated to climate projections (https://www.climatechange.vic.gov.au/adapting-to-climate-change-impacts/victorian-climate-projections-2019) provides projections specific for Victoria as both datasets and regional and technical reports, with guidance on how to use all of the information.

# MEXICO

## Instituto Nacional de Ecología y Cambio Climático (INECC; National Institute of Ecology and Climate Change)

INECC (http://www.gob.mx/inecc) is an institution within the Secretaria de Medio Ambiente y Recursos Naturales (SEMARNAT; the Mexican natural resources agency) that is dedicated to climate change and ecology research, the integration of technical and scientific knowledge into decision-making, and the formulation and evaluation of public policies related to ecology and climate change. Its website contains links to maps of vegetation, demographics, and watersheds, and the documents page provides links to many technical publications, including documents related to vulnerability and adaptation. The actions and programs link provides detailed information on all of the agency's projects, with links to associated references and tools. INECC also hosts another site, along with other federal agencies, with information on climate change in Mexico (https://cambioclimatico.gob.mx/).

## Comisión Nacional del Agua (CONAGUA; Mexican National Water Commission)

CONAGUA (http://www.gob.mx/conagua) is a federal agency charged with water-resources regulation and the operation of infrastructure for Mexico's surface and groundwater supplies. The agency operates chiefly from offices in the nation's capital, but also has regional offices throughout the U.S.-Mexico border region. The CONAGUA website is a valuable source of information on water-management policies and practices. It is also a resource for research, reports, and authoritative assessments of surface and groundwater resources. The latter include annual reports on water statistics for Mexico. The website provides an online map resource (Subgerencia de Información Geográfica del Agua [SIGA]), a digital water-information atlas, and a metadata exploration tool. Historical and real-time data are not readily available from the CONAGUA website but can be obtained by contacting CONAGUA's offices.

## Servicio Meteorológico Nacional (SMN; Mexican National Weather Service)

The SMN (https://smn.conagua.gob.mx/es/) is a branch of CONAGUA devoted to weather and climate observations and forecasts. The SMN website provides data on the average climate of Mexican states, analyses of recent conditions, seasonal summaries, and precipitation forecasts for the current month and the following two months. Monthly, state, and national-level data for the last ten years are available through the website. The SMN website also provides forecasts and information on extreme events, such as tropical cyclones and drought. The SMN does not conduct research on climate change; however, you can access climate change reports from SEMARNAT, Mexico's natural resources agency.

## Universidad Nacional Autónoma de México (UNAM; National Autonomous University of Mexico)

UNAM (https://www.unam.mx/), a public research university, has a comprehensive program on climate variability and climate change, with research spanning phenomena such as drought and the El Niño–Southern Oscillation, weather forecasting, impacts on society, and projections of future climate change. UNAM's website includes links to a Digital Climate Atlas for Mexico (http://uniatmos.atmosfera.unam.mx/ACDM/) and a Program for the Investigation of Climate Change (PINCC; http://www.pincc.unam.mx). The digital atlas displays maps of climate vulnerability and adaptation, as well as climate trends, average conditions, and projected future changes to several atmospheric and terrestrial parameters such as minimum, maximum, and average temperature, precipitation, drought, and climate extremes. The PINCC website's publications link points viewers to several publications on climate change, produced in conjunction with SEMARNAT and others, such as the three volumes of the 2015 Mexican Report on Climate Change. In recent years, UNAM researchers have collaborated with CONAGUA, SMN, and SEMARNAT on climate change research and outreach to stakeholders.

## Laboratorio Nacional de Modelaje y Sensores Remotos (LNMySR; National Laboratory of Modeling and Remote Sensing)

LNMySR (http://clima.inifap.gob.mx/lnmysr/) is an institution under the Instituto Nacional de Investigaciones Forestales, Agrícolas y Pecuarias (INIFAP; National Institute of Forestry, Agriculture, and Livestock Research) that produces seasonal and five-day forecasts of humidity and temperature conditions. From the website, users can access the forecasts and search for weather stations. The website's publications link provides links to scientific publications on forecasting and climate change, case studies, and videos. The website also contains a directory of researchers associated with LNMySR, and an information request form to obtain climate information from the National Stations Network.

## Centro de Investigación Científica y de Educación Superior de Ensenada, Baja California (CICESE; Center for Scientific Research and Higher Education in Ensenada, Baja California)

CICESE (http://www.cicese.edu.mx/) is a public research center of Consejo Nacional de Ciencia y Tecnología (CONACYT), Mexico's National Science Foundation, and is one of the main scientific centers in Mexico. The website provides links to publications, forecasts, and a weather data-management software system developed by CICESE, in collaboration with SMN, INECC, and others, known as CLICOM (CLImate COMputing project; http://clicom-mex.cicese.mx/). The associated web tool displays time series of temperature, precipitation, evaporation, and heat units (useful for the agricultural sector). The data used to inform the maps is also available to download, and CICESE plans to include extreme events data in the future.

## Red de Desastres Asociados a Fenómenos Hidrometeorológicos y Climáticos (REDESClim; Network of Disasters Associated with Hydrometeorological and Climatic Phenomena)

REDESClim (http://www.redesclim.org.mx/) is a research network of CONACYT focused on reducing the risk of impacts from natural disasters. The network brings together researchers, businessmen, decision makers, and the public to promote solutions for increasing resilience to natural disasters. The website provides various resources, including a list of national and international organizations and a catalog of national educational programs, including contact information, and is organized by topic. Sites of interest that can be accessed via the website include climate projections mapping tools. A database of publications and reports provide the reader with additional information on extreme events.

## Comisión Federal de Electricidad (CFE; Federal Electricity Commission)

The CFE (http://www.cfe.gob.mx/paginas/Home.aspx) is Mexico's Federal Electricity Commission, which generates electric power for about one hundred million people across the country. The CFE does not offer data directly to the public but will provide data upon request. The types of data it can provide include regional daily forecasts of precipitation, wind, and temperature; forecasts of precipitation and maximum and minimum temperature for fifteen days, one month, and six months; and regional projections of precipitation and temperature.

## Instituto Mexicano de Tecnología del Agua (IMTA; Mexican Institute of Water Technology)

IMTA (https://www.gob.mx/imta) addresses challenges associated with water management in Mexico. It has played a lead role in generating downscaled climate change projections for Mexico during the last two IPCC cycles. For the last two decades it has been conducting research related to climate change and water resources; its products include two national atlases on water vulnerability under climate change (2010 and 2015). These and other products related to climate change can be accessed in the open-access website https://www.gob.mx/imta#3032. IMTA also has a comprehensive graduate program on water sciences and technology.

## UNITED STATES

## National Integrated Drought Information System (NIDIS)

NIDIS (https://www.drought.gov/drought/) was authorized by the U.S. Congress in 2006 (Public Law 109–430) to improve the nation's capacity to proactively manage drought-related risks, by providing those affected with the best available information and tools to assess the potential impacts of drought and to better prepare for and mitigate the effects of drought, through a national drought early warning information system (DEWS). The system uses new and existing partner networks, made up of federal, tribal, state, local, and academic partners, to make climate and drought information understandable and accessible for decision makers and to improve their capacity to plan for and cope with the impacts of drought.

There are currently eight regional DEWS in the United States that can be accessed through the NIDIS website (https://www.drought.gov/drought/regions/dews). Each contains information on current drought conditions and water supply, as well as forecasts, reports, and partners for each region. The NIDIS website also provides links to maps of national drought conditions (current and forecasted), impacts, soil moisture conditions, vegetation indices, current and forecasted fire conditions, and water supply. Some of these maps, such as the Experimental Surface Water Monitor maps of soil moisture and snow-water equivalent, and the Evaporative Stress Index, show data for Mexico as well. The website's data search explains how to access the data used in the maps, and the resources page houses climate-outlook reports and other documents on climate and drought research and planning.

## Regional Integrated Sciences and Assessments (RISA; National Oceanic and Atmospheric Administration [NOAA])

For over a decade, the NOAA RISAs have conducted diverse, innovative, interdisciplinary, use-inspired, and regionally relevant research that informs resource management and public policy and builds the capacity of the United States to prepare for and adapt to climate variability and change. Working in sustained partnerships with local decision makers, RISA teams help stakeholders understand and adapt to regional climate impacts and are gaining insights into what is required to bridge the gap between research and applications. The types of products and management efforts undertaken by the RISAs vary widely but share the common feature of emerging from real-world challenges faced by stakeholders.

RISAs in the U.S.-Mexico border region include the California-Nevada Applications Program (CNAP; https://scripps.ucsd.edu/programs/cnap/), the Climate Assessment for the Southwest (CLIMAS; http://www.climas.arizona.edu/), and the Southern Climate Impacts Planning Program (SCIPP; http://www.southernclimate.org/). The Western Water Assessment RISA (WWA; http://wwa.colorado.edu/), although located outside of the border region, provides tools and resources for arid states, such as the Intermountain West Climate Dashboard (http://wwa.colorado.edu/climate/dashboard.html), as part of the NIDIS Intermountain West DEWS, and the NOAA Western Region Climate Service Providers Database (https://wrcc.dri.edu/ClimSvcProviders/). Other tools include TreeFlow (http://www.treeflow.info/), a comprehensive web resource that provides streamflow reconstructions, data access, and examples of the ways in which water managers are using streamflow reconstructions, and a tool devoted to winter precipitation reconstructions for Arizona and New Mexico (http://www.climas.arizona.edu/paleoclimate-tool). Scientists at these and other institutions can serve as a resource for developing site-specific analyses and reconstructions of past environmental parameters.

## Landscape Conservation Cooperatives (LCCs; Department of Interior [USDOI])

The LCCs (https://lccnetwork.org/) are management-science partnerships to provide applied science and coordinate efforts to develop science-based responses to climate change impacts to land, water, and wildlife resources. Landscape conservation is the main focus of LCC science delivery, with a goal of informing resource management decisions that pertain to climate change and other regional-scale stressors such as drought. The LCCs facilitate communication among scientists and land managers to create a mechanism for informed conservation planning, effective conservation delivery, and adaptive monitoring to evaluate the effects of management actions, and then modify actions as needed.

In 2019, the U.S. Department of the Interior stopped funding the LCCs, but several LCCs are still operating after receiving alternative support. For example, the California LCC is supported and led by the state and continues as the California Landscape Conservation Partnership. Additionally, the websites and resources of some of the LCCs are still operating. The Desert LCC was the primary LCC entity in the U.S.-Mexico border region. The Desert LCC (https://desertlcc.org/) covered a broad region, stretching from southern California and southern Nevada to west Texas, and south into central Mexico, covering the border states of eastern Baja California Norte, Sonora, Chihuahua, and Coahuila. Many of the maps developed by the Desert LCC, such as hydrography and springs maps and ecosystems maps, cover the entire Desert LCC region on both sides of the U.S.-Mexico border, and the website also provides some resources in Spanish.

## Climate Adaptation Science Centers (CASCs; United States Geological Survey [USGS])

CASCs (https://casc.usgs.gov/) are research collaborations that are guided by a U.S. Geological Survey (USGS) center director and hosted by a university or a consortium of universities. CASCs bring together expertise from university and federal scientists to support climate change research and collaborations. They work with natural resource managers and other providers of climate information to ensure that research by CASC-affiliated scientists contributes to robust decision-making. The southwestern United States region is served by the Southwest CASC (SW CASC) and the South Central CASC (SC CASC). Each CASC has a Stakeholder Advisory Committee, comprising representatives from federal and state resource-management agencies and tribes, which communicates science needs to the CASC, helps guide the science planning process, and conveys scientific results to end users to implement actions in the field.

The SW CASC and SC CASC websites contain information on funding opportunities, publications of work conducted by CASC affiliates, lists of partners (individual and institutional) and their areas of expertise, and links to resources, such as drought histories and precipitation outlooks. The SW CASC also links to the Southwest Climate and Environmental Information Collaborative (SCENIC), a tool by the SW CASC and the Western Regional Climate Center (WRCC) meant to support natural resource management decision makers throughout the southwestern United States. SCENIC provides access to observational, modeled, and remote sensing datasets as well as data analysis tools to assist users in visualizing the data.

## National Centers for Environmental Information (NCEI; National Oceanic and Atmospheric Administration [NOAA])

In 2015, NOAA (https://www.ncei.noaa.gov/) began consolidating its three existing National Data Centers (the National Climatic Data Center, National Geophysical Data Center, and National Oceanographic Data Center) into the NCEI to meet the demand for high-value environmental data and information by private industry and businesses, local to international governments, academia, and the general public. The NCEI hosts and provides access to oceanic, atmospheric, and geophysical data, and is one of the most significant archives of such data in the world. The NCEI website provides access to data in sixteen different categories, including natural hazards (e.g., extreme climate events), climate monitoring and extremes (e.g., monthly, seasonal, and annual climate reports), and land-based stations. The website also links to interactive maps, such as Climate Data Online (https://www.ncdc.noaa.gov/cdo-web/), which provides historical weather and climate information and a mapping tool to view the information.

The NCEI also provides regional services. Regional Climate Services Directors (https://www.ncdc.noaa.gov/rcsd) regularly communicate with stakeholders about climate information needs and bridge the gap between the NOAA and regional stakeholders.

Regional Climate Centers (see below) produce, and deliver to decision makers, climate data and information.

## Regional Climate Centers (RCC; National Centers for Environmental Information [NCEI])

Created in the 1980s, the RCC program specializes in the development of robust and efficient computer-based infrastructure for providing climate information, and seamless integration and storage of various streams of climate data. They have long partnered with regional stakeholders, including federal and state entities, and their mission is rooted in (1) user-centric services, and (2) active research, data stewardship, and effective partnerships. The RCCs developed the Applied Climate Information System (ACIS; http://www.rcc-acis.org/), which updates national climate data daily, and provides a suite of maps at a regional level. The RCCs perform applied climate research, such as on wildland fire and on how the El Niño–Southern Oscillation affects the region. They also develop decision-support tools and produce hundreds of climate summaries.

The centers that serve the southwestern United States are the Western Regional Climate Center (WRCC; https://wrcc.dri.edu/) and the Southern Regional Climate Center (SRCC; http://www.srcc.lsu.edu/). One of the centerpieces of both of these RCCs is that they have online tools to (a) provide easy access to historic monthly climate data and data quality information, and (b) allow users to visualize these data in graphs and maps (some tools are subscription services for analyzing many climate parameters). In addition, the WRCC has climate-tracker tools for visualizing climate change trends in observed climate records and provides links to forecasts and drought-monitoring tools.

## State Climate Offices

Each U.S. state has a State Climate Office, usually based at a state university, with a state climatologist appointed by the state to be a source of statewide climate information to support researchers, stakeholders, and the general public. State climatologists respond to inquiries and regularly give presentations throughout their states. They also work closely with other climate services partners, such as RCCs and RISAs, to improve communication and coordinate the referral of stakeholder inquiries. The websites of the state offices provide weather and climate information and links to resources, such as statewide water and climate data and seasonal outlooks. Some, such as the New Mexico Climate Center (https://weather.nmsu.edu/), also provide weather-station and other data (e.g., on droughts).

## U.S. Department of Agriculture (USDA) Regional Climate Hubs

The USDA Climate Hubs (https://www.climatehubs.oce.usda.gov/) were established in 2014 to deliver science-based knowledge, practical information, and program support to farmers, ranchers, forest landowners, and resource managers. In partnership with various entities such as universities, farm groups, and local to regional governments, they provide tools, strategies, management options, and technical support to help land managers adapt to climate change. They also work closely with RISAs and CASCs, which provide data, research, tools, and forecasts for the Hubs to integrate into services for the agricultural and forestry sectors, such as periodic assessments of climate risks and vulnerability in those sectors. The website contains a Climate Hubs Tool Shed (https://tools.taccimo.info/tbl_tools_list.php) that can be searched regionally in the sector of interest. Tools include the Adaptation Workbook, which guides land managers on how to integrate climate change into their conservation efforts, and The Nature Conservancy's Climate Wizard, which displays climate projections of temperature and precipitation for the country.

The two Climate Hubs that serve the U.S.-Mexico border region are the Southwest (https://www.climatehubs.oce.usda.gov/hubs/southwest) and Southern Plains (https://www.climatehubs.oce.usda.gov/hubs/southern-plains) Hubs. Each website provides regional and state climate assessments, and links to regional tools, data, and other educational materials, such as bulletins and fact sheets.

## U.S. Climate Resilience Toolkit

The U.S. Climate Resilience Toolkit (https://toolkit.climate.gov/), under the auspices of the U.S. Global Change Research Program and hosted by the NCEI, is a website that provides tools, information, and expertise on building climate resilience; it aims to reduce risk and vulnerability to climate hazards across the country. The website provides a framework for assessing risk and developing solutions, case studies to see how others are building resilience, climate resilience tools, a database of experts, and climate-relevant reports and training courses. The website also contains a research application, Climate Explorer, which offers graphs, maps, and data of observed and projected temperature, precipitation, and other parameters for every county in the contiguous United States.

## National Climate Assessment (NCA)

The NCA (http://www.globalchange.gov/what-we-do/assessment) is an assessment effort that culminates in regular reports to Congress and the citizens of the United States; the report is required every four years by the Global Change Research Act of 1990. The quadrennial report serves as a status report about climate change science and impacts for the United States, reporting on the effects of global change on the natural environment, agriculture, energy production and use, land and water resources, transportation, human health and welfare, human social systems, and biological diversity. The NCA also analyzes current trends in global change, both human-induced and natural, and projects major trends for the next twenty-five to one hundred years.

The third NCA report, or NCA3 (Melillo et al. 2014), was published online and in PDF. It included several innovations that added to the scope and relevance of previous assessments: (1) chapters devoted to cross-cutting issues, such as biogeochemical cycles, land-use and land-cover change, intersections between energy, water, and land use, indigenous and native lands and resources, decision support, mitigation, and adaptation; (2) interactive electronic publication as the main form of access; (3) links in the online material to metadata and data used in graphs and analyses related to key messages put forth in each chapter; (4) implementation of a substantial effort to institute a sustained assessment process (http://www.globalchange.gov/what-we-do/assessment/sustained-assessment), which supplements the four-year reports in order to facilitate the development and transfer of reports on special topics, case studies, and information that can easily be used as input to the next NCA (e.g., USGCRP 2016); (5) development and implementation of a set of indicators to track climate changes (http://www.globalchange.gov/browse/indicators); (6) implementation of a data and information system to support the NCA (https://data.globalchange.gov/); and (7) development of a network of cultivated partnerships between scientists and a broader audience of stakeholders to ensure the relevance and usability of future assessment reports and products (NCAnet: http://ncanet.usgcrp.gov/).

The fourth NCA report (NCA4; http://www.globalchange.gov/nca4), was published in two volumes, in 2017 and 2018 (USGCRP 2017; USGCRP 2018). NCA4 builds on the NCA3, with additional innovations, including the stand-alone Climate Science Special Report (USGCRP 2017) and new chapters on air quality and emerging multisectorial interactions. The NCA5 report is currently in its early stages and is expected in 2022 or later.

## U.S. National Phenology Network (USA-NPN)

Phenology is the study of the interactions between environmental change and seasonally recurring biological events (e.g. leaf-out, flowering, animal migrations), which can be very sensitive to hydrology and temperature changes and therefore provide valuable indicators of climate change. The USA-NPN (http://www.usanpn.org) is a consortium of scientists, citizens, organizations, government agencies, and nonprofit groups working together to monitor the impacts of climate change on plants and animals throughout the United States. The USA-NPN collects and shares phenology data and models with scientists, resource managers, and the public to help in decision-making and adapting to climate change. It is organized and directed by its National Coordinating Office (NCO), which facilitates communication between organizations, policymakers, and individuals who are interested in understanding the links between climate change and natural systems. The USA-NPN receives funding from several organizations, including the USGS, National Park Service, and the University of Arizona, where the NCO is located.

## Federal Government Agencies

Many other U.S. federal government agencies provide resources on climate change. For example, Reclamation's climate change page (https://www.usbr.gov/climate/) provides climate news, studies, risk assessments, and reports, such as a synthesis of climate change implications for water and environmental resources (USBR 2013). Reclamation's WaterSMART page (https://www.usbr.gov/watersmart/) allows the user to access basin studies and climate-risk assessments for river basins, and a tool (http://gdo-dcp.ucllnl.org/downscaled_cmip_projections/dcp Interface.html#Welcome), produced in collaboration with other agencies and organizations, provides data and maps for viewing climate and hydrology projections in U.S. river basins.

The USGS provides a National Climate Change Viewer (https://www2.usgs.gov/land resources/lcs/nccv.asp) through its Climate Research and Development Program. The interactive map displays parameters, such as temperature, precipitation, and runoff, over the next one hundred years, viewed by state or watershed. Users can delimit the displays for a variety of future time periods, emission scenarios, and models. The tool also provides tables and time-series graphs of the data.

The USFWS offers a five-month online course, Climate Academy, for natural resource managers and conservation professionals. The course brings in experts from various natural resources fields and teaches the fundamentals of climate science, provides tools and resources for climate adaptation, and increases climate literacy and communication. The course encourages networking among participants to stimulate collaboration on climate change response planning.

## Nonprofit Organizations

Nonprofit organizations also provide tools and resources related to climate information and assisting in climate adaptation. EcoAdapt (http://www.ecoadapt.org/) supports planning and management practitioners in climate change adaptation by making resources more accessible and building capacity in ways such as creating networks of projects and partners with similar adaptation concerns. The Climate Adaptation Knowledge Exchange (CAKE; http://www.cakex.org/), founded and managed by EcoAdapt, is an interactive online database that aims to build these networks by providing the best information available through a virtual library, a directory of practitioners to share knowledge, and data tools and information from other sites. Also available by CAKE is CRAVe (Climate Registry for the Assessment of Vulnerability), a searchable database and registry of vulnerability assessments across the country.

The Conservation Biology Institute (CBI) provides scientific support for conservation and recovery of biological diversity through research, education, planning, and outreach. Scientists, software engineers, and educators at CBI created Data Basin (https://databasin.org/), a mapping and analysis platform that allows users to explore over 2,700 datasets, create custom visualizations, publish datasets and maps, and develop decision-support tools.

# GLOSSARY

**adaptive capacity**: The ability of a plant species or system to reduce the impacts, or take advantage, of climate change.

**aggradation**\* (or channel aggradation): The raising or elevating of a bottomland surface through the process of alluvial deposition; it is the vertical component of accretion and is most frequently applied to sediment deposition on a channel bed, bar, or other near-channel surfaces, floodplain, or, less often, low-lying alluvial terrace.

**alluvial channel**\*: A channel formed within stream-deposited alluvium. *See also* channel.

**alluvial terrace**\*: An aggradational feature that is composed of unconsolidated to poorly consolidated alluvium and its weathering products, and generally reflects an abandoned floodplain surface.

**alluvium**\*: A general term for sediment deposited in a streambed, on a floodplain or other bottomland feature, in a delta, or at the base of a mountain during comparatively recent geological time.

**alpha (α)**: The confidence level of data analysis; it describes the probability of rejecting the null hypothesis when the null hypothesis is actually true.

**anabranch**\*: A separate channel of a stream that has diverged from a main stream channel and rejoins the stream at a downstream site; it is a discrete, semipermanent channel that may be of equal or smaller size as the main channel, thereby distinguishing it from channel braids that are not discrete and may be highly ephemeral.

**anastomosing stream**: A synonym for "braided stream."

**antidune**\*: A transient sand wave or dune in a fluvial setting or on a stream bed that moves upstream by processes in which the erosion of sand particles occurs on the downstream slope of the bedform followed by deposition of the sand particles on the next upstream slope.

**aquifer**\*: Any rock body or geological deposit of alluvium or similar rock debris that is partially or fully saturated with groundwater and has the properties of permeability (transmissivity) and porosity that enable it to yield the groundwater to a well or spring at a rate significantly high to fulfill a specified purpose; aquifers are grouped as unconfined, those controlled by near-surface gravitational and atmospheric-pressure conditions, and artesian, those that are poorly connected to the land surface due to an impermeable layer separating it from the land surface.

**aspect**\*: When used in a geomorphic context, the orientation or the compass direction in which a landform or surface faces (the north-facing slope of a mountain has a northerly aspect).

**assemblage**: A group of organisms of a number of different species that occur in the same area and interact through trophic and spatial relationships.

**avulsion**\*: A rapid change in the course or position of a stream channel, especially by the erosion of lowland alluvium, to bypass a meander and thereby shorten channel length and increase channel gradient; it commonly occurs during floods but also can occur by normal processes of lateral migration of a stream channel during nonflood discharges.

**bar**\*: In-channel sediment of relatively coarse bed material, typically coarse sand through cobbles in size, that is generally deposited during the recession of a high flow and is mostly exposed during periods of low flow; the upper surface of bars of perennial streams is typically equivalent to the stage of about 40 percent flow duration.

**basal-bark treatment**: A method of removing invasive species that involves the application of an herbicide to the lowest forty-six centimeters of young plants that have a basal diameter of less than ten centimeters.

**bathymetry**: The measurement of the depth of water in oceans, lakes, or streams. The term is the underwater equivalent to "topography."

**bed load** (or sediment discharged as bed load)\*: The sediment that is moved by sliding, rolling, or hopping (saltation) on or near the stream bed, essentially in continuous contact with it.

**belt transect**: A widening of a transect to form a quadrant that occupies a specific area. When vegetation is being measured, belt transects provide data on abundance (sometimes as density per unit area) as well as percent cover of the species found within the quadrant.

**benchmark**: A term to signify progress toward stated goals or objectives relative to the desired time frame for accomplishment.

**benthic invertebrates**: Organisms that live in or on the bottom sediment of rivers, streams, and lakes; they consume algae, biofilms, and organic matter, and therefore are important links in the food chain, including to fish.

**biological control**: A technique to control a pest by the introduction of a natural enemy or predator.

**blocked design**: In statistical design, the arrangement of experimental units in groups or blocks that are similar to one another—for example, a monitoring design that is quantifying survival of pole plantings versus seedling plantings along a stream reach that has a major elevation change. Revegetation areas are divided into two distinct stream reaches that contain both types of planting strategies: "high elevation" and "low elevation." In this instance, differences due to elevation are "blocked" as a factor that may account for variability in survival rates between plantings done via poles and seedlings.

**bottomland***: That part of an alluvial valley formed of and underlain by alluvium that has been transported and deposited by the stream flowing through the valley reach; bottomlands may include the channel bed and one or more terraces.

**bottomland ecosystems**: Biotic communities associated with streams, lakes, and other landscape features at which the availability of water is greater than in surrounding uplands; the availability of water at these areas may permit the establishment and growth of plant species not found on adjacent, drier uplands.

**braided stream***: A stream with a wide, relatively horizontal channel bed over which water during low flows forms an interlacing pattern of splitting into numerous small conveyances that again coalesce a short distance downstream; the conveyances, or sub-channels, lack channel characteristics, are highly ephemeral, and thereby are distinguishable from anabranches. A synonym is "anabranching stream," a biological term referring to the vein patterns of some plant leaves.

**capillary fringe***: A zone of the subsurface that is continuous with the underlying zone of saturation, contains capillary interstices (some or all of which are filled with water), and in which the pressure is less than atmospheric.

**channel***: A natural or constructed passageway or depression of perceptible linear extent containing continuously or periodically flowing water and sediment, or a connecting link between two bodies of water; an alluvial channel is one formed within stream-deposited alluvium.

**channel incision**: *See* incision.

**channel morphology**: The physical characteristics of a stream channel, principally the width, depth, and gradient.

**channel profile** (or longitudinal profile): Described by an elevation line determined by the lowest points in a stream channel as it decreases downstream; it is a major determinant of channel pattern.

**climate change adaptation**: Practices and adjustments by humans or natural systems that enhance resilience and/or reduce vulnerability to the impacts of global climate change.

**climate services**: Climate-related products and information that assist decision makers and the general public in understanding, and making decisions about, climate change.

**composition**: In biology, the contribution of a species being measured to the biological community of which it is a part.

**cone of depression**[†]: As applied to groundwater investigations, a depression in the potentiometric surface of a body of groundwater that has the shape of an inverted cone and develops around a well from which water is being withdrawn.

**confidence level**: The probability of rejecting the null hypothesis when it is actually true, or the percentage of all possible samples that can be expected to include the true population parameter; confidence level is typically represented by the Greek letter α (alpha).

**consumptive use**[†]: In hydrological studies, the difference between the total quantity of water withdrawn from a source for use and the quantity of water, in liquid (or, rarely, solid) form, returned to the source.

**curve number**[*]: An index of the runoff potential on a land surface in response to rainfall; curve numbers range from 0 (no runoff under any condition) to 100 (all rainfall of any event results in runoff).

**cut-stump treatment**: A method of removing invasive woody-plant species involving a combination of removing most of the above-ground biomass and immediately applying an herbicide on the cut stump.

**dendritic drainage pattern**[†]: Where parts of the stream branch irregularly in all directions and at almost any angle, resembling in plan the branching habit of some trees.

**deposition**[*]: The constructive process of accumulation into beds or irregular masses of loose sediment or other rock material by any natural agent; it is especially the mechanical settling of sediment from suspension or tractive movement in water.

**diadromous**: Species that spend portions of their life cycle partially in freshwater and partially in salt water.

**discharge**[*] (or flow rate): As a hydrological term of streamflow, expressed as the movement downstream per unit length of channel of a volume of water; water discharge is given in volume per unit time, typically cubic meters per second (m3 s-1). As a sedimentology term, discharge is the movement of a dry mass of sediment per unit length of channel in a specified time interval; technically, it is expressed in watts per meter (W m-1), but informally it is viewed as mass per unit time. Owing to theoretical considerations, the term sediment-transport rate is preferred to that of sediment discharge.

**dissolved load**[*]: The part of the total stream load that is carried in solution in surface or groundwaters; the amount of dissolved solids is a measure of the quality of the water and is generally expressed as milligrams per liter.

**dune**\*: Including dune field, an accumulation, or concentration, by depositional processes of water or wind as a low, small-scale mound, ridge—or more commonly, a complex (field, or zone) of mounds and ridges—of loose, well-sorted granular material (generally sand) that, if active, may be bare or, if inactive, partially to fully vegetated; dunes are subject to translocation, without a basic loss of scale or structure, by the action of streamflow, waves, or wind.

**ecosystem**\*: The complex of biotic populations, the biophysical (environmental) constraints on the biotic populations, and the ability of the complex to function as an ecological unit within a specified area or part of a watershed; within studies of bottomland ecology, an ecological unit is a specific site within a management unit along a stream reach where monitoring is conducted.

**ecosystem services**\*: The production of renewable natural resources through processes yielding clean water, soil, vegetation, and wildlife.

**effectiveness monitoring**: *See* outcome monitoring.

**effluent**: As applied to streamflow, the movement of groundwater into a stream channel from the underlying and adjacent alluvial deposits to become surface water.

**endemic species**: Plant or animal species unique to a defined geographical location.

**entrainment**\*: The process by flowing water or air, or by the mixing of water or air between opposing currents, of mobilizing sediment by picking up particles and transporting them in suspension, as suspended load, and along the channel or other surface of transfer, as bed (or traction) load; rates of hydrological entrainment depend on stream power (the product of discharge and water-surface slope) and the sizes of the sediment particles.

**environmental flow**: The quantity, timing, and quality of water flows required to sustain freshwater and estuarine ecosystems and the human livelihoods and well-being that depend upon these ecosystems.

**ephemeral stream**\*: Streamflow within a normally dry channel; the streamflow occurs inconsistently or infrequently and, except during periods when the ephemeral streamflows occur, the channel bed is directly underlain by unsaturated alluvium.

**erosion**: The process of detachment and transport of soil or rock particles and organic matter by the agents of raindrop impact and surface runoff from rainfall and snowmelt, along with the abrasive and transport action of moving fluids and solids such as wind, ice, soil, and rock debris. (WRO, 12-11-19)

**evaporation**\*: As applied to hydrology, the conversion of water to a gaseous or vapor state.

**evapotranspiration**\*: The loss of water from any surface by the combined processes of evaporation and transpiration.

**exposure**: In the context of biotic vulnerability to climate change, the amount of change either in climate or in climate-driven factors that a species or a biotic community experiences.

**facultative species**: Plants that are commonly found in terrestrial and/or riparian zones.

**flood**\*: Any climatically controlled, relatively high streamflow that overtops the natural or artificial banks in any reach of a stream, thereby being of geomorphic significance; where a floodplain exists, a flood is any flow that spreads over or inundates the floodplain.

**flood-frequency curve**\*: A graph showing recurrence intervals of floods plotted as the abscissa and the magnitudes of the floods plotted as the ordinate.

**floodplain**\*: A strip of relatively smooth land bordering a stream incision, built of sediment carried by the stream and dropped in slackwater beyond the influence of the swift current of the channel; the level of the floodplain is generally about the stage of the mean annual flood, and therefore one and only one floodplain level can occur in a limited reach of bottomland.

**floodplain reconstruction**: The processes of bottomland sediment deposition as floodplain alluvium, and the reestablishment of vegetation on the freshly deposited sediment following an erosive flood.

**flow duration**\*: The percentage of time that a specified discharge is equaled or exceeded.

**flow regime**: The amount, distribution, and movement of water along a stream's channel reach; more precisely, flow regime is characterized by the combined effects of discharge magnitude and frequency, flashiness, duration, and predictability of streamflow.

**fluvial**\*: Pertaining to streams.

**fluvial processes**: The interactions of flowing water in natural stream channels that result in the erosion of landforms or the transportation and deposition of sediment.

**functional diversity**: The elements of biodiversity that influence ecosystem function; it can be measured by the number of sets of species showing similar responses to the environment.

**global climate models**: Complex physical models that simulate the most important processes of the climate and include the atmosphere, ocean, land surface, and cryosphere.

**gradient**\*: As applied to stream channels, the rate of elevation change between two specified sites of horizontal distance measured along the thalweg of the channel; it is generally expressed as a nondimensional number (m m-1).

**growing degree days**: The number of days that occur in a time period during which the temperature exceeds a reference temperature that permits plant development; the reference temperature varies with plant species and is the temperature below which plant development ceases.

**habitat**\*: The living space for one or more organisms; it is described by the combined environmental parameters of the biotic and abiotic factors.

**hydraulic radius (R)**\*: Of a stream channel, the ratio of its cross-sectional area, A, to its wetted perimeter, WP: R = A/WP.

**hydrograph***: The graphical representation of a hydrological variable, such as the stage of a stream or the water level in a well, as a function of time; a hydrograph for runoff (streamflow) is a graph of the time-rate distribution of flowing water passing a site on the landscape, generally at a stream channel and often for a specific flow or runoff event.

**hyporheic zone***: The ill-defined volume of sediment, adjacent to and beneath an alluvial stream channel, through which groundwater moves roughly parallel to streamflow; water of the hyporheic zone generally is readily exchangeable with stream water, receiving water as bank storage through influent reaches of channel and yielding water as seepage through effluent reaches.

**in-channel revegetation**: The process within rehabilitation projects in which plants are placed to grow on bottomland surfaces where streamflow may occur.

**incision**†: The process whereby a downward-eroding stream deepens its channel or produces a narrow, steep-walled valley.

**influent**: As applied to streamflow, the movement of flowing surface water from a stream channel into the underlying and adjacent alluvial deposits to become groundwater.

**intermittent stream***: As a hydrological term, intermittently or seasonally flowing water in a natural, intermittent stream channel; the flow of an intermittent stream typically is derived from wet-season runoff or snowmelt, and the surface of an intermittent stream, or the bed of the channel upon which flow occurs, typically is higher than the level of the zone of saturation in the adjacent water-bearing alluvium or rocks.

**invasive species**: A species not native to a site or area; invasive, or introduced plant species, often spread rapidly and may cause environmental damage by overwhelming native vegetation.

**keystone species**: A species that has a pronounced effect on ecosystem processes; examples are large predators, beavers (by altering streamflow and sediment movement), and plants that disperse large numbers of seeds.

**line-point intercept technique**: A rapid and accurate technique for measuring plant cover along a linear transect; it is based on the number of points (of the total number) along the transect for which the target feature that is being measured is found. For use in riparian zones, point counts can include soil type, surface litter, rocks, biological soil crusts, and particular species of vegetation.

**lotic environment**: An area, such as that of a stream, that is influenced by water of unidirectional flow; lotic environments differ from lakes, ponds, and oceans (lentic environments) that have water and energy flow in more than one direction.

**management unit**: A geographically bounded area that is identified for practical purposes and targeted for management; in the context of stream-corridor restoration, it is the bottomland environment along a reach or stream segment at which restoration is occurring or where control sites have been established.

**mesic***: An environment (habitat) that is characterized by moist conditions; mesic species are adapted to that environment.

**mixed-effects model**: A statistical model that contains both fixed effects and random effects; such models are typically used when multiple correlated measurements are made on each unit of interest, such as survivorship, numbers of leaves and branches, and diameter at breast height (DBH) of vegetation planted as part of a revegetation project. Because of their advantage in dealing with missing values, mixed-effects models are often preferred over more traditional approaches.

**monoculture**: A stand or planting of a single plant species.

**multivariate analysis**: A type of statistical analysis that uses techniques permitting analysis and comparison of more than two statistical variables at once.

**nested plot design**: A technique for the sampling of vegetation using various sized plots nested within a frame; the design allows a variety of different species to be efficiently and effectively measured within the same plot or frame. For example, a larger plot area within a frame can be used to measure trees, whereas a smaller, "nested" plot within the same frame can be used to measure grasses.

**normal depth**: The depth of flow in a uniform channel for which the water surface is parallel to the channel profile and energy slope.

**obligate species**: Plants that under natural conditions are found and can survive only in a wetland environmental zone.

**off-channel revegetation**: A revegetation project at a riparian setting at which plantings are on surfaces beyond (above) the bottomland environment and are separated from most stream processes.

**outcome monitoring**: A question-driven, science-based approach that tracks and documents the success of a stream corridor restoration project over time; it is synonymous with effectiveness monitoring.

**paired t-test**: A statistical procedure to determine whether the mean difference between two sets of observations on samples is zero.

**peak discharge**: The highest flow or discharge of a water/sediment mixture that occurs during a specific flood event or time period.

**perched aquifer**: An unconfined aquifer, generally of limited thickness and areal extent, that occurs above, or on top of, a stratigraphic layer of low permeability that inhibits or prevents the downward movement of groundwater in the unsaturated zone from reaching the areal zone of unconfined saturation; thus, the term refers to a condition by which water accumulates and is stored in the alluvium or rock overlying the impermeable layer.

**perennial stream***: As a hydrological term, continuously flowing water in a natural stream channel; the surface of a perennial stream fluctuates at or near the upper level of the zone of saturation in the adjacent water-bearing alluvium or rocks.

**phenology**\*: As a general concept, the branch of science that treats of relations between short-term climatic variations and periodic biological phenomena; as a part of ecology, phenology concerns principally short-term changes in biological processes, such as bird migration, plant flowering and seed dispersal, and tree growth, that vary with season; and seasonal variations such as air and water temperatures and water availability.

**phreatophyte**\*: A plant dependent on water in the zone of saturation (groundwater of the saturated zone), either directly or through the capillary fringe. Fremont cottonwood, Goodding willow, and Arizona walnut are phreatophytes common to the arid southwestern United States.

**point-centered quarter method**: A dimensionless method for measuring composition and size-class distribution of stands of woody plants.

**point transect**: Readings taken, for surveying vegetation, at random or systematic locations along a transect or tape that is extended across a small area or site of interest.

**pole**: As related to stream corridor rehabilitation, a branch or trunk of a woody plant without roots and above-ground growth; pole planting is a technique that places unrooted tree branches or stems in direct contact with saturated soil or sediment. Branches and trunks of some species can be cut into usable sections for propagation. Cottonwood (*Populus fremontii*) and willow (*Salix gooddingii*) yield poles that are commonly used in revegetation projects in the arid Southwest.

**pool-riffle sequence**\*: In an alluvial channel, refers to a succession of one or more combinations of pools and riffles along the channel in the downstream direction.

**prescribed burn or fire**: A planned, controlled burn conducted to meet specific management objectives.

**propagule**: A vegetative structure that can become detached from a plant and give rise to a new plant; examples include poles, cuttings, seedlings, and seeds.

**quadrat**\*: A small area, possibly a square with one-meter sides, within which samples or observations of plants, rock fragments, soil conditions, or other land-surface characteristics are taken; most commonly quadrats are used in ecological studies to document the species occurring in the area. Often quadrats are positioned along a transect to obtain detailed knowledge or develop a statistical analysis of vegetation or other characteristics on a landform or plant community.

**reach**\*: An uninterrupted part of a stream channel between two points; generally the two points are where readily recognizable tributary inflows occur, but can also include features such as meander bends, gorges, or a significant change in geology.

**recruitment**: As an ecological term, when juvenile organisms (generally plants) survive to be added to a population, by propagation, immigration, or birth.

**regime**\* (or regime theory): The concept that alluvial stream channels are self-forming and self-adjusting; the term applies only to channels that make at least part of their boundaries from their transported load, carrying out the process at different places and times in any one stream channel in a balanced or alternating manner that prevents unlimited growth or removal of boundaries. Thus, a stream channel is said to be "in regime" when it has achieved an approximate equilibrium between matter and energy entering a stream reach and matter and energy leaving the reach.

**repeated-measures analysis**: A technique applied to data that are collected at the same site over time; it can help account for variation caused by the passage of time or year-to-year variation.

**replicate**: One of multiple sample points of data collected within a monitoring unit type.

**restoration\***: As applied to this guidebook, any action meant to shift a stream's deteriorated hydroecological condition toward a state characterized by enhanced resilience, decreased vulnerability to the impacts of stressors, improved habitat for native flora and fauna, and enhanced ecosystem services for people. As applied more generally to stream corridors that have been altered through human activity, restoration is the attempt to recreate the adjusted physical and biological conditions that were present prior to the alteration; a goal of restoration, therefore, is to minimize and eliminate the effects of human-induced alterations, thus promoting stable landforms, bioproductivity, and species diversity.

**riffle\***: As applied to alluvial stream channels, a short, relatively shallow and coarse-bedded length over which the stream flows at ordinarily higher velocity and greater turbulence than it does through upstream and downstream pooled reaches where cross-sectional areas of the channel are greater, bed material is smaller, and velocities and turbulence are less.

**riparian\***: Pertaining to the banks of a stream; within ecology the term has been broadened to refer to biota and other characteristics of alluvial bottomlands.

**riparian zone\***: As applied to the study of fluvial systems, an ecological term referring to that part of the fluvial landscape inundated or saturated by flood flows; it consists of all surfaces of active fluvial landforms inundated or saturated by flood flows.

**runoff\*** (or rainfall excess): That part of precipitation that appears in surface streams ($m^3$ $s^{-1}$) and is the amount of rainfall input minus hydrological abstractions, or losses, of interception, depression storage, infiltration, and evapotranspiration; it is more restricted than streamflow as it does not include stream channels affected by artificial diversions, storage, or other works of man.

**sample point**: The site of a single sample of data collected at a specific location on the ground; multiple sample points within a monitoring unit are replicates.

**sand bar**: Bar formed primarily of sand. *See* bar.

**saturated zone**: *See* zone of saturation.

**sediment budget\***: An accounting, or inventory, of the sediment-transport rate, generally as components based on particle-size ranges entering and leaving a specified area or stream reach; when the fluxes of sediment that enter and leave are unequal, the assumption follows that the differences signify the net amounts of sediment that are stored or taken from storage within the area or reach.

**sediment load**: The quantity of sediment, as measured by dry weight or by volume, that is transported during a given period of time; it typically is reported in watts per meter (W m-1) or in mass per unit time.

**sediment-rating curve***: A line (curve) averaging concentrations of fluvial sediment in transport, generally as measured from suspended-sediment samples, collected through the range of discharges typical of a stream; it shows mean variation in sediment concentration with variation in discharge for the period of data collection.

**sediment-transport rate*** (or sediment discharge; or sediment flux): The rate at which a dry weight of sediment passes a section of a stream channel in a given period of time. *See* discharge.

**seepage study**: The quantification of inflows and outflows in a reach of stream channel through the use of flow measurements, streamflow gauges, or related methods to estimate discharge.

**sensitivity**: In the context of climate change, the innate characteristics of an organism or ecosystem, such as tolerance to temperature change, that predispose it to being more or less susceptible to climate change and other environmental variations.

**size-class**: In the context of this guidebook, the size and life development of biota; size classes of trees are often described as seedling, sapling, and adult/mature, with each class having a designated height, which varies by the species being measured.

**slope-area method**: A technique to estimate flood discharge using only survey equipment and an estimation of channel roughness.

**species-area curve**: An approach to determine the number of transects to use in monitoring, especially for individual species.

**spherical densitometer**: An instrument that uses a spherical mirror that reflects the sky to determine the amount of tree cover; it is used to determine thinning of a canopy or spacing for replanting, or to gauge habitat suitability for a particular area.

**stage***: The height (gauge height) of a water surface above an established datum plane, generally at a gauging station.

**stage-discharge curve*** (or rating curve): A graph showing the relation between the gauge height and the amount of water flowing in a channel.

**stage-discharge relation***: The relation between stage and discharge expressed by the stage-discharge curve.

**stream impoundment**: The structure, typically a dam and reservoir, that alters the natural flow of water and stores water for agriculture, power generation, flood protection, or various uses related to human needs.

**streamflow***: The discharge (m3 s-1) that occurs in (and, during floods, adjacent to) a natural channel. The term "streamflow" is more general than "runoff" and can be applied to discharge regardless of whether it is affected by diversion or regulation; streamflow is the water remaining after losses of precipitation or snowmelt to evaporation or sublimation and after available groundwater has satisfied the needs of vegetation and replenishment of soil moisture.

**suspended sediment*** (or suspended load): Sediment moved in suspension in water and is maintained in suspension by the upward component or turbulent currents or by colloidal suspension; the regulatory term for suspended sediment is Total Suspended Solids (TSS).

**taxa (plural of taxon)**: Taxonomic groups of any rank, such as a species, family, or class.

**tectonic processes**: Physical processes associated with the structure of the earth's crust and the large-scale dynamics that take place within it and which may cause erosion of landforms and the transportation and deposition of sediment.

**terrace***: A valley-contained surface that typically is expressed as a long, narrow, nearly level or gently inclined landform bounded along the lower edge by a steeper ascending slope; a terrace is always topographically higher than the floodplain, and is inundated by floods of greater magnitude than the mean annual flood.

**thalweg***: For studies of stream-channel dynamics and the purposes of this guidebook, the line within a stream channel connecting the lowest points of the channel.

**transpiration***: The process by which water in living organisms, primarily plants, passes into the atmosphere.

**trophic**: Relating to nutrition and the processes of nutrition.

**vegetative propagation**: The propagation of a plant by the growth of root-crown buds and root suckers.

**vegetation structure**: The organization of individual plants in space that constitutes the morphology and architecture of a plant community; included are the vertical layers of plants of different heights in a particular plant community, the presence or absence of gaps in the forest canopy, and the horizontal spacing of individual plants.

**vulnerability**: In the context of climate change, the extent to which a species, ecosystem, or vegetated area is likely to be harmed as a result of climate change and associated stresses.

**water balance***: An accounting of the volumes of water entering, leaving, and stored in a hydrological area or unit, typically a drainage basin or aquifer, during a specified time period in which the amount of water entering the area or unit equals the amount leaving.

**watershed***: A drainage divide or a "water parting," but usage of the term has commonly been altered to signify a drainage-basin area contributing water to a network of stream channels, a lake, or other topographical lows where water can collect.

**well field**: A collection of water wells in a small area.

**wetland**\*: A bottomland or low-lying area, including ephemeral-lake floors, at which water either is shallowly ponded on the surface or has a persistent (weeks or longer) near-surface condition of groundwater saturation adequate to support hydrophytic vegetation.

**wetted perimeter (WP)**\*: Of a channel section, the length of which water is in contact with the channel bed and banks; wetted perimeter is a hydraulic parameter in the computation of streamflow from physical properties of the channel.

**xeric**\*: An environment (habitat) that is characterized by deficient moisture; xeric species, or xerophytes, are plants that are adapted to a habitat of low moisture availability.

**zone of saturation**\*: That part of the subsurface in which the interstices of porous and permeable rocks are saturated with water under pressure equal to or greater than atmospheric pressure.

# CONTRIBUTORS

DR. **ANDREW BARTON** is a forest ecologist, science writer, and professor of biology at the University of Maine at Farmington. His research focuses on responses of plant communities to environmental change, especially climate and wildfire. His current projects include monitoring contemporary ecological trends in forests in Maine and the U.S.-Mexico Borderlands, management and restoration of fire-prone plant species and vegetation on federal lands, and the employment of the new ECOSTRESS instrument on the International Space Station to predict vegetation drought stress and recovery after wildfires. He is co-author of *The Changing Nature of the Maine Woods and Ecology* and co-editor of *Recovery of Old-growth Forests in Eastern North America*.

**JEFFERY BENNETT** is a professional geologist and has an MS in geology from the Northern Arizona University and a BS in earth science from Sul Ross State University. He is currently employed by the American Bird Conservancy as a conservation delivery specialist for the Rio Grande Joint Venture, a binational, public-private habitat conservation partnership. Prior to starting work with the Rio Grande Joint Venture in 2018, he worked for the National Park Service for fifteen years as a physical scientist and hydrologist. While with the park service, he coordinated science and resource-management projects related to Big Bend National Park and the Rio Grande Wild and Scenic River, its tributaries, and related groundwater systems. His work has focused on building strong binational and regional partnerships for the conservation of aquatic and grassland habitats. He serves on the Far West Texas Regional Water Planning Group and the Expert Science team for the establishment of instream flows for the upper Rio Grande.

**MARK K. BRIGGS** is a stream ecologist with over twenty-five years restoring numerous rivers in the western United States and northern Mexico. His passion for stream restoration is driven by the strong belief that bringing back damaged streams provides a wealth of benefits for both native species and people. Highlights of his career in restoration include working with the Zuni Nation and the Tohono O'ohdam Nation to restore native riparian ecosystems sacred to their culture. He also helped lead binational restoration efforts along the Rio Grande/Bravo and the Colorado River Delta. His past publications on stream restoration include books, book chapters, and numerous peer-reviewed articles. Mark currently works for RiversEdge West on stream restoration in the southwestern United States. He is an adjunct professor at the University of Arizona and sits on the editorial board of the international journal *Restoration Ecology*. When not working to restore streams, he is often found hiking and biking.

DR. **DANIEL BUNTING** is a geospatial biologist working for the Science Applications program within the U.S. Fish and Wildlife Service. In this capacity, he provides science support within the Service while strengthening relationships with state partners to facilitate landscape-scale conservation across boundaries. Daniel studied watershed management and ecohydrology at the University of Arizona and over the last decade has provided consulting for the rehabilitation and restoration of dryland rivers across the southwestern United States. He has previously worked six years as a project manager for the Harris Environmental Group where he managed numerous projects ranging from biological evaluations of state and federally protected species to large-scale habitat restoration. Currently residing in Austin, Texas, he enjoys hiking, camping, and mountain biking.

**BROOKE BUSHMAN** is the water projects coordinator for The Nature Conservancy in Arizona, facilitating hydrological assessments, design, and monitoring of groundwater recharge projects, with a focus on benefiting the Upper San Pedro River in southeastern Arizona. Brooke also coordinates the annual San Pedro Watershed Wet-Dry Mapping, a multidecadal citizen-science dataset that establishes the extent of the river's base flow during the driest time of year. Trained as an anthropologist and water-resource planner, much of her work focuses on delivering scientific, financial, and capacity resources to advance water projects that benefit both people and nature.

**TODD CAPLAN** is an Albuquerque-based ecologist and natural resources manager with twenty-five years of experience addressing watershed restoration and habitat conservation issues in the southwestern United States. His work has included performing burned-area rehabilitation on wildlife refuges in Arizona and New Mexico, developing tropical forest restoration programs in Papua New Guinea, and leading large-scale floodplain habitat restoration programs throughout the southwest. Much of Todd's professional experience has centered on managing interdisciplinary teams of scientists with applying existing data and implementing new research to advance endangered species habitat restoration practices along the Middle Rio Grande in New Mexico.

DR. **BARRY CHERNOFF** joined the Wesleyan Faculty in 2003 where he holds the Robert K. Schumann Chair of Environmental Studies. He currently chairs the environmental studies program and is director of the College of the Environment. Chernoff's research centers on the freshwater fishes of the neotropical region, primarily those in South America in the Amazon. His research includes aquatic ecology, evolutionary biology, and conservation. Chernoff has published ninety-three peer-reviewed scientific works, including six books and edited volumes. He has led international teams on expeditions designed to conserve large watersheds of the world, having made more than thirty-three expeditions in thirteen countries. He holds visiting positions at Universidad Central de Venezuela and the Museu Zoologia de Universidade de Sao Paulo, Brasil. From 1993 to 1999, he served on the U.S. National Committee for the International Union of Biological Sciences elected by the National Academy of Sciences / National Research Council and served as vice-chairman from 1995 to 1997 and chairman from 1997 to 1999. He serves on the Inland Fish Commission's Endangered and Threatened Fishes Panel for the Connecticut Department of Environmental Protection.

**KELON CRAWFORD** was born in Durango, Colorado, but grew up in the Hill Country near Kerrville, Texas. After high school, she attended the University of Texas at Austin and received a BA in geography with an earth science concentration in 2009. Upon completion of her degree, she accepted an ecology internship with the Student Conservation Association at Bandelier National Monument and went on to work at Los Alamos National Laboratory as a hydrologic technician. These opportunities led to numerous seasonal technician jobs with the National Park Service, the Bureau of Land Management, and Utah State University. Kelon currently lives in Terlingua, Texas, and is the lead restoration ecologist for Río Bravo Restoration, where she has worked on a variety of restoration and long-term monitoring projects along the Rio Grande/Bravo over the last decade. Additionally, she is a certified yoga instructor and part-time river guide. In her spare time, you will find her on the yoga mat and enjoying the outdoors camping, hiking, boating, or a combination of all three.

**DAVID J. DEAN** is a research hydrologist with the U.S. Geological Survey Southwest Biological Science Center, Grand Canyon Monitoring and Research Center. David holds a bachelor's degree in geology and geography from the University of St. Thomas in Minnesota, and a master's degree in watershed sciences from Utah State University. David's research focuses on the geomorphology of streams with high suspended-sediment loads, sediment transport, surrogate methods for measuring sediment transport, anthropogenic driven changes in river morphology, the effects of vegetation on channel form, and stream restoration. Most of David's research has been conducted on river systems of the American Southwest, including the Rio Grande, the Green River, the Colorado River, the Little Colorado River, and Moenkopi Wash. He is currently based in Flagstaff, AZ.

**DR. MEGAN FRIGGENS** is a research ecologist with the USFS Rocky Mountain Research Station. Her research focuses on measuring and predicting the impacts of natural and anthropogenic disturbances on wildlife habitat, wildlife-disease interactions, and other natural and cultural resources. As part of this work, Megan has developed several decision support tools to help resource managers estimate the vulnerability of species and landscapes to changing climate and fire regimes.

**DR. GREGG GARFIN** is associate professor and associate extension specialist in the University of Arizona's School of Natural Resources and the Environment, and university director of the Southwest Climate Science Center—a partnership between the USGS and a consortium of research institutions in the southwestern U.S. He has worked extensively with water and natural resource managers, planners and concerned citizens in the southwestern United States and northern Mexico, to examine resource and community resilience and to improve preparedness for drought and other climate-related hazards. He was chapter lead for the Southwest chapter in the Third and Fourth National Climate Assessments.

DR. **GARY GARRETT** is a research biologist in the Department of Integrative Biology at the University of Texas at Austin. From 2009 to 2013, he was the director of the Watershed Conservation Program at the Texas Parks and Wildlife Department. From 1982 to 2008, he worked as a conservation biologist in the Research Division of TPWD. His interests are centered on the conservation of aquatic natural resources and he has authored more than one hundred scientific publications on the subject. He has a long history of working cooperatively with private landowners, local communities, NGOs, universities, and other state and federal agencies to protect, restore, and enhance critical habitats. In particular, he has worked on conservation in the Chihuahuan Desert region over a period spanning five decades.

DR. **RANDY GIMBLETT** is a professor in the School of Natural Resources and the Environment of the University of Arizona. Randy has engaged in research work studying human-landscape interactions and their associated conflicts and public policies related to the protection of special environments and environmental experiences for nearly three decades. He has published over 120 refereed papers in the field of human behavior, recreation planning and management, and ecological modeling of complex adaptive systems, and specialized in building models that couple human-landscape systems across space and time at multiple scales. This work has culminated in three books: *Integrating GIS and Agent Based Modeling Techniques for Understanding Social and Ecological Processes* (2002), *Monitoring, Simulation and Management of Visitor Landscapes* (2008), and *Sustainable Wildlands: A Prince William Sound Case Study of Science-Based Strategies for Managing Human Use* (forthcoming). Much of his research and applications have focused on the development of field-based methods for collecting dispersed human- or pedestrian-use data coupled with agent-based modeling and its application to solve spatial dynamic problems.

DR. **EDUARDO GONZÁLEZ** is a research scientist at Colorado State University. Eduardo has authored more than fifty publications on riparian and wetland plant ecology, with a focus on evaluating the effectiveness of restoration projects. Eduardo completed his PhD at Alcalá University (Spain) in 2010 and has spent much time since then studying river and wetland systems in Spain, France, eastern Canada, and the western United States. His current work mainly focuses on the response of plant communities to management of invasive *Tamarix* along dryland rivers.

DR. **REBECA GONZÁLEZ VILLELA** is chief of Environmental Flows Projects at the Instituto Mexicano de Tecnología del Agua (IMTA—Mexican Institute of Water Technology) in Morelos, México. Rebeca received her PhD in biology from La Universidad Nacional Autónoma de México (UNAM). She also taught bachelor's and master's degree classes in teaching and research at UNAM, Universidad Simón Bolivar (USB), and Universidad Autónoma del Estado de Morelos (UAEM). Just prior to her work at IMTA, Rebeca was an invited professor at the faculty of Agriculture, Food and Natural Resources in Australia in 2010. At IMTA, Rebeca focuses her research on improving methodologies for managing streamflow in Mexico. She has published several books, chapters in books, and numerous articles in peer-reviewed national and international journals.

**COLLIN HAFFEY** is a conservation manager with The Nature Conservancy in New Mexico, where he works to support the Rio Grande Water Fund (RGWF). The goal of the RGWF is to restore a natural fire regime to forests, build resilient communities, and provide water for one million people. His other conservation work focuses on forest restoration, reversing post-fire ecological type-conversion, and helping communities prepare for and respond to climate change–driven disturbances.

DR. **MICHAEL HAMMER** is an authority on the taxonomy and ecology of Australian fishes, particularly freshwater species. His work involves biodiversity assessment and species discovery, linking the taxonomy and ecology of fishes into wider conservation issues, providing key advice to management, promoting awareness of fishes and their plight, and engaging in specific recovery action. He has published widely and has described several new species. His work is achieved through a collaborative base with partner research agencies and museums, along with government and community stakeholders. Michael is currently the curator of fishes at the Museum and Art Gallery of the Northern Territory in Darwin and focuses on fish conservation in northern Australia and its near neighbors.

**JEANMARIE HANEY** is principal hydrologist with Haney Hydrologic and has thirty-five years of experience conducting hydrogeological investigations in Arizona and across the American Southwest, Mexico, and Chile. She spent eighteen years as a hydrologist for The Nature Conservancy in Arizona, where she worked to expand understanding of interactions between humans and natural ecosystems, especially in the realm of water management. Jeanmarie has worked across sectors and boundaries to provide science for solutions that benefit people while maintaining or increasing natural ecosystem function and services. She focuses on developing, implementing, and assessing monitoring programs that provide high-quality data for assessing attainment of goals; interpreting data and information for decision makers and the public; and seeking solutions that maintain water supply for rivers and wetlands, while meeting human needs. She has specialized in groundwater investigations and the impact of pumping on surface water. Jeanmarie is an Arizona native devoted to conservation in the American Southwest and spends her free time exploring wild places and her own backyard.

**SUE HARVISON** is the owner of 3 Bar Ranch, Terlingua, Texas. Along with an educational background in social work, Sue has had a lifelong love of nature and the beauty of natural landscapes. Her interest in conservation and preservation of native plants and wildlife came from years of serving on the boards of the Lady Bird Johnson Wildflower Center and the Fort Worth Nature Center. Sue became involved with the Terlingua Creek Project through the efforts of the Texas Parks and Wildlife Department, World Wildlife Fund, and Río Bravo Restoration. Restoring the Terlingua Creek watershed to its natural state—before the combined actions of mining and ranching depleted the native trees and grasses—is vital for the habitat of wildlife and ecosystems dependent on it.

DR. **KATHARINE HAYHOE** is an atmospheric scientist whose research focuses on under-standing what climate change means for people and the places where we live. She is a professor at Texas Tech University and has served as lead author on the Second, Third, and Fourth National Climate Assessments. She also hosts and produces the PBS Digital Series *Global Weirding*, and has been named one of TIME's 100 Most Influential People, Fortune's 50 World's Greatest Leaders, and a UN Champion of the Earth. Katharine has a BSc in physics from the University of Toronto and an MS and PhD in atmospheric science from the University of Illinois.

DR. **OSVEL HINOJOSA-HUERTA** has been working in conservation and research projects in northwestern Mexico since 1997, in particular in wetland areas of the Sonoran Desert. His activities include the evaluation and recovery of birds and their habitats, the implementation of community-based restoration projects, and the creation of partnerships with governments and stakeholders for the conservation of nature. He was the director of the Water and Wet-lands Program for Pronatura Noroeste in Mexico, where he led the efforts to restore the Colorado River Delta for over twenty years, including the restoration of river flows and the facilitation of binational negotiations between Mexico and the United States for the Colorado River. Currently, he is the director of the Coastal Solutions Fellowship Program at the Cornell Lab of Ornithology in Ithaca, New York, where he is working to develop capacity and cross-collaborative projects to protect threatened coastal habitats for communities and shorebirds along the Pacific Flyway from Mexico to Chile.

DR. **CHRIS HOAGSTROM** is presently a professor in the Department of Zoology, Weber State University. He has co-authored several publications on ecology and conservation of fishes in arid and semiarid environments as well as on the biogeography of North American fishes. Previous employment includes a decade with the U.S. Fish & Wildlife Service, New Mexico Fishery Resources Office, Albuquerque, New Mexico. He presently works and lives in Ogden, UT.

**ROBERT ITAMI**, at the time of contributing to the guidebook, was director of Geo-Dimensions Pty Ltd. in Melbourne, Australia. His firm specializes in research focused on visitor management in outdoor recreation, visitor monitoring and simulation, vessel traffic monitoring simulation and management, community consultation, visitor carrying capac-ity, level of sustainable activity, and multicriteria evaluation. Bob has published extensively in this area and has worked throughout the United States, Canada, Europe, Australia, and New Zealand on numerous studies with a variety of federal land-management agencies. His contribution to this book was the development of computer simulation models of river vessel traffic and developing methods for determining river user perceptions of different levels of vessel traffic to safety, enjoyment, and environmental impact. Bob is now happily retired in Cuenca, Ecuador.

**SARAH R. LEROY** is assistant staff scientist with the Arizona Institute for Resilient Environment and Societies at the University of Arizona, and the science communications coordinator with the Southwest Climate Adaptation Science Center—a partnership between the USGS and a consortium of research institutions in the southwestern United States. She works to bridge the gap between science and society by working closely with decision makers to get them the scientific information they need in order to make decisions to prepare for the impacts of climate change.

DR. **SHARON LITE** is an ecologist, scientific writer, and professional editor. With a MA and PhD in geography from Arizona State University, she has contributed to the study of riparian systems in the desert and has her work published in multiple scientific journals. Her research has focused on the relationships between fluvial geomorphology and riparian plant biogeography, with an emphasis on the restoration of riparian ecosystems. Currently, Sharon focuses her energy on editing, raising a family, and working as an administrator at Shearim Torah High School, an Orthodox Jewish girls' school in Phoenix, Arizona. On her own time, Sharon enjoys long runs, riding her bicycle, and playing the piano.

DR. **MAURICIO DE LA MAZA-BENIGNOS** has a PhD with maximum honors, summa cum laude, in biology and sustainable development from Universidad Autónoma de Nuevo León; an MBA from the University of Lancaster, United Kingdom, with exchange program with ESC-Lyon (today, EMLYON Business School), France, where Mauricio specialized in strategic planning; a degree in agricultural engineering, agronomy and animal science from Monterrey Tech; and a law degree with Honors for Excellence from the TecMilenio, Mexico. Mauricio is currently part of Mexico's National System of Researchers (SNI 1). In 2015, he was awarded the UANL Research Prize for the best research project in the natural sciences and received the Dr. José Álvarez del Villar Award from the Mexican Ichthyological Society for the best doctoral dissertation in 2014. Mauricio has international experience in conflict resolution, project management, and coordinating multicultural and multidisciplinary teams. He also develops legal analysis, advocacy, and policy work in the defense of water, the natural environment, and sustainable development. Since 2015, Mauricio has been the chief executive officer at Pronatura Noreste A.C.

**LAURA MCCARTHY** is the State Forester in New Mexico. She was previously with The Nature Conservancy where she held three different positions: New Mexico Associate State Director, New Mexico Director of Conservation, and U.S. Senior Policy Advisor for Forests and Fire. Laura launched the Rio Grande Water Fund as a pilot project, and in the process, she discovered a passion for connecting forest and water managers to help them see watersheds through one another's point of view.

DR. **AMY MCCOY** is a founding partner of Martin & McCoy LLC, where she addresses vexing water challenges through the lens of science, data analysis, and policy. She is also an adjunct research scientist with the University of Arizona Southwest Center and participates in ecological and water security research collaborations. She has previously worked as a partner at AMP Insights and as an ecologist with the Sonoran Institute, and was a Captain and athlete in the U.S. Air Force. She graduated with a BS in environmental biology from Yale University, an MA in environmental studies from the University of Southern California, and a PhD in arid lands resource sciences from the University of Arizona.

DR. **KIRK MCDANIEL**, now retired, joined the New Mexico State University (NMSU) faculty as a professor in 1978 with a joint appointment in the NMSU Extension Service as a range brush and weed control specialist. With forty years of research and over two hundred publications, he is nationally and internationally recognized as a leader in the development of vegetation management applications in natural ecosystems. He has been honored with several distinguished research awards and has always emphasized collaboration with others, including landowners and ranchers, agency conservationists, private industry, students, and university scientists.

DR. **MARTÍN JOSÉ MONTERO MARTÍNEZ** received a BS in physics from Autonomous University of Puebla in 1989, an MS in geophysics from the National Autonomous University of Mexico in 1993, and a PhD in Atmospheric Sciences from the University of Arizona in 1998. Martín was head of Atmospheric Environmental Monitoring at Servicio Meteorológico Nacional from 2011 to 2014. From 2014 to date, he has worked as a researcher at the Mexican Institute of Water Technology (IMTA). Over his professional career, Martín has published fourteen papers in indexed journals, ten book chapters, and two books (co-editor), and has one copyright product registration to his name. His current lines of research include climate change, downscaling techniques, quality control, and homogenization.

DR. **JOAQUIN MURRIETA-SALDIVAR** is a cultural ecologist who specializes in building resilience in diverse communities, creating the connections between people, culture, and natural resources. Joaquin has vast experience on the multicultural border region between the United States and Mexico, where he has implemented several community-based approaches to watershed management, river restoration, geotourism, conservation with native peoples, and best-management practices with the ranching community. He successfully implemented several community-based river restoration practices in the binational rivers of the Colorado, San Pedro, and Santa Cruz (a community-based project along the Santa Cruz River with the town of San Lazaro is highlighted in chapter 6) in the Arizona-Sonora region. With the Kwapa Community in the Colorado River Delta, Joaquin developed community mapping strategies that are elevating the long-term role of the Kwapa in river restoration—cultural knowledge, management, and policy in the Colorado River Delta. By finding common ground with the ranching community of the Santa Cruz and San Pedro Rivers, he continues to expand best-management practices for river restoration and range management with community benefits as well as binational watershed health. In Tucson, Arizona, within the Santa Cruz River basin, he currently is part of the Watershed Management Group team, where the focus is on developing and implementing a fifty-year vision for the restoration of the Santa Cruz River.

DR. **JOHN NIELSEN-GAMMON** has been on the faculty at Texas A&M University since 1991. John is currently a Regents Professor of Atmospheric Sciences and also serves as the

Texas State Climatologist. He received a PhD from the Massachusetts Institute of Technology in 1990. He does research on various types of extreme weather, from droughts to floods, as well as air pollution and computer modeling. As Texas State Climatologist, he helps the state of Texas make the best possible use of weather and climate information through applied research, outreach, and service on state-level committees. He is a fellow of the American Meteorological Society.

DR. **JAVIER OCHOA-ESPINOZA** studied agronomy and animal science at Antonio Narro University in the northern Mexican state of Coahuila, where he fell in love with natural resources and biological sciences. He earned a master's degree in ecology and natural resources management from La Universidad de Chihuahua, and a doctorate in the same topic from Nuevo Leon University. Javier has spent the last fourteen years working in the protected areas of Santa Elena, Ocampo, and Maderas del Carmen, which are located along the RRGB in the northern states of Chihuahua and Coahuila. In this capacity, he worked on a wide variety of conservation topics, including wildlife and habitat monitoring, community development, environmental education, and ecosystem restoration. Along the RGB, Javier focused on river restoration efforts that included invasive species management, human dimensions, and community water use.

DR. **W. R. OSTERKAMP** was a research hydrologist, emeritus, with the U.S. Geological Survey. His research included bottomland dynamics, interactions of vegetation with water and sediment, channel islands, and fluvial processes of arid areas. Waite's publications addressed geomorphology, hydrology, geology, sediment movement, ecology, water quality, archaeology, and the economic benefits of earth-science and sediment networks. Waite retired several decades ago and spent much of the last years of his life writing, hiking, and being with his two dogs—Sophie and Pepper—and happily divided their time during the year between Arizona and Wisconsin.

**JENNIFER PITT** is Colorado River Program Director for the National Audubon Society, where she works with water users and governments to protect and restore bird habitat while ensuring reliable water supplies for rural and urban communities. Previously, Jennifer worked for nearly two decades with the Environmental Defense Fund as part of a binational team that spearheaded the release of the pulse flow from the United States through the Colorado River Delta in Mexico (the pulse flow is highlighted in several chapters of the guidebook).

**HELEN M. POULOS** is a plant ecologist and adjunct assistant professor at Wesleyan University's College of the Environment. Her research examines the effects of natural and anthropogenic disturbances on plant community structure and function. Helen's work is applied and focuses on evaluating the effectiveness of vegetation management activities on plant community dynamics across large landscapes in southwestern North America and elsewhere. Science delivery and decision support are critical components of her research focus. Helen works directly with managers on both federal and private lands on both sides of the U.S.-Mexico border to examine temporal trends in plant community structure in response to restoration as a tool for guiding adaptive management.

**JON RADTKE** leads the Sustainable Water and Sustainable Agriculture programs for Coca-Cola North America. In this role, he works across functions to lead strategy, policy, and stakeholder engagement across these strategic sustainability priorities. Primary areas of focus include water conservation in manufacturing, source water protection, community water partnerships, sustainable agriculture initiatives, and marine litter strategies. One goal of The Coca-Cola Company is to return to nature and communities an amount of water equivalent to the water used in Coca-Cola's beverages and their production by 2020. Jon's leadership helped the company to reach this goal five years early and to position The Coca-Cola Company as an industry leader in water stewardship. Jon holds a bachelor's degree in geology and a master's in hydrogeology with over twenty-five years of professional experience managing water resources.

**JEFF RENFROW** has spent over thirty years working on the Big Bend National Park reach of the Rio Grande / Río Bravo as a river guide and for the last twelve years has devoted his professional life to restoration work on the river and its tributaries. He is the founder or Rio Grande Scientific Support Services and Río Bravo Restoration, an NGO working on river restoration and environmental monitoring. Jeff has a bachelor's degree in fisheries management from Texas A&M University. His passion is working to enhance binational cooperation along the RGB. When he's not working, you can often find him out on the river having fun.

DR. **HOLLY RICHTER** is the Arizona Water Projects Director for The Nature Conservancy. She works closely with Conservancy staff and external partners in the development of projects that enhance water supplies for western rivers, including the San Pedro River, Verde River, and the Colorado River/Delta in Mexico. Holly initiated the binational wet/dry mapping program for the San Pedro River—now in its twenty-first year—was a founding member of both the Upper San Pedro Partnership and the Cochise Conservation and Recharge Network, and served as vice chair of the Organizing Board for the Upper San Pedro Water District.

DR. **DAVID RISSIK** is a senior principal climate change adaptation at BMT in Australia. He works on a variety of environmental and climate change risk projects, including several in the area of habitat restoration. David has worked in government, academia and the private sector, and he was deputy director of the National Climate Change Adaptation Research Facility (NCCARF). He is an adjunct professor at Griffith University and affiliated with NCCARF and the Australian Rivers Institute. He is a non-executive director of Green Cross Australia and a past president of the Society of Wetlands Scientists Australian Chapter.

**AIMEE MICHELLE ROBERSON** is employed by the American Bird Conservancy as the coordinator for the Rio Grande Joint Venture, a binational, public-private habitat conservation partnership. Prior to starting work with the Rio Grande Joint Venture in 2013, Aimee worked for the U.S. Fish and Wildlife Service for fifteen years. Aimee's work has emphasized conservation of endangered species and migratory birds, along with their habitats, as well as coordination of large, complex initiatives and partnerships at cross-regional and international scales. Aimee's experience has centered around building and nurturing conservation partnerships, including collaborating with stakeholders, facilitating decision making, fostering effective linkages between science and management, building consensus, and developing joint strategies for conserving wildlife, water, and ecosystems in the United States and Mexico. Aimee has an MS in conservation biology from the University of Minnesota and a BA in geology from Macalester College.

DR. **ALFREDO RODRÍGUEZ** earned his bachelor's degree in engineering and master's degree in water resources in arid zones at Autonomous University of Chihuahua. He received his doctoral degree from the University of Texas in El Paso. Nineteen years ago, his interest in the topic of water deepened when he began to research drought in the state of Chihuahua and groundwater contamination by arsenic with the Institute of Ecology, AC. During those years, he was part of the National Research System (SNI). Over the last fourteen years, he has worked for the World Wildlife Fund (WWF-México) as coordinator of the Water Program, where he develops and implements strategies for water conservation and its associated natural resources, mainly in the Río Conchos basin. Their results have been published in scientific articles and book chapters dedicated to the themes of water, soil, and other natural resources. His life's work revolves around the conservation of nature for the benefit of society, wildlife, and his family. "With water, everything, without it, nothing."

DR. **JULIO SERGIO SANTANA SEPÚLVEDA** received a BS in civil engineering from the Universidad Michoacana de San Nicolás de Hidalgo in 1981 and a master's degree in science in computational systems from the Instituto Tecnológico y de Estudios Superiores de Monterrey in 1987. Julio earned his PhD in computer sciences at the University of Salford, UK, in 1999. He has numerous papers in indexed journals, eight book chapters, one book, and two copyright product registrations to his credit. Julio has been a researcher at the Mexican Institute of Water Technology (IMTA) since 2000, where he focuses on data science, statistical data analysis, and global climate change.

**KAREN J. SCHLATTER** is a research coordinator at the University of Florida Water Institute where she facilitates water-related research and education initiatives in Florida. Previously, Karen was the associate director of the Colorado River Delta Program at the Sonoran Institute and in that capacity worked in the Delta from 2010 to 2019. Karen led ecological restoration and monitoring efforts for the Sonoran Institute and was a co-manager of the Minute 319 and 323 binational monitoring and restoration teams. Karen specializes in riparian restoration ecology and vegetation monitoring and promotes science-based policy through interdisciplinary partnerships with diverse stakeholders. She has an MS in environmental studies from the University of Colorado Boulder and a BS in biology from McGill University, Montreal.

DR. **PATRICK B. SHAFROTH** is a research ecologist with the U.S. Geological Survey, Fort Collins Science Center. Since 1991, Pat has conducted research on riparian ecosystems, primarily in arid and semiarid regions of the western United States. He and colleagues have focused their work on understanding relationships between surface and groundwater hydrology, fluvial processes, and the dynamics of native and introduced riparian vegetation. This research has often been conducted in the applied contexts of riparian ecosystem restoration and the management of nonnative species.

**CARLOS A. SIFUENTES** received a degree in Forestry Agronomy at the University of Antonio Narro Autonomous and worked with the national nongovernmental organization Profauna on a variety of natural resource conservation issues from 1990 to 1996. Carlos was a park ranger with the Protected Area of Maderas del Carmen Flora and Fauna Protection (CONANP) from 1996 to 1999 and deputy director of that same protected area from 2000 to 2006. Between 2006 and 2013, Carlos was the director of Cañon de Santa Elena, Ocampo, Maderas del Carmen and El Monumento (aka Río Bravo Natural Monument), which constitute the four protected areas managed by CONANP in the greater Big Bend region along the Rio Grande/Bravo and international border between the United States and Mexico. Carlos is currently CONANP's regional director and is based in the city of Saltillo, Coahuila. In this capacity, he manages seventeen Natural Protected Areas of the Northeast and Sierra Madre Oriental regions of Mexico.

**DR. JOE SIROTNAK** spent his career as a botanist and ecologist for the National Park Service (1999–2016) and the Bureau of Land Management (2016–2019). The common theme and focus of Joe's professional life was the restoration of wildlands. Throughout his career he worked on projects as diverse as restoring native cutthroat trout populations in Great Basin National Park, managing exotic plants on the Big Bend National Park reach of the Rio Grande, and working on the revegetation of native sagebrush habitat on the Morley Nelson Snake River Birds of Prey National Conservation Area. As the botanist at Big Bend National Park, Joe was central to the development of a strong binational team that is working to bring back this critical international reach of the RGB. Joe's work is being carried on by dedicated friends and colleagues.

**DR. JULIE STROMBERG** is a plant ecologist and recently retired from Arizona State University, where she specialized in understanding how human actions have influenced riparian ecosystems of the American Southwest. She and her partner currently live on a four-acre habitat garden in urban Phoenix, where she is having great fun putting into practice some of the principles she taught in her restoration and conservation biology courses. She also devotes considerable time to helping shelter dogs ("domesticated nature") lead their best lives.

**PAUL TASHJIAN** joined Audubon in March 2018 to establish a comprehensive water and restoration strategy for the Rio Grande in New Mexico. He has expanded Audubon's role in the Rio Grande by building capacity through federal and state grants, representation on federal water teams, and coordination with many partners from irrigation districts to other nonprofit environmental organizations. Prior to joining Audubon, Paul spent twenty-seven years working as a regional hydrologist for the U.S. Fish and Wildlife Service in the southwestern United States. His expertise includes water management and water protection for wildlife, river restoration, water law, and wetland workshop coordination. Paul was the founder and coordinator of the Bosque Hydrology Group, an interagency, interuniversity think tank that focused on the physical restoration of the Middle Rio Grande. He coordinated the quantification and protection of water rights on the National Wildlife Refuge with Department of Interior solicitors and federal and state water-management agencies, conducted numerous studies and workshops to improve wetland management on refuges, and implemented river restoration projects throughout the southwestern United States. Paul enjoys fishing, photography, and family. He holds degrees from Colorado College (BA) and Temple University (MS).

**DALE TURNER** is a conservation scientist for The Nature Conservancy in Arizona, preparing science assessments and developing regional and local plans to guide conservation actions. Trained as a wildlife biologist, his recent work has included maintaining and restoring streams and riparian areas, with a particular focus on the restoration of the Colorado River Delta in Mexico.

DR. **KEVIN URBANCZYK** is a professor in the Department of Biology, Geology and Physical Sciences and director of the Rio Grande Research Center at Sul Ross State University, where he has been teaching since 1991. His academic background includes BS and MS degrees from Sul Ross, and a PhD from Washington State University. His environmental research includes groundwater and surface water studies in west Texas, and studies of the geomorphology of the Rio Grande. He utilizes modern survey technology for developing elevation models and multispectral imagery (using LiDAR and UAVs). At the state level, he has contributed to the development of Groundwater Availability Models for the Texas Water Development Board, has been a member of the Basin to Bay Expert Science Team for the Upper Rio Grande for the Texas Commission on Environmental Quality, and is a director of the Brewster County Groundwater Conservation District. He is actively involved in a binational research and conservation projects on the Rio Grande/Bravo along the U.S.-Mexico border in the Big Bend Region and is also involved in a multinational twelve-river monitoring project coordinated out of Hungary.

DR. **LINDSAY WHITE** is with the Commonwealth Environmental Water Office and has over twenty-five years of experience working in stream management and restoration in Australia as a researcher, government official, and consultant. Lindsay has worked on stream restoration from the scale of the Murray-Darling Basin, right down to small-scale projects adjacent to paddocks. Highlights of Lindsay's career include designing stream restoration projects, operating releases from Murray River dams, conducting doctoral research on the hydraulics of fishways and the biology of native fish species using them, managing multidisciplinary technical analyses that informed the Living Murray First Step decision, piloting the buyback of water entitlements in three states concurrently, being part of a team that managed billions of dollar of environmental water holdings at the Commonwealth Environmental Water Office as part of the implementation of the Basin Plan, and advising a past Governor-General of Australia on river management issues during a tour of the Darling River.

DR. **NICK WHITEROD** is a senior aquatic ecologist at the not-for-profit organization Aquasave–Nature Glenelg Trust, with two decades experience monitoring and researching to assist environmental management across southern Australia. His background is diverse, with expertise in freshwater fish and crayfish species (including invasive fish species), bioenergetics, phytoplankton, and invasive species across river, wetland and terrestrial environments. He specializes in small-bodied freshwater fish and freshwater crayfish conservation and management, with a focus on conservation translocation. In 2018, Nick received a Churchill Fellowship to explore conservation translocation approaches for freshwater species across the world. He has published on varied topics, including population genetics, population modelling, spatial and temporal dynamics, reintroduction ecology, and bioenergetics.

DR. **BART WICKEL** is a senior scientist and conservation hydrologist with the Stockholm Environment Institute (SEI), which is based in Davis, California. Bart's work focuses on spatial analyses and tool development for sustainable water-resource freshwater ecosystem conservation in a changing climate. He has worked in a broad range of development contexts in river basins around the world, focusing on conservation and water-resource development, remote sensing and GIS, spatial analysis, climate change adaptation, land-use and hydropower development impacts, and ecosystem services. Prior to joining SEI he was the lead hydrologist with the WWF Conservation Science Program in Washington, D.C. Bart holds degrees from the Vrije Universiteit Amsterdam (MSc, 1998) and the University of Bonn (PhD, 2004, magna cum laude).

DR. **FRANCISCO ZAMORA** is a scientist and a conservationist with more than twenty years of experience working in natural resource management, particularly in the Delta of the Colorado River in Mexico. He is responsible for integrating community stewardship, applied science, and local values in an alliance to reform water policy, conserve and restore priority areas, and build knowledge and capacity for collaboration between water managers and local leaders. Francisco obtained his PhD in resource geography from Oregon State University in October 2002, with a minor in natural resource economics. He also holds a master's degree in marine resource management from Oregon State University and a bachelor's degree in oceanography from Autonomous University of Baja California. He has published his work on ecological restoration, policy, and binational collaboration issues in the Colorado River Delta.

# INDEX